Metal Oxide Catalysis

Edited by
S. David Jackson and
Justin S. J. Hargreaves

Related Titles

C. N. R. Rao, A. Müller,
A. K. Cheetham (eds.)

Nanomaterials Chemistry

Recent Developments and New Directions
2007
ISBN: 978-3-527-31664-9

C. N. R. Rao, A. Müller,
A. K. Cheetham (eds.)

The Chemistry of Nanomaterials

Synthesis, Properties and Applications
2004
ISBN: 978-3-527-30686-2

R. A. van Santen, M. Neurock

Molecular Heterogeneous Catalysis

A Conceptual and Computational Approach
2006
ISBN: 978-3-527-29662-0

H. U. Blaser, E. Schmidt (eds.)

Asymmetric Catalysis on Industrial Scale

Challenges, Approaches and Solutions
2004
ISBN: 978-3-527-30631-2

J. Hagen

Industrial Catalysis

A Practical Approach
2006
ISBN: 978-3-527-31144-6

Metal Oxide Catalysis

Volume 2

Edited by
S. David Jackson and Justin S. J. Hargreaves

WILEY-VCH

WILEY-VCH Verlag GmbH & Co. KGaA

The Editors

Prof. S. David Jackson
University of Glasgow
Department of Chemistry
WestCHEM
Joseph Black Building
Glasgow, G12 8QQ
United Kingdom

Dr. Justin S. J. Hargreaves
University of Glasgow
Department of Chemistry
WestCHEM
Joseph Black Building
Glasgow, G12 8QQ
United Kingdom

■ All books published by Wiley-VCH are carefully produced. Nevertheless, authors, editors, and publisher do not warrant the information contained in these books, including this book, to be free of errors. Readers are advised to keep in mind that statements, data, illustrations, procedural details or other items may inadvertently be inaccurate.

Library of Congress Card No.: applied for

British Library Cataloguing-in-Publication Data
A catalogue record for this book is available from the British Library.

Bibliographic information published by the Deutsche Nationalbibliothek
Die Deutsche Nationalbibliothek lists this publication in the Deutsche Nationalbibliografie; detailed bibliographic data are available on the Internet at <http://dnb.d-nb.de>.

© 2009 WILEY-VCH Verlag GmbH & Co. KGaA, Weinheim

All rights reserved (including those of translation into other languages). No part of this book may be reproduced in any form – by photoprinting, microfilm, or any other means – nor transmitted or translated into a machine language without written permission from the publishers. Registered names, trademarks, etc. used in this book, even when not specifically marked as such, are not to be considered unprotected by law.

Cover Design Schulz Grafik-Design, Fußgönheim
Composition SNP Best-set Typesetter Ltd., Hong Kong
Printing Strauss GmbH, Mörlenbach
Bookbinding Litges & Dopf GmbH, Heppenheim

Printed in the Federal Republic of Germany
Printed on acid-free paper

ISBN: 978-3-527-31815-5

Contents to Volume 2

Contents to Volume 1 XI

11	**Oxidation Reactions over Supported Metal Oxide Catalysts: Molecular/Electronic Structure–Activity/Selectivity Relationships** 487	
	Israel E. Wachs and Taejin Kim	
11.1	Overview 487	
11.2	Introduction 487	
11.3	Monolayer Surface Coverage 488	
11.4	Molecular and Electronic Structures 490	
11.5	Number of Exposed Catalytic Active Sites (N_s) 491	
11.6	Surface Reactivity 492	
11.7	Steady-State Reactivity (TOF) 493	
11.8	Number of Participating Catalytic Active Sites in Oxidation Reactions 494	
11.9	Role of Surface Acid Sites on Oxidation Reactions 495	
11.10	Other Supported MO_x Redox Active and Acidic Systems 496	
11.11	Conclusions 496	
	References 497	
12	**Vanadium Phosphate Catalysts** 499	
	Johnathan K. Bartley, Nicholas F. Dummer, and Graham J. Hutchings	
12.1	Introduction 499	
12.2	The Active Catalyst 500	
12.2.1	The Oxidation State of the Catalyst 502	
12.2.2	The Phosphorus-to-Vanadium Ratio of the Catalyst 504	
12.2.3	The Role of Amorphous Material 505	
12.2.4	The Disordered Plane 507	
12.2.5	Acid–Base Properties 507	
12.3	Preparation of VPP Precursors 508	
12.3.1	The Preparation of Novel Vanadium Phosphates 513	
12.4	Activation of the Catalyst Precursors 514	
12.4.1	Activation Procedures 514	
12.4.2	Structural Transformations during Activation 517	

Metal Oxide Catalysis. Edited by S. David Jackson and Justin S. J. Hargreaves
Copyright © 2009 WILEY-VCH Verlag GmbH & Co. KGaA, Weinheim
ISBN: 978-3-527-31815-5

12.5	Promoted Catalysts 519
12.6	Mechanism of n-Butane Partial Oxidation 524
12.6.1	Consecutive Alkenyl Mechanism 524
12.6.2	Consecutive Alkoxide Mechanism 527
12.6.3	Concerted Mechanism 529
12.6.4	Redox Couple Mechanism 530
12.7	Concluding Comments 530
	References 531

13	**Heterogeneous Catalysis by Uranium Oxides** 539
	Stuart H. Taylor
13.1	Introduction 539
13.2	Structure of Uranium Oxides 540
13.3	Historical Uses of Uranium Oxides as Catalysts 543
13.4	Catalysis by Uranium Oxides 547
13.4.1	Total Oxidation 547
13.4.2	Selective Oxidation 528
13.4.3	Reduction 554
13.4.4	Steam Reforming 556
13.5	Conclusions 558
	References 559

14	**Heteropolyoxometallate Catalysts for Partial Oxidation** 561
	Jacques C. Védrine and Jean-Marc M. Millet
14.1	Introduction 561
14.2	History of Polyoxometallates 565
14.3	Properties and Applications of Polyoxometallates 566
14.4	Catalytic Applications in Partial Oxidation Reactions 568
14.4.1	Oxidation with Molecular Oxygen 570
14.4.2	Oxidation by Hydrogen Peroxide 575
14.5	Characterization: Redox and Acid–Base Properties 578
14.5.1	IR Spectroscopy 581
14.5.2	Photoacoustic Spectroscopy 582
14.5.3	UV-Visible Spectroscopy 583
14.5.4	Nuclear Magnetic Resonance Spectroscopy 583
14.5.5	Electron Spin Resonance (ESR) Spectroscopy 584
14.5.6	Electrochemistry of Keggin Heteropoly Compounds 585
14.5.7	Thermal Analysis 586
14.5.8	Microcalorimetry of Acid or Basic Probe Adsorption 586
14.6	Conclusions and Perspectives in Polyoxometallate Application in Heterogeneous Oxidation Catalysis 587
	References 589

15	**Alkane Dehydrogenation over Vanadium and Chromium Oxides**	595
	S. David Jackson, Peter C. Stair, Lynn F. Gladden, and James McGregor	
15.1	Introduction 595	
15.2	Commercial LPG Dehydrogenation Process 596	
15.3	Lummus/Houdry CATOFIN® Process 596	
15.4	Chromia 596	
15.5	Vanadia 601	
15.6	Conclusions 610	
	References 610	
16	**Properties, Synthesis and Applications of Highly Dispersed Metal Oxide Catalysts** 613	
	Juncheng Hu, Lifang Chen, and Ryan Richards	
16.1	Introduction 613	
16.2	Properties 614	
16.2.1	Structure and Bonding 614	
16.2.2	Defects 616	
16.2.3	Acid–Base Properties of Metal Oxides 617	
16.2.4	Redox Property of Metal Oxides 619	
16.3	Synthesis 619	
16.3.1	Sol–Gel Technique 620	
16.3.1.1	Hydrolysis and Condensation of Metal Alkoxides 622	
16.3.1.2	Solvent Removal and Drying 623	
16.3.2	Co-precipitation Methods 627	
16.3.2.1	Co-precipitation from Aqueous Solution at Low Temperature 628	
16.3.2.2	Sonochemical Co-precipitation 630	
16.3.2.3	Microwave-Assisted Co-precipitation 631	
16.3.3	Solvothermal Technique 633	
16.3.4	Micro-Emulsion Technique 636	
16.3.5	Combustion Methods 638	
16.3.6	Others 639	
16.3.6.1	Vapor Condensation Methods 639	
16.3.6.2	Spray Pyrolysis 640	
16.3.6.3	Templated/Surface Derivatized Nanoparticles 640	
16.4	Applications in Catalysis 641	
16.4.1	Oxygenation of Alkanes 641	
16.4.2	Biodiesel Production 643	
16.4.3	Methanol Adsorption and Decomposition 645	
16.4.4	Destructive Adsorption of Chlorocarbons 649	
16.4.5	Alkene Metathesis 650	
16.4.6	Claisen–Schmidt Condensation 652	
16.5	Conclusions 653	
	References 654	

17	**Preparation of Superacidic Metal Oxides and Their Catalytic Action** 665
	Kazushi Arata
17.1	Introduction 665
17.2	Preparation 669
17.2.1	Sulfated Metal Oxides of Zr, Sn, Ti, Fe, Hf, Si, and Al 669
17.2.1.1	Preparation of Zirconia Gel 669
17.2.1.2	Preparation of Stannia Gel 669
17.2.1.3	Preparation of H_4TiO_4 670
17.2.1.4	Preparation of $Fe(OH)_3$ 670
17.2.1.5	Preparation of $Hf(OH)_4$ 670
17.2.1.6	Sulfation, Calcination, and Catalytic Action 670
17.2.1.7	Preparation of Sulfated Silica 671
17.2.1.8	Preparation of Sulfated Alumina 671
17.2.1.9	Property and Characterization 671
17.2.1.10	One-Step Method for Preparation of SO_4/ZrO_2 672
17.2.1.11	Commercial Gels for Preparation of SO_4/ZrO_2 and SO_4/SnO_2 672
17.2.1.12	Effect of Drying and Calcination Temperatures on the Catalytic Activity of SO_4/ZrO_2 673
17.2.2	Tungstated, Molybdated, and Borated Metal Oxides 674
17.2.2.1	Preparation of WO_3/ZrO_2 and MoO_3/ZrO_2 674
17.2.2.2	Preparation of WO_3/SnO_2, WO_3/TiO_2, and WO_3/Fe_2O_3 674
17.2.2.3	Preparation of B_2O_3/ZrO_2 674
17.2.2.4	Property and Characterization 675
17.3	Determination of Acid Strength 675
17.3.1	Hammett Indicators 676
17.3.2	Test Reactions 677
17.3.3	Temperature-Programmed Desorption (TPD) 677
17.3.4	Temperature-Programmed Reaction (TPRa) 678
17.3.5	Ar-TPD 678
17.3.6	Ar-Adsorption 680
17.4	Nature of Acid Sites 682
17.5	Isomerization of Butane Catalyzed by Sulfated Zirconia 685
17.6	Isomerization of Cycloalkanes 686
17.7	Structure of Sulfated Zirconia 687
17.8	Promoting Effect 689
17.8.1	Effect of Addition of Metals to Sulfated Zirconia on the Catalytic Activity 689
17.8.2	Effect of Mechanical Mixing of Pt-Added Zirconia on the Catalytic Activity 690
17.9	Friedel–Crafts Acylation of Aromatics 692
17.10	Ceramic Acid 695
17.10.1	Tungstated Stannia 696
17.10.2	Tungstated Alumina 697
17.11	Application to Sensors and Photocatalysis 698
	References 698

18	**Titanium Silicalite-1** *705*
	Mario G. Clerici
18.1	Synthesis and Characterization *706*
18.2	Hydroxylation of Alkanes *707*
18.2.1	Titanium Silicalite-1 *708*
18.2.2	Other Ti-Zeolites *712*
18.3	Hydroxylation of Aromatic Compounds *712*
18.3.1	Hydroxylation of Phenol *713*
18.3.1.1	Titanium Silicalite-1 *713*
18.3.1.2	Other Ti-Zeolites *715*
18.3.2	Hydroxylation of Benzene *716*
18.3.3	Oxidation of Substituted Benzenes *717*
18.4	Oxidation of Olefinic Compounds *717*
18.4.1	Epoxidation of Simple Olefins *717*
18.4.1.1	Titanium Silicalite-1 *718*
18.4.1.2	Other Ti-Zeolites *722*
18.4.2	Epoxidation of Unsaturated Alcohols *724*
18.4.3	Epoxidation of Allyl Chloride and other Substituted Olefins *726*
18.4.4	Epoxidation with Solvolysis/Rearrangement of Intermediate Epoxide *726*
18.5	Oxidation of Alcohol and Other Oxygenated Compounds *727*
18.6	Ammoximation of Carbonyl Compounds *730*
18.7	Oxidation of N-Compounds *732*
18.8	Oxidation of S-Compounds *734*
18.9	Industrial Processes Catalyzed by TS-1 *734*
18.9.1	Hydroxylation of Phenol to Catechol and Hydroquinone *734*
18.9.2	Salt-Free Production of Cyclohexanone Oxime *734*
18.9.3	Propene Oxide Synthesis (HPPO) *735*
18.10	Problems in the Use of H_2O_2 and Possible Solutions *736*
18.10.1	Direct Synthesis of Hydrogen Peroxide *737*
18.10.2	In Situ Production of Hydrogen Peroxide *737*
18.10.3	Process Integration *738*
18.10.4	Miscellanea *739*
18.11	Adsorption, Active Species and Oxidation Mechanisms *740*
18.11.1	Adsorption and Catalytic Performances *740*
18.11.2	The Structure of Ti–OOH Species *742*
18.11.3	Reactive Intermediates and Oxidation Mechanisms *743*
18.11.4	Proposal for a General Mechanistic Scheme *746*
18.12	Conclusions *748*
	References *749*
19	**Oxide Materials in Photocatalytic Processes** *755*
	Richard P.K. Wells
19.1	Introduction *755*
19.2	Basic Principles of Heterogeneous Photocatalysis *756*

19.3	Traditional Photocatalysts 757
19.4	Improving Photocatalytic Activity 760
19.4.1	Visible Light Sensitization by Adsorption of Organic and Inorganic Dyes 760
19.4.2	Visible Light Sensitization by Anion Doping 760
19.4.3	Visible Light Sensitization by Metal Ion Implantation Techniques 761
19.4.4	Physical Methods to Enhance Photocatalytic Activity 763
19.4.5	Potential-Assisted Photocatalysis 766
19.5	Conclusions 766
	References 767

20 Catalytic Ammoxidation of Hydrocarbons on Mixed Oxides 771
Fabrizio Cavani, Gabriele Centi, and Philippe Marion

20.1	Introduction 771
20.2	Propene Ammoxidation to Acrylonitrile 775
20.3	Propane Ammoxidation to Acrylonitrile 778
20.3.1	Mo/V/Te/Sb/(Nb)/O Catalysts 782
20.3.2	Rutile-Type Antimonate Catalysts 786
20.4	Alkylaromatic Ammoxidation 791
20.4.1	Alkylbenzenes and Substituted Alkylbenzenes 791
20.4.2	Alkylaromatics Containing Hetero-Groups 795
20.4.3	Ammonolysis vs Ammoxidation 796
20.5	Ammoxidation of Unconventional Molecules 797
20.5.1	The Ammoxidation of C_4 Hydrocarbons 797
20.5.2	The Ammoxidation of Cyclohexanol and Cyclohexanone 800
20.5.3	The Ammoxidation of Cyclohexane and n-Hexane 802
20.5.4	The Ammoxidation of Benzene 805
20.5.5	Ammoxidation of C_2 Hydrocarbons 807
20.5.6	Conclusions on the Ammoxidation of Unconventional Molecules 808
20.6	Use of Other Oxidants for Ammoxidation Reactions 810
20.7	Conclusions 810
	References 811

21 Base Catalysis with Metal Oxides 819
Khalaf AlGhamdi, Justin S. J. Hargreaves, and S. David Jackson

21.1	Introduction 819
21.2	Catalysts and Catalytic Processes 825
21.2.1	Alkali Metal Oxides 826
21.2.2	Alkaline Earth Metal Oxides 830
21.2.3	Hydrotalcites 835
21.2.4	Rare Earth Oxides 836
21.2.5	Basic Zeolites 837
21.2.6	Zirconia Superbases 837
21.3	Outlook 838
	References 840

Index 845

Contents to Volume 1

1   **EPR (Electron Paramagnetic Resonance) Spectroscopy of Polycrystalline Oxide Systems**   *1*
    Damien M. Murphy

2   **The Application of UV-Visible-NIR Spectroscopy to Oxides**   *51*
    Gianmario Martra, Enrica Gianotti, and Salvatore Coluccia

3   **The Use of Infrared Spectroscopic Methods in the Field of Heterogeneous Catalysis by Metal Oxides**   *95*
    Guido Busca

4   **Resonance Raman Spectroscopy – Θ-Al_2O_3-Supported Vanadium Oxide Catalysts as an Illustrative Example**   *177*
    Zili Wu, Hack-Sung Kim, and Peter C. Stair

5   **Solid-State NMR of Oxidation Catalysts**   *195*
    James McGregor

6   **Photoelectron Spectroscopy of Catalytic Oxide Materials**   *243*
    Detre Teschner, Elaine M. Vass, and Robert Schlögl

7   **X-ray Absorption Spectroscopy of Oxides and Oxidation Catalysts**   *299*
    Michael Stockenhuber

8   **Theory: Periodic Electronic Structure Calculations**   *323*
    Rudy Coquet, Kara L. Howard, and David J. Willock

9   **Thermal Analysis and Calorimetric Methods**   *391*
    Simona Bennici and Aline Auroux

10 **Transmission Electron Microscopy**   *443*
    Wuzong Zhou

11
Oxidation Reactions over Supported Metal Oxide Catalysts: Molecular/Electronic Structure–Activity/Selectivity Relationships

Israel E. Wachs and Taejin Kim

11.1
Overview

Supported vanadium oxide catalysts find wide application as oxidation catalysts and are the focus of this chapter. The highly dispersed nature of the surface vanadium oxide phases on oxide supports allows their detailed characterization and the establishment of molecular/electronic structure–activity/selectivity relationships. The surface vanadium oxide species exist on oxide supports under dehydrated conditions as isolated and polymeric surface VO_4 species with one terminal V=O bond and three bridging V–O–M bonds, in which M represents either an adjacent V or an oxide support cation. Fundamental oxidation studies with the CH_3OH chemical probe molecule over the well defined supported vanadium oxide catalysts show that the bridging V–O–V and terminal V=O bonds do not play a critical role in this redox reaction. However, kinetic studies reveal that the bridging V–O–Support bond dominates the overall reaction by controlling the number of chemisorbed methanol molecules present on the catalyst surface during steady-state reaction conditions. The number of chemisorbed methanol molecules is determined by the electron density on the oxygen atom in the bridging V–O–Support bond, which increases with decreasing electronegativity of the support cation. This general trend is also observed for other oxidation reactions involving one surface VO_4 site. For oxidation reactions involving multiple redox or redox–acidic sites, both the specific oxide support, via the bridging V–O–Support bond, and the surface density of catalytic active sites control the surface intermediate coverage during steady-state reaction. These observations are outlined in more detail in this chapter.

11.2
Introduction

Supported metal oxide catalysts consist of dispersed surface metal oxide species, the catalytic active sites, which are supported on high-surface-area oxides [1–3]. The

Metal Oxide Catalysis. Edited by S. David Jackson and Justin S. J. Hargreaves
Copyright © 2009 WILEY-VCH Verlag GmbH & Co. KGaA, Weinheim
ISBN: 978-3-527-31815-5

surface coverage of the supported metal oxide phase can vary from below to above monolayer coverage. Below monolayer surface coverage, isolated and polymerized surface metal oxide species are generally present on the support. Above monolayer surface coverage, crystalline nanoparticles (NPs) are also present and reside on top of the surface metal oxide monolayer. Although this class of catalysts has been referred to as "monolayer oxide catalysts," it should be recognized that the surface coverage of the supported metal oxide phase can vary widely over all coverage (sub-monolayer, monolayer and above monolayer). Monolayer surface coverage is a critical parameter for supported metal oxide catalysts, since different molecular structures of the catalytically active supported metal oxide phase are present below and above monolayer coverage (surface species and crystalline NPs, respectively).

Supported metal oxide catalysts have been extensively investigated since the 1950s. They are employed in numerous large-volume industrial catalytic applications [4–6]: olefin metathesis over supported Re_2O_7/Al_2O_3 and WO_3/SiO_2 [7, 8], ethylene polymerization over supported CrO_3/SiO_2 [9, 10], o-xylene oxidation to phthalic anhydride over supported V_2O_5/TiO_2 [11, 12], oxidation/ammoxidation of alkyl pyridines for production of pharmaceuticals over supported V_2O_5/TiO_2 and V_2O_5/Al_2O_3 catalysts [13], selective catalytic reduction (SCR) of NO_x with ammonia or other reducing agents to N_2 over supported V_2O_5-WO_3/TiO_2 and V_2O_5-MoO_3/TiO_2 catalysts [14–20], H_2S oxidation to elemental sulfur over supported Fe_2O_3/SiO_2 and MnO_x/SiO_2 catalysts [21], combustion of automotive exhaust over supported PtO_x/CeO_2 and NO_x storage over supported $PtO_x/BaO_x/Al_2O_3$ [22, 23]. Supported metal oxide catalysts represent model well defined catalysts for fundamental studies since the surface coverage of the catalytic active sites can be varied in the sub-monolayer region and their molecular and electronic structures readily characterized by many available spectroscopic techniques, even under reaction conditions [3]. Furthermore, combination of the molecular and electronic structural information with corresponding catalytic performance data allows the establishment of molecular/electronic structure–activity/selectivity relationships that facilitate the design of advanced catalytic materials.

The focus of this chapter will be on supported vanadia catalysts since they represent the most examined and applied supported metal oxide catalyst systems. The characterization and catalysis data will emphasize the work of the Wachs group and collaborators since both characterization and examination of their catalytic performance has been conducted on the same samples. For the development of structure–activity relationships it is critical that the characterization and catalytic performance data be performed on the same catalyst samples so that any discrepancies are minimized.

11.3
Monolayer Surface Coverage

As indicated in the introduction, it is critical to determine the surface coverage corresponding to monolayer coverage because of the different metal oxide species present above and below monolayer coverage. Many methods have been developed to determine monolayer surface coverage of supported metal oxide catalytic materials and all methods essentially give the same results, as should be the case if the methodologies employed are correct. The discussion below assumes that the supported metal oxide catalysts were well prepared in the sub-monolayer region so that crystalline NPs are not present. Monolayer surface coverage can be determined with (i) X-ray Photoelectron Spectroscopy (XPS) by plotting the intensity ratio of surface MO_x/support signals as a function of MO_x content and establishing where there is a break or "knee" in the curve when crystalline NPs are present [1, 24], (ii) by plotting the Low Energy Ion Scattering Spectroscopy (LEISS) intensity ratio of surface MO_x/support signals as a function of MO_x content and establishing where there is a break in the curve [25], (iii) by plotting the Raman intensity ratio of surface MO_x/support signals as a function of MO_x content and establishing where there is a break in the curve (in this method, the support Raman band must be employed as an internal reference) [24, 26], (iv) determining the appearance of crystalline MO_x NPs as a function of MO_x content [1, 25, 27], (v) monitoring the consumption of surface hydroxyls vs MO_x content with IR spectroscopy since the surface metal oxides anchor to the support by titrating the surface hydroxyls [1, 24], (vi) CO_2 chemisorption on the unreacted surface hydroxyls, but care must be taken if there is selective titration of surface hydroxyls by the supported MO_x [1, 24], and (vi) CH_3OH-Temperature Programmed Surface Reaction (TPSR) spectroscopy which monitors the number of exposed surface MO_x sites, since this parameter decreases above monolayer coverage as crystalline metal oxide NPs are introduced into the catalyst [26]. The catalytic activity (moles $(gs)^{-1}$) generally also levels off above monolayer coverage because of the reduced number of exposed MO_x sites above monolayer coverage, but this approach only works if the catalytic reaction being employed is not sensitive to the presence of the crystalline metal oxide NPs [11–13]. Consequently, it is risky to rely solely on the catalytic reaction to determine monolayer coverage, and independent spectroscopic determination of monolayer coverage is the proper method for monolayer determination. As already mentioned, the monolayer surface coverage usually has the same value independent of the method of determination.

The Raman spectra of a series of supported V_2O_5/Al_2O_3 catalysts as a function of vanadium oxide loading are presented in Figure 11.1. Monolayer surface coverage is readily determinable from this series of Raman spectra since below monolayer coverage only the broad Raman band of dehydrated surface VO_x species are present at 1008–1031 cm^{-1} and the sharp crystalline V_2O_5 NPs Raman band at 995 cm^{-1} as well as accompanying broader bands at 690 and 525 cm^{-1} appear above monolayer surface coverage. Thus, monolayer surface coverage of vanadium oxide on Al_2O_3 corresponds to 7.9 V atoms/nm^2. For supported vanadium oxide catalysts, monolayer surface coverage has been determined to be ~8 V atoms/nm^2 [25] on all oxide supports (Al_2O_3, TiO_2, ZrO_2, Nb_2O_5, CeO_2, etc.) with the exception of SiO_2. The lower reactivity of the silica surface hydroxyls results in a maximum surface vanadium oxide dispersion of ~3 V atoms/nm^2 on SiO_2 [27]. The bulk V_2O_5 phase has a surface vanadium oxide density of ~10 V atoms/nm^2 on the V_2O_5 (010) basal plane, which shows that surface density of bulk metal oxides is not always quantitatively representative of surface metal oxide phases on oxide supports.

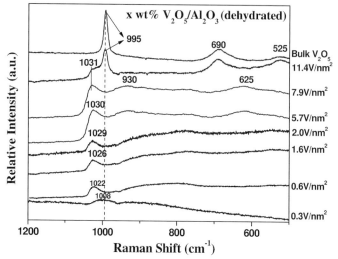

Figure 11.1 Raman spectra of supported V_2O_5/Al_2O_3 catalysts under dehydrated conditions as a function of vanadium oxide loading.

11.4
Molecular and Electronic Structures

The molecular structures of the surface vanadium oxide species have been determined in recent years with the application of numerous *in situ* spectroscopic techniques. *In situ* solid-state ^{51}V NMR, X-ray Absorption Near-Edge Spectroscopy (XANES) and Extended X-ray Absorption Fine Structure (EXAFS) characterization studies have demonstrated that the surface vanadium oxide species possess VO_4 coordination under dehydrated conditions [28, 29]. Furthermore, both isolated and polymerized surface VO_4 species are present on oxide supports. The extent of polymerization of the surface VO_4 species has recently been estimated with *in situ* UV-Vis Diffuse Reflectance Spectroscopy (DRS) from the optical edge energy, E_g, which reflects the extent of polymerization (E_g decreases monotonically with increasing extent of polymerization and is related to the electron mobility). The extent of polymerization of the surface VO_4 species as a function of surface vanadium oxide density in the sub-monolayer region is presented in Figure 11.2 from UV-Vis DRS measurements [30]. Below ~15% of monolayer coverage, the surface VO_4 species are isolated and possess one terminal V=O bond (Raman bands at 1008–1031 cm^{-1} in Figure 11.1) and three bridging V–O–Support bonds (weak and broad Raman band at ~900 cm^{-1} in Figure 11.1). At intermediate surface coverage, both isolated and polymeric surface VO_4 species coexist on the Al_2O_3 support and the extent of polymerization increases monotonically with surface coverage. At monolayer surface coverage, essentially all the surface VO_4 species are polymerized and possess one terminal V=O bond (1031 cm^{-1}) together with bridging

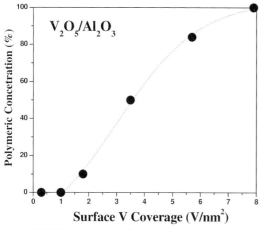

Figure 11.2 Concentration of polymeric surface VO_4 species for dehydrated supported V_2O_5/Al_2O_3 catalysts in the sub-monolayer region as a function of surface vanadium oxide density.

V–O–Support (930 cm^{-1}) and V–O–V (625 cm^{-1}) bonds. Similar distributions of monomeric and polymeric surface VO_4 species are found on other oxide supports with the exception of SiO_2 [30]. For the supported V_2O_5/SiO_2 catalyst system, only isolated surface VO_4 species are present below the maximum dispersion limit (<3 V atoms/nm^2). For all supported vanadium oxide catalysts, crystalline V_2O_5 NPs are also present above the monolayer surface coverage or maximum dispersion limit [31].

11.5
Number of Exposed Catalytic Active Sites (N_s)

Below monolayer surface coverage, the number of exposed catalytic active sites is just the number of surface VO_4 species present on the support, and increases linearly with surface vanadium oxide coverage until monolayer. Above monolayer surface coverage, both exposed surface VO_4 and crystalline V_2O_5 NP sites are present and the contribution of each component must be considered. The number of exposed vanadium oxide catalytic active sites can be determined with CH_3OH-TPSR spectroscopy, since a monolayer of surface CH_3O^* is formed during CH_3OH chemisorption at ~100 °C [32, 33]. The redox vanadium oxide sites yield HCHO as the reaction product and the exposed acidic Al_2O_3 sites form dimethyl ether. The area under the HCHO TPSR curves reflects the number of redox sites. The number of redox sites for the supported V_2O_5/Al_2O_3 catalyst system as a function of vanadium oxide loading is shown in Figure 11.3. In the sub-monolayer region, the number of catalytic redox sites increases linearly with the vanadium oxide

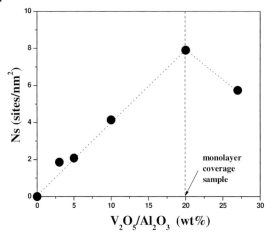

Figure 11.3 Number of exposed redox catalytic active sites for supported V_2O_5/Al_2O_3 catalysts as a function of vanadium oxide loading. Monolayer coverage corresponds to the 20% V_2O_5/Al_2O_3 (~7.9 V atoms/nm^2) catalyst sample.

loading and then decreases with increasing vanadium oxide loading above 20% V_2O_5/Al_2O_3 (7.9 V atoms/nm^2) when crystalline V_2O_5 NPs are also present. Note that the maximum number of catalytic active redox sites corresponds to monolayer surface VO_4 coverage. This suggests that the crystalline V_2O_5 NPs are covering up more surface VO_4 sites than their contribution to the number of catalytic active sites. The low number of exposed redox catalytic sites for the supported V_2O_5 NPs is a consequence of their platelet morphology, which results in an inactive (001) basal plane that terminates with the V=O functionality and only the minor edge sites terminating with V–O–V and V-OH sites are able to chemisorb CH_3OH [34, 35].

11.6
Surface Reactivity

The surface reactivity of the isolated and polymeric surface VO_4 species present in supported V_2O_5/Al_2O_3 catalysts was also determined from CH_3OH-TPSR experiments. As mentioned above, CH_3OH chemisorbs as surface $CH_3O^{\bullet}$ on supported vanadium oxide catalysts and the rate determining step involves breaking the surface methoxy C–H bond to form the HCHO redox product. The CH_3OH-TPSR spectra revealed that the peak temperature for breaking the C–H bond of the surface methoxy intermediate was ~190 °C (corresponding to a first-order rate constant of ~0.2 HCHO molecules/(site sec) at 230 °C) and is independent of surface vanadium oxide coverage in the sub-monolayer region. The constant surface kinetics with coverage indicates that there is no difference in the reactivity between monomeric and polymeric surface VO_4 species on Al_2O_3. Furthermore, this also suggests that the bridging V–O–V bonds in the polymeric surface VO_4 species are not critical in the surface $CH_3O^\cdot$ dehydrogenation to HCHO reaction step. The TPSR peak temperature was also not affected by the presence of crystalline V_2O_5 NPs since they contributed too few sites and the peak temperature value for bulk crystalline V_2O_5 corresponds to ~188 °C, which is not very different.

Similar findings were also observed for other supported vanadium oxide catalysts as a function of surface vanadium oxide coverage. Furthermore, with the exception of the supported V_2O_5/SiO_2 catalyst, the CH_3OH-TPSR peak temperature of all the supported vanadium oxide catalysts were not very different and varied within the narrow temperature range 182–192 °C, which corresponds to a factor of ~2 in k_{rds}. Thus, the surface methoxy C–H breaking kinetics are relatively constant for all the supported vanadium oxide catalysts and not significantly affected by the oxide support.

11.7
Steady-State Reactivity (TOF)

The methanol oxidation Turnover Frequency, (TOF) (the number of HCHO molecules/(site sec)), for the supported V_2O_5/Al_2O_3 catalysts were determined from steady-state methanol oxidation studies at 230 °C and knowledge of N_s (see Figure 11.3 above). The TOF for methanol oxidation over the supported V_2O_5/Al_2O_3 was found to be essentially constant at ~7×10^{-3} HCHO molecules/(site sec) for all vanadium oxide loading. The constant TOF demonstrates that the TOF is the same for surface VO_4 monomers and polymers, and that the bridging V–O–V bonds in the polymeric surface VO_4 species are not critical for methanol oxidation to formaldehyde. It is not easy to determine the TOF for the crystalline V_2O_5 NPs above monolayer coverage since their contribution to the overall TOF was minor.

Changing the specific oxide support, however, dramatically affected the steady-state TOF values by orders of magnitude: V_2O_5/CeO_2 (~1.0 HCHO molecules/(site sec)) > V_2O_5/ZrO_2 (~0.22 HCHO molecule/(site sec)) > V_2O_5/TiO_2 (~0.17 HCHO molecules/(site sec)) > V_2O_5/Al_2O_3 (~0.007 HCHO molecules/(site sec)). To fully understand the origin of this significant variation in TOF, it is necessary to examine the overall kinetics for methanol oxidation. The steady-state kinetics is expressed by [36]

$$\text{TOF} = K_{ads} \times k_{rds} \times P_{CH3OH} \tag{11.1}$$

in which k_{rds} represents the rate determining step of methanol oxidation to HCHO (the surface methoxy C–H bond-breaking step), K_{ads} is the CH_3OH equilibrium adsorption (the CH_3OH O–H bond breaking step) and P_{CH3OH} is the methanol partial pressure. The methanol oxidation reaction is zero order in O_2 partial pressure because the surface VO_4 sites are fully oxidized under the methanol oxidation reaction conditions in excess oxygen [37]. The CH_3OH-TPSR findings demonstrate that k_{rds} is relatively constant as the oxide support is changed, which means that changes in K_{ads} are responsible for the dramatic TOF variations since the methanol partial pressure is constant in all the studies. Increases in K_{ads} result in higher surface $CH_3O^{\bullet}$ concentrations under reaction conditions. This means that the dramatic changes in TOF with the specific oxide support is related to the number of surface methoxy intermediates on the catalyst surface during methanol oxidation and not variations in the rate-determining-step first-order k_{rds}.

Recent density functional theory calculations for CH_3OH chemisorption found that the most energetically favorable chemisorption site on supported VO_4 species is the bridging V–O–Support bond [38–40]. The density functional theory methanol oxidation calculation over supported V_2O_5/SiO_2 also concluded that the changes in reactivity for the different supports must be in the chemisorption step since the first-order energetics and kinetics for breaking the C–H bond of the surface $CH_3O^{\bullet}$ intermediate is very similar for the gas phase $O=V-(OCH_3)_3$ molecular complex and the surface methoxy intermediate on the model V_2O_5/SiO_2 cluster [39].

11.8
Number of Participating Catalytic Active Sites in Oxidation Reactions

Below monolayer coverage, the redox VO_4 catalytic active sites are 100% dispersed on the oxide supports and their surface coverage can be systematically varied (see Figure 11.3). Thus, the kinetic Equation 11.1 can be rewritten as

$$\text{Rate} = K_{ads} \times k_{rds} \times P_{CH3OH} \times N_s^n \tag{11.2}$$

in which n represents the number of participating catalytic active sites. Furthermore, for constant temperature and methanol partial pressure the log of Equation (11.2) results in

$$\log \text{Rate} = \log(K_{ads} \times k_{rds} \times P_{CH3OH}) + n \log N_s \tag{11.3}$$

and a plot of log Rate vs log N_s yields a line with slope n, the number of participating redox surface VO_4 catalytic active sites in the reaction, and the constant intercept of $\log(K_{ads} \times k_{rds} \times P_{CH3OH})$. The value of n for CH_3OH oxidation is 1 since the TOF is constant with coverage in the sub-monolayer region. Equation 11.3 can also be generalized to other oxidation reactions over supported vanadium oxide catalysts where the reaction is zero order in oxygen partial pressure, which corresponds to excess oxygen, as

$$\log \text{Rate} = \log(K_{ads} \times k_{rds} \times P_{Reactant}) + n \log N_s \tag{11.4}$$

The values of n have been determined for a number of oxidation reactions over supported vanadium oxide catalysts and are given in Table 11.1.

Table 11.1 Number of participating catalytic active sites for different oxidation reactions. Determined by application of Equation (11.4).

Catalytic oxidation reaction	Number of participating sites	Ref.
$CH_3OH \rightarrow HCHO$ (formaldehyde)	1	[41]
$SO_2 \rightarrow SO_3$ (sulfur trioxide)	1	[42]
$C_2H_6 \rightarrow C_2H_4$ (ethylene)	1	[43]
$C_3H_8 \rightarrow C_3H_6$ (propylene)	1	[30, 44, 45]
$C_3H_6 \rightarrow C_3H_4O$ (acrolein)	2	[46]
$n\text{-}C_4H_{10} \rightarrow C_4H_2O_3$ (maleic anhydride)	multiple[a]	[47]
$NO_x/NH_3 \rightarrow N_2$ (molecular N_2)	multiple[a]	[15]

a Exact values of n have not been determined, but TOF increases with surface coverage, reflecting the participation of multiple catalytic active sites.

All the reactions involving the participation of one surface VO_4 site are 2 e⁻ reactions that require only one O atom. All the reactions involving multiple surface VO_4 sites are 4 to 8 e⁻ reactions requiring the involvement of multiple O atoms. Note that each surface VO_4 species can only be reduced from V^{+5} to V^{+3}, which corresponds to the release of one O atom. Similar to the CH_3OH oxidation to HCHO, varying the oxide support changes the TOF values by a couple of orders of magnitude for all the oxidation reactions listed in Table 11.1, which again demonstrates the pronounced effect of the bridging V–O–Support bond on all oxidation reactions.

11.9
Role of Surface Acid Sites on Oxidation Reactions

The concentration of surface Brønsted acids increases monotonically with surface VO_4 coverage and qualitatively follows the presence of polymeric surface VO_4 species [43, 48]. The presence of the surface Brønsted acid sites, however, has no effect on the TOF values for oxidation reactions involving only one surface VO_4 site. Surface Brønsted acid sites, however, can facilitate oxidation reactions requiring both surface redox and acid sites. For example, (i) the oxidation of *n*-butane to maleic anhydride does require the presence of acid sites and is probably related to the ring-closure step [47, 49], (ii) the presence of surface acid sites, Brønsted as well as Lewis, enhances the NO_x SCR reaction because the acid sites facilitate the chemisorption of basic NH_3 [14, 20] and (iii) the presence of surface acid sites enhances the chemisorption of propane via a precursor state [50]. Thus, surface Brønsted acid sites do not facilitate oxidation reactions involving one redox surface VO_4 site, but may facilitate oxidation reactions involving multiple redox surface VO_4 sites, especially if the oxidation reaction requires the participation of dual redox–acid sites.

11.10
Other Supported MO$_x$ Redox Active and Acidic Systems

Although this chapter focuses on oxidation reactions involving redox supported vanadium oxide catalysts, similar trends with surface coverage and specific oxide support also apply for other redox supported transition metal oxide catalysts, such as supported MoO$_3$ [51], CrO$_3$ [52] and Re$_2$O$_7$ [53]. The redox supported vanadium oxide catalytic system was chosen for this review because of the extensive studies that these catalysts have received in recent years as well as their widespread industrial applications.

Supported Nb$_2$O$_5$ [54], Ta$_2$O$_5$ [55] and WO$_3$ [26] catalysts typically possess almost no redox potential and exclusively behave as surface acid sites. Other than their acidic properties, these supported metal oxides possess similar molecular and electronic structural characteristics as the redox surface sites discussed above.

11.11
Conclusions

Supported metal oxide catalysts are a new class of catalytic materials that are excellent oxidation catalysts when redox surface sites are present. They are ideal catalysts for investigating catalytic molecular/electronic structure–activity selectivity relationships for oxidation reactions because (i) the number of catalytic active sites can be systematically controlled, which allows the determination of the number of participating catalytic active sites in the reaction, (ii) the TOF values for oxidation studies can be quantitatively determined since the number of exposed catalytic active sites can be easily determined, (iii) the oxide support can be varied to examine the effect of different types of ligand on the reaction kinetics, (iii) the molecular and electronic structures of the surface MO$_x$ species can be spectroscopically determined under all environmental conditions for structure–activity determination and (iv) the redox surface sites can be combined with surface acid sites to examine the effect of surface Brønsted or Lewis acid sites. Such fundamental structure–activity information can provide insights and also guide the molecular engineering of advanced hydrocarbon oxidation metal oxide catalysts such as supported metal oxides, polyoxo metallates, metal oxide supported zeolites and molecular sieves, bulk mixed metal oxides and metal oxide supported clays.

Acknowledgments

The support of the United States Department of Energy–Basic Energy Sciences grant DEFG02–93ER14350 during the writing of this manuscript is gratefully acknowledged.

References

1 Wachs, I.E. (1996) *Catalysis Today*, **27**, 437.
2 Wachs, I.E. (2005) *Catalysis Today*, **100**, 79.
3 Weckhuysen, B.M. and Wachs, I.E. (2001) Catalysis by supported metal oxides, in *Handbook of Surfaces and Interfaces of Materials* (ed. H.S. Nalwa), Academic Press, New York, p. 63.
4 Thomas, C.L. (1970) *Catalytic Processes and Proven Catalysts*, Academic Press, New York.
5 Weissermel, K. and Arpe, H.-J. (1978) *Industrial Organic Chemistry*, Verlag Chemie, Weinheim.
6 Wachs, I.E. (ed.) (1999) Special issue on *Supported Metal Oxide Catalysts. Catalysis Today*, **51**(2), 271.
7 Xiaoding, X., Boelhouwer, C., Vonk, D., Benecke, J.I. and Mol, J.C. (1986) *Journal of Molecular Catalysis*, **36**, 47.
8 Banks, R.L. (1984) Olefin metathesis: technology and application, in *Applied Industrial Catalysi* (ed. B.E. Leach), Academic Press, New York, p. 215.
9 McDaniel, M.P. (1986) *Advanced Catalysis*, **33**, 4.
10 Weckhuysen, B.M., Wachs, I.E. and Schoonheydt, R.A. (1996) *Chemical Reviews*, **96**, 3327.
11 Wachs, I.E., Saleh, R.Y., Chan, S.S. and Chersich, C.C. (1985) *Applied Catalysis*, **15**, 339.
12 Grabowski, R., Grzybowska, B., Haber, J. and Sloczynski, J. (1975) *Reaction Kinetics and Catalysis Letters*, **2**, 81.
13 Heinz, D., Hoelderich, W.F., Krill, S., Boeck, W. and Huthmacher, K. (2000) *Journal of Catalysis*, **192**, 1.
14 Bosch, H. and Janssen, F. (1988) *Catalysis Today*, **2**, 369.
15 Wachs, I.E., Deo, G., Weckhuysen, B.M., Andreini, A., Vuurman, M.A., de Boer, M. and Amiridis, M.D. (1996) *Journal of Catalysis*, **161**, 211.
16 Amiridis, M.D., Wachs, I.E., Deo, G., Jehng, J.-M. and Kim, D.S. (1996) *Journal of Catalysis*, **161**, 247.
17 Ramis, G., Busca, G., Lorenzelli, V. and Forzatti, P. (1990) *Applied Catalysis*, **64**, 243.
18 Lietti, L., Svachula, J., Forzatti, P., Busca, G., Ramis, G. and Bregani, F. (1993) *Catalysis Today*, **17**, 131.
19 Topse, N.-Y., Dumesic, J. and Topsoe, H. (1995) *Journal of Catalysis*, **151**, 241.
20 Amiridis, M.D., Duevel, R.V. and Wachs, I.E. (1999) *Applied Catalysis B Environmental*, **20**, 111.
21 Visser, R.J.A.M., Terörde, P.J., van den Brink, L.M., van Dillen, A.J. and Geus, J.W. (1993) *Catalysis Today*, **17**, 217.
22 Weber, W.H. (2000) Raman applications in catalysts for exhaust-gas treatment, in *Raman Scattering in Materials Science* (eds W.H. Weber and R. Merlin), Springer, New York, p. 233.
23 Takahashi, N., Shinjoh, H., Iijima, T., Suzuki, T., Yamazaki, K., Yokota, K., Suzuki, H., Miyoshi, N., Matsumoto, S., Tanizawa, T., Tanaka, T., Tateishi, S. and Kasahara, K. (1996) *Catalysis Today*, **27**, 63.
24 Wachs, I.E. and Segawa, K. (1992) Supported metal oxides, in *Characterization of Catalytic Materials*, Butterworth-Heinemann, Stoneham, MA, p. 69.
25 Briand, L.E., Tkachenko, O.P., Gurya, M., Gao, X., Wachs, I.E. and Gruenert, W. (2004) *Journal of Physical Chemistry B*, **108**, 4823.
26 Kim, T.J., Burrows, A., Kiely, C.J. and Wachs, I.E. (2007) *Journal of Catalysis*, **246**, 370.
27 Gao, X., Bare, S.R., Weckhuysen, B.M. and Wachs, I.E. (1998) *Journal of Physical Chemistry B*, **102**, 10842.
28 Eckert, H. and Wachs, I.E. (1989) *Journal of Physical Chemistry*, **93**, 6796.
29 Yoshida, S., Tanaka, T., Nishimura, Y., Mizutani, H. and Funabiki, T. (1998) *Proc. 9th Intern. Congress Catal., Calgary, Canada*, Vol. 3 (eds M.J. Phillips and M. Ternan), p. 1473.
30 Tian, H., Ross, E.I. and Wachs, I.E. (2006) *Journal of Physical Chemistry B*, **110**, 9593.
31 Vuurman, M.A., Hirt, A.M. and Wachs, I.E. (1991) *Journal of Physical Chemistry*, **95**, 9928.
32 Briand, L.E., Farneth, W.E. and Wachs, I.E. (2000) *Catalysis Today*, **62**, 219.

33 Burcham, L.J., Briand, L.E. and Wachs, I.E. (2001) *Langmuir*, **17**, 6164.
34 Briand, L.E., Jehng, J.-M., Cornaqglia, L., Hirt, A.M. and Wachs, I.E. (2003) *Catalysis Today*, **78**, 257.
35 Badlani, M. and Wachs, I.E. (2001) *Catalysis Letters*, **75**, 137.
36 Holstein, W. and Machiels, C. (1996) *Journal of Catalysis*, **162**, 118.
37 Burcham, L.J., Deo, G., Gao, X. and Wachs, I.E. (2000) *Topics in Catalysis*, **11/12**, 85.
38 Khaliullin, R.Z. and Bell, A.T. (2002) *Journal of Physical Chemistry B*, **106**, 7832.
39 Doebler, J., Pritzsche, M. and Sauer, J. (2005) *Journal of the American Chemical Society*, **127**, 10861.
40 Calatayud, M. and Minot, C. (2004) *Journal of Physical Chemistry B*, **108**, 15679.
41 Deo, G. and Wachs, I.E. (1994) *Journal of Catalysis*, **146**, 323.
42 Dunn, J.P., Koppula, P.R., Stenger, H.G. and Wachs, I.E. (1998) *Applied Catalysis B Environmental* **19**, 103.
43 Martinez-Huerta, M.V., Gao, X., Tian, H., Wachs, I.E., Fierro, J.L.G. and Banares, M.A. (2006) *Catalysis Today*, **118**, 279.
44 Gao, X., Jehng, J.-M. and Wachs, I.E. (2002) *Journal of Catalysis*, **209**, 43.
45 Zhao, Z., Gao, X. and Wachs, I.E. (2001) *Journal of Physical Chemistry B*, **107**, 6333.
46 Zhao, C. and Wachs, I.E. (2006) *Catalysis Today*, **118**, 332.
47 Wachs, I.E., Deo, G., Jehng, J.-M., Weckhuysen, B.M., Guliants, V.V., Benziger, J.B. and Sundaresan, S. (1997) *Journal of Catalysis*, **170**, 75.
48 Turek, A.M., Wachs, I.E. and DeCanio, E. (1992) *Journal of Physical Chemistry*, **96**, 5000.
49 Deo, G. and Wachs, I.E. (1994) *Journal of Catalysis*, **146**, 335.
50 Mitra, B., Wachs, I.E. and Deo, G. (2006) *Journal of Catalysis*, **240**, 151.
51 Hu, H. and Wachs, I.E. (1995) *Journal of Physical Chemistry*, **99**, 10911.
52 Kim, D.S. and Wachs, I.E. (1993) *Journal of Catalysis*, **142**, 166.
53 Kim, D.S. and Wachs, I.E. (1993) *Journal of Catalysis*, **141**, 419.
54 Jehng, J.-M. and Wachs, I.E. (1990) *Catalysis Today*, **8**, 37.
55 Chen, Y., Fierro, J.L.G., Tanaka, T. and Wachs, I.E. (2003) *Journal of Physical Chemistry B*, **107**, 5243.

12
Vanadium Phosphate Catalysts

Jonathan K. Bartley, Nicholas F. Dummer, and Graham J. Hutchings

12.1
Introduction

Vanadium phosphates have been of considerable interest since the mid-1960s, when Bergman and Frisch [1] found them to be effective catalysts for the oxidation of *n*-butane to maleic anhydride (Equation 12.1).

$$\text{butane} + 3.5\, O_2 \xrightarrow{\text{VPO}} \text{maleic anhydride} + 4 H_2O \quad (12.1)$$

Prior to this, maleic anhydride had been industrially manufactured by the oxidation of benzene over supported V_2O_5-MoO_3 catalysts. In the late 1970s, when pollution laws that restricted benzene emissions came into effect, industry began to use the *n*-butane route. The reaction is of great importance as it is the only industrial large-scale selective oxidation of alkanes currently in operation. It also involves the functionalization of an alkane, providing a use for this rather unreactive oil fraction.

Maleic anhydride is used as a chemical intermediate in the synthesis of fumaric and tartaric acids, certain agricultural chemicals, resins in numerous products, dye intermediates and pharmaceuticals [2]. It is primarily used as a co-monomer for unsaturated polyester resins, which are used in the production of bonding agents for plywood manufacture and when mixed with glass fibres for reinforced plastics. Annual production of maleic anhydride is estimated to be over one million tonnes [3].

The selective oxidation of C_4 hydrocarbons leads to various commercially desirable products, which encouraged the development of the route involving the oxidation of *n*-butane to maleic anhydride. This new route was also less expensive owing to the lower cost of butane compared with benzene.

The *n*-butane oxidation process is interesting as it is an extensive oxidation, with the cleaving of eight C—H bonds and the introduction of three oxygen atoms, yet it still manages to occur selectively. The process is a fourteen-electron oxidation, in comparison to other selective oxidation processes, which only require a maximum of four electrons.

Many well characterized, crystalline vanadium phosphate phases have been identified, whose structure and catalytic properties have been well documented. Some of the most widely studied are the V^{5+} vanadyl orthophosphates (α-, β-, γ-, δ-, ε- and ω-$VOPO_4$, and $VOPO_4 \cdot 2H_2O$), and the V^{4+} vanadyl hydrogen phosphates ($VOHPO_4 \cdot 4H_2O$, $VOHPO_4 \cdot \frac{1}{2}H_2O$, $VO(H_2PO_4)_2$), vanadyl pyrophosphate (($VO)_2P_2O_7$) and vanadyl metaphosphate ($VO(PO_3)_2$). Of these compounds, $VOHPO_4 \cdot \frac{1}{2}H_2O$ (vanadyl hydrogen phosphate hemihydrate) is of particular interest as a catalyst precursor, which after activation gives a catalyst mainly composed of $(VO)_2P_2O_7$ (vanadyl pyrophosphate, hereafter VPP).

In this chapter we will discuss how the behavior of the catalyst is influenced by a number of factors including: the method of preparation of the precursor, the oxidation state of the catalyst, the phosphorus/vanadium ratio of the catalyst and the activation conditions. A variety of techniques have been used to characterize the morphology and nature of the active sites of the catalysts and gain insight into the mechanism. Furthermore, the effect of these preparatory techniques will be discussed, with emphasis on the partial oxidation of *n*-butane. We will pay particular attention to industrially relevant examples where possible and attempt to describe the current state of the art.

Although there is a great deal of debate about certain aspects of vanadium phosphate catalysts, Hodnett [4] has laid out a set of statements that most researchers generally agree upon:

- The most active and selective catalysts comprise mainly $(VO)_2P_2O_7$.
- During testing in *n*-butane lean conditions, the oxidation state of the catalyst is close to +4.
- The surface has some phosphorus enrichment. Only the surface layers are directly involved in catalysis.
- The rate determining step is butane activation by hydrogen abstraction

12.2
The Active Catalyst

The structure of vanadium phosphate catalysts is dependent on a number of factors. The P/V stoichiometry, thermal treatment time, activation temperature and gas phase composition can all affect catalyst composition. By varying these factors a variety of crystalline phases can be identified (by high-resolution transmission electron microscopy (HRTEM) [5] Figure 12.1a and X-ray diffraction Figure 12.1b) in the freshly activated catalyst [6]. It is widely accepted that VPP plays an important role in the oxidation of butane to maleic anhydride and most hypotheses are based on the (100) face (Figure 12.2). Additionally, this phase has been reported to be an efficient catalyst for the oxyfunctionalization of light paraf-

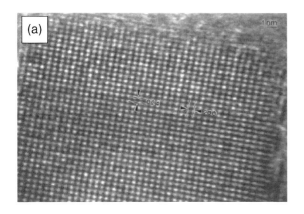

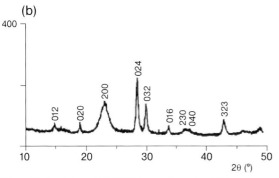

Figure 12.1 (a) An axial HREM image from the (021) projection of $(VO)_2P_2O_7$, (b) XRD pattern of an activated $(VO)_2P_2O_7$ catalyst. (Reproduced with permission).

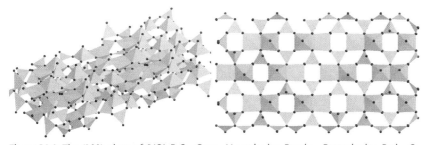

Figure 12.2 The (100) plane of $(VO)_2P_2O_7$. Grey = V octahedra, Purple = P tetrahedra, Red = O.

fins: (a) for the oxidation of ethane to acetic acid [6, 7], (b) for the oxidation and ammoxidation of propane to acrylic acid [8] and acrylonitrile [9, 10], respectively, and (c) for the oxidation of n-pentane to maleic and phthalic anhydrides [10–19].

The catalytic behavior of the different crystal faces has been investigated by Inumaru and coworkers by exposing individual planes [20]. VPP was deactivated by the surface deposition of SiO_2. The crystallites were then fractured to expose the side faces, for example the (021) and (001) faces. The side faces were found to be non-selective, with maleic anhydride formed only on the (100) face. There is debate as to whether VPP is indeed the active catalyst, or if a combination of phases are responsible for the reaction.

Transient studies by Ballarini and coworkers showed that the active surface of equilibrated catalysts is different depending on the reaction conditions and the P/V ratio of the catalyst [21]. At low temperature (320 °C) an active surface forms that is selective and probably is more like $VOPO_4$ than VPP. However, as the temperature is increased to 380 °C this material becomes less selective. The active phase formed at $T > 380$ °C was found to be less active than the low-temperature phase ($T < 380$ °C) but has increased selectivity at this temperature. At these temperatures the active site is found to hydrolyze and oxidize and Ballarini and coworkers propose that the active surface is a VO_x/polyphosphoric acid mixture. The authors speculate that the different phase evolutions at different temperatures, which are also dependent on very minor changes in the P/V ratio, could be the cause for the very differing surfaces observed by both *in situ* and *ex situ* studies of the active catalyst.

12.2.1
The Oxidation State of the Catalyst

The final oxidation state of the activated catalysts varies between +4.00 and +4.40, depending on the amount of V^{5+} present in the catalyst. There is extensive discussion as to whether V^{5+} and V^{3+} phases are important in the reaction mechanism.

Ebner and Thompson have postulated that the V^{5+} phases that are formed during the activation period, are unimportant and do not contribute to the oxidation mechanism [22]. They have found that after several hundred hours on stream, the V^{5+} orthophosphate phases are reduced to pyrophosphate, giving an active catalyst with a final oxidation state of +4.00 to +4.04. The concentration of oxygen and butane in the reactant feedstock determines the time needed to equilibrate the catalyst. Based on this study, they suggest that other researchers (who find V^{5+} phases in the active catalyst) have not performed the activation process fully, or have an unfavorable redox potential in the gas stream.

A series of vanadium phosphate catalysts prepared by different routes and containing different phases were examined by Guliants and coworkers [23]. From this study it was concluded that the catalytically active phase is an active surface layer on VPP. Their experimental results showed $VOPO_4$ phases to be detrimental to the performance of the catalyst. This was confirmed by Cavani and Trifirò, who suggested that V^{5+} sites are responsible for the over-oxidation of maleic anhydride to carbon oxides [24].

A number of groups dispute this one-phase hypothesis. They suggest that V^{5+} phases are important in the active catalyst and are formed as the result of a redox mechanism (Figure 12.3) [25–27]. Bordes cites the apparent need for two contradictory conditions during the oxidation of butane as evidence for a multi-phase active catalyst [19]. It was stated that oxygen associated with V^{4+} activates butane

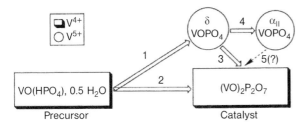

1: Oxydehydration 3,5: Reduction V^{5+} to V^{4+}
2: Topotactic transformation 4: Isovalence transformation

Figure 12.3 Scheme of the proposed evolution of the VPO catalyst with time [32].

while the oxygen associated with V^{5+} is incorporated later. V^{4+} gives a very active catalyst that has poor selectivity, while V^{5+} gives a very selective catalyst that has a low activity. This mechanism has been supported by the X-ray Photoelectron Spectroscopy (XPS) studies of Coulston and coworkers [28] and Temporal Analysis of Products (TAP) studies by Rodemerck and coworkers [29, 30] and Lorences and coworkers [31]. These studies report that in the absence of V^{5+} sites no maleic anhydride was observed and the main reaction product was furan. When V^{5+} sites were present maleic anhydride was produced.

Experimental results on pure vanadium phosphate phases and active catalysts suggest that the active catalyst is VPP with domains of V^{5+} on the (100) face [33] The lack of selectivity of side faces found by Inumaru and coworkers [34, 35] is attributed to the difficulty of the re-oxidation to V^{5+} of these planes. Hutchings and coworkers propose a V^{4+}/V^{5+} couple, which can be present on the surface of a range of vanadium phosphate phases, as the active site [36]. The active phase is suggested to comprise a well dispersed micro-crystalline $VOPO_4$ on a VPP matrix. Centi and coworkers proposed that the reaction proceeds via a series of redox couples [37], the activation of butane requiring a V^{4+}–V^{3+} couple, while the subsequent conversion to maleic anhydride requires a V^{5+}–V^{4+} couple.

There has been speculation about the source of the V^{5+} in these two-phase systems. α_{II}-$VOPO_4$, γ-$VOPO_4$ and δ-$VOPO_4$ are commonly found in $(VO)_2P_2O_7$-based catalysts [38, 39]. More recently, ω'-$VOPO_4$ has been identified as a surface species [40–42] in the final active catalysts, especially those activated in slightly reducing conditions. Koyano and coworkers [43, 44] propose the X_1 phase as the V^{5+} source, although they admit this phase may be δ-$VOPO_4$. It should also be noted that a recent *in situ* X-ray Diffraction (XRD) study by Conte and coworkers [45] showed that metastable $VOPO_4$ phases can exist at the elevated temperatures of the catalytic process but that these can be undetectable in the final catalyst as they disorder into an amorphous material at room temperature.

Agaskar and coworkers [46] proposed that the catalyst surface contains clusters of four active dimeric sites that can each exist in one of four states. These states differ in the number of oxygen atoms associated with them and in the oxidation state of the vanadium ions present (Scheme 12.1). The V^{3+} site [S_0] acts as an

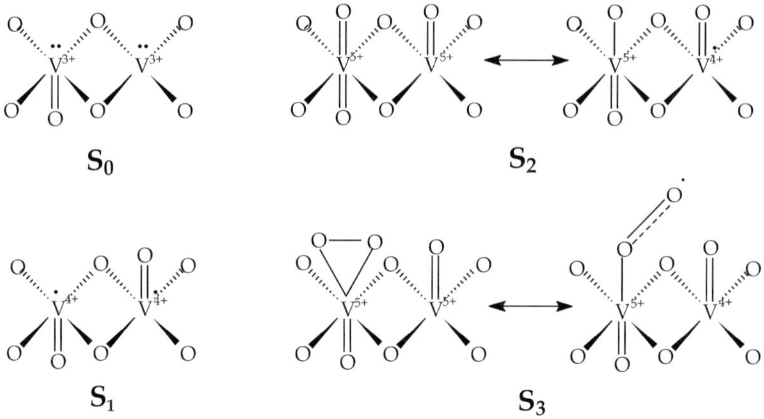

12 Vanadium Phosphate Catalysts

Scheme 12.1

oxygen acceptor. It can either react with gaseous oxygen to give [S_2], or accept an oxygen atom from [S_2] to give [S_1]. The active site for the butane adsorption and reaction is thought to be one associated with two vanadyl groups ([S_2] or [S_3]). Rodemerck and coworkers [29] have investigated the importance of V^{3+} using TAP experiments. They propose that only V^{4+} and V^{5+} sites are important, and demonstrate that maleic anhydride was not formed on catalysts with an oxidation state below +3.96 [29]. However, this is the bulk oxidation state and so does not discount the possibility of isolated V^{3+} sites being present at the catalyst surface.

12.2.2
The Phosphorus-to-Vanadium Ratio of the Catalyst

Trifirò has proposed that the active catalyst is pure VPP and found that the catalyst had a slight increase in oxidation state after the equilibrium period [47]. The small increase from +4.00 to +4.03 was reproducible and attributed to isolated V^{5+} surface sites being formed. The P/V ratio is proposed to be a key factor in the stabilization of V^{4+} within the catalyst, as $VOPO_4$ formation becomes very difficult at P/V ratios above 2.0. Trifirò had stated that a very high surface P/V ratio is required for an active and selective catalyst, and experimentally has found surface P/V ratios of 10:1.

The P/V ratio was proposed by Ruiz and coworkers to be an important factor in determining the activity and selectivity of the catalyst [48]. They acknowledge that the majority of industrial catalysts have two or more different phases present, and suggest that the active catalyst is a mixture of two phases with different P/V ratios. By mixing a low P/V ratio catalyst and a high P/V ratio catalyst, they found a 5.7-fold increase in yield and a 3.8-fold increase in selectivity, compared to the sum of the individual phases. Ruiz suggests different roles for the two phases. The high P/V ratio phase acts as an oxygen acceptor, whereas the low P/V ratio phase acts as an oxygen donor. Experiments with Sb_2O_4 and $BiPO_4$, which are known to be good oxygen donors, support this assignment of roles for the different phases. Ruiz found mechanical mixtures of a high P/V ratio catalyst with Sb_2O_4 and $BiPO_4$ gave a 6.0-fold increase in yield and a 5.4-fold increase in selectivity, over the standard $(VO)_2P_2O_7$ catalyst.

Garbassi and coworkers [49] observed that a P/V ratio of 1.05–1.1 is necessary for a high catalytic performance.

This slight excess of phosphorus is proposed to stabilize the V^{4+} phase, as V^{5+} is only detected in catalysts with a bulk P/V ratio less than 1.0. The findings of Garbassi and coworkers were confirmed by Hodnett and coworkers [50, 51] who suggested that a surface enrichment of phosphorus (P/V = 2.4) is the important factor for a selective catalyst. They propose that the excess surface phosphorus is responsible for the selectivity of the catalyst by isolating the V^{4+} active site on the surface.

Morishige and coworkers [52–54] found that although the industrial catalyst had a bulk P/V ratio of approximately 1.0, the surface had a high phosphorus concentration. Their studies showed that the surface P/V ratio was 1.6–1.8, when the bulk P/V ratio was 0.95–1.2. They suggested that a phosphorus-rich layer, supported on the surface of $(VO)_2P_2O_7$, is responsible for the oxidation of butane to maleic anhydride. For catalysts with a P/V ratio of between 1.0 and 2.0, Satsuma and coworkers also found catalytic improvements [55]. They proposed that at P/V ratios less than 1.0 a number of vanadium sites remain inactive but at higher P/V ratios all surface sites are active.

However, it should be noted that Coulston and coworkers [28] reported contradictory findings. In summary, they found that the surface P/V of a number of pure vanadium phosphate phases and activated catalysts are all closer to that of the bulk than has been previously reported, leading to the claim that phosphorus enrichment reported by other groups was based on incorrect calibration of XPS instrumentation.

12.2.3
The Role of Amorphous Material

There is considerable discussion in the literature as to whether amorphous material, commonly found in vanadium phosphate catalysts, plays a role in the catalysis of *n*-butane to maleic anhydride. Industrial catalysts undergo a lengthy activation process (often >1000 h) before they are equilibrated. During the activation, the crystallinity of the catalysts increases, leading some researchers to suggest that the more crystalline the catalyst, the better the performance [25]. Guliants and coworkers observed that although an amorphous surface layer was found to be present on fresh catalysts this disappeared on-line, leading to an increase in catalytic activity [56]. Most researchers consider that the (100) plane of crystalline VPP is the catalytically active plane, and the mechanisms that have been proposed use this as their active site [46, 57–69]. However, other researchers consider that an amorphous material supported on a $(VO)_2P_2O_7$ matrix plays an important role in catalysis.

As catalysts commonly contain both amorphous and crystalline material, it cannot be stated with any certainty whether or not the amorphous phase is catalytically active. However, experimental observations have added weight to the theory that amorphous material is the catalytically active phase in vanadium phosphate catalysts.

Research by Ruiz and coworkers [48] into high and low P/V ratio catalysts, led to the hypothesis that the active catalyst is made up of VPP in conjunction with an amorphous phase with a high oxidation state near +5.

Morishige and coworkers also suggested that the active catalyst is an amorphous phase with excess phosphorus on the surface of $(VO)_2P_2O_7$ [52]. Evidence was found by extracting the amorphous phase from the bulk. Catalytic testing yielded the same activity and selectivity as $(VO)_2P_2O_7$, suggesting that the amorphous phase is catalytically active. It has been proposed that the amorphous phase is phosphorus rich, accounting for the high surface P/V ratio that is commonly observed experimentally [28, 48, 50–52, 70].

In situ Raman spectroscopy has been carried out on precursors prepared in aqueous solution (referred to as VPA materials see Section 12.3.1), as they were converted to the active catalyst [38]. They found that during the activation, there is a structural disordering at 370 °C, which corresponds to the appearance of maleic anhydride. The disordering was found to occur at a lower temperature (300 °C), when maleic anhydride was added to the butane/air reaction mixture. This demonstrates that the presence of the products is important in controlling the structural transformations, and that a highly disordered structure can be important in selective butane oxidation.

Further evidence for the catalytic importance of amorphous material comes from experiments using cobalt-doped catalysts. Hutchings and coworkers found that doping the catalysts with cobalt improved their performance [36]. Additionally, Sajip and coworkers [71] found that the cobalt-promoted catalysts are far more disordered than the undoped catalysts. In the doped catalysts, the promoter is dispersed in the amorphous phase and cobalt is not found in the VPP crystals. It is thought that one of the properties of the cobalt promoter is the stabilization of the disordered phase and V^{5+} phases in the final catalysts, which leads to improved performance. This implies that the disordered material is the catalytically active vanadium phosphate phase.

Hutchings and coworkers [72–74] have prepared vanadium phosphate catalysts using supercritical precipitation methods. These materials were found to be amorphous by XRD and electron diffraction, but showed activity comparable to standard vanadium phosphate catalysts. This demonstrates that an amorphous surface layer can be the active phase in these catalysts and that the crystalline VPP that has been so well studied may be nothing more than an elaborate support.

However, Schimoda and coworkers [75] and others [6, 76] have dismissed the V^{4+} amorphous phases prepared from the precursor $VO(H_2PO_4)_2$ as less selective than the crystalline $(VO)_2P_2O_7$ catalyst.

12.2.4
The Disordered Plane

The selective oxidation of *n*-butane and 1-butene on vanadium phosphate catalysts prepared via different routes was investigated by Cavani and coworkers [77] Precursors prepared in aqueous medium were found to have greater crystallinity than those prepared in organic solvents (the activity and selectivity of which was the same for 1-butene oxidation). However, for butane activation, the crystalline catalyst was considerably less active than the organically prepared catalyst, which had an XRD pattern showing some disorder in the (100) plane.

It has been suggested that organic compounds occluded between the vanadium phosphate layers cause this disorder [78]. The disorder may be derived from a number of structural modifications: a missing oxygen atom, an inversion from a *trans*- to a *cis*- vanadyl position, or from the modification of the V—O bond strength. It is proposed that these defects can all cause the creation of new active centers for butane activation. Furthermore, Cornaglia and coworkers [79] also report an increase in selectivity to maleic anhydride as the disorder in the (100) plane decreases.

The opposite effect has been reported for a series of organically prepared catalysts [25]. The catalytic performance to maleic anhydride was found to improve with the increase of crystallinity of VPP. A number of studies on catalysts that have been on-line for several hundred hours have provided supporting evidence that crystalline compounds are the most active and selective [79, 80].

12.2.5
Acid–Base Properties

Some research groups consider the acidity of the catalysts to be an important aspect in controlling the catalyst performance. An infrared study of the acid sites using NH_3, pyridine and acetonitrile as probe molecules showed the existence of Lewis and Brønsted acid sites [24, 65, 79, 81–83]. A correlation was observed between the selectivity to maleic anhydride and the number of strong Lewis acid sites. It has been suggested that hydrogen abstraction occurs on Lewis acid sites (V^{4+}), and that butane C—H bond cleavage results from interactions between Lewis acid sites and Brønsted acid sites (P—OH) [83].

Centi and coworkers [84] have suggested that, in addition to Lewis acidity, Brønsted acidity plays an important role in the selective oxidation of butane to maleic anhydride. The surface phosphorus enrichment means that a number of P—OH groups are present on the catalyst surface. Centi offers three hypotheses for the role of Brønsted acidity: the stabilization of reaction intermediates, the stabilization of an adsorbed oxygen species, or the generation of an organic surface species that is involved in oxygen activation or transport.

Ai measured the acid site concentration by the adsorption of ammonia [85]. No correlation was found between the P/V ratio, the acidity and the catalytic activity.

This has been attributed to the use of ammonia as a probe molecule, since this cannot distinguish between Lewis and Brønsted acidity. Cornaglia and coworkers [79] measured the acid sites using pyridine and acetonitrile. However, the pyridine results showed no correlation between the activity and selectivity to maleic anhydride and either the Lewis-to-Brønsted acid sites ratio (L/B acidity ratio) or the Lewis acid site concentration.

Adsorption of acetonitrile enabled the strength of the Lewis sites to be measured and a greater number of strong Lewis sites were found to be present in organically made catalysts than in those prepared in an aqueous medium. Thus, the greater the concentration of strong Lewis acid sites in the catalyst, the higher the maleic anhydride yield. Cornaglia and coworkers suggest that the strong Lewis sites are responsible for butane dehydrogenation.

12.3
Preparation of VPP Precursors

Vanadium phosphate catalysts are obtained by activating the catalyst precursor in the reaction feedstock. After pre-treatment, the catalyst is equilibrated and catalytic activity remains consistent throughout the lifetime of the catalyst. The activated catalysts are formed topotactically from the precursors [86]. For this reason, a great deal of research is based around the preparation of catalyst precursors with well defined, favorable morphologies.

VOHPO$_4 \cdot \frac{1}{2}$H$_2$O (Figure 12.4) is the catalytic precursor for (VO)$_2$P$_2$O$_7$. A number of preparation methods are commonly used to prepare VOHPO$_4 \cdot \frac{1}{2}$H$_2$O. These usually involve reacting V$_2$O$_5$ and H$_3$PO$_4$ in the presence of a reducing agent.

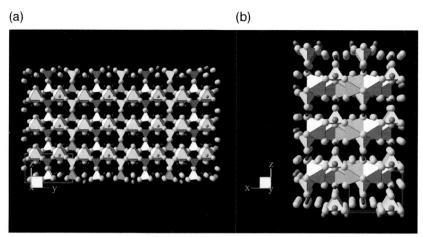

Figure 12.4 Schematic diagram showing the layered structure of hemihydrate precursors in the (a) (100) and (b) (010) directions (both orthogonal to the (001) layer normal direction) [87].

Originally, catalyst precursors were prepared in aqueous media, most commonly using hydrochloric acid as the reducing agent [80, 88–91]. This is commonly referred to as the VPA route:

$$V_2O_5 + HCl \xrightarrow[\text{2. }H_3PO_4]{\text{1. }\Delta,\text{ 2h}} VOHPO_4 \cdot \tfrac{1}{2}H_2O \qquad (12.2)$$

Alternative aqueous routes have been used by a number of groups to prepare $VOHPO_4 \cdot \tfrac{1}{2}H_2O$. Oxalic acid [88, 92], lactic acid [93], phosphorous acid [93] and $NH_2OH \cdot HCl$ [75] have all been investigated as reducing agents in place of hydrochloric acid.

There has also been investigation into alternative vanadium sources. Poli and coworkers [88, 92] used NH_4VO_3 as the vanadium source in conjunction with H_3PO_4 and oxalic acid, and Harouch Batis and coworkers [91] used a VCl_3/V_2O_5 mixture instead of vanadium pentoxide. Mizuno and coworkers [95] reported a

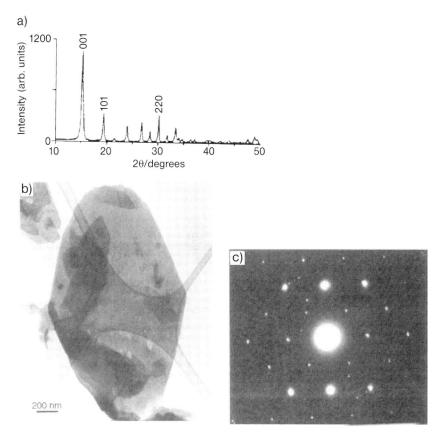

Figure 12.5 (a) XRD pattern, (b) bright-field image and (c) selected area diffraction pattern from the hemihydrate precursor material [94]. (Reproduced with permission).

preparative route using vanadium metal to reduce vanadium pentoxide. They heated a mixture of phosphoric acid, cetyltrimethylammonium chloride, vanadium and vanadium pentoxide in an autoclave, at 200 °C for 48 h. Schimoda and coworkers [75] have also reported the direct reaction of V_2O_4 and H_3PO_4.

In the 1970s, catalyst precursors prepared in organic media became increasingly popular. The most common route (the VPO route, Figure 12.5), is a one-pot method using alcohol as both the solvent and the reducing agent:

$$V_2O_5 + H_3PO_4 + \text{alcohol} \xrightarrow{\Delta, 16h} VOHPO_4 \cdot \tfrac{1}{2}H_2O \qquad (12.3)$$

A number of alcohols have been used in this preparation, isobutanol being the most common [96]. Another common organic route uses a mixture of isobutanol and benzyl alcohol. V_2O_5 is refluxed in the alcohol for an hour before H_3PO_4 is added, and the mixture refluxed for a further hour [96, 97].

Johnson and coworkers [86] described a method for the preparation of $VOHPO_4 \cdot \tfrac{1}{2}H_2O$ by reduction of $VOPO_4 \cdot 2H_2O$ with alcohol. This is known as the VPD route. This route was investigated more fully by Horowitz and coworkers [98] for short-chain alcohols, and Ellison and coworkers [99, 100] for longer chain alcohols. $VOPO_4 \cdot 2H_2O$ is prepared by heating an aqueous solution of V_2O_5 and H_3PO_4 under reflux conditions for 16 h. This is then reduced with alcohol to yield $VOHPO_4 \cdot \tfrac{1}{2}H_2O$:

$$V_2O_5 + H_3PO_4 \xrightarrow[\Delta, 16h]{H_2O} VOPO_4 \cdot 2H_2O \xrightarrow[\Delta, 16h]{\text{alcohol}} VOHPO_4 \cdot \tfrac{1}{2}H_2O \qquad (12.4)$$

Alternative vanadium sources have been investigated using organic solvents. Doi and Miyake [101, 102] used V_4O_9 as the vanadium source. V_2O_5 was initially reduced to V_4O_9 by isobutanol. The V_4O_9 was then reacted with H_3PO_4 with a range of alcohols as the solvent. As with their aqueous preparations Harouch Batis and coworkers [91] used a VCl_3/V_2O_5 mixture instead of vanadium pentoxide.

Guilhaume and coworkers [103] investigated the effect of the reactant V_2O_5 morphology on the morphology of the final catalysts. They found a correlation between the size of the V_2O_5 grains and the preferentially exposed planes of $(VO)_2P_2O_7$. This study was only carried out on a few V_2O_5 samples, and further research on this aspect of the preparation is needed.

Investigations comparing organically and aqueously prepared $VOHPO_4 \cdot \tfrac{1}{2}H_2O$, found that the final catalysts had very similar specific activities [89, 91] However, the organic preparation routes tend to give precursors with a higher surface area than those prepared in aqueous solution. As the morphology of the precursor is retained in the active catalyst, the organically prepared catalysts show the better catalytic performance (Figure 12.6).

Hutchings and coworkers [89] investigated VPA, VPO and VPD catalysts, prepared with a range of alcohols (Figure 12.7). VPA catalysts were found to have a cubic morphology and a low surface area and VPO catalysts were found to have a platelet morphology. However, VPD catalysts prepared with primary alcohols gave a high surface area catalyst with a rosette structure, characterized by an X-ray dif-

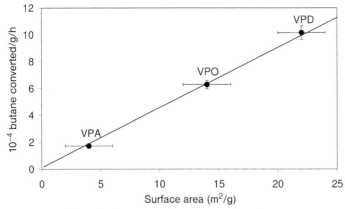

Figure 12.6 Relationship between catalyst activity and surface area for standard vanadium phosphate catalysts for the oxidation of n-butane [89]. (Reproduced with permission).

fraction pattern with only one peak corresponding to the (220) reflection. VPD catalysts prepared with secondary alcohols had a similar morphology and surface area to VPO catalysts and have a characteristic X-ray diffraction pattern containing many peaks, with the (001) reflection as the dominant feature. The catalysts with rosette morphology were found to have a considerably higher activity than the platelets. This is probably due to the increased surface area, as all the VOHPO$_4$·½H$_2$O catalysts are reported to have similar specific activities (the activity to maleic anhydride per unit area).

Horowitz and coworkers [98] investigated a range of preparations in organic solvents. They reported the rosette structure for catalysts prepared by a VPD method with isobutanol and straight-chain alcohols, as well as a VPO type preparation using a 1:10 mixture of benzyl alcohol and isobutanol or 1-butanol. This study found that the platelet catalysts were more selective than the rosettes. They suggest this is due to the rosette structure obscuring the active plane. The greater selectivity of thick platelets has been confirmed by other researchers [26].

Okuhara and coworkers [104–110] have modified the preparations using intercalated and exfoliated VOPO$_4$·2H$_2$O. Using this technique, intercalating compounds such as amines, amides, alcohols or carboxylic acids can replace the water between the vanadium phosphate layers. These materials can then be delaminated in a polar organic solvent and the exfoliated VOPO$_4$ reduced to give V^{4+} vanadium phosphates with unusual morphologies. It is thought that this process can occur in an alcohol, which leads to the formation of the rosette structures found by reduction of VOPO$_4$·2H$_2$O with a primary alcohol [106].

The crystallization of VOHPO$_4$·½H$_2$O from V$_2$O$_5$ and H$_3$PO$_4$ has been studied in detail by O'Mahony and coworkers [111, 112] using time-resolved X-ray diffraction. They found that an intermediate phase was formed initially but this then disappeared as VOHPO$_4$·½H$_2$O was detected. Concurrent focused ion beam

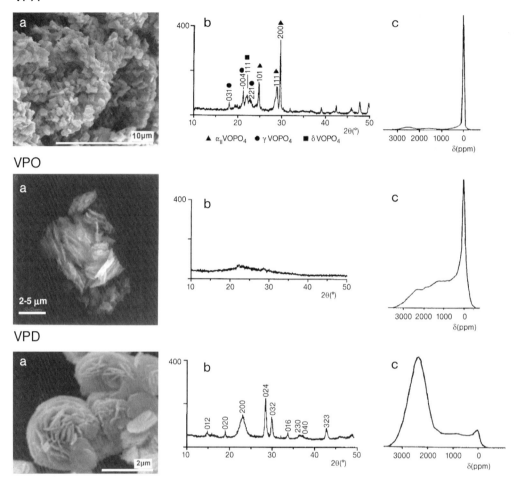

Figure 12.7 (a) An SEM micrograph, (b) an XRD pattern, (c) ^{31}P NMR spin echo mapping spectrum from: activated VPA, activated VPO, and activated VPD [5]. (Reproduced with permission).

microscopy showed rosette structures forming from delaminated plates as the reaction proceeded (Figure 12.8).

A number of groups have also studied $VO(H_2PO_4)_2$ as a catalyst precursor [113–118]. Mount and Raffleson [119] prepared $VO(H_2PO_4)_2$ by heating V_2O_5 and H_3PO_4 with H_3PO_3 in an autoclave at 150 °C. They found this material decomposed at 360 °C to yield $VO(PO_3)_2$. Hannour and coworkers [120–122] prepared $VO(H_2PO_4)_2$ by heating an aqueous solution of V_2O_5, H_3PO_4 and oxalic acid. This was calcined in air to give β-$VO(PO_3)_2$. They also prepared α-$VO(PO_3)_2$ by heating V_2O_5 in a large excess of H_3PO_4. Both of these catalysts were poorly selective to maleic anhydride, with carbon oxides being the major products.

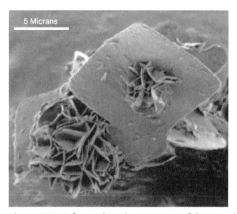

Figure 12.8 A focused ion beam image of the samples recovered 2 min after reaction of V_2O_5 and H_3PO_4 in alcohol. Rosette-shaped hemihydrate particles appear to grow out from the basal plane of the platelets [112]. (Reproduced with permission).

This is consistent with previous studies that have shown $VO(PO_3)_2$ is not as catalytically active as $(VO)_2P_2O_7$. Hutchings and Higgins [123] found beneficial results by removing $VO(H_2PO_4)_2$ by solvent extraction, from the $VOHPO_4 \cdot \frac{1}{2}H_2O$. This yielded a higher surface area precursor and a more active catalyst after activation. Most $VOHPO_4 \cdot \frac{1}{2}H_2O$ preparations include boiling in water as a final step to remove water soluble impurities.

12.3.1
The Preparation of Novel Vanadium Phosphates

Bordes and Courtine [124] prepared a number of precursors that required calcination in air, oxygen or nitrogen to give the final catalyst. $NH_4(VO_2)_2PO_4$, $(NH_4)_2[(VO_2)_2C_2O_4(HPO_4)_2]\cdot 5H_2O$ and NH_4HVPO_6 gave final catalysts mainly comprising $(VO)_2P_2O_7$, $VO(PO_3)_2$ or $V(PO_3)_3$, depending on the activation conditions.

The synthesis of new vanadium phosphate precursors as reported by Benziger and coworkers [125] consisted of intercalated n-alkyl amine pillars inserted between the layers of $VOHPO_4 \cdot \frac{1}{2}H_2O$. The $(VO)_2P_2O_7$ catalysts derived from these precursors show an increase in selectivity which is attributed to stacking faults created by the pillars. Vanadyl phosphonates with the formula $VOC_nH_{2n+1}PO_3 \cdot xH_2O$, ($n = 0$ to 4, $x = 1$ or 5) were also synthesized. These could be converted into $(VO)_2P_2O_7$ at considerably lower temperatures than $VOHPO_4 \cdot \frac{1}{2}H_2O$ and produced catalysts with higher surface areas and increased yields of maleic anhydride.

The gas-phase synthesis of $VOPO_4 \cdot 2H_2O$ has been reported [126]. A gas stream of $VOCl_3$, $POCl_3$ and H_2O with N_2 as carrier, was passed through a furnace where the powdered $VOPO_4 \cdot 2H_2O$ was collected. This was converted to α- and β-$VOPO_4$ by calcination in nitrogen, before *in situ* activation gave the $(VO)_2P_2O_7$ catalyst.

Michalakos and coworkers [127] prepared vanadium phosphate catalysts using an aerosol process. The aerosol was created with aqueous solutions of NH_4VO_3 and H_3PO_4 (with air as the carrier) and sprayed into a furnace. The solid was collected at the reactor exit on a cooled filter. The compound was found to be $VOPO_4 \cdot nH_2O$ which was converted to α_I-$VOPO_4$ with a small amount of $VO(H_2PO_4)_2$ by calcining. The catalyst was found to be more active than VPA catalysts, despite having a lower surface area.

A number of groups have prepared vanadium phosphate catalysts using hydrothermal synthesis [92, 93, 128–130]. Using standard reaction mixtures, Dong and coworkers [128] showed that at elevated temperatures and pressures different materials are synthesized from those obtained under reflux conditions. Pressure did not seem to affect the product formed, but as the temperature increased to >200 °C further reductions occurred and V^{3+} products formed. However, these materials were not found to have enhanced catalytic activity compared to traditionally prepared materials. At lower temperatures, hydrothermal syntheses have produced catalysts with comparable activity to those prepared under standard conditions [92, 93, 129, 130]. Taufiq-Yap and coworkers [129] found an enhancement in activity for hydrothermally prepared catalysts and suggested this was due to a modification in the redox behavior of the catalysts evidenced by TPO/TPR experiments.

Hydrothermal syntheses have also been used to prepare new porous materials. Doi and Miyake [131] obtained a mesoporous vanadium phosphate compound by intercalating a surfactant (*n*-tetradecyltrimethyl ammonium chloride) between the layers of $VOHPO_4 \cdot \frac{1}{2}H_2O$. Furthermore, Bu and coworkers [132] have also reported a new mesoporous vanadium phosphate compound synthesized with an organic template molecule. $NaVO_3$, V, H_3PO_4, H_2O and the template piperazine were mixed in a molar ratio of $1.0:0.5:5.17:476:0.71$. A dark blue gel was formed after 15 minutes stirring, and the mixture was then heated at 170 °C for seven days in an autoclave. Light blue, needle-like crystals were observed and then recovered by filtration. The catalytic activity of these mesoporous vanadium phosphates have not been reported.

12.4
Activation of the Catalyst Precursors

Research into the activation of precursors falls into two categories: the study of structural and morphological changes during the activation period, and the effect of different activation methods on the final catalytic behavior.

12.4.1
Activation Procedures

The catalyst precursor $VOHPO_4 \cdot \frac{1}{2}H_2O$ must be activated to the $(VO)_2P_2O_7$ catalyst. This is usually done *in situ* with the reaction feedstock of 1.5% butane in air.

Table 12.1 Influence of activation conditions on unpromoted V–P–O catalysts [135].

Entry	Activation atmosphere	Temp (°C)	Time (h)	MA$_{selc}$ @400°C (%)	Conv @400°C (%)	Ref.
1	1% Bu/Air	380	100	50	20	[133]
2	1% Bu/Air	380	1000	80	20	[133]
3	O_2,[a]	500	1	84	ca. 12	[134]
4	30% O_2 in N_2	400	3	62.3	98.2	[135]

a Sample heated in N2 (750°C, 72h) then heated in O_2.

A number of pre-treatments have been claimed to speed up the activation process. These can involve heating the catalyst in an inert atmosphere, a reducing environment or an oxidizing environment (Table 12.1).

The effects of standard activation procedures (adopted by industry) and fast activation procedures (often reported in the literature) have been investigated by Lombardo and coworkers [80]. In the standard activation procedure, the catalyst was heated in air up to reaction temperature, followed by introduction of the butane in three steps, up to 1.5%. The gas hourly space velocity (GHSV) was increased in four steps to 2500 h^{-1}. This procedure could take up to 380 hours.

In the fast activation procedure the catalyst was heated under 1.0% butane in air up to reaction temperature, at a GHSV of 900 h^{-1}. This was held for 3 h before the butane concentration was increased to 1.5% and the GHSV to 2500 h^{-1}. Lombardo and coworkers found that the standard activation procedure gave final catalysts that were more crystalline, had less V^{5+} phases present and were far more active than the fast activated catalysts.

Albonetti and coworkers [133] have compared equilibrated and non-equilibrated catalysts. Non-equilibrated catalysts, which had only been on-line for 100 hours, were found to be poorly crystalline, with an oxidation state of +4.36 and a surface area of 11 m^{2}g^{-1}. The equilibrated catalysts that had been on-line for 1000 hours were more crystalline, had an increased surface area (23 m^{2}g^{-1}), and an oxidation state of +4.00. The equilibrated catalyst was found to be considerably more active and selective than the non-equilibrated catalyst.

The effect of oxidation pre-treatments on the catalyst have been investigated by Aït-Lachgar and coworkers [134]. Pure VPP catalysts were obtained by heating VOHPO$_4$·½H$_2$O in a flow of N$_2$ at 750°C for 72 h. This was then oxidized in a flow of O$_2$ for between 0.5 and 24 h. The catalyst oxidized for 1 h showed the highest selectivity to maleic anhydride, although all the pre-treated catalysts were more selective than pure (VO)$_2$P$_2$O$_7$. This is thought to be due to the introduction of V^{5+} phases. Previous studies have found V^{5+} to be detrimental to catalysts, many researchers suggesting they may be responsible for the total oxidation of butane [22–24, 80, 133].

Cheng and Wang [135] have studied the effect of calcining catalysts in air, N$_2$ and CO$_2$. VOHPO$_4$·½H$_2$O could be converted into an amorphous V^{5+} phase, crystalline $VOPO_4$ or $(VO)_2P_2O_7$ with varying crystallinity, depending on the calcining temperature and atmosphere. The calcining environment that resulted in the best catalyst was different for promoted and unpromoted catalysts, depending on how easily they were oxidized. It was found that for unpromoted $VOHPO_4 \cdot \frac{1}{2}H_2O$ the best environment was 30% O_2 in N_2. This gave a catalyst that was composed of dispersed V^{5+} species on a $(VO)_2P_2O_7$ matrix.

The addition of water to the reaction feed was investigated [136]. This led to two significant effects being noted. The selectivity to maleic anhydride increased (with increased yields of acetic and acrylic acids). There was also an increase in the surface area of the catalysts activated with the *n*-butane/water/air feed compared to the dry activated catalyst. Arnold and Sundaresan excluded the possibility of the water vapor acting as a diluent by performing experiments with an *n*-butane/N_2/air feed. Instead they proposed that water is adsorbed onto the surface, blocking sites that are responsible for over-oxidation of the products.

Contractor and coworkers [137] confirmed the effects of water vapor in the gas feed, using a riser reactor. They proposed that the water and oxygen are in direct competition for catalytic sites. Adsorbed water decreases the amount of activated oxygen available, which decreases the activity, but prevents over-oxidation, which in turn increases the selectivity. It is suggested that lattice oxygen is responsible for the selective oxygen, while adsorbed or gaseous oxygen (which is inhibited by the steam) forms the total oxidation products. Furthermore, Vedrine and coworkers [138] proposed that water plays an important role in maintaining a hydrated catalyst surface, which allows a redox mechanism to occur easily.

Centi and Perathoner [139] have discussed the benefits of small amounts of sulfur dioxide in the reaction feed. The SO_2 is thought to adsorb on the surface to form stable $VOSO_4$, blocking the redox activity of surface V^{5+} sites (considered to be responsible for over-oxidation). This results in an increase in selectivity to maleic anhydride, particularly at high conversions.

Recently a very detailed study of activation conditions was carried out by Patience and coworkers [140]. A number of parameters were investigated including temperature, pressure, time and gas-phase composition with respect to O_2 and H_2O. They found that standard conditions (390 °C and atmospheric pressure) gave the best performance although the activation time could be shortened by increasing the pressure. Water had a deleterious effect on the catalyst even at low concentrations and this was proposed to be due to an increase in the V^{5+} observed.

Mechanochemistry has been proposed as an activation method [141–146]. This involves milling the catalyst precursor in a solvent, usually ethanol, prior to its conversion to the active catalyst. Experiments have shown that this procedure promotes the exposure of the (100) plane in the catalyst, which is considered to be the active crystal face [141–143, 146], and can reduce the particle size thus increasing the surface area [141, 144–146]. Experiments with promoters show that this procedure gives more active and selective catalysts than those prepared by either chemical means or mechanical mixing [93, 141, 146].

12.4.2
Structural Transformations during Activation

The importance of precursor morphology in determining catalyst performance was reported by Johnson and coworkers [86]. This is due to the topotactic transformation (Figure 12.3), so the $(VO)_2P_2O_7$ catalyst retains the morphology of the $VOHPO_4 \cdot \tfrac{1}{2}H_2O$ precursor. Insight into how the transformation occurs can help in the design of pre-treatments and activation procedures that will produce catalysts with improved properties.

Kiely and coworkers [94] investigated the differences in activated catalysts which were obtained from $VOHPO_4 \cdot \tfrac{1}{2}H_2O$ precursors prepared via VPA, VPO and VPD routes. The VPA catalyst XRD pattern shows a mixture of $VOPO_4$ phases, with $(VO)_2P_2O_7$ present as a minority phase, and there is a weak V^{4+} peak seen in the ^{31}P NMR spin echo mapping spectrum. TEM revealed α_{II}-$VOPO_4$, δ-$VOPO_4$, $(VO)_2P_2O_7$ and some amorphous material (Figure 12.9). The $(VO)_2P_2O_7$ widely considered to be the active phase constituted only 10% of the catalyst.

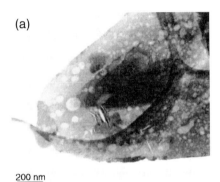

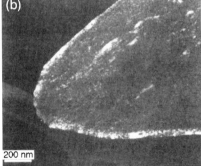

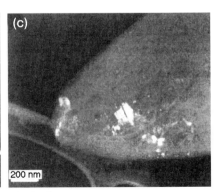

Figure 12.9 Bright-field image showing a typical platelet from the VPO-0.1 sample. Corresponding dark-field micrographs taken in (b) the $\mathbf{g}^{pyro}$ = (024) reflection of $(VO)_2P_2O_7$ and (c) the $\mathbf{g}^{delta}$ = (022) reflection of δ-$VOPO_4$ ($\mathbf{g}$ = diffraction vector) [94]. (Reproduced with permission).

No crystalline phases were detected in the VPO catalyst using XRD. The ^{31}P NMR spin echo mapping spectrum showed two peaks that were assigned to disorganized and crystalline $(VO)_2P_2O_7$. TEM showed large plates, proposed to be amorphous material with small rectangular crystals of $(VO)_2P_2O_7$, preferentially exposing the (100), (021) and (012) planes. About 20% of the sample was composed of δ-$VOPO_4$ plates containing cracks, which may be formed by the initial loss of water from $VOHPO_4 \cdot \frac{1}{2}H_2O$.

The VPD catalyst appears more crystalline than the VPO catalyst. XRD and ^{31}P NMR both show crystalline $(VO)_2P_2O_7$, with a small amount of $VOPO_4$ and disorganized $(VO)_2P_2O_7$ also visible in the spin echo mapping spectrum. The TEM study shows the characteristic rosettes make up 95% of the catalyst, with a few flat platelets. Diffraction patterns showed the rosettes to be made up of $(VO)_2P_2O_7$ (100) planes, while the flat plates can be indexed to α_{II}-$VOPO_4$.

It is clear that the method of preparation, as well as the activation procedure, can have an effect on the structure of the active catalyst. This must be taken into account when comparing activation procedures carried out on catalyst precursors prepared by different means.

Research has been carried out to investigate the changes observed during the activation process. This takes the form of heating a catalyst precursor up to the reaction temperature in an n-butane/air feedstock, and leaving it on-line for varying times (the standard times that have been adopted are 0.1 hour, 8 hours, 84 hours and 132 hours). The samples are then taken off-line and characterized as fully as possible.

Abon and coworkers [32] report that during the activation period the catalyst goes through a number of changes. Initially, $VOHPO_4 \cdot \frac{1}{2}H_2O$ is transformed into poorly crystalline $(VO)_2P_2O_7$ and δ-$VOPO_4$. The initial δ-$VOPO_4$ is transformed into α_{II}-$VOPO_4$ and $(VO)_2P_2O_7$, which become more crystalline with time on-line. The increase in crystallinity and surface area are proposed to be responsible for the increase in activity observed. A decrease in V^{5+} phases (responsible for total oxidation) is observed with activation time, which accounts for the increase in selectivity to maleic anhydride.

Additionally, a similar study was conducted by Kiely and coworkers [94] on VPO catalysts. Prior to activation, rhomboidal plates of the precursor $VOHPO_4 \cdot \frac{1}{2}H_2O$ were observed by TEM. As the sample is heated in the reaction feedstock, cracks are formed in the plates, due to the loss of water of crystallization (0.1 hour) (Figure 12.10a). Even after a short time on-line $(VO)_2P_2O_7$ and δ-$VOPO_4$ are seen in the diffraction pattern. The δ-$VOPO_4$ and $VOHPO_4 \cdot \frac{1}{2}H_2O$ phases gradually diminish with time, until they can no longer be observed in the 132 hour sample (Figure 12.10d).

As the activation proceeds, a crystalline rim starts to form around the edge of the plate, while in the center of the plate a more disordered phase is formed, with dispersed crystals of δ-$VOPO_4$ (8 hours) (Figure 12.10b). After 84 hours the rim (made up of small oblong crystallites of $(VO)_2P_2O_7$) has thickened and large holes start to appear in the center of the plates (Figure 12.10c). After 132 hours the inside of the platelet has also become more crystalline $(VO)_2P_2O_7$ (Figure 12.10d).

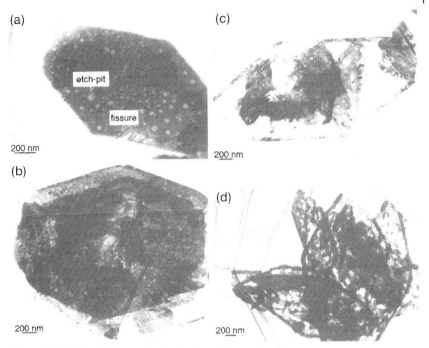

Figure 12.10 Bright-field transmission electron micrographs of platelet morphologies in (a) VPO-0.1, (b) VPO-8, (c) VPO-84 and (d) VPO-132 activated catalysts [94]. (Reproduced with permission).

12.5
Promoted Catalysts

Industrial catalysts for oxidation reactions rarely use a single bulk phase. A number of promoter elements are added that can act purely as textural promoters, or enhance the activity and selectivity of the bulk catalyst. The role of promoters on vanadium phosphate catalysts has been addressed mainly in the patent literature and Hutchings [147] has provided an extensive review of these patents.

A number of groups have tested a wide range of promoter elements and compounds. Hutchings and Higgins [148] found that chromium, niobium, palladium, antimony, ruthenium, thorium, zinc and zirconium had very little effect on the specific activity of $(VO)_2P_2O_7$. A significant increase in surface area was observed with zirconium, zinc and chromium, which could be of use as structural promoters. Iron-, cesium- and silver-doped catalysts showed a decrease in the specific activity, while cobalt and molybdenum were the only promoters found to increase the specific activity.

The selectivity decreased for catalysts doped with cesium, palladium, ruthenium, zinc and zirconium. This was thought to be due to these metals promoting the over-oxidation of maleic anhydride to carbon oxides. However, molybdenum

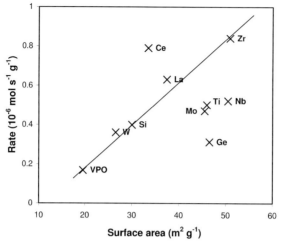

Figure 12.11 Plot of the activity against surface area for a number of promoted catalysts [155]. (Reproduced with permission).

was found to poison the over-oxidation reaction. From this work Hutchings and Higgins concluded that only cobalt and molybdenum act as promoters. Other elements reported as promoters are only responsible for an increase in surface area of $(VO)_2P_2O_7$.

Ye and coworkers [149–154] tested a large number of promoter elements and found the activity to maleic anhydride changes in the order:

$$Zr > Ce > La > Fe > Co > Cu > Nb > Ti > Mo > Ca > Si > W > Ni > Ge > K$$

However, the activity reported by Ye and coworkers is not the specific activity, and the surface areas of the promoted catalysts show a large variation (26.3 to 50.8 $m^2 g^{-1}$). Hutchings [155] has plotted (Figure 12.11) the activity against the surface area for a number of promoted catalysts and deduced that most of the catalysts conform to a linear correlation. The only enhancement of the specific activity was given by the Ce-promoted catalyst. This shows that care must be taken in the interpretation of results, particularly when catalysts prepared by different methods are compared. A review of the promoter literature revealed that Ce, Co, Cr, Cu, Fe, Hf, La, Mo, Nb, Ni, Ti and Zr are commonly reported to enhance the activity [155]. These cations are suggested to form solid solutions, $[(VO)_xM_{1-x}]_2P_2O_7$ (where M is a promoter cation). The inclusion of cations of different size or charge in the VPP lattice is likely to cause defects, which could then function as active sites for butane oxidation. In this section we shall discuss the common promoter elements that have been shown to give an increase in activity compared to undoped catalysts.

A number of other groups have also found that zirconium enhances the activity of vanadium phosphate catalysts [11, 56, 146, 148, 150, 154, 156–163]. Zeyss and coworkers [158] investigated catalysts doped with 5 to 15% zirconium. Unlike the observations of Hutchings and Higgins [148], the zirconium was not incorporated into the $(VO)_2P_2O_7$ lattice, but was found in an amorphous phase, which is proposed to be the catalytically active phase. This is probably the reason that zirconium was not found to increase the surface area as Ye and coworkers [153] observed. Zeyss suggested that the zirconium influences the amount of $VOPO_4$ phases formed during the activation of the catalyst and that these are the cause of the increased activity. The positive or negative effect of $VOPO_4$ in the active catalyst is the subject of considerable discussion (see Section 12.2), so this explanation of the promotional effect of zirconium is quite controversial. Sant and Varma [164] also studied the role of zirconium as a promoter. They found that low concentrations of zirconium lowered the temperature required to reach the maximum yield. Various reasons for this observation have been put forward. The increase in surface area and the increase in oxygen transport rates can be sufficiently altered by the zirconium to result in high yields of maleic anhydride at lower temperatures. The roles of zirconium, zinc and titanium have been studied as catalytic and structural promoters [156]. The results showed that all these promoters had a significant effect, if added in the correct proportion. Confirming previous findings, 1.5% zirconium had the most beneficial effect on the activity; good catalytic performance could be achieved at lower temperatures.

It is proposed that zirconium and titanium both create acidic surface sites on the vanadium phosphate surface. This prevents the desorption of reaction intermediates (butene, butadiene and furan), while facilitating the desorption of the acidic maleic anhydride. A large amount of zinc promoter resulted in a loss of surface acidity, leading to over-oxidation of strongly adsorbed maleic anhydride. However, a small addition of 0.6% zinc enhanced the catalyst performance by creating basic sites, which increase the rate of butane activation. At low zinc concentrations the slight loss in surface acidity does not have a great effect. Takita and coworkers [157, 165] studied the effects of zinc oxide on the catalyst. They found an increase in catalytic performance that was ascribed to the increased rate of re-oxidation of catalyst. Additionally, a range of transition metals and transition metal oxides were tested to determine if they could act as promoters. It was found that the conversion is significantly increased over catalysts containing manganese, cobalt and zirconium, but decreased over the catalysts containing TiO_2 and MoO_2. Also the selectivity to maleic anhydride showed an increase of 8 to 10% for catalysts containing TiO_2, copper and zinc.

The conclusions of their study related the specific activity to maleic anhydride to the electronegativity of the promoter elements added. They found that the V=O stretching mode has a larger wavenumber as the electronegativity of the promoter increases, and that the larger the wavenumber of the V=O stretching mode, the smaller the specific activity becomes. The stronger the V=O bond is, the higher the frequency of the V=O stretching mode will be, hence the larger the wavenumber will be. So the more electronegative the promoter, the stronger the V=O bond, and the lower the specific activity.

Bej and Rao [166–170] conducted a detailed study of molybdenum- and cerium-promoted vanadium phosphate catalysts. They found an increase in the selectivity of these catalysts, compared to the unpromoted catalyst, albeit with a slight decrease in activity. They attribute this finding to the promoters preventing over-oxidation of the maleic anhydride to carbon oxides. They also found that the promoted catalyst could withstand more severe reaction conditions, which was again attributed to less carbon oxides being formed, which can poison the catalyst.

In common with Hutchings and Higgins [148], Bej and Rao suggest that the molybdenum prevents the reduction of the V^{4+} ions to V^{3+}, a species that is considered to be responsible for the formation of total oxidation products. Cerium is proposed to increase the conversion of butane.

The promotional effects of cobalt [71, 74, 150, 152, 154, 157, 162, 171–184] and iron [71, 110, 148, 151, 152, 160–163, 173, 175, 176, 182, 185–189] have been widely studied. Ben Abdelouahab and coworkers [173] looked at the effect of various promoters on the structure of organically prepared catalysts. Both cobalt and iron promoters were found to increase the selectivity to maleic anhydride, but butane conversion was found to decrease with cobalt promoters and increase with iron promoters.

As with the zirconium study by Zeyss and coworkers, cobalt and iron were found to promote the formation of $VOPO_4$ phases during activation of the precursor to the active catalyst. The difference in activity is considered to be due to the redox potentials of the promoters. As the V^{4+}/V^{5+} ratio decreases the butane conversion is stabilized by iron (as the Fe^{3+}/Fe^{2+} redox potential is lower than the V^{5+}/V^{4+} redox potential). As the Co^{3+}/Co^{2+} redox potential is higher than the V^{4+}/V^{5+} redox potential, the conversion of butane decreases when the V^{4+}/V^{5+} ratio decreases.

A similar promotional effect was observed for catalysts prepared using an aqueous route [174]. The iron- and cobalt-promoted catalysts are associated with an increase in selectivity. The iron-doped catalyst showed an increase in activity while the cobalt-doped catalyst activity decreased. The decrease in activity of the cobalt-promoted catalyst is attributed to the formation of $VOPO_4 \cdot 2H_2O$ in the final catalyst. The $VOPO_4 \cdot 2H_2O$ is formed by the oxidation of $VOHPO_4 \cdot \frac{1}{2}H_2O$ during the introduction of the promoters using the incipient wetness technique.

The method of preparation of the catalyst was found to alter the effect of the promoter [177]. With standard organically prepared VPO, the effect of cobalt and iron was found to be the same as previously described [133, 173–176, 183]. The increase in catalytic performance is proposed to be due to the stabilization of V^{4+}–V^{5+} dimers; the proposed active site. However, with catalysts prepared from $VOPO_4 \cdot 2H_2O$ in organic solvents, iron has no promotional effect. This is proposed to be due to the loss of crystallinity and surface area of the rosette crystals formed by this preparative route. Similarly the increase in activity due to cobalt is thought to be a structural effect, influencing the development of the (100) plane of $(VO)_2P_2O_7$.

Zazhigalov and coworkers [143] investigated cobalt-doped VPO catalysts prepared by co-precipitation and impregnation methods. The performance of catalysts prepared by both methods was increased, compared to the unpromoted catalyst. The cobalt is thought to be present as cobalt phosphate, which is considered to stabilize excess phosphorus at the surface, which has previously been found to be an important feature of active catalysts (Section 12.2).

Oxygen donors (Sb_2O_4 and $BiPO_4$) were used by Ruiz and coworkers [48], and it was found these increased the activity and selectivity when used to promote a high P/V ratio catalyst. Tamaki and coworkers [190] tested the promotional affects of magnesium, manganese, lanthanum and bismuth. They found bismuth to be the most effective promoter, increasing the selectivity at conversions up to 90 mol%, compared with unpromoted VPP. In line with studies on the active catalyst by Morishige and coworkers [52], there is speculation that the bismuth is incorporated into a phosphorus-rich amorphous surface species. Again, the additive is thought to reduce the over-oxidation of products and the total combustion of butane to carbon oxides.

The promotional effects of alkali and alkali earth metals, were investigated by Zazhigalov and coworkers [143]. The promoters can easily donate electrons to the $(VO)_2P_2O_7$. This was reported to lead to an increased negative charge on the oxygen atoms and an increase in the basic properties of the catalyst. The activation of butane by dehydrogenation occurs more readily on a basic catalyst and so the rate is increased. Acid sites are also proposed to be important to enable desorption of products to prevent over-oxidation. It has been suggested that surface species must be tuned by the promoters, to have a mix of acid and base sites in appropriate amounts for butane activation to be enhanced while allowing the selectivity to maleic anhydride to remain undiminished.

Centi and coworkers [84] have examined the effect of potassium doping, which was found to inhibit the P—OH Brønsted acid sites. The lack of Brønsted sites was thought to have two unfavorable effects on the catalyst. Firstly; the formation of lactones and maleic anhydride from furan was inhibited and, secondly, carbon-containing residues were strongly adsorbed onto the surface, deactivating the catalyst.

An increased performance has been reported with catalysts doped with indium and tetraethylorthosilicate (TEOS) [185]. The increase in catalytic performance is only observed with both promoters in the catalyst. It is proposed that the promoters work by facilitating the oxidation of the catalyst during activation, giving rise to $VOPO_4$ phases, and drastically decreasing the thickness and size of the $(VO)_2P_2O_7$ crystallites leading to a higher surface area.

Harouch Batis and coworkers [91] investigated the effect of chromium, which Hutchings and Higgins [148] observed had no effect on specific activity or selectivity. In this study, a surface enrichment of chromium was found to give a decreased maleic anhydride yield, while at high conversion, the catalyst was deactivated by surface coking. It is thought that the active site is different on the doped catalyst and the undoped vanadium phosphate. This leads to the formation of butene and furan at low conversions, which cannot usually desorb from vanadium phosphate catalysts, and at high conversions to over-oxidation to carbon oxides. Matsuura and coworkers [191] report that using niobium phosphate as promoter leads to an increase in activity of vanadium phosphate catalysts. It is thought that an increase in Lewis acid sites is responsible for the enhanced performance.

Zazhigalov and coworkers [142, 143] have reported the incorporation of bismuth compounds into vanadium phosphate catalysts using mechanochemistry. This involves milling the catalyst precursor and the promoter in ethanol. The mechanochemistry preparation yields catalysts with a higher activity and greater selectivity to maleic anhydride than those prepared by chemical means, or mechanical mixtures.

As illustrated by the work by Sananes-Schulz and coworkers [177] the preparation method of the catalyst can alter the effect of the promoter, as can the method of doping. Hutchings and Higgins have warned against misinterpretation of promotional effects for catalysts prepared by incipient wetness and co-precipitation using acid solutions. Acidic solutions can cause the formation of $VOPO_4 \cdot 2H_2O$, which has a detrimental effect on the catalytic performance that can be mistakenly attributed to the promoter. They recommend that these methods of introducing promoters must be used with care, and in particular that the acidity of the impregnation solution should be carefully monitored. Further confusion is caused by different groups reporting contrasting results for the same promoters.

The effect of promoters on VPO performance has been summarized by Ballarini and coworkers (Table 12.2) [192].

12.6
Mechanism of n-Butane Partial Oxidation

The oxidation of *n*-butane to maleic anhydride is a 14-electron oxidation. It involves the abstraction of eight hydrogen atoms, the insertion of three oxygen atoms, and a multi-step polyfunctional reaction mechanism that occurs entirely on the adsorbed phase. No intermediates have been observed under standard continuous flow conditions, although mechanisms for this process have been proposed based on a variety of experimental and theoretical findings. The description of the active site is linked to the mechanism and is the subject of considerable debate in the literature. The mechanisms are linked to the researchers' hypotheses of the active site, which will be discussed in a separate section in this chapter. It is widely accepted that the (100) plane of vanadyl pyrophosphate, $(VO)_2P_2O_7$, (referred to as the (020) plane by certain authors) plays an important role in the selective oxidation of butane.

The structure has been determined by XRD and consists of edge-sharing VO_5 units linked by pyrophosphate tetrahedra (Figure 12.2). This is viewed as the active surface for most of the proposed mechanisms. Here we will discuss several of the mechanisms thought to account for the production of maleic anhydride, which have been debated in the literature.

12.6.1
Consecutive Alkenyl Mechanism

A consecutive alkenyl mechanism has the widest support in the literature [58–64, 83, 200]. Once butane has adsorbed onto the vanadium phosphate surface, it is transformed via adsorbed alkenyl intermediates into maleic anhydride. A summary of the mechanism is shown in Scheme 12.2.

12.6 Mechanism of n-Butane Partial Oxidation

Table 12.2 Summary of recent achievements on the effect of promoters on VPO performance [192].

Dopant, optimal amount (%w.r.t. V)	Promotional effect[a]	Reasons for promotion	Ref.
Co, 0.77	C 15–25%, S 0–11%, under hydrocarbon-rich conditions	Control of the optimal V^{5+}/V^{4+} surface ratio; stabilization of an amorphous Co/V/P/O compound	[71, 193, 194]
Co, 13%	C 55–79%, S 43–35%, at 653K	Optimal surface Lewis acidity	[179, 183, 195, 196]
Ce + Fe	C 44–60%, S63–66% in the absence of O_2	Improvement of redox properties	[182]
Fe, 8%	Increase of catalytic activity	Fe replaces V^{4+} in VPP. The re-oxidation rate is increased	[105, 197]
Ga, 10%	C 22–73%, S 55–51%	Increase of surface area + increase of intrinsic activity (electronic effect)	[198]
Nb	C 20–17%, S 35–53%	Increase of surface acidity promotes desorption of MA	[159]
Nb, 1%	C 58–75%, S 70–70%	Nb concentrates at the surface, where defects are generated. Nb acts as an n-type dopant; development of a more oxidized surface	[199]

a C = conversion; S = selectivity for the undoped compared with the doped catalyst, under fixed reaction conditions.

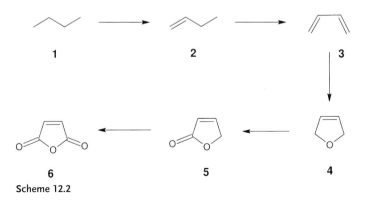

Scheme 12.2

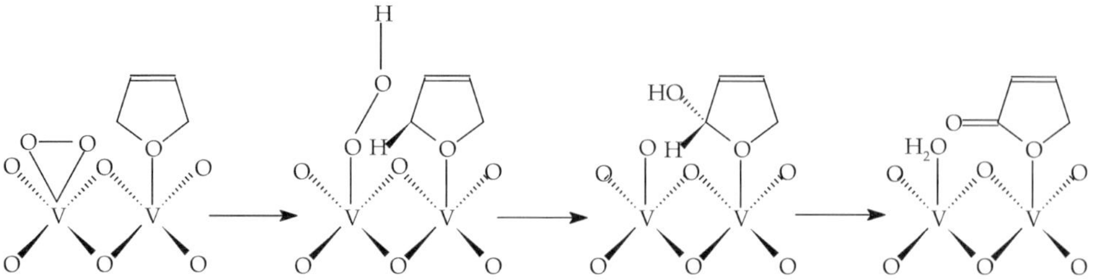

Scheme 12.3

The initial step is thought to be hydrogen abstraction from *n*-butane (**1**), giving 1-butene (**2**), followed by a further hydrogen abstraction to form 1, 3-butadiene (**3**). A 1, 4 insertion of an electrophilic surface oxygen atom occurs, producing dihydrofuran (**4**). Dihydrofuran is then oxidized to the asymmetric lactone (**5**) from which maleic anhydride (**6**) is formed by a final oxidation of the remaining CH_2 group.

There are different ways in which gaseous oxygen can adsorb onto the surface of the catalyst. In their theoretical study, Schiøtt and Jørgensen [58, 59] suggest that the gaseous oxygen is adsorbed in an η^2-peroxo coordination mode (Scheme 12.3) as this leads to a favorable overlap of the ϕ_{C-H} and ϕ^*_{O-O} orbitals. Furan is formed by oxygen insertion into adsorbed 1, 3-butadiene (**7**). The ϕ_{C-H} orbital donates electron density into the ϕ^*_{O-O} orbital, weakening the C–H bond and forming an O–H bond (**8**). Then there is a favorable carbon–oxygen interaction to give intermediate (**9**). The asymmetric lactone intermediate (**10**) is finally formed by the loss of water. This process is repeated on the reverse side of the lactone to give maleic anhydride.

A consecutive reaction mechanism was also proposed by Gleaves and Centi [61]. This was based on experimental work to back up the theoretical calculations of Schiøtt and Jørgensen. Although the proposed intermediates are not detected under reaction conditions they have been seen with fuel-rich gas feeds and under temporal conditions. Using a TAP reactor, the products are detected in the order: butane → butene → butadiene → furan. However, these conditions differ significantly from standard continuous flow reaction conditions. Taufiq-Yap and coworkers [64] surmised the same mechanism from temperature programmed reaction (TPR) and temperature programmed desorption (TPD) experiments on *n*-butane, 1-butene and 1, 3-butadiene. Temperature programmed oxidation (TPO) experiments suggest that the active oxygen species for selective oxidation is lattice oxygen, and that the replenishment of the surface oxygen from the bulk is the rate determining step.

The active oxygen species was investigated by Abon and coworkers [57], using isotopic labeling experiments. Initially, they found the products contained only lattice ^{16}O. As the reaction proceeded, more ^{18}O atoms were incorporated into the products. They also concluded that lattice oxygen was the active oxygen species, and that it was replenished by the gas-phase oxygen. This is widely accepted to be the case and has been confirmed by numerous studies [31].

Although Misono and coworkers [63] agreed with the consecutive mechanism, they suggested a different rate determining step to Taufiq-Yap and coworkers [64]. By determining kinetic data for the reaction of n-butane, 1-butene and 1, 3-butadiene over VPP, they concluded that the initial dehydrogenation of butane was rate determining. The mechanism for this H abstraction has been studied in more detail by Millet [201].

Other research [65, 67, 202] has shown that butene oxidation can produce many selective products (furan, acetaldehyde and methyl vinyl ketone) which are not detected during butane oxidation. It cannot be assumed that the oxidation of butane and the unsaturated reactants proceed along the same pathway. The kinetic data must be viewed with this in mind, although butane activation is widely accepted to be the rate determining step. The intermediates are capable of desorbing from the surface, (as seen in the TAP studies) but do not do so, indicating that the further reactions occur more readily than desorption.

12.6.2
Consecutive Alkoxide Mechanism

A consecutive mechanism was proposed by Zhang-Lin and coworkers [68, 69]. The mechanism was based on kinetic data calculated for the oxidation of butane, 1-butene, 1, 3-butadiene and furan over $(VO)_2P_2O_7$ and $VOPO_4$ phases. Unlike TAP studies, the kinetic data suggested that furan is not an intermediate for butane oxidation, but is an intermediate for butadiene oxidation. The differences observed in the oxidation of butane and the unsaturated hydrocarbons questions the validity of applying butene and butadiene oxidation results to the butane system.

The consecutive alkenyl mechanism (Scheme 12.4) was put forward by Zhang-Lin and coworkers [68, 69] as the route for oxidation of unsaturated reactants such as 1-butene. The weakly adsorbed intermediates are in equilibrium with the gas phase, which enables furan to be seen as a product for butene oxidation [65–67].

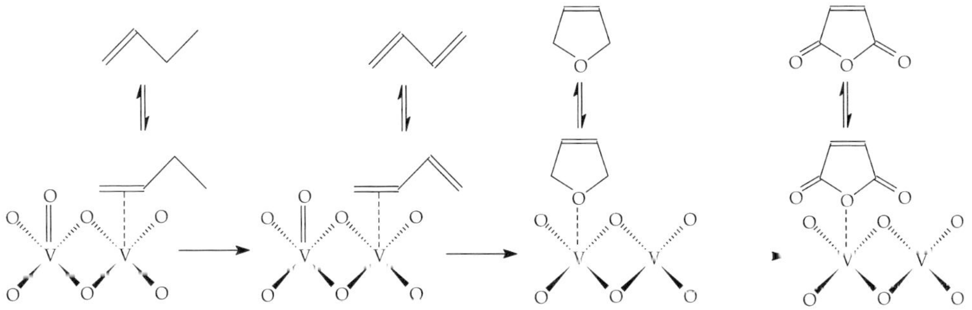

Scheme 12.4

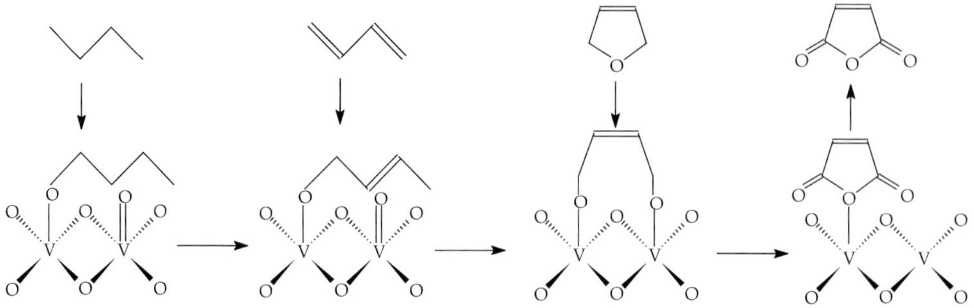

Scheme 12.5

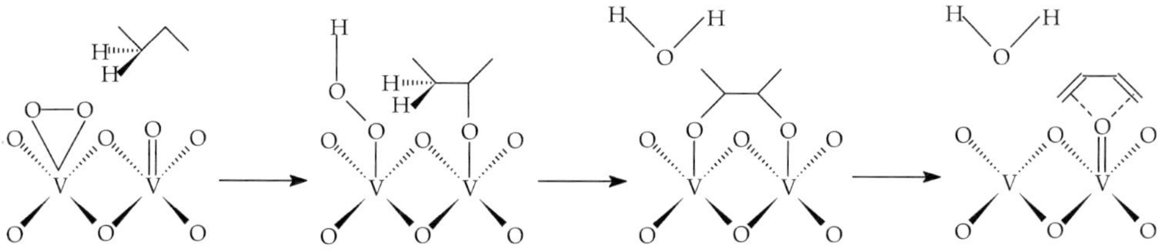

Scheme 12.6

Unlike the previous work, this study examined the fact that none of the alkene intermediates desorb from the surface. Zhang-Lin and coworkers proposed that the reaction proceeded via more strongly adsorbed alkoxide intermediates that would remain on the surface for the whole oxidation scheme (Scheme 12.5).

Agaskar and coworkers [46] have proposed a different consecutive alkoxide mechanism, based on theoretical and experimental results. Unlike other proposals, they suggest that adsorbed oxygen is the active oxygen species.

Theoretical calculations imply that gaseous oxygen is adsorbed onto the surface as a η^2-peroxo species. However, it is thought that a η^1-superoxo species may be the dominant form, if the oxide species is stabilized by interactions with neighboring metal-oxo atoms (**11**) (Scheme 12.6).

Gaseous butane is activated by hydrogen abstraction by the adsorbed superoxo species, to give a surface-bound hydroperoxy group (**12**). Simultaneously, the butene is adsorbed onto a surface vanadyl group. This is followed by a further hydrogen abstraction by the hydroperoxy group, to give a water molecule and a bridging alkoxide surface species (**13**). The next step involves a Brønsted acid site, where the alkoxide undergoes an acid-catalyzed reaction, to give 1,3-butadiene and another water molecule (**14**). The final steps to maleic anhydride then proceed as described by Schiøtt and Jørgensen [58, 59] (Scheme 12.3).

12.6.3
Concerted Mechanism

The concerted mechanism proposed by Ziółkowski and coworkers [203, 204] is based on theoretical calculations on the (100) plane of $(VO)_2P_2O_7$. The butane is adsorbed via hydrogen bonds to the active site. The active site has the following properties:

The site has three routes of relatively easy movement of hydrogen along the surface. Abstraction of the first two hydrogen atoms (from carbons C_1 and C_4) forms very strong bonds between the molecule and the catalyst (C_1–V_C, C_4–V_D). These bonds enable the molecule to be anchored to the surface long enough for the reaction to occur (Figure 12.12).

The molecule must have contact with five active oxygen atoms: P_A–O_A, P_B–O_B, V_E–O_E, and weakly adsorbed oxygen atoms O_D and O_C. O_E, O_A and O_B make bridges from which hydrogen atoms from all four carbons may move away. The oxygen atoms O_E and O_C are responsible for anchoring the terminal carbons, and P_A–O_A, V_E–O_E and V_C–O_C are in convenient positions to be incorporated into the molecule to form maleic anhydride (Scheme 12.7).

The adsorbed butane is activated (by hydrogen abstraction) to give butadiene before the concerted step to form maleic anhydride. The formation of maleic anhydride creates seven oxygen vacancies on the surface. The re-oxidation of the surface is proposed to be the rate determining step.

However, there is no experimental evidence for this mechanism. It is based on the deduction that the idealized surface structure has the right configuration for

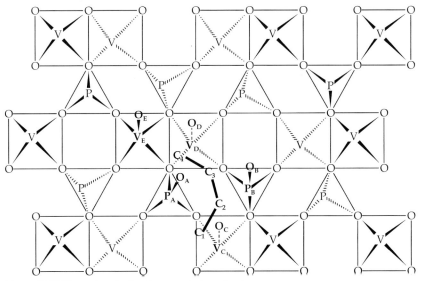

Figure 12.12 The active site for the concerted mechanism proposed by Ziółkowski and coworkers [203, 204]. (Reproduced with permission).

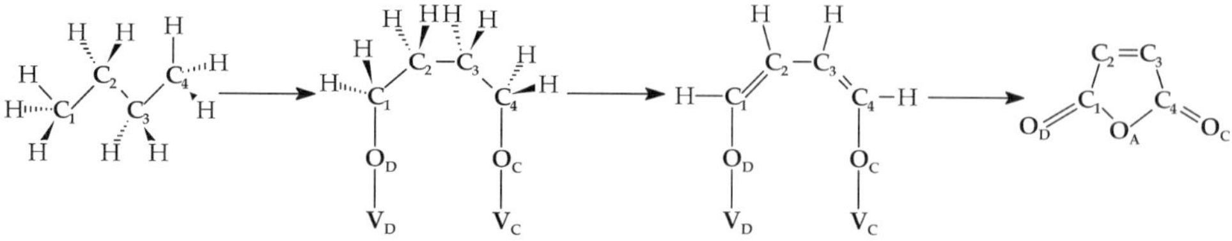

Scheme 12.7

the reaction to occur in this way. It is proposed that the reaction occurs in one step after the adsorption of butane on the active site.

12.6.4
Redox Couple Mechanism

Centi and coworkers [37] have proposed that the reaction proceeds via a series of redox couples. The mechanism is based on experimental data obtained at high butane concentrations and total oxygen conversions. Under these conditions the catalyst undergoes a partial reduction and many V^{3+} sites are formed. The activation of butane requires a V^{4+}–V^{3+} couple, while the subsequent conversion to maleic anhydride needs a V^{5+}–V^{4+} couple. The alkenes from butane dehydrogenation can desorb if no oxidizing sites are available. When the oxidizing sites are available, the alkenes are adsorbed and react very quickly to give maleic anhydride.

Under normal reaction conditions with lower concentrations of butane, there will be fewer V^{3+} sites. Following butane activation there will be numerous oxidizing sites available and so desorbed alkenes are not detected. Apart from the distribution and availability of the different sites, the authors consider that the mechanism at low and high butane concentrations is the same.

12.7
Concluding Comments

Vanadium phosphates have been applied to a number of selective oxidation and ammoxidation reactions, although the partial oxidation of n-butane to maleic anhydride remains the most widely studied reaction for these catalysts. Even though the first patent for this reaction was filed over 40 years ago, there are still a large number of papers published on this system every year.

This may, in part, be due to this reaction being the only industrial large scale functionalization of an alkane currently in operation, and knowledge gleaned from vanadium phosphate systems can provide valuable information that can be applied to different alkane activation reactions. To this end a number of the publications have focused and continue to focus on a fundamental understanding of the catalyst active surface and the active site. In recent years, transient techniques such as the use of TAP reactors and *in situ* characterization studies have advanced the understanding of the system, although this has led to considerable debate as to the exact nature of the active surface and the active site. Despite the fact that model studies and mechanistic proposals have been postulated using crystalline VPP as the active surface, increasingly this is being disputed, with evidence pointing to the importance of V^{5+} and/or amorphous phases in the catalyst.

It is no surprise that a large number of publications and patents describing the addition of promoter elements, as these can be easily introduced into the industrial catalyst formulation. It is perhaps more surprising that a number of publications are focused on new preparation methods for a catalyst that is used industrially with high selectivity and activity. These investigations may be driven by engineering requirements, with the advances in new technology such as riser reactors or membrane reactors requiring catalysts with different characteristics to those used in fixed-bed reactors.

It is clear that vanadium phosphate catalysts are still widely studied although not widely understood. Advances in research methodology, particularly the number of complementary *in situ* characterization techniques now available, may be able to further the understanding of alkane activation and selective oxidation. This fundamental understanding will be beneficial in the design of new catalyst systems for alkane functionalization and provide new uses for this relatively unreactive, under-utilized feedstock.

References

1. Bergman, R.I. and Frisch, N.W. (1966) US patent 3 293 268.
2. Trivedi, B.C. and Culbertson, B.M. (1982) *Maleic Anhydride*, Plenum, New York.
3. Varma, R.L. and Saraf, D.N. (1979) *Industrial & Engineering Chemistry Product Research and Development*, **18**, 7.
4. Hodnett, B.K. (1993) *Catalysis Today*, **16**, 131.
5. Kiely, C.J., Burrows, A., Sajip, S., Hutchings, G.J., Sananes, M.T., Tuel, A. and Volta, J.-C. (1996) *Journal of Catalysis*, **162**, 31.
6. Bordes, E. (1987) *Catalysis Today*, **1**, 499.
7. Ai, M. (1986) *Journal of Catalysis*, **101**, 389.
8. Centi, G., Tosarelli, T. and Trifiro, F. (1993) *Journal of Catalysis*, **142**, 70.
9. Centi, G., Pesheva, D. and Trifiro, F. (1987) *Applied Catalysis*, **33**, 343.
10. Centi, G. and Trifiro, F. (1990) *Chemical Engineering Science*, **45**, 2589.
11. Chen, S., Lao, L. and Shao, H. (2006) *Petroleum Science*, **3**, 65.
12. Sobalik, Z., Carrazan, S.G., Ruiz, P. and Delmon, B. (1999) *Journal of Catalysis*, **185**, 272.
13. Zanthoff, H.W., Kubias, B. and Hutchings, G.J. (1997) *DGMK Tagungsbericht*, **9705**, 307.
14. Lopez Granados, M., Fierro, J.L.G., Cavani, F., Colombo, A., Giuntoli, F. and Trifiro, F. (1998) *Catalysis Today*, **40**, 251.
15. Sobalik, Z., Gonzalez, S., Ruiz, P. and Delmon, B. (1997) *Studies in Surface Science and Catalysis*, **110**, 1213.
16. Sobalik, Z., Ruiz, P. and Delmon, B. (1997) *Studies in Surface Science and Catalysis*, **110**, 481.
17. Sobalik, Z., Gonzalez, S. and Ruiz, P. (1995) *Studies in Surface Science and Catalysis*, **91**, 727.
18. Centi, G., Lopez-Nieto, J., Pinelli, D. and Trifiro, F. (1989) *Industrial & Engineering Chemistry Research*, **28**, 400.
19. Bordes, E. (1993) *Catalysis Today*, **16**, 27.

20 Inumaru, K., Misono, M. and Okuhara, T. (1997) *Applied Catalysis, A: General*, **149**, 133.
21 Ballarini, N., Cavani, F., Cortelli, C., Ricotta, M., Rodeghiero, F., Trifiro, F., Fumagalli, C. and Mazzoni, G. (2006) *Catalysis Today*, **117**, 174.
22 Ebner, J.R. and Thompson, M.R. (1993) *Catalysis Today*, **16**, 51.
23 Guliants, V.V., Benziger, J.B., Sundaresan, S., Wachs, I.E., Jehng, J.M. and Roberts, J.E. (1996) *Catalysis Today*, **28**, 275.
24 Cavani, F. and Trifiro, F. (1997) *Applied Catalysis, A: General*, **157**, 195.
25 Volta, J.-C. (1996) *Catalysis Today*, **32**, 29.
26 Okuhara, T. and Misono, M. (1993) *Catalysis Today*, **16**, 61.
27 Schuurman, Y. and Gleaves, J.T. (1994) *Industrial & Engineering Chemistry Research*, **33**, 2935.
28 Coulston, G.W., Thompson, E.A. and Herron, N. (1996) *Journal of Catalysis*, **163**, 122.
29 Rodemerck, U., Kubias, B., Zanthoff, H.W., Wolf, G.U. and Baerns, M. (1997) *Applied Catalysis, A: General*, **153**, 217.
30 Rodemerck, U., Kubias, B., Zanthoff, H.W. and Baerns, M. (1997) *Applied Catalysis, A: General*, **153**, 203.
31 Lorences, M.J., Patience, G.S., Cenni, R., Diez, F. and Coca, J. (2006) *Catalysis Today*, **112**, 45.
32 Abon, M., Bere, K.E., Tuel, A. and Delichere, P. (1995) *Journal of Catalysis*, **156**, 28.
33 Zhang, Y., Sneeden, R.P.A. and Volta, J.C. (1993) *Catalysis Today*, **16**, 39.
34 Okuhara, T., Inumaru, K. and Misono, M. (1993) *Catalytic Selective Oxidation* (eds J. Hightower and T. Oyama), ACS Symposium Series, **523**, ACS: Washington, DC, p. 156.
35 Inumaru, K., Okuhara, T. and Misono, M. (1992) *Chemistry Letters*, 1955.
36 Hutchings, G.J., Kiely, C.J., Sananes-Schulz, M.T., Burrows, A. and Volta, J.C. (1998) *Catalysis Today*, **40**, 273.
37 Centi, G., Fornasari, G. and Trifiro, F. (1984) *Journal of Catalysis*, **89**, 44.
38 Hutchings, G.J., Desmartin-Chomel, A., Oller, R. and Volta, J.C. (1994) *Nature (London)*, **368**, 41.
39 Abdelouahab, F.B., Olier, R., Guilhaume, N., Lefebvre, F. and Volta, J.C. (1992) *Journal of Catalysis*, **134**, 151.
40 Amoros, P., Marcos, M.D., Roca, M., Alamo, J., Beltran-Porter, A. and Beltran-Porter, D. (2001) *Journal of Physics and Chemistry of Solids*, **1393**, 62.
41 Kubias, B., Wolf, H., Wolf, G.-U., Duvauchelle N. and Bordes E. (2000) *Chemie-Ingenieur-Technik*, **72**, 249.
42 Amoros, P., Marcos, M.D., Alamo, J., Beltran, A. and Beltran, D. (1994) *Materials Research Society Symposium Proceedings*, **346**, 391.
43 Koyano, G., Okuhara, T. and Misono, M. (1995) *Catalysis Letters*, **32**, 205.
44 Koyano, G., Okuhara, T. and Misono, M. (1998) *Journal of the American Chemical Society*, **120**, 767.
45 Conte, M., Budroni, G., Bartley, J.K., Taylor, S.H., Carley, A.F., Schmidt, A., Murphy, D.M., Girgsdies, F., Ressler, T., Schloegl, R. and Hutchings, G.J. (2006) *Science (Washington, DC)*, **313**, 1270.
46 Agaskar, P. A., DeCaul, L. and Grasselli, R.K. (1994) *Catalysis Letters*, **23**, 339.
47 Trifiro, F. (1993) *Catalysis Today*, **16**, 91.
48 Ruiz, P., Bastians, P., Caussin, L., Reuse, R., Daza, L., Acosta, D. and Delmon, B. (1993) *Catalysis Today*, **16**, 99.
49 Garbassi, F., Bart, J.C.J., Montino, F. and Petrini, G. (1985) *Applied Catalysis*, **16**, 271.
50 Hodnett, B.K. and Delmon, B. (1984) *Journal of Catalysis*, **88**, 43.
51 Hodnett, B.K., Permanne, P. and Delmon, B. (1983) *Applied Catalysis*, **6**, 231.
52 Morishige, H., Tamaki, J., Miura, N. and Yamazoe, N. (1990) *Chemistry Letters*, 1513.
53 Yamazoe, N., Morishige, H. and Teraoka, Y. (1989) *Studies in Surface Science and Catalysis*, **15**, 44.
54 Morishige, H., Teraoka Y. and Yamazoe N. (1988) *Shokubai*, **30**, 480.
55 Satsuma, A., Hattori, A., Mizutani, K., Furuta, A., Miyamoto, A., Hattori, T. and Murakami, Y. (1989) *Journal of Physical Chemistry*, **93**, 1484.
56 Guliants, V.V., Benziger, J.B., Sundaresan, S., Wachs, I.E. and Hirt, A.M. (1999) *Catalysis Letters*, **62**, 87.

57 Abon, M., Bere, K.E. and Delichere, P. (1997) *Catalysis Today*, **33**, 15.
58 Schiott, B. and Jorgensen, K.A. (1993) *Catalysis Today*, **16**, 79.
59 Schiott, B., Jorgensen, K.A. and Hoffmann, R. (1991) *Journal of Physical Chemistry*, **95**, 2297.
60 Gleaves, J.T., Ebner, J.R. and Kuechler, T.C. (1988) *Catalysis Reviews – Science and Engineering*, **30**, 49.
61 Gleaves, J.T. and Centi, G. (1993) *Catalysis Today*, **16**, 69.
62 Gleaves, J.T., Yablonskii, G.S., Phanawadee, P. and Schuurman, Y. (1997) *Applied Catalysis, A: General*, **160**, 55.
63 Miyamoto, K., Nitadori, T., Mizuno, N., Okuhara, T. and Misono, M. (1988) *Chemistry Letters*, 303.
64 Taufiq-Yap, Y.H., Sakakini, B.H. and Waugh, K.C. (1997) *Catalysis Letters*, **48**, 105.
65 Busca, G., Ramis, G. and Lorenzelli, V. (1989) *Journal of Molecular Catalysis*, **50**, 231.
66 Cavani, F., Centi, G., Manenti, I., Riva, A. and Trifiro, F. (1983) *Industrial & Engineering Chemistry Product Research and Development*, **22**, 565.
67 Lischke, G., Hanke, W., Jerschkewitz, H.G. and Oehlmann, G. (1985) *Journal of Catalysis*, **91**, 54.
68 Zhang-Lin, Y., Forissier, M., Sneeden, R.P., Vedrine, J.C. and Volta, J.C. (1994) *Journal of Catalysis*, **145**, 256.
69 Zhang-Lin, Y., Forissier, M., Vedrine, J.C. and Volta, J.C. (1994) *Journal of Catalysis*, **145**, 267.
70 Garbassi, F., Bart, J.C.J., Tassinari, R., Vlaic, G. and Lagarde, P. (1986) *Journal of Catalysis*, **98**, 317.
71 Sajip, S., Bartley, J.K., Burrows, A., Sananes-Schulz, M.-T., Tuel, A., Claude Volta, J., Kiely, C.J. and Hutchings, G.J. (2001) *New Journal of Chemistry*, **25**, 125.
72 Hutchings, G.J., Bartley, J.K., Webster, J.M., Lopez-Sanchez, J.A., Gilbert, D.J., Kiely, C.J., Carley, A.F., Howdle, S.M., Sajip, S., Caldarelli, S., Rhodes, C., Volta, J.C. and Poliakoff, M. (2001) *Journal of Catalysis*, **197**, 232.
73 Hutchings, G.J., Lopez-Sanchez, J.A., Bartley, J.K., Webster, J.M., Burrows, A., Kiely, C.J., Carley, A.F., Rhodes, C., Haevecker, M., Knop-Gericke, A., Mayer, R.W., Schloegl, R., Volta, J.C. and Poliakoff, M. (2002) *Journal of Catalysis*, **208**, 197.
74 Lopez-Sanchez, J.A., Bartley, J.K., Burrows, A., Kiely, C.J., Haevecker, M., Schloegl, R., Volta, J.C., Poliakoff, M. and Hutchings, G.J. (2002) *New Journal of Chemistry*, **26**, 1811.
75 Shimoda, T., Okuhara, T. and Misono, M. (1985) *Bulletin of the Chemical Society of Japan*, **58**, 2163.
76 Guliants, V.V., Benziger, J.B., Sundaresan, S., Yao, N. and Wachs, I.E. (1995) *Catalysis Letters*, **32**, 379.
77 Cavani, F., Centi, G. and Trifiro, F. (1985) *Journal of the Chemical Society, Chemical Communications*, 492.
78 Busca, G., Cavani, F., Centi, G. and Trifiro, F. (1986) *Journal of Catalysis*, **99**, 400.
79 Cornaglia, L.M., Lombardo, E.A., Anderson, J.A. and Garcia Fierro, J.L. (1993) *Applied Catalysis, A: General*, **100**, 37.
80 Lombardo, E.A., Sanchez, C.A. and Cornaglia, L.M. (1992) *Catalysis Today*, **15**, 407.
81 Cornaglia, L.M. and Lombardo, E.A. (1994) *Studies in Surface Science and Catalysis*, **90**, 429.
82 Ben Abdelmalek, S., Batis, H. and Ghorbel, A. (1993) *Journal de la Societe Algerienne de Chimie*, **3S**, 265.
83 Abon, M. and Volta, J.-C. (1997) *Applied Catalysis, A: General*, **157**, 173.
84 Centi, G., Golinelli, G. and Busca, G. (1990) *Journal of Physical Chemistry*, **94**, 6813.
85 Ai, M. (1986) *Journal of Catalysis*, **100**, 336.
86 Johnson, J.W., Johnston, D.C., Jacobson, A.J. and Brody, J.F. (1984) *Journal of the American Chemical Society*, **106**, 8123.
87 Sartoni, L., Delimitis, A., Bartley, J.K., Burrows, A., Roussel, H., Herrmann, J.-M., Volta, J.-C., Kiely, C.J. and Hutchings, G.J. (2006) *Journal of Materials Chemistry*, **16**, 4348.
88 Poli, G., Resta, I., Ruggeri, O. and Trifiro, F. (1981) *Applied Catalysis*, **1**, 395.
89 Hutchings, G.J., Sananes, M.T., Sajip, S., Kiely, C.J., Burrows, A., Ellison, I.J. and

Volta, J.C. (1997) *Catalysis Today*, **33**, 161.
90 Cornaglia, L.M., Sanchez, C.A. and Lombardo, E.A. (1993) *Applied Catalysis, A: General*, **95**, 117.
91 Batis, N.H., Batis, H., Ghorbel, A., Vedrine, J.C. and Volta, J.C. (1991) *Journal of Catalysis*, **128**, 248.
92 Meisel, M., Wolf, G.-U., Worzala, H., Eichele, K. and Grimmer, A.-R. (2000) *Phosphorus Research Bulletin*, **11**, 81.
93 Hutchings, G.J. and Higgins, R. (1997) *Applied Catalysis, A: General*, **154**, 103.
94 Kiely, C.J., Burrows, A., Hutchings, G.J., Bere, K.E., Volta, J.-C., Tuel, A. and Abon, M. (1997) *Faraday Discussions*, **105**, 103.
95 Mizuno, N., Hatayama, H. and Misono, M. (1997) *Chemistry of Materials*, **9**, 2697.
96 O'Connor, M., Dason, F. and Hodnett, B.K. (1990) *Applied Catalysis*, **64**, 161.
97 Cavani, F., Centi, G. and Trifiro, F. (1984) *Applied Catalysis*, **9**, 191.
98 Horowitz, H.S., Blackstone, C.M., Sleight, A.W. and Teufer, G. (1988) *Applied Catalysis*, **38**, 193.
99 Ellison, I.J., Hutchings, G.J., Sananes, M.T. and Volta, J.C. (1994) *Journal of the Chemical Society, Chemical Communications*, 1093.
100 Sananes, M.T., Ellison, I.J., Sajip, S., Burrows, A., Kiely, C.J., Volta, J.C. and Hutchings, G.J. (1996) *Journal of the Chemical Society–Faraday Transactions*, **92**, 137.
101 Doi, T. and Miyake, T. (1997) *Applied Catalysis, A: General*, **164**, 141.
102 Miyake, T. and Doi, T. (1995) *Applied Catalysis, A: General*, **131**, 43.
103 Guilhaume, N., Roullet, M., Pajonk, G., Grzybowska, B. and Volta, J.C. (1992) *Studies in Surface Science and Catalysis*, **72**, 255.
104 Hiyoshi, N., Yamamoto, N., Ryumon, N., Kamiya, Y. and Okuhara, T. (2004) *Journal of Catalysis*, **221**, 225.
105 Kamiya, Y., Ueki, S., Hiyoshi, N., Yamamoto, N. and Okuhara, T. (2003) *Catalysis Today*, **78**, 281.
106 Yamamoto, N., Hiyoshi, N. and Okuhara, T. (2002) *Chemistry of Materials*, **14**, 3882.
107 Yamamoto, N., Okuhara, T. and Nakato, T. (2001) *Journal of Materials Chemistry*, **11**, 1858.
108 Hiyoshi, N., Yamamoto, N. and Okuhara, T. (2001) *Chemistry Letters*, 484.
109 Nakato, T., Furumi, Y., Terao, N. and Okuhara, T. (2000) *Journal of Materials Chemistry*, **10**, 737.
110 Nakato, T., Furumi, Y. and Okuhara, T. (1998) *Chemistry Letters*, 611.
111 O'Mahony, L., Curtin, T., Zemlyanov, D., Mihov, M. and Hodnett, B.K. (2004) *Journal of Catalysis*, **227**, 270.
112 O'Mahony, L., Henry, J., Sutton, D., Curtin, T. and Hodnett, B.K. (2003) *Catalysis Letters*, **90**, 171.
113 Bartley, J.K., Wells, R.P.K. and Hutchings, G.J. (2000) *Journal of Catalysis*, **195**, 423.
114 Hannour, F.K., Martin, A., Bruckner, A., Wolf, G.U. and Lucke, B. (1998) *Reaction Kinetics and Catalysis Letters*, **63**, 225.
115 Bethke, G.K., Wang, D., Bueno, J.M.C., Kung, M.C. and Kung, H.H. (1997) *3rd World Congress on Oxidation Catalysis*.
116 Sananes, M.T., Ellison, I.J., Sajip, S., Burrows, A., Kiely, C.J., Volta, J.C. and Hutchings, G.J. (1996) *Journal of the Chemical Society-Faraday Transactions*, **92**, 137.
117 Sananes, M.T., Hutchings, G.J. and Volta, J.C. (1995) *Journal of Catalysis*, **154**, 253.
118 Sananes, M.T., Hutchings, G.J. and Volta, J.C. (1995) *Journal of the Chemical Society-Chemical Communications*, 243.
119 Mount, R.A., Raffelsonand, H. and Robinson, W.D. (1978) Monsanto Co., USA, US Patent 4116868.
120 Martin, A., Hannour, F.K., Bruckner, A. and Lucke, B. (1998) *Reaction Kinetics and Catalysis Letters*, **63**, 245.
121 Hannour, F.K., Martin, A., Kubias, B., Lucke, B., Bordes, E. and Courtine, P. (1998) *Catalysis Today*, **40**, 263.
122 Martin, A., Steinike, U., Rabe, S., Lucke, B. and Hannour, F.K. (1997) *Journal of the Chemical Society–Faraday Transactions*, **93**, 3855.
123 Higgins, R. and Hutchings, G.J. (1980) Imperial Chemical Industries Ltd., UK. Application: US Patent 4222945.
124 Bordes, E. and Courtine, P. (1979) *Journal of Catalysis*, **57**, 236.
125 Benziger, J.B., Guliants, V. and Sundaresan, S. (1997) *Catalysis Today*, **33**, 49.

126 Takita, Y., Hashiguchi, T. and Matsunosako, H. (1988) *Bulletin of the Chemical Society of Japan*, **61**, 3737.

127 Michalakos, P.M., Bellis, H.E., Brusky, P., Kung, H.H., Li, H.Q., Moser, W.R., Partenheimer, W. and Satek, L.C. (1995) *Industrial & Engineering Chemistry Research*, **34**, 1994.

128 Dong, W.-S., Bartley, J.K., Dummer, N.F., Girgsdies, F., Su, D., Schloegl, R., Volta, J.-C. and Hutchings, G.J. (2005) *Journal of Materials Chemistry*, **15**, 3214.

129 Taufiq-Yap, Y.H., Hasbi, A.R.M., Hussein, M.Z., Hutchings, G.J., Bartley, J. and Dummer, N. (2006) *Catalysis Letters*, **106**, 177.

130 Griesel, L., Bartley, J.K., Wells, R.P.K. and Hutchings, G.J. (2005) *Catalysis Today*, **99**, 131.

131 Doi, T. and Miyake, T. (1996) *Chemical Communications (Cambridge)*, 1635.

132 Bu, X., Feng, P. and Stucky, G.D. (1995) *Journal of the Chemical Society, Chemical Communications*, 1337.

133 Albonetti, S., Cavani, F., Trifiro, F., Venturoli, P., Calestani, G., Granados, M.L. and Fierro, J.L.G. (1996) *Journal of Catalysis*, **160**, 52.

134 Aït-Lachgar, K., Abon, M. and Volta, J.C. (1997) *Journal of Catalysis*, **171**, 383.

135 Cheng, W.-H. and Wang, W. (1997) *Applied Catalysis, A: General*, **156**, 57.

136 Arnold, E.W. III and Sundaresan, S. (1988) *Applied Catalysis*, **41**, 225.

137 Contractor, R.M., Horowitz, H.S., Sisler, G.M. and Bordes, E. (1997) *Catalysis Today*, **37**, 51.

138 Vedrine, J.C., Millet, J.M.M. and Volta, J.-C. (1996) *Catalysis Today*, **32**, 115.

139 Centi, G. and Perathoner, S. (1998) *Catalysis Today*, **41**, 457.

140 Patience, G.S., Bockrath, R.E., Sullivan, J.D. and Horowitz, H.S. (2007) *Industrial & Engineering Chemistry Research*, **46**, 4374.

141 Taufiq-Yap, Y.H. and Goh, C.K. (2005) *Eurasian Chemico-Technological Journal*, **7**, 73.

142 Haber, J., Zazhigalov, V.A., Stoch, J., Bogutskaya, L.V. and Batcherikova, I.V. (1997) *Catalysis Today*, **33**, 39.

143 Zazhigalov, V.A., Haber, J., Stoch, J., Bogutskaya, L.V. and Bacherikova, I.V. (1996) *Applied Catalysis, A: General*, **135**, 155.

144 Taufiq-Yap, Y.H., Goh, C.K., Hutchings, G.J., Dummer, N. and Bartley, J.K. (2006) *Journal of Molecular Catalysis A: Chemical*, **260**, 24.

145 Wang, X., Xu, L., Chen, X., Ji, W., Yan, Q. and Chen, Y. (2003) *Journal of Molecular Catalysis A: Chemical*, **206**, 261.

146 Ji, W., Xu, L., Wang, X., Hu, Z., Yan, Q. and Chen, Y. (2002) *Catalysis Today*, **74**, 101.

147 Hutchings, G.J. (1991) *Applied Catalysis*, **72**, 1.

148 Hutchings, G.J. and Higgins, R. (1996) *Journal of Catalysis*, **162**, 153.

149 Ye, D., Satsuma, A., Hattori, T. and Murakami, Y. (1991) *Applied Catalysis*, **69**, L1.

150 Ye, D., Satsuma, A., Hattori, T. and Murakami, Y. (1990) *Journal of the Chemical Society, Chemical Communications*, 1337.

151 Ye, D., Fu, M., Cheng, S. and Liang, H. (2003) *Abstracts of papers*, 226th ACS National Meeting. New York, NY, September 7–11, 2003, COLL.

152 Ye, D., Fu, M., Tian, L., Liang, H., Rong, T., Cheng, S. and Pang, X. (2003) *Research on Chemical Intermediates*, **29**, 271.

153 Ye, D., Satsuma, A., Hattori, A., Hattori, T. and Murakami, Y. (1993) *Catalysis Today*, **16**, 113.

154 Ye, D., Fu, M., Hong, L., Cheng, S., Rong, T. and Pang, X. (2004) *Huanan Ligong Daxue Xuebao Ziran Kexueban*, **32**, 49.

155 Hutchings, G.J. (1993) *Catalysis Today*, **16**, 139.

156 Sananes, M.T., Petunchi, J.O. and Lombardo, E.A. (1992) *Catalysis Today*, **15**, 527.

157 Takita, Y., Tanaka, K., Ichimaru, S., Mizihara, Y., Abe, Y. and Ishihara, T. (1993) *Applied Catalysis, A: General*, **103**, 281.

158 Zeyss, S., Wendt, G., Hallmeier, K.-H., Szargan, R. and Lippold, G. (1996) *Journal of the Chemical Society – Faraday Transactions*, **92**, 3273.

159 Guliants, V.V., Benziger, J.B., Sundaresan, S. and Wachs, I.E. (2000) *Studies in Surface Science and Catalysis*, **130B**, 1721.

160 Kourtakis, K. and Gai, P.L. (2005) US Patent 6903047.
161 Xu, L., Chen, X., Ji, W. and Yan, Q. (2002) *Reaction Kinetics and Catalysis Letters*, **76**, 335.
162 Liang, H. and Ye, D. (2005) *Journal of Natural Gas Chemistry*, **14**, 177.
163 Liang, H., Ye, D. and Lin, W. (2006) *Yunnan Daxue Xuebao, Ziran Kexueban*, **28**, 68.
164 Sant, R. and Varma, A. (1993) *Journal of Catalysis*, **143**, 215.
165 Takita, Y., Tanaka, K., Ichimaru, S., Ishihara, T., Inoue, T. and Arai, H. (1991) *Journal of Catalysis*, **130**, 347.
166 Bej, S.K. and Rao, M.S. (1992) *Industrial & Engineering Chemistry Research*, **31**, 2075.
167 Bej, S.K. and Rao, M.S. (1992) *Applied Catalysis, A: General*, **83**, 149.
168 Bej, S.K. and Rao, M.S. (1991) *Industrial & Engineering Chemistry Research*, **30**, 1819.
169 Bej, S.K. and Rao, M.S. (1991) *Industrial & Engineering Chemistry Research*, **30**, 1824.
170 Bej, S.K. and Rao, M.S. (1991) *Industrial & Engineering Chemistry Research*, **30**, 1829.
171 Hodnett, B.K. and Delmon, B. (1983) *Applied Catalysis*, **6**, 245.
172 Abdelouahab, F.B., Herrmann, J.-M., Volta, J.-C. and Ziyad, M. (1995) *Journal of the Chemical Society–Faraday Transactions*, **91**, 3231.
173 Abdelouahab, F.B., Olier, R., Ziyad, M. and Volta, J.C. (1995) *Journal of Catalysis*, **157**, 687.
174 Hutchings, G.J., Ellison, I.J., Sananes, M.T. and Volta, J.C. (1996) *Catalysis Letters*, **38**, 231.
175 Sananes-Schulz, M.T., Abdelouahad, F.B., Hutchings, G.J. and Volta, J.C. (1996) *Journal of Catalysis*, **163**, 346.
176 Cavani, F., Colombo, A., Trifiro, F., Sananes Schulz, M. T., Volta, J.C. and Hutchings, G.J. (1997) *Catalysis Letters*, **43**, 241.
177 Sananes-Schulz, M.T., Tuel, A., Hutchings, G.J. and Volta, J.C. (1997) *Journal of Catalysis*, **166**, 388.
178 Cornaglia, L.M., Carrara, C.R., Petunchi, J.O. and Lombardo, E.A. (1999) *Applied Catalysis, A: General*, **183**, 177.
179 Cornaglia, L., Carrara, C., Petunchi, J. and Lombardo, E. (2000) *Studies in Surface Science and Catalysis*, **130B**, 1727.
180 Carrara, C., Irusta, S., Lombardo, E. and Cornaglia, L. (2001) *Applied Catalysis, A: General*, **217**, 275.
181 Sajip, S., Bartley, J.K., Burrows, A., Rhodes, C., Claude Volta, J., Kiely, C.J. and Hutchings, G.J. (2001) *Physical Chemistry Chemical Physics*, **3**, 2143.
182 Shen, S., Zhou, J., Zhang, F., Zhou, L. and Li, R. (2002) *Catalysis Today*, **74**, 37.
183 Cornaglia, L., Irusta, S., Lombardo, E.A., Durupty, M.C. and Volta, J.C. (2003) *Catalysis Today*, **78**, 291.
184 Taufiq-Yap, Y.H. (2006) *Journal of Natural Gas Chemistry*, **15**, 144.
185 Cheng, W.-H. (1996) *Applied Catalysis, A: General*, **147**, 55.
186 McCormick, R.L., Alptekin, G.O., Herring, A.M., Ohno, T.R. and Dec, S.F. (1997) *Journal of Catalysis*, **172** (1), 160.
187 Boudin, S., Guesdon, A., Leclaire, A. and Borel, M.M. (2000) *International Journal of Inorganic Materials*, **2**, 561.
188 Taufiq-Yap, Y., Goh, C.K., Waugh, K.C. and Kamiya, Y. (2005) *Reaction Kinetics and Catalysis Letters*, **84**, 271.
189 Dobner, C., Duda, M., Raichle, A., Wilmer, H., Rosowski, F. and Hoelzle, M. (2007) BASF Aktiengesellschaft, Germany. Application: WO 2007012620.
190 Tamaki, J., Morishita, T., Morishige, H., Miura, N. and Yamazoe, N. (1992) *Chemistry Letters*, 13.
191 Matsuura, I., Ishimura, T. and Kimura, N. (1995) *Chemistry Letters*, 769.
192 Ballarini, N., Cavani, F., Cortelli, C., Ligi, S., Pierelli, F., Trifiro, F., Fumagalli, C., Mazzoni, G. and Monti, T. (2006) *Topics in Catalysis*, **38**, 147.
193 Mota, S., Volta, J.C., Vorbeck, G. and Dalmon, J.A. (2000) *Journal of Catalysis*, **193**, 319.
194 Abon, M., Herrmann, J.M. and Volta, J.C. (2001) *Catalysis Today*, **71**, 121.
195 Cornaglia, L.M., Carrara, C.R., Petunchi, J.O. and Lombardo, E.A. (2000) *Catalysis Today*, **57**, 313.

196 Pries de Oliveira, P.G., Eon, J.G., Chavant, M., Riche, A.S., Martin, V., Caldarelli, S. and Volta, J.C. (2000) *Catalysis Today*, **57**, 177.
197 Satsuma, A., Kijima, Y., Kamiya, S.-I., Komai, Y., Nishikawa, E. and Hatttori, T. (2001) *Catalysis Today*, **71**, 161.
198 Sartoni, L., Bartley, J.K., Wells, R.P.K., Kiely, C.J., Volta, J.C. and Hutchings, G.J. (2004) *Journal of Molecular Catalysis, A: Chemical*, **220**, 85.
199 Duarte de Farias, A.M., Gonzalez, W.D.A., Pries de Oliveira, P.G., Eon, J.-G., Herrmann, J.-M., Aouine, M., Loridant, S. and Volta, J.-C. (2002) *Journal of Catalysis*, **208**, 238.
200 Busca, G. and Centi, G. (1989) *Journal of the American Chemical Society*, **111**, 46.
201 Millet, J.-M.M. (2006) *Topics in Catalysis*, **38**, 83.
202 Cavani, F., Centi, G. and Trifiro, F. (1983) *Industrial & Engineering Chemistry Product Research and Development*, **22**, 570.
203 Ziółkowski, J., Bordes, E. and Courtine, P. (1993) *Journal of Molecular Catalysis*, **84**, 307.
204 Ziółkowski, J., Bordes, E. and Courtine, P. (1990) *Journal of Catalysis*, **122**, 126.

13
Heterogeneous Catalysis by Uranium Oxides
Stuart H. Taylor

13.1
Introduction

The element uranium has been used, in the form of its natural oxide, since ancient times. It was used as an orange-yellow coloring agent for ceramics dating from at least 79 AD. The element was identified by Martin Heinrich Klaproth in 1789 when it was discovered in natural minerals. The element was named in honor of the recently discovered new planet Uranus. In 1841 Peligot showed that Klaproth's *substance*, previously believed to be the metal, was in fact the oxide UO_2. Shortly thereafter Peligot showed that it was possible to produce the metal by reduction of uranium tetrachloride by the metals sodium and potassium. The construction of the periodic table by Mendeléev in 1872 focused attention on uranium as it was then the heaviest of the known elements. This stimulated much research on the element but it was not until 1896 that Antoine Becquerel recognized the radioactive properties of uranium. A research program starting in 1934 and led by Enrico Fermi resulted in the fissile properties of uranium being used for power generation and production of nuclear weapons. The isotope ^{235}U is still of primary importance and is used for power generation in both metallic and oxide forms. It is the only naturally occurring nuclide that undergoes nuclear fission with thermal neutrons.

The average concentration of uranium in the earth's crust is somewhere in the range 2–4 ppm, which is very similar to elements such as molybdenum and approximately 40 times greater than silver. Estimates of uranium reserves suggest that there are 4.7 million tons of easily accessible uranium minerals, and a further 35 million tons that could be recovered with further investment. In addition, a further 4.6 billion tons is present in sea water. Uranium is obtained mainly from the mineral uraninite, also called pitchblende, which consists largely of UO_2. Uranium-rich mineral resources are mined by a combination of open-cast and underground extraction methods, whilst poorer grade deposits are often recovered by leaching with acid or alkali.

Heterogeneous catalysis by compounds of uranium, and in particular the oxides of uranium, is well established and has a long history. The versatility of uranium

Metal Oxide Catalysis. Edited by S. David Jackson and Justin S. J. Hargreaves
Copyright © 2009 WILEY-VCH Verlag GmbH & Co. KGaA, Weinheim
ISBN: 978-3-527-31815-5

oxide based catalysts is related to the rich and diverse properties of the wide variety of phases and mixed phases that can be synthesized. The aim of this chapter is to introduce the reader to the chemistry of uranium compounds and highlight their uses as heterogeneous catalysts.

13.2
Structure of Uranium Oxides

Before examining the efficacy of uranium oxides as catalysts it is beneficial to consider the structures of the oxides. The three main oxides of uranium are UO_2 (brown-black), UO_3 (orange-yellow) and U_3O_8 (green-black). In addition to these three compounds, a considerable number of oxides exist within the stoichiometric range UO_2–UO_3. The range of stoichiometries and structures that are possible for the uranium–oxygen system make it the most complex of the Actinide elements, and one of the most complicated of all the elements. Changing the oxidation state of a given uranium ion is often accompanied by a modification of the structure. For example, within a limited temperature range, the uranium oxide structure and the uranium oxidation state can be influenced by the atmosphere to which it is exposed [1]. The complexity of the uranium–oxygen system is illustrated in Figure 13.1.

Allen and Holmes [2] have investigated the mechanism of transformation of UO_2 to UO_3 and they have also summarized the intermediate phases that have been identified (Table 13.1).

UO_2 has the fluorite structure; it is a face centered cubic based structure with each uranium ion coordinated to eight oxygen ions (Figure 13.2). The fact that there are vacant coordination positions in the lattice is critical for the catalytic, and other, properties of the oxide. This is because ion exchange is more efficient by

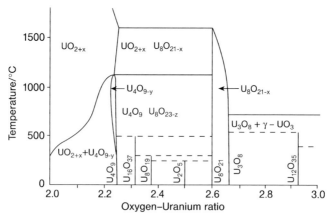

Figure 13.1 Phase diagram of uranium oxides. (*Journal of the Chemical Society, Dalton Transactions* (1982), 2169; reproduced by permission of The Royal Society of Chemistry).

Table 13.1 Crystallographic data for the known uranium oxides [2].

Phase	O:U ratio	Structure	Crystal class	Unit cell dimensions (Å)	Ref.
UO_2	2.00	fluorite	cubic	$a = 5.470$	[3]
UO_{2+x}	2.00–2.25	fluorite	cubic	$a = 5.470–5.445$	[4]
α-U_4O_9	2.45–2.25	fluorite	rhombohedral	$a = 4 \times (5.441–5.444)$; $\alpha = 90.078°$ ($T = 20°C$)	[5]
β-U_4O_9	2.25	fluorite	cubic	$a = 4 \times 5.438$ ($T = 65°C$)	[6]
γ-U_4O_9	2.25	fluorite	cubic	$a = 4 \times (5.47–5.50)$ ($T > 600°C$)	[6]
α-U_3O_7	2.27–2.33	fluorite	tetragonal	$a = 5.472$; $c = 5.397$	[7]
β-U_3O_7	2.33	fluorite	tetragonal	$a = 5.363$; $c = 5.531$	[7]
$U_{16}O_{37}$ (γ-U_3O_7)	2.31	fluorite	tetragonal	$a = 5.407$; $c = 5.497$	[8]
U_8O_{19} (δ-U_3O_7)	2.375	fluorite	monoclinic	$a = 5.378$; $b = 5.559$; $c = 5.378$; $\beta = 90.29°C$	[9]
γ-U_2O_5	2.50	fluorite	monoclinic	$a = 5.410$; $b = 5.481$; $c = 5.410$; $\beta = 90.49°$	[8]
α-U_2O_5	2.50	layered	hexagonal	$a = 3.885$; $c = 4.082$	[8]
β-U_2O_5	2.50	layered	hexagonal	$a = 3.813$; $c = 13.18$	[8]
U_2O_5	2.50	layered	orthorhombic	$a = 8.29$; $b = 31.71$; $c = 6.73$	[10]
α-U_3O_8	2.660–2.667	layered	orthorhombic	$a = 6.715$; $b = 11.96$; $c = 4.146$	[11, 12]
β-U_3O_8	2.67	layered	orthorhombic	$a = 7.07$; $b = 11.45$; $c = 8.30$	[13]
$U_{12}O_{35}$	2.92	layered	orthorhombic	$a = 6.91$; $b = 3.92$; $c = 4.12$	[14]
α-UO_3	3.00	layered	orthorhombic	$a = 6.84$; $b = 43.45$; $c = 4.12$	[15]
β-UO_3	3.00	layered	monoclinic	$a = 10.34$; $b = 14.33$; $c = 3.91$; $\beta = 99.03$	[16]
γ-UO_3	3.00		tetragonal	$a = 6.013$; $c = 19.975$	[17]
δ-UO_3	3.00	ReO_3-type	cubic	$a = 4.16$	[18]
ε-UO_3	3.00	layered	triclinic	$a = 4.002l$; $\alpha = 98.10°$; $b = 3.841$; $\beta = 90.20°$; $c = 4.165$; $\gamma = 120.17$	[19]
η-UO_3	3.00		orthorhombic	$a = 7.511$; $b = 5.466$; $c = 5.224$	[20]

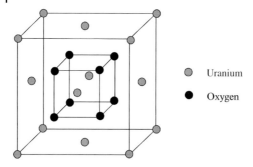

Figure 13.2 Fluorite structure of UO_2.

exchange through lattice vacancies. Another important factor is that the UO_2 fluorite structure is readily able to accommodate up to 10% additional oxygen in the lattice without any change of the structure [21]. The study of the oxidation of uranium oxides can be divided into two distinct regions: the oxidation of stoichiometric cubic UO_2 to orthorhombic U_3O_8 and the subsequent oxidation of this phase to UO_3 [22].

In the first region the initial stage is the oxidation step of UO_2 to $UO_{2.25}$, where gradual addition of oxygen leads to displacement of the ideal lattice positions until the structure of U_4O_9 ($UO_{2.25}$) is reached. The addition of this oxygen has no effect on the uranium sub-lattice as the oxygen is incorporated in interstitial sites. Increasing the stoichiometry to $UO_{2.12}$, the oxygen is distributed randomly in the interstitial sites; however, as the O/U ratio increases further towards $UO_{2.25}$, the formation of clusters in the lattice takes place. The clusters develop to form ordered cluster chains. The clusters are known as 2:2:2 clusters and contain two interstitial oxygen atoms in the (110) direction, two vacancies in the oxygen sub-lattice and two interstitial oxygen atoms in the (111) direction [23, 24].

For the stoichiometry $UO_{2.25}$, the structure can be simplified to an arrangement of 4:3:2 clusters [25]. These clusters are composed of four interstitial oxygen atoms in the (110) direction with three oxygen vacancies and two interstitial oxygen atoms in the (111) direction. The addition of this oxygen causes the expansion of the cubic structure so that the cell dimension for U_4O_9 is approximately four times that of UO_2 [6], although the cubic structure is retained.

Increasing the O/U ratio further from the stoichiometry U_4O_9 to U_3O_7, the structure of the oxide changes from the cubic crystal system to a variety of tetragonal structures [7, 8]. Further oxidation from U_3O_7 to U_2O_5 results in a further change of structure from the fluorite structure to a layered structure that is close to that observed for U_3O_8 [8, 10]. The U_2O_5 phase was first shown conclusively to exist by Rundle and coworkers in 1948 [10]. This work showed that the structure was orthorhombic like U_3O_8, but the actual cell dimensions were larger [11–13]. It is interesting that U_2O_5 phases have been reported with a monoclinic fluorite structure [8], with a hexagonal layered structure [8] and with an orthorhombic layered structure similar to that for U_3O_8 [10]. Allen and Holmes [2] suggest that the β-U_2O_5 could represent an intermediate bridging structure, since it shows neither fluorite-type nor U_3O_8-type structure.

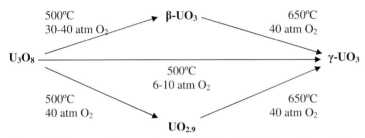

Figure 13.3 Summary of preparation of γ-UO$_3$ from U$_3$O$_8$. (Adapted from [14]).

The second region of oxidation covers the addition of oxygen from U$_3$O$_8$ to UO$_3$. Numerous phases have been identified for UO$_3$, with the majority showing layered structures similar to U$_3$O$_8$ [14–19]. UO$_3$ can be prepared from U$_3$O$_8$ by heating between 400 and 700 °C; however, it must be prepared under pressures of up to 40 atmospheres [13]. The transformation of U$_3$O$_8$ to UO$_3$ phases is summarized in Figure 13.3.

In UO$_3$ phases the uranium atom may be coordinated to six, seven or eight oxygen atoms, leading to at least five known modifications. For example the γ-UO$_3$ phase has two independent uranium atoms U (1) and U (2) in the structure [17]. The coordination polyhedron around U (2) is a slightly distorted octahedron, whereas U (1) is surrounded by eight oxygen atoms forming a somewhat distorted dodecahedron. This chapter is not intended to provide a detailed understanding of the multiple and complex phases of the uranium oxides, but it is clear even from a brief survey that the structures and chemistry are diverse.

13.3
Historical Uses of Uranium Oxides as Catalysts

Uranium compounds have historically been used as catalysts for many years, dating back to the initial development of catalysis as a recognized scientific discipline. In 1922, James published a paper detailing the vapor-phase low-temperature catalytic oxidation of fuel oil and other crude petroleum fractions [26]. It was concluded that to obtain satisfactorily high yields of sufficient quality for industrial use a catalyst must be employed. An apparatus was developed to test catalysts on a pilot-plant scale and it was found that the most effective configuration consisted of three catalyst layers, called catalyst screens. The most effective catalyst system for the reaction was an initial bed composed of uranium oxide and two subsequent beds containing molybdenum oxide. The uranium oxide was the best catalyst for the oxidation to aldehyde-type compounds and was particularly favored when acids were the desired products, as the higher yields of aldehyde were converted to acids over the molybdenum oxide screens. Typically the catalyst was supported on asbestos, although it is unclear whether the asbestos was a support in the way conventionally associated with a catalyst or merely acted as gauze to hold the catalyst bed in place. Air was used as oxidant and this was fed independently before each catalyst bed, which ensured that the oxygen concentration laterally through the catalyst screens was relatively low and probably helped to reduce overoxidation. The production of acids was typically carried out at 280 °C, whilst increasing the reaction temperature to ca. 400 °C resulted in the production of hydrocarbons and lower molecular weight oxygenates, which were suitable for use as fuels. The data presented are relatively scant; however, the publication is one of the first to indicate that uranium oxide is a potentially important oxidation catalyst.

Early work has also demonstrated that uranium oxide catalysts show promising activity for the oxidation of hydrocarbons in the liquid phase. A patent granted to the Dow Chemical Company describes a process for the manufacture of phenol from the partial oxidation of benzene [27]. The Dow Chemical patent claims that oxides of vanadium, molybdenum, tungsten and uranium were all effective catalysts for the oxidation of benzene to phenol in the presence of aqueous sodium hydroxide. The process was operated at approximately 200 atm under air in the temperature range 320–400 °C with an alkali solution of 20–25%. Sodium benzoate was formed in the aqueous phase and the unreacted benzene remained as an immiscible organic phase. Phenol was liberated by acidification of the aqueous phase. The process was 100% selective to the mono phenol product, and the unreacted benzene was easily recycled. It is stated that the uranium oxide catalyst gave the best results, although these results are not specified. Furthermore, the nature of the uranium oxide catalyst is not clear, but the patent acknowledges the existence of several different oxides and it is implied that all have been investigated and there is no differentiation in their activity.

The vapor phase oxidation of aromatic hydrocarbons using uranium oxide catalysts is also discussed in a patent filed shortly after the liquid-phase process and was assigned to the Barrett Corporation of New Jersey [28]. Studies concentrated mainly on the oxidation of toluene, and a large range of metal oxides were investigated. Air was used as the oxidant and it was pre-mixed with toluene before passing over the catalyst maintained at standard test conditions. The conditions are not specifically stated although it is thought that they are similar to 14/1 air/toluene by weight, pressure slightly elevated above atmospheric and a temperature of 500 °C. The reactivity was classed into four groups, these are described below and the oxides contained in them are also listed:

Group 1: Relatively high benzaldehyde production and relatively low combustion oxides of tantalum, tungsten, zirconium and molybdenum.
Group 2: Relatively high benzaldehyde production and relatively high combustion oxides of manganese, chromium, copper, nickel, thorium and uranium.
Group 3: Relatively low benzaldehyde production and relatively high combustion oxides of cobalt and cerium.
Group 4: Relatively low benzaldehyde production and relatively low combustion oxides of titanium, bismuth and tin.

Table 13.2 Catalytic data for toluene oxidation using uranium oxide based catalysts [6].

Catalyst	GHSV (h^{-1})	Toluene:air	Temp (°C)	O$_2$ consumed in total oxidation (%)	O$_2$ consumed in partial oxidation (%)
UO$_2$WO$_4$	1455	0.72	445	7.8	4.8
			480	23.8	14.2
			545	50.5	26.2
UO$_2$WO$_4$ + Al$_2$O$_3$	1455	0.72	405	30.9	21.4
			545	62.8	28.8
UO$_2$MoO$_4$ (no support)	414	0.63	406	6.7	8.6
			430	15.2	12.4
			524	74.3	23.8
U(MoO$_4$)$_2$	500	0.28	385	7.1	7.1
			415	27.6	16.2
			450	68.6	29.5
			510	73.3	24.7
U(MoO$_4$)$_2$	492	0.16	390	15.2	8.6
			420	36.1	12.4
			470	75.3	16.2

GHSV, gas hourly space velocity.

The activity demonstrated by uranium and molybdenum was significantly better than the other catalysts in the respective groups, whilst vanadium oxide was not classified in any of the groups as it produced quantities of maleic acid and benzoic acid in addition to benzaldehyde and carbon oxides. The catalyst performance was considerably enhanced by synthesizing catalysts containing mixtures of the oxides. The catalysts were prepared by a type of impregnation technique which involved placing a support, usually pumice or asbestos, in a solution of the metal salts before evaporating the solution to dryness.

Parks and Katz [29] also carried out early studies into the oxidation of toluene to benzaldehyde and benzoic acid over a range of catalysts, concentrating on uranium, tungsten and molybdenum oxides. Reactions used an air/toluene mixture with a gas hourly space velocity ranging from ca 400–1500 hour^{-1} and reaction temperatures 350–520 °C. Catalytic activity was expressed in terms of oxygen consumption to total and partial oxidation products. Selected data are presented in Table 13.2. A considerable range of catalysts were tested, and although specific yields of products are not available it is clear that the catalysts are active for oxidation. Appreciable yields of selective oxidation products were obtained when uranium oxide was used in combination with other oxides that are recognized as selective oxidation components.

In another patent also granted to the Barrett Corporation of New Jersey [30], the partial oxidation of ethanol to acetaldehyde by oxide catalysts was investigated. The patent details a process for reducing combustion by controlling the reaction temperature by means of efficient heat removal from the functioning catalyst. This was achieved by packing the catalyst in a series of tubular reactors, and is the precursor to the multi-tubular design that is used so effectively by the modern chemical industry. The majority of results are concerned with vanadium oxide as the catalyst, which at 300 °C and 0.39 sec contact time produced 70 parts acetaldehyde per 100 parts of ethanol. Acetic acid (10 parts) was also produced and only approximately 3% of the ethanol feed was combusted to carbon dioxide. No specific data were presented but it was acknowledged that many oxides, including those of uranium, were active. Cobalt, tin, cerium and titanium oxides only showed low acetaldehyde yields; all the other oxides showed reasonable yields although they were all lower than vanadium oxides. It was also highlighted that oxides of uranium, chromium, manganese and copper showed higher levels of combustion. The study was extended to include mixed oxide catalysts: one such system consisted of 93% uranium oxide and 7% molybdenum oxide, which yielded almost exclusively acetaldehyde with virtually no carbon dioxide and only a small amount of acetic acid products.

In 1932 Wietzel and Pfaundler [31] described the use of a uranium oxide based catalyst to produce *valuable hydrocarbons of low boiling point* from various sources including coal, tar and mineral oils. One example describes the use of uranium oxide in a process in which a fraction of mineral oil (BPt > 270 °C) was passed with excess hydrogen over the catalyst at 450 °C and 200 at. The actual catalytic material was fine aluminum granules activated with 1–2% of uranyl nitrate. The catalyst produced aromatics of boiling point less than 200 °C in a yield of over 80%. Uranium oxide was also cited for an example to convert bituminous coal tar (BPt 300–420 °C). The pressure was the same as in the experiments described above, with the temperature raised to 480 °C. The catalyst in this example was formed by heating aluminum gauze in ammonium vanadate and uranyl nitrate in hydrochloric acid. The resulting uranium–vanadium–aluminum catalyst was able to convert the coal tar to 70% oil.

Thus, historically, uranium oxides have been used as catalysts, and more often they have been used as catalyst components in combination with other metal oxides. Often it is difficult to identify the catalysts unambiguously: there is little characterization data in the studies, and it is most likely that the specific stoichiometries of uranium oxides quoted as catalysts are not correct. There are many other examples of the use of uranium oxides for heterogeneous catalysis and the few examples presented in this section are typical of some of the earliest uses. It is interesting to note that, although some of the work highlighted was carried out over 80 years ago, some of the aims, such as selective hydrocarbon oxidation, are still major research aims for heterogeneous catalysis today.

13.4
Catalysis by Uranium Oxides

13.4.1
Total Oxidation

One of the earlier studies investigating the total oxidation of uranium oxides concentrated on the oxidation of carbon monoxide [32]. The oxide U_3O_8 was the most active catalyst probed for the formation of carbon dioxide by oxidation by molecular oxygen. Comparison was made with V_2O_5, MoO_3 and WO_3; although these oxides are not recognized as high-activity catalysts for carbon monoxide oxidation the results indicated that U_3O_8 was a potential catalyst for total oxidation.

The oxidative destruction of volatile organic compounds (VOCs) over uranium oxides has been studied by Hutchings, Taylor and coworkers [33, 34]. Studies have shown that U_3O_8 is a highly active catalyst for the destruction of a wide range of chemically diverse VOCs. In the case of benzene oxidation over U_3O_8 a conversion of 100% at 400°C was reached, with selectivities of 27% and 73% for CO and CO_2 respectively. Comparing these results with the total oxidation activity of Co_3O_4, a well known active combustion catalyst, it was found that even at 450°C the conversion of benzene over Co_3O_4 was only 90%. The uranium oxide U_3O_8 was particularly active for the total oxidation of chlorinated VOCs [33]. For example, investigating the oxidation of chlorobenzene, 99.7% conversion was reached at only 350°C with selectivity to CO_x of 100%. Furthermore, oxidation of chlorobutane at 350°C led to 100% selectivity to CO_x with a conversion higher than 99.5%. No catalyst deactivation was observed for the oxidation of the VOCs, even for prolonged oxidation of chlorinated compounds [35].

Short-chain linear alkanes are amongst the most difficult of VOCs to destroy. A study has investigated the catalytic activity of uranium oxide catalysts for the destruction of alkanes in the C_1–C_4 range [36]. Uranium oxide, U_3O_8, showed relatively low activity for the combustion of methane and ethane and moderate activity for propane and n-butane. Catalyst activity was improved by supporting the uranium oxide on silica and further improvements were achieved by the addition of chromium. X-ray Diffraction (XRD), X-ray Photoelectron Spectroscopy (XPS) and Temperature-Programmed Reaction (TPR) characterization data indicated that supporting the U_3O_8 phase and adding chromium modified the structure and chemistry of the oxide. This modification may culminate in an increase of the detect structure of the oxide, resulting in the increased oxidation activity.

The effect of water addition on the complete oxidation of benzene and propane VOCs by uranium oxide catalysts has been investigated [37]. Benzene oxidation was studied using a silica-supported U_3O_8 catalyst. Complete oxidation was promoted by the addition of 2.6% water compared with the reactivity when no water was added to the reactant feed. Increasing the water concentration to 12.1% resulted in a suppression of oxidation activity. Investigation of propane oxidation using U_3O_8 showed a dramatic promotion of activity. Propane conversion was ca 50% at 600°C without added water, whilst it increased to 100% at 400°C with the addition of 2.6% water. Comparison of total oxidation activity with Mn_2O_3 showed that any level of water addition suppressed conversion, and this was in clear contrast to the U_3O_8 catalyst. *In situ* powder XRD studies showed that the bulk U_3O_8 structure was stable under all the reaction conditions. The origin of the increased activity is not clear but it may be due to modification of the catalyst surface, possibly aiding the activation of the VOCs by increased hydroxylation.

A Temporal Analysis of Products (TAP) reactor has been used to investigate the mechanism of oxidation by uranium oxide catalysts [35, 38]. A combination of TAP pulse experiments with oxygen present and absent in the gas phase indicated that the lattice oxygen from the catalyst was responsible for the total oxidation activity. It was proposed that the catalyst operates by a redox mechanism using lattice oxygen and the high activity shown by U_3O_8 was due to the facile uranium redox couple and the non-stoichiometry of the oxide. Isotopically labeled oxygen studies of carbon monoxide oxidation confirmed that lattice oxygen was the active oxidant.

In a much earlier patent, the removal of organics from exhaust gases by oxidation over a supported uranium oxide catalyst was reported by Hofer and Anderson [39]. The catalyst was 4% U_3O_8 supported on alumina spheres. The authors used the incipient wetness technique to impregnate alumina with uranyl nitrate solution. In this case the catalyst precursors were calcined at 700 °C for 3 h to decompose the uranium salt. The use of other uranium compounds as starting materials was mentioned and these included uranyl acetate, uranium ammonium carbonate and uranyl chloride. The alumina-supported catalyst had a surface area of ca 400 $m^2 g^{-1}$ and further added components, such as copper, chromium and iron, were highlighted as efficient additives to increase activity.

The catalysts were evaluated by exposure to a simulated automobile exhaust gas stream composed of 0.2% isopentane, 2% carbon monoxide, 4% oxygen and a balance of nitrogen. The temperature required to oxidize the isopentane and carbon monoxide was used to compare catalyst performance. The chromium-promoted catalyst oxidized isopentane at the lowest temperature, and a mixed chromium/copper-promoted catalyst proved the most efficient for oxidizing carbon monoxide and isopentane. It is interesting to note that the test rig used a stationary engine with 21 pounds of catalyst. Although the catalyst was very effective it is difficult to envisage uranium oxide catalysts employed for emission control of mobile sources.

13.4.2
Selective Oxidation

Uranium oxides have been investigated as catalysts and catalyst components for selective oxidation. They are more commonly used as catalyst components, but there are also reports of uranium oxide alone as a selective oxidation catalyst. The oxidation of ethylene over UO_3 has been studied by Idriss and Madhavaram [40] using the technique of temperature programmed desorption (TPD). Table 13.3 shows the desorption products formed during TPD after ethylene adsorption at room temperature on UO_3. The production of acetaldehyde from ethylene indicates

Table 13.3 Products from TPD of ethylene adsorbed on UO_3 [33].

Product	Desorption temperature (°C)	Carbon yield (%)	Carbon selectivity (%)
Ethylene	127–427	85.7	–
Acetaldehyde	207	8.3	58
Furan	277	6.0	42
CO_2	above 527	not calculated	–

the ability of UO_3 to oxidize olefins in a relatively facile manner owing to the lability of lattice oxygen. However, in this case it must be taken into account that during TPD there is no regeneration of surface reduced sites in contrast to steady-state oxidation with oxygen present in the gas phase. On the other hand, the identification of furan among the products shows that uranium oxides are active for carbon–carbon bond formation in addition to carbon–oxygen selective bond formation.

Idriss and Madhavaram [41] have also studied the partial oxidation of ethanol to furan over UO_3. The oxides U_3O_8 and UO_2 were also studied; UO_2 was inactive whilst U_3O_8 only exhibited very low activity. Using UO_3, the maximum selectivity to furan was 23% at 150 °C with a conversion of 81%, with acetaldehyde being the other major product. Comparing the reactions of ethylene and ethanol, the authors highlight two points. Firstly, no ethylene was formed during the reaction of ethanol. Secondly, acetaldehyde was formed from both ethylene and ethanol, and ethanol was formed as a trace product from ethylene oxidation. These facts were used to propose a mechanism in which the formation of furan was via an ethoxide intermediate [41].

More commonly, uranium has been used as a catalyst component for mixed-metal oxide catalysts for selective oxidation. Probably the most well known of these mixed oxide catalysts are those based on uranium and antimony. The uranium–antimony catalysts are exceptionally active and selective and they have been applied industrially. An interpretation of the catalyst structure and reaction mechanism has been reported by Grasselli and coworkers [42, 43] who discovered the catalyst. The USb_3O_{10} mixed oxide has been extensively used for the oxidation/ammoxidation reaction of propylene to acrolein and acrylonitrile. The selective ammoxidation of propylene was investigated by Grasselli and coworkers [44], and it has been demonstrated that at 460 °C a 62.0% selectivity to acrolein with a conversion of 65.2% can be achieved. Furthermore, Delobel and coworkers [45] studied the selective oxidation of propylene over USb_3O_{10}, which at 340 °C gave a selectivity to acrolein of 96.7%.

The structure of USb_3O_{10} is complex and it is crucial for selective oxidation activity. USb_3O_{10} contains one type of uranium, two types of antimony and four different types of oxygen. The unit cell is composed of eight formula weight units ($Z = 8$). The structure is composed of layers that contain heavy atoms and oxygen alternating with layers of oxygen only. Five layers of planes containing heavy metal atoms are required to completely describe the unit cell of USb_3O_{10}. It was found that there are two types of lattice oxygen in the structure, one giving high selectivity

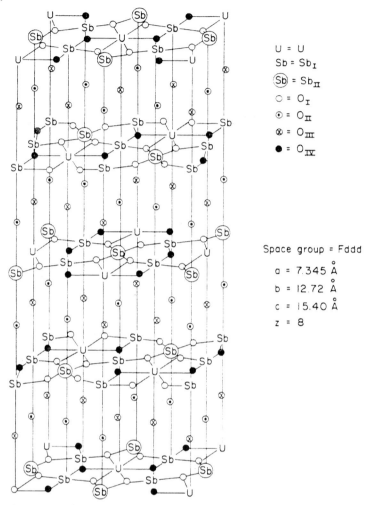

Figure 13.4 USb$_3$O$_{10}$ structure. (*Journal of Catalysis*, 25 (1972), 273; reproduced by permission of Academic Press).

for partial oxidation, and the other, which was less active and tended to be less selective, producing carbon oxides. Grasselli and coworkers [43] suggested that the active oxygen was O(IV) and the less active one was O(I), as shown in Figure 13.4. The USbO$_5$ structure is very similar to that of USb$_3$O$_{10}$. USbO$_5$ has the heavy atoms at positions close to those of USb$_3$O$_{10}$, but with a slight distortion of all atoms, leading to a lower symmetry sub-group.

The USbO$_5$ phase was proposed as a precursor to the USb$_3$O$_{10}$ phase, which at high temperature decomposes to give USbO$_5$. The decomposition of UO$_2$(NO$_3$)$_2$ and oxidation of the Sb$_2$O$_3$, from which the catalyst was prepared, led to the formation of UO$_3$ and Sb$_2$O$_4$ respectively. On thermal treatment the antimony oxide penetrates the lattice of UO_3 and forms the $USbO_5$ phase, according to the reactions

$$2UO_3 + Sb_2O_4 \rightarrow [2UO_3 \cdot Sb_2O_4] \rightarrow 2USbO_5$$

$$2U_3O_8 + 3Sb_2O_4 + O_2 \rightarrow [6UO_3 \cdot 3Sb_2O_4] \rightarrow 6USbO_5$$

The Sb_2O_5 phase is stabilized by the presence of the $USbO_5$ phase and further reaction proceeds to form the desired USb_3O_{10} catalytically active phase.

$$USbO_5 + Sb_2O_5 \rightarrow [USbO_5 \cdot Sb_2O_5] \rightarrow USb_3O_{10}$$

$$USbO_5 + Sb_2O_4 + \frac{1}{2}O_2 \rightarrow [USbO_5 \cdot Sb_2O_5] \rightarrow USb_3O_{10}$$

Powder X-ray diffraction studies of $USbO_5$, USb_3O_{10} and Sb_2O_4 as a function of calcination time and temperature allowed investigation of the mechanism of formation of the phases, $USbO_5$ and USb_3O_{10}. Accordingly, $USbO_5$, which was formed at about 675 °C, was considered to be the precursor of the USb_3O_{10} phase [43]. At this temperature Sb_2O_5 was also initially present and since it was not stable it reacted with $USbO_5$ to produce USb_3O_{10}. The USb_3O_{10} phase increased in concentration as the temperature was increased. At 980 °C the USb_3O_{10} phase was decomposed to re-form $USbO_5$.

The uranium–antimony oxide system remains as a basis of interest for catalysts. The preparation of a new uranyl antimonate has been described and it was prepared by hydrothermal synthesis from UO_3, Sb_2O_3 and KCl [46]. A detailed structural analysis was reported, but more importantly the $UO_2Sb_2O_4$ was selective for the oxidation of propylene to acrolein.

Some further mixed-metal oxides containing uranium, such as bismuth uranates, are known to be effective catalysts in certain selective oxidation reactions, for instance toluene partial oxidation. The oxidative demethylation of toluene using bismuth uranate as oxidant has been studied by Van deer Baan and coworkers [47] in a non-steady-state reactor. The reaction conditions were: temperature, 480 °C; pulse volume, 0.534 cm^3; mole fraction of toluene, 0.023; gas flow, 25 cm^3 min^{-1} and catalysts used were samples with different Bi to U atomic ratios. Results indicated that at the reaction temperature the most active catalyst was Bi_2UO_6 with a selectivity of 70% to benzene in the absence of gas-phase oxygen. The experiments carried out showed that bismuth uranate is a very active but unselective oxidant in the very early stages of the toluene oxidation reaction. However, as soon as the catalyst starts to become depleted of oxygen the activity decreased and the selectivity increased. The reduced uranate was easily reoxidized by molecular oxygen in a second step. The mechanism proposed for this reaction was a redox mechanism, which occurred via a benzoate-like intermediate. It was stated that the oxidation of toluene gave two main products: benzaldehyde, which dissociated to benzene and carbon monoxide and benzoic acid, which also dissociated to produce benzene and carbon dioxide.

In a more detailed study of bismuth uranate reduction by toluene using pulsed flow and steady-state flow, further insight into the mechanism of oxidation was obtained [48]. In the temperature range 455–527 °C, Bi_2UO_6 was reduced by toluene to metallic bismuth and UO_2. The reaction rate was related to the surface reaction rate and the rate of diffusion of oxygen through the lattice. It was apparent that Bi_2UO_6 contained two types of reactive lattice oxygen. The first was a nonselective species that was ca. 2% of the total lattice oxygen and it showed activation for diffusion of ca. 107 kJ mol^{-1}. The second type was a species responsible for selective oxidation and had a diffusion energy of ca. 261 kJ mol^{-1} at 470 °C. Evidence for the formation of shear planes in the Bi_2UO_6 structure was presented and the relatively small concentration of nonselective oxygen was thought to be from positions in the shear planes.

The oxidation of carbon monoxide to carbon dioxide using similar bismuth uranate catalysts has been reported by Derouane and coworkers [49]. The work on carbon monoxide oxidation confirmed that the bismuth uranate catalyst operated by a redox mechanism. These studies on bismuth uranates highlight the important role played by oxygen transfer via the lattice, and reinforce the importance of the ability of uranium to exhibit relatively facile redox behavior.

The selective oxidation of toluene has been studied over a number of catalysts based on metal oxides, with the U/Mo oxide system being one of the most active and selective[50, 51]. The main products in the oxidation of toluene, excluding the non-oxidative coupling products, were benzaldehyde, benzoic acid, maleic anhydride, benzene, benzoquinone, CO and CO_2. Under the same reaction conditions toluene may also yield coupling products such as phthalic anhydride, methyldiphenylmethane, benzophenone, diphenylethanone and anthraquinone, as shown by Zhu and coworkers [51]. A range of different uranium-based oxides were tested [51] and the results obtained are shown in Table 13.4.

Table 13.4 Conversion and selectivity data for toluene oxidation using U-based mixed oxides at 500 °C [51].

Catalyst (atomic ratio)	Contact time (s)	Conversion (%)	Benzaldehyde selectivity (%)	Benzaldehyde yield (mol%)
U alone	0.065	21.4	90	19.3
	0.200	47.6	69	32.1
	0.320	66.0	49	32.0
	0.400	77.4	47	35.8
U:P (8:2)	0.032	37.8	47	17.6
	0.065	54.2	38	15.4
U:W (9:1)	0.032	38.4	69	26.4
U:V (9:1)	0.032	48.4	54	26.4
U:Sb (1:4)	1.3	11.0	50	5.6

Under the reaction conditions used, a U_3O_8 catalyst demonstrated appreciable selective oxidation activity. The best results, in terms of both activity and selectivity to benzaldehyde, were obtained with the mixed oxides with U:Mo atomic ratios in the range 8:2 to 9:1. The maximum yield of benzaldehyde was 40 mol%. On the other hand, antimony-based uranium oxides were not found to be effective as catalyst for this reaction. U–Mo and Bi–Mo mixtures also exhibited promising activity and selectivity to benzaldehyde. Bi–Mo and Bi–Mo–P–Si catalysts were also tested. Qualitatively there was little difference between the product distributions from the two catalysts. The major products formed were benzaldehyde, benzene and carbon oxides, as well as traces of anthraquinone and benzoic acid.

Perhaps one of the most demanding selective oxidation reactions is the oxidation of methane to the oxygenates formaldehyde and methanol. Although the use of uranium as a catalyst component for selective methane partial oxidation is relatively rare, there is one notable attempt. Dowden and Walker reported the activity of a range of two-component catalysts, one component of which was molybdenum [52]. They developed these catalysts based on the principles outlined by the proposal of a virtual mechanism [53]. Results were reported for MoO_3/ZnO, $(MoO_3)_4/Fe_2O_3$, MoO_3/VO_2 and MoO_3/UO_2 supported on 1:3 $Al_2O_3:SiO_2$ with a low area of ca. $0.1 \, m^2 g^{-1}$, containing ca. 5% active oxide. Experimental conditions were 30 bar pressure with a $CH_4:O_2$ ratio of 97:3 in the temperature range 430–500 °C. In order to maintain high selectivity to CH_3OH the reactor effluent was cooled to below 200 °C within 0.03 s of leaving the heated catalyst bed, by the injection of cooling water. Although the most successful catalyst was the mixed oxide of molybdenum and iron, which gave yields of 869 g (kg cat)$^{-1}$h^{-1} and 100 g (kg cat)$^{-1}$h^{-1} of CH_3OH and $HCHO$ respectively, these were only marginally greater than the yields from the molybdenum–uranium oxide catalyst.

A study has been undertaken to compare the effectiveness of molybdenum and uranium oxide and iron sodalite catalysts with the homogeneous gas-phase oxidation of methane [54]. Catalyst performance was evaluated in a high-pressure annular reactor and data were compared to the reactivity of the empty reactor. It was concluded that none of the catalysts gave any advantage over the homogeneous reaction. Indeed, using a catalyst only reduced the selectivity to the desired partial oxidation products. Similar conclusions have been reached for many catalysts used for the partial oxidation of methane, and therefore it is perhaps not surprising that uranium oxide catalysts are no different.

Catalysts using molybdenum and uranium oxides have also been used for selective oxidation of other alkanes. The partial oxidation of isobutane over MoO_3–UO_3–SiO_2 was studied by Corma and coworkers [55], and led to the formation of a range of oxygenated hydrocarbons. The primary products formed in this reaction were methallyl alcohol, methacrolein, acetone and biacetyl. In addition, a large number of secondary products were obtained, such as acetic acid, ethanal, formic acid, methanal, and methacrylic acid. Total oxidation to carbon oxides was also determined. The best performance of the silica-supported MoO_3/UO_3 catalyst was obtained at 380 °C, with an oxygen:isobutene ratio of 12:1, to produce a 45% selectivity to methacrolein and 10% to acetone.

Taylor and coworkers demonstrated that catalysts of iron and uranium oxide were effective for the selective oxidation of propane and propene to formaldehyde [56]. Catalysts were prepared by co-precipitation and were most effective when prepared with Fe:U ratios of 0.5:3 and 1:3. It was possible to achieve 44% selectivity towards formaldehyde at 450 °C with a 42% conversion. The balance of the products was carbon oxides. Characterization of the catalysts showed that they were composed of iron oxide highly dispersed on UO_3, and it was a combination of the highly dispersed iron oxide and the UO_3 that was responsible for the selective oxidation. The mechanism of formaldehyde production is not clear, but there are other studies in the literature reporting similar observations. Oxidation of propane over highly dispersed iron oxide supported on SiO_2 showed the formation of formaldehyde [57]. Lee and coworkers have also shown that formaldehyde is produced during the oxidation of isobutene using a urania–titania catalyst [58]. Supporting the uranium oxide on titania modified the redox behavior of the uranium component and this modification was responsible for the enhanced selective oxidation function.

In recent years, interest in the use of gold-based catalysts has expanded considerably. In particular, the potential of gold catalysts for selective oxidation now receives significant attention. Gold-based catalysts supported on a wide range of materials have been prepared and tested for a wide range of reactions. Choudhary and coworkers have used nanoparticles of gold supported on U_3O_8 for the selective oxidation of benzyl alcohol to benzaldehyde by molecular oxygen in solvent-free conditions [59]. Variations of catalyst preparation conditions and reaction conditions were studied. The best catalysts were those with high gold loadings and small gold particles. Increasing the reaction temperature or time increased benzyl alcohol conversion, whilst benzaldehyde selectivity decreased and benzyl benzoate selectivity increased. The addition of a range of solvents was deleterious for catalyst performance.

13.4.3
Reduction

The versatility of uranium oxides in functioning as catalysts for a range of reactions is demonstrated by their ability to catalyze a range of reduction reactions, although primarily they have been used for oxidation. The reduction of acetaldehyde on UO_2, prepared by reduction of UO_3 with hydrogen, has been studied by Idriss and coworkers [40] using TPD. The formation of C_4 olefins indicated the ability of UO_2 surfaces to abstract large amounts of oxygen from carbonyl species bound to the surface. This functionality was due to the ability of the fluorite structure of UO_2 to readily abstract and accommodate a considerable quantity of oxygen in vacant sites in the oxide lattice. As discussed in Section 13.2, this behavior is well known and is a specific feature of the structure of UO_2.

More recently, Madhavaram and Idriss studied the reactions of acetaldehyde over the oxides UO_2, α-U_3O_8 and β-UO_3 [60]. The products were strongly dependent on the U:O ratio of the oxide. In agreement with earlier work, UO_2 showed activity for reductive coupling to produce C_4 olefins, whilst U_3O_8 produced predominantly crotonaldehyde by aldol condensation. It was possible to produce both furan and crotonaldehyde over UO_3, and stoichiometric reactions using TPD demonstrated that the type of product was related to the surface coverage. At low surface coverage of acetaldehyde, furan was predominant and the product distribution shifted to furan and crotonaldehyde as the surface coverage increased. The mode of acetaldehyde adsorption was studied by FTIR, and distinctly different modes were identified depending on the uranium oxide catalyst. The differences in products over the oxides have been explained in terms of the possible adsorption modes and the variation in the semiconductor properties of the different oxides.

Two further reduction reactions of interest are the TPD of acetone over U_3O_8, the principal product being isobutene, and over UO_2 forming mainly propylene [61]. Again a significant difference between the two uranium oxides used for these reactions resides in the fact that, in the first case, U_3O_8 reacted leading to C–C bond formation to give a C_4 olefin. However, over UO_2 the main product was propylene, once again showing the ability of UO_2 to accommodate excess oxygen in its fluorite structure.

The dehydrogenation of ethylbenzene is an important process used for styrene manufacture, and uranium oxide catalysts have been investigated for this reaction. A catalyst of uranium dioxide supported on alumina showed high selectivity to styrene of 96% at high conversion [62, 63]. The catalyst was synthesized as a higher oxide of uranium and initially it was not UO_2. Consequently, over the initial on-stream period only carbon dioxide and water were observed, as the catalyst produced total oxidation products. However, as the reaction preceeded the uranium oxide was reduced *in situ* by the ethylbenzene and hydrogen to form the active UO_2 phase. It was only when the uranium oxide was fully reduced to UO_2 that styrene was produced with high selectivity.

Nickel catalysts supported on uranium oxides have been reported for the hydrogenation of carbon dioxide to methane [64, 65]. The catalysts were selective below 500 °C, as CO was the major product at higher temperatures. The nickel was deposited on the catalysts by evaporation and the reduction characteristics of the catalysts were complex, depending on the calcination conditions and the metal content. The uranium oxide support had a crucial role in maintaining the high dispersion of the active nickel by preventing sintering. The most active catalysts were those with the highest stable nickel surface areas.

Uranium oxide catalysts have largely been employed for the reduction of organic species but, in a series of interesting studies, a uranium oxide catalyst has also been used for the reduction of NO_x and simultaneous oxidation of CO [66]. Studies showed that NO_x was converted to N_2 with 100% selectivity under favorable reaction conditions. Using a mixture of 4%NO, 4%CO with a balance of He, different uranium oxides were tested in a fixed bed micro-reactor. The results obtained are shown in Table 13.5, and compared with a conventional supported Pt catalyst.

At lower temperatures, reduction of NO produced N_2O as the major product, whereas an increase of reaction temperature not only enhanced NO conversion but generally improved selectivity, as the only product obtained was N_2.

More detailed studies of uranium-based catalysts for NO_x reduction and CO oxidation have been published and concentrate on catalyst characterization [67]

Table 13.5 NO conversion and selectivity to N_2 over uranium oxide catalysts [66].

Catalyst	Conversion (%)	Selectivity (%)	Temperature (°C)
U_3O_8	100	100	800
$U_3O_8/\gamma\text{-}Al_2O_3$ 800 °C[a]	100	100	400
$U_3O_8/\gamma\text{-}Al_2O_3$ 450 °C[a]	100	100	400
$Pt/\gamma\text{-}Al_2O_3$	10	100	250
	100	35	400

a Calcination temperature.

and catalyst performance [68]. Catalysts were prepared from the precursors uranyl nitrate and uranium(IV) chloride, which were supported on $\gamma\text{-}Al_2O_3$, SiO_2 and mesoporous SiO_2. Both the support and the uranium oxide precursor were found to influence the nature of the catalyst. Calcination of the mesoporous SiO_2 supported material at 800 °C resulted in significant extrusion of the uranium from the support, resulting in the formation of large orthorhombic domains of U_3O_8. The formation of a U_3O_8 phase was promoted on the mesoporous support and by the presence of chloride. On the silica and alumina supports hexagonal U_3O_8 was formed, with the presence of chloride inhibiting the growth of larger uranium oxide domains on all the catalysts.

Calcination of the uranium oxide mesoporous SiO_2 supported catalyst resulted in sintering of the active uranium oxide phase into larger particles and this was detrimental to catalytic activity. However, preparing the mesoporous supported catalysts from uranyl nitrate using thermal treatment in dilute CO/O_2 or CO/NO resulted in the best catalysts with high activity comparable to a Pt/Al_2O_3 catalyst. The rate expression for the best catalyst was zero order with respect to NO and showed an order of 1.4 with respect to CO. This was in clear contrast to the reaction over bulk U_3O_8, which was dependent solely on the NO concentration. For the majority of catalysts the presence of residual chloride resulted in lower activity. The exception was the $\gamma\text{-}Al_2O_3$ supported catalyst prepared from the chloride precursor and thermally treated at 600 °C in dilute CO/O_2. It was possible to correlate the catalytic activity with the residual chloride content and the average crystallite size of the supported uranium oxide.

13.4.4
Steam Reforming

Nicklin, with others, filed several early patents describing the use of uranium oxides as steam reforming catalysts [69]. U_3O_8 was used along with nickel oxide as the basis of a steam reforming catalyst, and it was modified with potassium species (potassium hydroxide, potassium oxide and/or potassium carbonate), all supported on either alumina or a mix of alumina and magnesium oxide. The uranium and nickel catalysts proved to be extremely efficient for steam reforming.

In later work Nicklin describes a reduction–oxidation cycle that was used to treat the catalysts prior to use [70]. Four to six cycles were performed, with the oxidation taking place at 600–650 °C and the reduction at temperatures no higher than 600 °C (preferably not above 550 °C). The gases used for these processes were hydrogen and oxygen for the reduction and oxidation steps respectively. The catalyst studied in this patent differed from the previous work, as additional UO_3 was present in the catalyst. The reduction–oxidation step was added to enhance catalyst activity compared to the earlier work. The authors observed that during the oxidation part of the process cycle the amount of oxygen used must be reduced with each successive oxidation to prevent the catalyst becoming highly pyrophoric.

A later patent presents more details of catalyst performance and specific details of the preparation and composition of the catalyst [71]. The supported catalyst used was composed of nickel (23.20%), uranium (11.45%) and potassium (0.23%). A naphtha stream was used to determine catalyst activity. A steam:naphtha ratio of 3.8:1 was employed at a pressure of ca. 11 bar and a gas hourly space velocity of 2058 h^{-1}. The temperature of the inlet gas was 465 °C, whilst the temperature of the outlet gas was 756 °C. The outlet of the reactor contained methane, carbon monoxide, carbon dioxide and hydrogen in the proportions $CH_4:CO:CO_2:H_2 = 6.7:12.5:14.6:66.0$.

Results of experiments varying the ratio of uranium to nickel showed that the ratios giving the largest surface area and catalyst volume were in the range 0.45–0.76 (U:Ni). These two characteristics were the most important for activity for these reactions. The catalysts were in a reduced state, which could explain the addition of the reduction–oxidation step in the previous patent. A further reason for using catalysts in the 0.45–0.74 U:Ni range was that the catalyst demonstrated greatest resistance to coke deposition at a ratio of 0.71:1.

Gavin [72] has studied similar catalysts to Nicklin and coworkers for steam reforming. It has been suggested that the nickel and uranium combine to form a nickel uranate phase (NiO · $3UO_3$). Once the catalysts were reduced, the active components were nickel from excess nickel oxide, nickel from the nickel uranate and tetra-uranium oxide (U_4O_9). Gavin described how the composition of the catalyst affected the amount of NiO · $3UO_3$. The relationship between the catalyst composition and activity for steam reforming is shown in Table 13.6.

Table 13.6 Activity of nickel–uranium oxide catalysts for steam reforming of naphtha [72].

Catalyst composition (%)				Relative amount of NiO · $3UO_3$	Conversion (%)
Ni	U	Ba	K		
11.0	7.0	0	0	9.1	100.0
11.7	7.6	1.5	0	4.6	100.0
11.7	7.7	1.6	0	5.2	99.0
11.3	7.4	2.11	0	5.0	98.5
10.9	6.7	2.12	0	3.6	95.6
10.6	7.2	0	0.59	3.6	93.0
12.0	7.9	1.72	0.50	2.0	90.0

The presence of the alkaline components is essential and they are converted into carbonates by heating in carbon dioxide during the catalyst preparation. It is argued that as the concentration of the alkaline component increased there was increased likelihood that the nickel uranate phase would react to produce barium uranate (BaU_2O_7) and NiO. The barium and potassium were also thought to reduce the tendency of coking on the catalyst surface, and their concentration is a balance between their efficacy for producing the most active phases and reducing coke formation.

Despite the earlier patent reports of the suitability of nickel-based catalysts incorporating uranium oxide for steam reforming, uranium did not become a component of commercial steam reforming catalysts. Nevertheless, interest has continued in assessing the efficacy of uranium oxide as a steam reforming catalyst component. Gordeva and coworkers prepared relatively porous oxides of UO_2 and U_3O_8 as supports for nickel and ruthenium as catalysts for methane steam reforming [73]. The catalysts were designed so that they could be used for conversion of nuclear energy to chemical energy by the production of hydrogen. High production rates of hydrogen were observed at 1 bar pressure and temperatures of 600–700 °C. Under operating conditions in a nuclear reactor, fissile products would be expected to contaminate the syngas. In order to try and limit contamination, studies also investigated containing the uranium oxide within a thin coating of MgO/Al_2O_3. The influence of the coating on catalyst activity is not clear and catalysts to be used for such advanced processes would clearly need further development. However, uranium oxide based catalysts are ideal for applications of this type, owing to their fissile properties and the knowledge base that is already in place because of the use of uranium oxides by the nuclear industry.

13.5
Conclusions

Uranium oxides have been used as catalysts and catalyst components for a relatively wide range of reactions. The oxides of uranium are numerous, with the main oxides being UO_2, U_3O_8 and UO_3. The structures of the oxides can be complex, as can the relationship between the phases. However, the oxides have many properties that make them versatile catalysts and catalyst components. Uranium oxides have been most widely applied as catalysts for oxidation reactions, and these include total oxidation and partial oxidation. Uranium oxide based catalysts have demonstrated excellent performance for selective oxidation and have been used commercially, although this is no longer the case. Uranium oxides have also been employed as catalysts for a range of reduction reactions and for steam reforming. The use of uranium oxides as catalysts may be a controversial issue; however, depleted uranium oxide is relatively widely available. The main concern in using uranium as a catalyst component is associated with its toxicity, which is comparable with that of lead, and it must be handled accordingly.

Acknowledgments

The author would like to thank Dr Maria-Luisa Palacios and Dr Richard Harris.

References

1. Allen, G.C. and Tempest, P.A. (1982) *Journal of the Chemical Society, Dalton Transactions*, 2169.
2. Allen, G.C. and Holmes, N.R. (1995) *Journal of Nuclear Materials*, **223**, 231.
3. Cordfunke, E.H.P. (1969) *The Chemistry of Uranium*, Elsevier, New York.
4. Lynds, L., Young, W.A., Mohl, J.S. and Liebowitz, G.G. (1962) *Nonstoichiometric compounds*, ACS Advances in Chemistry Series (ed. R.F. Gould), American Chemical Society, vol. 39, p. 58.
5. Vanlierde, W., Pelsmaekers, J. and Lecocq-Robert, A. (1970) *Journal of Nuclear Materials*, **37**, 276.
6. Belbeoch, B., Boivineau, J.C. and Perio, P. (1967) *Journal of Physics and Chemistry of Solids*, **28**, 1267.
7. Westrum, E.F. and Gronvold, F. (1962) *Journal of Physics and Chemistry of Solids*, **23**, 39.
8. Hoekstra, T.H.R., Siegel, S. and Gallagher, F.X. (1970) *Journal of Inorganic and Nuclear Chemistry*, **32**, 3237.
9. Hoekstra, H.R., Siegel, S. and Charpin, P. (1968) *Journal of Inorganic and Nuclear Chemistry*, **30**, 519.
10. Rundle, R.E., Baeziger, N.C., Wilson, A.S. and MacDonald, R.A. (1948) *Journal of the American Chemical Society*, **70**, 99.
11. Loopstra, B.O. (1964) *Acta Crystallographia*, **17**, 651.
12. Ball, R.G.J. and Dickens, P.G. (1991) *Journal of Materials Chemistry*, **1**, 105.
13. Loopstra, B.O. (1970) *Acta Crystallographia*, **B26**, 656.
14. Hoekstra, H.R. and Siegel, S. (1961) *Journal of Inorganic and Nuclear Chemistry*, **18**, 154.
15. Greaves, C. and Fender, B.E.F. (1972) *Acta Crystallographia*, **28**, 3609.
16. Debets, P.C. (1966) *Acta Crystallographia*, **21**, 589.
17. Loopstra, B.O., Taylor, J.C. and Waugh, A.B. (1977) *Journal of Solid State Chemistry*, **20**, 9.
18. Wait, E. (1955) *Journal of Inorganic and Nuclear Chemistry*, **1**, 309.
19. Kovba, L.M., Viadavskii, L.M. and Labut, E.L. (1963) *Zhurnal Strukturnoi Khimii*, **4**, 627.
20. Siegel, S., Hoekstra, H.R. and Sherry, E. (1966) *Acta Crystallographia*, **20**, 292.
21. Allen, G.C. and Tempest, P.A. (1983) *Journal of the Chemical Society, Dalton Transactions*, **267**, 7.
22. Colmenares, C.A. (1984) *Progress in Solid State Chemistry*, **12**, 257.
23. Willis, B.T.M. (1964) *Proceedings of the British Ceramic Society*, **1**, 9.
24. Willis, B.T.M. (1978) *Acta Crystallographia*, **A34**, 88.
25. Catlow, C.R.A. (1977) *Proceedings of the Royal Society of London*, **A353**, 533.
26. James, J.H. (1922) *Chemical and Metallurgical Engineering*, **26**, 209.
27. Hall, W.J. (1926) US Patent 1,595,299, August 10, 1926.
28. Craver, A.E. (1927) US Patent 1,636,954, July 26, 1927.
29. Parks, W.G. and Katz, J. (1933) *Industrial and Engineering Chemistry*, **28**, 3193.
30. Craver, A.E. (1927) US Patent 1,636,952, July 26, 1927.
31. Wietzel, R. and Pfaundler, C. (1932) US Patent 1,844,998, February 16, 1932.
32. Nozaki, F. and Ohki, K. (1972) *Bulletin of the Chemical Society of Japan*, **45**, 3473.
33. Hutchings, G.J., Heneghan, C.S., Hudson, I.D. and Taylor, S.H. (1996) *Nature*, **384**, 341.
34. Hutchings, G.J., Heneghan, C.S., Hudson, I.D. and Taylor, S.H. (1996) *Heterogeneous Hydrocarbon Oxidation* (eds B.K. Warren and S.T. Oyama), ACS Symposium Series, ACS, Washington, DC, vol. 638, p. 58.

35 Taylor, S.H., Heneghan, C.S., Hutchings, G.J. and Hudson, I.D. (2000) *Catalysis Today*, **59**, 249.
36 Taylor, S.H. and O'Leary, S.R. (2000) *Applied Catalysis B*, **25**, 137.
37 Harris, R., Hutchings, G.J., Boyd, V.J. and Taylor, S.H. (2002) *Catalysis Letters*, **78**, 369.
38 Heneghan, C.S., Hutchings, G.J., O'Leary, S.R., Taylor, S.H., Boyd, V.J. and Hudson, I.D. (1999) *Catalysis Today*, **54**, 3.
39 Hofer, L.J.E. and Anderson, R.B. (1964) US Patent 3,140,148, July 7, 1964.
40 Madhavaram, H. and Idriss, H. (1997) *Studies in Surface Science and Catalysis*, **110**, 265.
41 Madhavaram, H. and Idriss, H. (1999) *Journal of Catalysis*, **184**, 553.
42 Grasselli, R.K., Suresh, D.D. and Knox, K. (1970) *Journal of Catalysis*, **18**, 356.
43 Grasselli, R.K. and Suresh, D.D. (1972) *Journal of Catalysis*, **25**, 273.
44 Grasselli, R.K. and Callahan, J.L. (1969) *Journal of Catalysis*, **14**, 93.
45 Delobel, R., Baussart, H., Le Bras, M., Le Maguer, D. and Leroy, J.M. (1982) *Journal of the Chemical Society, Faraday Transactions*, **78**, 485.
46 Sykora, R.E., King, J.E., Illies, A.J. and Albrecht-Schmitt, T.E. (2004) *Journal of Solid State Chemistry*, **177**, 17170.
47 Van der Baan, H.S., Steenhof, J.G., De Jong, J.G. and Guffens, C.H.E. (1972) *Journal of Catalysis*, **26**, 401.
48 Steenhof, J.G., Guffens, C.H.E. and Van der Baan, H.S. (1973) *Journal of Catalysis*, **32**, 149.
49 Collette, H., Derouane, E.G., Verbist, J.J., Deremince-Mathieu, V. and Nagy, J.B. (1987) *Journal of the Chemical Society, Faraday Transactions*, **83**, 1263.
50 Ai, M. (1991) *Heterogeneous Catalysis and Fine Chemicals II*, Elsevier Science Publishers B.V, Amsterdam, pp. 423–30.
51 Zhu, J. and Anderson, S.L.T. (1989) *Journal of the Chemical Society, Faraday Transactions*, **85**, 3629.
52 Dowden, D.A. and Walker, G.T. (1971) UK Patent 1,244,001, August, 1971.
53 Dowden, D.A., Schnell, C.R. and Walker, G.T. (1968) *Proceedings of the 4th International Congress on Catalysis, Moscow*, Paper 62, pp. 201–15.
54 Walker, G.S., Lapszewicz, J.A. and Foulds, G.A. (1994) *Catalysis Today*, **21**, 519.
55 Corma, A., Corberán, V.C. and Kremeric, G. (1984) *Industrial & Engineering Chemistry Product Research and Development*, **23**, 546.
56 Taylor, S.H., Hutchings, G.J., Palacios, M. and Lee, D.F. (2003) *Catalysis Today*, **81**, 171.
57 Teng, Y. and Kobayashi, T. (1998) *Catalysis Letters*, **55**, 33.
58 Lee, A.F., Sears, P.J., Pollington, S.D., Overton, T.L., Wells, P.B. and Lee, D.F. (2000) *Catalysis Letters*, **70**, 183.
59 Choudhary, V.R., Jha, R. and Jana, P. (2007) *Green Chemistry*, **9**, 267.
60 Madhavaram, H. and Idriss, H. (2004) *Journal of Catalysis*, **224**, 3589.
61 Madhavaram, H., Buchanan, P. and Idriss, H. (1997) *Journal of Vacuum Science and Technology A*, **15**, 1685.
62 Heynen, H.W.G. and Van der Baan, H.S. (1974) *Journal of Catalysis*, **34**, 167.
63 Heynen, H.W.G., Van der Baan, H.S. and Camp-van Berkel, C.G.M.M. (1977) *Journal of Catalysis*, **48**, 386.
64 Berry, F.J., Murray, A. and Parkyns, N.D. (1993) *Applied Catalysis A*, **100**, 131.
65 Berry, F.J., Murray, A. and Steel, A.T. (1998) *Journal of the Chemical Society, Faraday Transactions*, **84**, 2783.
66 Pollington, S.D., Lee, A.D., Overton, T.L., Sears, P.J., Wells, P.B., Hawley, S.E., Hudson, I.D., Lee, D.F. and Ruddock, V.J. (1999) *Journal of the Chemical Society, Chemical Communications*, 725.
67 Campbell, T., Newton, M.A., Boyd, V., Lee, D.F. and Evans, J. (2005) *Journal of Physical Chemistry B*, **109**, 2085.
68 Campbell, T., Newton, M.A., Boyd, V., Lee, D.F. and Evans, J. (2006) *Journal of Molecular Catalysis A*, **245**, 62.
69 Nicklin, T. and Burgess, K.H. (1970) GB Patent 1,198,991, July 15, 1970.
70 Nicklin, T., Clack, J. and Burges, K.H. (1971) US Patent 3,630,967, January 1, 1971.
71 Nicklin, T. and Farrington, F. (1974) US Patent 3,847,836, November 12, 1974.
72 Gavin, D.G. (1974) US Patent 3,974,098, 1974.
73 Gordeva, L.G., Moroz, Y.I. Aristov, E.M., Rudina, N.A., Zaikovskii, V.I., Tanashev, Y.Y. and Parmon, V.M. (1995) *Journal of Nuclear Materials*, **218**, 202.

14
Heteropolyoxometallate Catalysts for Partial Oxidation
Jacques C. Védrine and Jean-Marc M. Millet

14.1
Introduction

Polyoxometallates (POMs) are a large class of nanosized transition metal–oxygen clusters [1]. Their structures, sizes and properties correspond to compounds intermediate between molecules and oxides. They are composed of anions having metal–oxygen octahedra as the basic structural unit. The octahedra are linked together to yield a stable and compact skeleton of polymeric oxoanions formed by the condensation of mononuclear oxoanions (isopolyanions) or of more than two different mononuclear oxoanions (heteropolyanions) in acidic media as shown in the equation:

$$p[XO_r]^{x-2r} + q[MO_n]^{m-2n} + zH^+ \rightarrow X_pM_qO_s^{(p \cdot x + m \cdot q - z \cdot s)} + (z/2)H_2O$$

with M = metal (designated as addenda atom), X = heteroatom, x = the valency of the heteroatom, m = the valency of the addenda atom and $s + (z/2) = n \cdot q + r \cdot p$ (oxygen balance). The elements M that can act as addenda atoms in heteropoly or isopolyanions are limited to those with both a favorable combination of ionic radius and charge and the ability to form $d\pi$–$p\pi$ M—O bonds. However, there are no such restrictions for the heteroatom X. The most common addenda atoms are molybdenum or tungsten and, less frequently, tantalum, vanadium and niobium, or mixtures of these elements in their highest oxidation state (d^0) [2]. In principle all the elements of the periodic table could act as heteroatoms, though the most usual are P^V, As^V, Si^{IV}, Ge^{IV}, Ce^{IV}, Th^{IV}, B^{3+}, Co^{3+}, Al^{3+}, Cr^{3+}, and so on. The counter ions, required to charge balance the anions, may be protons, alkali metals or other metal cations.

More than twenty types of structure, incorporating from four to forty metal atoms and from one to nine heteroatoms are known. There are five general types of POM, namely (i) the "Lindqvist" structure (O_h symmetry of six MoO_6 edge-shared octahedra) $[M_6O_{19}]^{n-}$; (ii) the "Keggin" structure, α-$[(XO_4)M_{12}O_{36}]^{n-}$ (T_d symmetry of four M_3O_{13} groups of three MO_6 octahedra), (iii) the "Dawson" structure

---

Metal Oxide Catalysis. Edited by S. David Jackson and Justin S. J. Hargreaves
Copyright © 2009 WILEY-VCH Verlag GmbH & Co. KGaA, Weinheim
ISBN: 978-3-527-31815-5

14 Heteropolyoxometallate Catalysts for Partial Oxidation

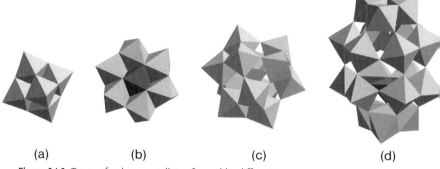

Figure 14.1 Types of polyoxometallates formed by different arrangements of MO_6 octahedra. (a) Lindqvist, (b) Anderson, (c) Keggin, (d) Dawson.

α-$[(XO_4)_2M_{12}O_{54}]^{n-}$ (D_{3h} symmetry of two "Keggin" fragments α-XM_9O_{34}), (iv) the "Anderson" structure $[H_x(XO_6)M_6O_{18}]^{n-}$ (D_{3h} symmetry planar arrangement of seven edge-shared MO_6 octahedra), (v) the $[(XO_{12})M_{12}O_{30}]$ structure (I_h symmetry with an XO_{12} icosahedron surrounded by six equivalent M_2O_9 groups of face-shared MO_6 octahedra, linked together by corner-sharing tetravalent cations). A schematic drawing of the first four types is presented in Figure 14.1.

The most common compounds, particularly for catalytic application, belong to the 12 series (M/X = 12) and are Keggin-type heteropolyacids (HPA), as they are the most stable, are more easily available and have been studied in more detail. The well defined Keggin structure comprises a central XO_4 tetrahedron surrounded by twelve edge- and corner-sharing metal–oxygen octahedra (MO_6) units as shown in Figure 14.2. They contain heterododecametallate anions with the formula

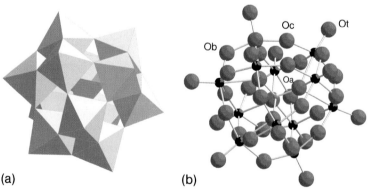

Figure 14.2 Schematic representations of the Keggin anion showing the four distinct oxygen sites. (a) Polyhedral, (b) ball and stick.

$[XM_{12}O_4]^{x-8}$ where x is the oxidation degree of the central heteroatom X. The addenda atoms, M (Mo^{VI}, W^{VI}), can be partly replaced by other metals, in particular by Ta^V, V^V and Nb^V, resulting in a change of the anionic charge (one per substituted metal atom) and thus also in the number of counter-cations. These compounds are nearly always negatively charged and their negative charge density is variable depending on their elemental composition and molecular structure.

Solid POMs have a hierarchical structure and can be divided into three sub-structures [3], namely: primary, secondary and tertiary, as shown in Figure 14.3. These structures are important for understanding their heterogeneous catalytic properties. The primary structure is the structure of the heteropolyanion itself, that is the metal oxide cluster. The secondary structure is the three-dimensional arrangement consisting of polyanions, counter-cations and additional molecules (in particular organic molecules or water clusters such as $H_5O_2^+$, hydrated species, or dimethyl sulfoxide (DMSO) in $H_4SiW_{12}O_{40} \cdot 9DMSO$ and pyridine in $PW_{12}O_{40} \cdot [(C_5H_5N)_2H]_3$). The tertiary structure is the arrangement by which the secondary structure assembles into solid particles, and this relates to properties

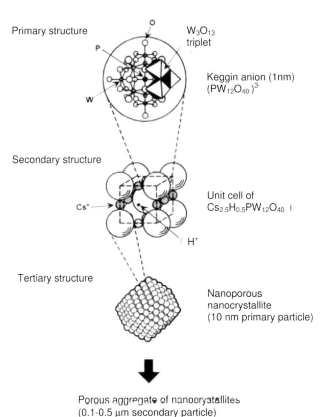

Figure 14.3 Primary, secondary and tertiary structures. (Taken from figure 14.2 in Ref. [3]).

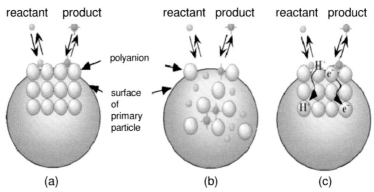

Figure 14.4 Types of catalysis in solid heteropoly compounds. (a) Surface, (b), pseudo-liquid bulk type (I), (c) bulk-type (II). (Taken from Ref. [3]).

such as particle size, surface area and porosity. Based on this hierarchical structure, it was demonstrated that there are three different modes of catalysis: surface-type catalysis, pseudo-liquid bulk-type catalysis and bulk-type catalysis, as represented in Figure 14.4. Surface-type catalysis corresponds to ordinary heterogeneous catalysis, which takes place on the outer surface of the solid. The other two types of catalysis are bulk-type and occur when the diffusion of reactant molecules in the solid (diffusion into the lattice rather than into the pores) is faster than the reaction. The solid bulk forms a pseudo-liquid phase in which catalytic reaction can proceed. Thus the catalyst appears solid but behaves like a liquid (solvent). As the active sites in the bulk, for example protons or transition metals, take part in catalysis, very high catalytic activity is observed. Owing to the flexible nature of the solid structure of some POMs (variable secondary structure), reactants having polarity or basicity are readily absorbed into the solid lattice, between the polyanions in the lattice, sometimes causing expansion of the lattice, and reaction occurs therein. In other words, the reaction becomes three-dimensional as in a solution. Owing to this behavior, POM catalysts often exhibit high catalytic activities and unique selectivity.

Heteropolyacids generally have a small surface area ($10\,m^2\,g^{-1}$) and no porosity. However, microporosity has been observed in the presence of alkali metal countercations, especially in cases involving large cations such as K^+ and Cs^+ as well as with NH_4^+, and so on. A full chapter detailing the microporosity of POMs can be found in Ref. [4]. For instance, the pore size of acidic Cs^+ salts ($Cs_xH_{3-x}PW_{12}O_{40}$) has been shown to be controlled by the Cs^+ content, for example $Cs_{2.2}H_{0.8}PW_{12}O_{40}$ has micropores in the range 0.62–0.72 nm diameter compared to less than 0.59 nm for $Cs_{2.1}H_{0.9}PW_{12}O_{40}$ and >0.85 nm for $Cs_{2.5}H_{0.5}PW_{12}O_{40}$. Correspondingly, there is a great difference in surface area values, for example ~$10\,m^2\,g^{-1}$ for $H_3PW_{12}O_{40}$ and $130\,m^2\,g^{-1}$ for its $Cs_{2.5}$ salt. The microporosity of NH_4^+ and Cs^+ salts of H_3PW_{12}, H_3PMo_{12} and H_4SiW_{12} Keggin-type heteropolyacids (HPA) has been investigated

[5] by ^{129}Xe NMR. It has been observed that ammonium salts had a pore size of 0.9 nm regardless of the anion, while for Cs$^+$ salts pore size varies between 0.7 and 1.0 nm depending upon the nature of the anion. However, this porosity has frequently been observed to be lost upon catalytic application.

Control of the shape and size of pores (micro- and meso-pores) in polyoxometallates is an important objective for the future. Pioneering studies have shown that such control may be applied to the primary structure with generation of pores between the anionic species [6] or to the tertiary structure with generation of pores between self-assembled $(NH_4)_3PW_{12}O_{40}$ micro-crystallites [7]. However, attempts to control the pore size in primary structures with the use of organic templates have failed so far, mainly because of the difficulties encountered in the template removal step, during which the structure collapses. Success has been reported in the control of pore size in tertiary structures for applications at low temperatures; however, for applications at high temperature, procedures need to be improved, as the aggregates are thermally unstable. Alkali and alkaline earth substituted Keggin-type phosphomolybdic or phosphotungstic acids have been shown to form nano-crystallites composed of alkaline salt particles (10–20 nm) thickly covered by the acid after precipitation [8]. Upon heat treatment, these particles are converted to particles having a size similar to those before heat treatment with a more or less uniform composition. These particles are connected epitaxially to each other to form aggregates that can develop an important internal porosity when the ratio of alkaline earth cations to protons is greater than or equal to 2.5 [8]. The complete homogeneity of the composition of this type of polyoxometallate after heat treatment is rather difficult to prove, since the acid phase coating the salt particles is not detectable by X-ray diffraction and can only be observed by X-ray photoelectron spectroscopy or Raman spectroscopy [9].

14.2
History of Polyoxometallates

POMs have been known for about two centuries. The discovery of the first ammonium salt of the dodecamolybdophosphate $(NH_4)_3PMo_{12}O_{40}$ (12:1 composition) was reported by Berzelius [10] in 1826 and its tungsto counterpart by Galissard de Marignac [11] in 1864. However, the field remained undeveloped. The first attempts to understand the composition of heteropolyanions were based on Werner's coordination theory [12]. Structure determination for heteropoly compounds was not possible until the advent of X-ray diffraction techniques. A hypothesis proposed in 1908 by Miolati and Pizzighelli [13], based upon sharing metal–oxygen polyhedra, was adopted and developed by Rosenheim [14] and Pauling [15], during the following 25–30 years. Rosenheim was probably the most productive and influential researcher in the field of polyanion chemistry. According to the Miolati–Rosenheim theory, heteropoly acids are based on six coordinate heteroatoms bound with MO_4. It was Keggin [16] who solved the structure of the most important 12:1 heteropoly compound $H_3PW_{12}O_{40} \cdot 5H_2O$ by X-ray diffraction. Evans [17] determined the structure of the Anderson compound (6:1 series for $Te^{6+}[Mo_6O_{24}]^{6-}$) and Dawson [18] determined that of the 18:2 heteropoly compound $[P_2W_{18}O_{62}]^{6-}$.

Souchay and his group brought clarity in the field and have provided a major contribution to polyoxometallate chemistry [1a, 19]. Since the early 1990s, polyoxometallate chemistry has expanded tremendously, following fundamental discoveries by Müller and his group at Bielefeld University [20]. These scientists have shown that aqueous solutions of molybdates result in a huge variety of linkable units and a large variety of clusters, among them amazing wheels and spheres such as the giant anions $[La_{16}As_{12}W_{148}O_{524}]^{76-}$ (mass ~4000, diameter 4 nm) or $\{Mo_{132}\}$, an inorganic superfullerene built up from 12 pentagonal (Mo)Mo$_5$ units (one pentagonal MoO$_7$ unit sharing edges with five MoO$_6$ octahedra) and 30 different Mo_2^V linkers or $[Mo_{72}^{VI}Mo_{60}^{V}O_{372}(CH_3COO)_{30}(H_2O)_{72}]^{42-}$ or $[Mo_{126}^{VI}Mo_{28}^{V}O_{462}H_{14}(H_2O)_{70}]^{14-}$, and so on. These wheels can be considered as nanostructures mimicking oxide surfaces and allowing reactions to occur on well defined sites. The history of polyoxometallate chemistry and its applications for nano-chemistry has been published [21]. Today, POMs still represent an exciting area of research and technology, and new compounds are discovered every month. For instance, the development of functionalized polyoxometallate [22] and polyoxometallate-based polymers [23] with creation of giant two-dimensional networks has appeared recently and constitute new perspectives in the field of organic–inorganic frameworks.

14.3
Properties and Applications of Polyoxometallates

There are more commercial applications of polyoxometallates than of any other class of cluster compound and there is also much potential for additional applications in areas ranging from catalysis and medicine to proton conductors, materials and micro-device technology [24]. These applications stem from the wide range of molecular architectures, surface charge density and chemical and electronic properties exhibited by POMs, as described above. Varying in size from one to a few nanometers, they are used as electrode functionalization agents [25], anti-retroviral agents [26] and for detection, separation and quantification. The latter uses are based on properties such as their high molecular weight (>2000), their solubility in water or organic solvents (depending on the size of the counter-ion size – solubility decreases with size), the fact that a wide variety of elements can be incorporated into the polyanion framework changing the properties, their electrochemical activity and their reducibility to form colored species depending on the substitution element and extent (e.g. from yellow to orange when Mo is substituted by V, while reduction leads to the famous "heteropoly blues"). A vast range of applications has been described in fields as different as synthesis of organic and organometallic derivatives of POMs, absorbents (ion-exchange materials and alcohol or carboxylic acid radical detectors), electron-transfer reactions, photo- and electro-chromism, flame retardants and smoke suppressants, together with applications in metallurgy (corrosion inhibitors and coatings on steel and aluminum), medicine and catalysis. Nevertheless, catalysis is by far the most important field of application (representing > 80% in terms of patent applications). Most of the pioneering work was performed in Japan (Izumi, Misono, Mizuno, Ono, Otake, Yoneda and coworkers), in Russia (Matveev, Kozhevnikov, Kholdeeva and coworkers) and in France (Souchay, Hervé and coworkers).

POMs can display very strong Brønsted acidity and are efficient oxidants, exhibiting fast and reversible multi-electron redox transformations under mild conditions. This acidity is very strong, as the negative charge is delocalized over much larger anions than for mineral acids or solids such as zeolites. Consequently, the electrostatic interaction between protons and the anion is much less for POMs. The general trend in acidity among the most common HPAs is a function of the hetero atom [27] with Co < B < Si, Ge < P, falling into the following order [28] determined in an ammonia desorption study:

$$H_3PW_{12}O_{40} > H_4SiW_{12}O_{40} \sim H_4GeW_{12}O_{40} > H_3PMo_{12}O_{40} > H_4SiMo_{12}O_{40} \sim H_4GeMo_{12}O_{40}$$

with NH_3 desorption maxima being at 865, 805, 736 and 696 K, respectively. Recent studies report a decrease in the quantity of acid sites, but not in their strength, when dispersing POMs on silica-type supports [29–31] such as HMS, MCM41 and Kenyaite.

The acid–base and redox properties can be controlled across a broad spectrum by varying the chemical composition, the extent of hydration, the type of support for supported samples, the thermal treatment, and so on. For instance, the substitution of M^{VI} by M^{n+} of close valence ($n = 5, 4, 3$ or 2) has been studied and leads to additional negative charges on the HPA and thus to additional protons (from one to four respectively) for charge balance, thus generating a higher acid density. An example is $H_{3+x}PV_x^V M_{12-x}^{VI}O_{40}$, which can be synthesized with $x =$ between 1 and 3. Other metal elements can also be substituted. The problem is that during synthesis one could have either framework substitution of the M element or cationic exchange between protons and the added metal cation, as is well known for zeolite materials. It follows that the number of protons could be less than $3 + x$ in the above formula, so that the balance between redox and acid properties of such materials is difficult to control during synthesis, since they depend on temperature, pH conditions, and so on.

POMs are promising catalysts for acid, redox and bifunctional catalysis. In many structures, the transition metal addenda atoms such as Mo or W exist in two oxidation states, which results in different redox properties as determined by polarography. The exceptional ability of heteropolyanions to act as electron reservoirs has been demonstrated by the preparation and characterization of numerous reduced derivatives [32]. They also exhibit high solubility in polar solvents, which means that they can be used in homogeneous catalysis. The wide range of applications of heteropoly compounds are based on their unique properties which include size, mass, electron and proton transfer (and hence storage) abilities, thermal stability, lability of "lattice oxygen" and the high Brønsted acid strength of the corresponding acids. An added advantage is that the HPAs can be separated and can be extracted into organic solvents.

Transition metal substituted polyoxometallates (TMSP) are synthesized using transition metals as addenda ions. Most commonly, first row transition metals are incorporated into the framework. The transition metal element fits into the lacunary vacancy created in the Keggin structural unit by the loss of an MO_6 octahedron, which is equivalent to the stoichiometric loss of an MO^{n+} unit, resulting in the formation of the $[XM_{11}O_{39}]^{n-}$ anion. This involves the partial degradation of the Keggin unit by lowering the pH of the solution upon the addition of a suitable buffer. The solubilities of the TMSP complexes can be controlled by selecting an appropriate counter-cation. Usually alkali metals and tetra-alkyl ammonium groups are the counter-cations of choice. There are two methods for the synthesis of a TMSP. The first method involves the addition of a transition metal salt to the already prepared isolated lacunary salt under appropriate synthetic conditions, while the second method corresponds to the generation of *in situ* lacunary vacancies followed by the addition of the transition metal salt.

Applications of POMs to catalysis have been periodically reviewed [33–40]. Several industrial processes were developed and commercialized, mainly in Japan. Examples include liquid-phase hydration of propene to isopropanol in 1972, vapor-phase oxidation of methacrolein to methacrylic acid in 1982, liquid-phase hydration of isobutene for its separation from butane–butene fractions in 1984, biphasic polymerization of THF to polymeric diol in 1985 and hydration of *n*-butene to 2-butanol in 1989. In 1997 direct oxidation of ethylene to acetic acid was industrialized by Showa Denko and in 2001 production of ethyl acetate by BP Amoco.

Since the application of POMs is so widespread across areas such as homogeneous and heterogeneous catalysis, as well as acid and redox catalysis, it is not possible to exhaustively review all the applications. Thus the rest of this chapter focuses only on the catalytic properties of the polyoxometallates in heterogeneous gas- or liquid-phase oxidation reactions, and reviews the most recent progress in the knowledge of their properties and working process, underlining both their potential and their limitations.

14.4
Catalytic Applications in Partial Oxidation Reactions

It is known [41] that partial oxidation reactions in heterogeneous catalysis involves redox properties of the solid catalysts, allowing the well known Mars–van Krevelen mechanism [42] to occur, or at least to be facilitated. Acid–base properties are also an important feature, as they play a determining role in the activation of the reactants and in the desorption of the intermediate compounds. For instance, an acid surface will favor desorption of acid products, thus avoiding further over-oxidation, while a basic surface will favor desorption of basic products as olefins. It follows that heteropolyoxometallate compounds, in particular TMSP, appear as potential catalysts for partial oxidation reactions [43]. Oxidation catalysis by polyoxometallates, especially of the Keggin type, is a rapidly expanding area owing to their unusual versatility and compatibility with environmentally friendly conditions (employing oxidants such as O_2 and H_2O_2) and reactions. Research in the area of oxidation using these compounds has been intense since the 1990s, as existing catalytic processes leave ample margin for improvement with scope to develop new catalysts working under more stringent reaction conditions. A general overview of applications of POMs in oxidation reactions has been presented by Centi and coworkers [44]. Extensive applications have been found in areas ranging from fine chemical synthesis and alkane up-grading, to the degradation of toxic materials.

As described above, substitution of the addenda atoms by either transition metals (TMSP) or other addenda atoms modifies the redox features of the heteropolyoxometallates. The synthesis of TMSP associated with metalloporphyrins was studied thoroughly by Hill and his group [45]. These systems behave as oxidation catalysts by transferring oxygen from a typical donor to the organic substrate and exhibit the attractive features of metalloporphyrins, namely ability for dioxygen binding, formation of high valence species with stoichiometric oxygen transfer, reaction with oxidants such as iodosobenzene, sodium periodate and *tert*-butyl hydroperoxide. The strong binding of the metalloporphyrin with the d electrons of the transition metal ions permits it to be retained within the structure during the catalytic cycle, although changes in oxidation state of the metal occur, which influence its mobility. If during the oxidation process structural degradation of the catalyst takes place, the transition metal species is lost and precipitates as metal oxide. On their own, metalloporphyrins are organic molecules that are thermodynamically unstable in the presence of strong oxidizing agents and hence can undergo oxidative degradation during reaction. However, the TMSP complexes are more thermally robust than the uncomplexed metalloporphyrin and are not susceptible to oxidative degradation. Consequently, they are able to catalyze oxidation reactions for longer periods than metalloporphyrins alone.

Mixed addenda complexes are those in which one or more of the addenda atoms in the framework are substituted by other addenda-type atoms. The attractiveness of this class of compounds as catalysts for oxidation is their high oxidation potential, low cost, thermal stability and oxidative robustness, ease of preparation and solubility in media ranging from water to hydrocarbons. The redox potential of a POM depends on the negative charge density and on the elemental composition. Both factors can be controlled to a great extent during the synthesis process. The relationship between redox potential and elemental composition is dictated by the presence or absence of highly oxidizing addenda metal atoms. The decreasing order of redox potentials is $V^V > Mo^{VI} > W^{VI}$. Hence, the molybdovanadophosphate system has been the most extensively studied for oxidation reactions. All the types of Keggin compounds mentioned above have been found to be compatible in operation with environmentally friendly oxidants, such as oxygen and hydrogen peroxide, and also with oxidants like *tert*-butyl hydroperoxide, iodosobenzene, sodium periodate and potassium persulfate. Many examples are given below so that the readers can appreciate the breadth of the field of application of POMs for catalytic oxidation reactions using either oxygen (air) or hydrogen peroxide (H_2O_2) as oxidizing agents.

14.4.1
Oxidation with Molecular Oxygen

For TMSP materials, the efficacy of oxidation using molecular oxygen is influenced by their oxidation potential. They act as catalysts by oxygen transfer from a typical donor to a typical TMSP followed by transfer of this activated form of oxygen to an organic substrate. TMSPs are potential catalysts for the epoxidation of olefins in the presence of aldehydes and molecular oxygen or air. Mizuno and coworkers [46] have demonstrated the significant catalytic activity of $[PW_{11}CoO_{39}]^{5-}$ for the epoxidation of alkenes such as cyclohexene, 1-decene and styrene by molecular oxygen at 303 K in the presence of aldehydes such as isobutyraldehyde and pivaldehyde. They proposed that this reaction involves peracids as intermediates, and the formation of per-isobutyric acid has been confirmed by ^1H NMR [47].

Alkene epoxidation by dioxygen in the presence of isobutyraldehyde and of the tetrabutylammonium salts of transition metal substituted heteropolyanions $[PW_{11}MO_{39}]^{n-}$ (M = Co^{2+}, Mn^{2+}, Cu^{2+}, Pd^{2+}, Ti^{IV}, Ru^{3+}, V^V) has been studied by Kholdeeva and coworkers [48]. In this work, *trans*-stilbene was used as the model substrate in an acetonitrile medium. Selectivity of epoxidation reached 95% at complete alkene conversion. The reaction was inhibited by 2,6-di-*tert*-butyl-4-methylphenol indicating a chain radical mechanism, and the acyl peroxy radical was the active species for epoxidation. Oxidation of olefins and ketones by molecular oxygen/aldehyde on a V heteropolyoxometallate system has been studied by Hamamoto and coworkers [49]. Olefins were epoxidized with dioxygen in the presence of two equivalents of 2-methyl propanal under the influence of $(NH_4)_6[PMo_6V_6O_{40}]$ to give the corresponding epoxides in moderate to good yields. This system was also extended to allylic and homo allylic alcohols. Baeyer–Villiger oxidation of cyclic ketones was achieved using benzaldehyde instead of 2-methylpropanal.

Kuznetsova and coworkers have shown that $[PW_{11}Fe(H_2O)O_{39}]$ in the pH range 3.5–5 at 293 K is an active catalyst for the oxidation of H_2S with O_2 to produce elemental sulfur [50]. Harrup and coworkers have found that polyoxometallate catalysts such as $K_5[ZnPW_{11}O_{39}]$, α-$K_8[SiW_{11}O_{39}]$, α-$K_6[ZnSiW_{11}O_{39}]$ and $K_4[NaP_5W_{30}O_{110}]$ are active for the oxidation of H_2S to elemental sulfur at 60 °C under 1.1 atmospheres pressure of O_2 [51]. Khenkin and Hill have observed that Cr^{3+} heteropolytungstate and its corresponding oxo-form Cr^V are efficient catalysts for the oxidation of alkenes, alkanes, alcohols and triphenylphosphines by a variety of oxidants such as ClO, H_2O_2 or PhIO [52]. Kuznetsova and coworkers have studied complexes of Pd^{2+} and Pt^{2+} with $[PW_{11}O_{39}]^{7-}$ and have found that they are active for the oxidation of benzene to phenol in a mixture of O_2 and H_2 gases in a two-phase water–benzene system at a temperature of 283–313 K [50]. Iron heteropolyacid has also been found by Seo and coworkers to be active for phenol synthesis by liquid phase oxidation of benzene with molecular oxygen [53].

The vanadium-substituted heteropolyanions have a fairly high oxidation potential (0.7 V relative to the normal hydrogen electrode) and are capable of oxidizing substrates ranging from organic to inorganic compounds. They act as reversible oxidants, that is, their reduced forms can be reoxidized to the original form by oxygen under mild conditions. The $V^{IV} \leftrightarrow V^{V}$ transformation is actually responsible for the redox activity. Neumann and coworkers have successfully carried out the oxidative dehydrogenation of α-terpinene to p-cymene by mixed addenda compounds of the type $H_5[PMo_{10}V_2O_{40}]$ [54]. The reaction mechanism involves the formation of a stable substrate complex in the catalyst reduction (substrate oxidation state) stage and the formation of a μ-peroxo intermediate in the catalyst reoxidation stage. Oxidation of trialkyl-substituted phenols such as 2,3,6-trimethyl phenol in the presence of phosphomolybdovanadium heteropolyacids has been reported by Kholdeeva and coworkers [55]. The product obtained was the 2,3,5-trimethyl-1,4-benzoquinone, an intermediate in Vitamin E synthesis, with 86% yield at 100% conversion, with 2,2'-3,3'-6,6'-hexamethyl-4,4'-biphenol being isolated as an intermediate. The divanadium-substituted phosphomolybdates have been found to catalyze the oxidation of dialkylphenols to diphenoquinones. The rate is highly dependent on the oxidation potential of the substrate, and the reaction proceeds by electron transfer from the substrate to the heteropolyanion catalyst. The divanadium-substituted heteropolyanion has been found by Lissel and coworkers to catalyze aerobic oxidation of dialkyl phenols to diphenoquinones and the oxidation of 2,3,5-trimethylphenol to 2,3,5-trimethyl-1,4-benzoquinone [56]. The reaction rate has been found to be dependent on the oxidation potential of the substrate and to proceed by electron transfer from the substrate to the heteropolyanion catalyst. These catalysts are equally efficient for oxybromination in organic media. For instance, oxybromination of phenol, anisole, o-cresol, p-cresol, 1-naphthol, N,N-diethylaniline, toluene, cumene, acetone, cyclohexanone and 1-octene to the corresponding bromides has been achieved under ambient conditions by Neumann and coworkers [57]. The oxidation of 2-methylcyclohexanone and cyclohexanone by O_2 to 6-oxo-heptanoic acid and adipic acid respectively has been observed on molybdovanadophosphoric acids [58].

The oxidation of benzylic derivatives with oxygen has been studied using $(NH_4)_6[PMo_6V_6O_{40}]$ as catalyst [59], as well as the oxidative dehydrogenation of benzylic amines to the corresponding Schiff base amines with oxygen in toluene solution at 373 K and the oxidation of isochroman and indan to 3,4-dihydroisocoumarin and 1-indanone with high selectivity. Similarly, oxidative cleavage of ketones, such as substituted cycloalkanones, 1-phenylalkanones and open-chain ketones to the corresponding acids was observed [60]. For instance, substrates such as 2,4-dimethyl cyclopentanone were oxidized to 5-oxo-3-methyl hexanoic acid and 1-phenylpropan-1-one, and open-chain ketones such as pentan-3-one was oxidized to the corresponding carboxylic acid.

The oxyfunctionalization of low molecular weight alkanes has attracted much attention because of their low cost and chemical stability as feedstock. Their oxidation over POM catalysts has been widely studied by controlling redox properties upon substituting M addenda by transition metal elements [61–63]. For example,

TMSP catalysts have been studied for the oxidation of propane with M = Co^{2+}, Fe^{3+}, Ga^{3+}, Ni^{2+}, Sb^{3+} and Zn^{2+} incorporated into $Cs_{2.5}H_{1.5}(M)PV_1Mo_{11}O_{40}$ in an M : V = 1 : 1 atomic ratio [36a]. Propene was the main product with about 80% selectivity at 5% propane conversion with Ni > Co > Fe > Zn at T_{react} = 595, 618, 646 and 673 K respectively, carbon oxides being the other main products. About 60% selectivity has been obtained for Ga- and Sb-substituted POMs at 613 and 628 K respectively and 26% at 578 K for the starting $Cs_{2.5}H_{1.5}$ material, which also led to CO_x (46%), acetic acid (24%) and acrylic acid (4%). It has been shown that the acidity of all samples was different (about 1.6 H^+/KU (Keggin unit) for the starting $Cs_{2.5}H_{1.5}$ sample and 2.4, 2.3, 1.9, 1.4, 2.1 and 3.0 H^+/KU for M = Ni, Co, Fe, Zn, Ga and Sb respectively). It was then clear that the balance between the redox and acid properties of the POMs are important features in determining their catalytic properties. For Ga samples, by changing the relative amount of Ga [43d], oxygenates, propene and CO_x have been found to be formed with a maximum for 0.16 Ga/KU (about 45% acrolein, acetic and acrylic acids against about 33% CO_x and 22% propene at 573 K), while the number of protons has been found to be similar (3.0–3.3 per KU). Substituting W for Mo in a $Cs_{2.5}H_{1.5}PV_1Mo_{11}O_{40}$ led to stronger acidity of the sample and more CO_x and acetic acid in propane oxidation reaction at the expense of propene [64].

Substituting Mo by V (x = 1, 2 and 3 per KU) in $PMo_{12-x}V_xO_{40}$, Centi and coworkers have shown that pentane is oxidized mainly to maleic anhydride while the VPO catalyst gives both maleic and phthalic anhydrides [65]. Such V-substituted $PMo_{12-x}V_xO_{40}$ POMs are the best known catalysts for the oxidation of isobutane to methacrolein and methacrylic acid [66]. Insertion of Ni into the $Cs_{2.5}$ of $Cs_{2.5}Ni_{0.08}H_{1.34}PVMo_{11}O_{40}$ has been demonstrated to improve the oxidation of isobutane to methacrolein and methacrylic acid with molecular oxygen [67, 68]. At 613 K the yield of methacrylic acid reached 9.0%. The optimal content of Cs and V was found to be equal to 2.5 and 1, respectively, and the addition of Ni enhanced the yield of methacrylic acid. In agreement with the statements above, it has been clearly shown that for $Cs_xH_{3-x}PMo_{12}O_{40}$ catalysts the factors that control the catalytic activity are the oxidizing ability and the protonic acidity of the catalysts [69].

Iron and copper have been the most widely studied addenda elements. It has been suggested in a study of $Cs_{2.5}M_{0.08}H_{1.5-0.08n}PVMo_{11}O_{40}$ (M = Cu^{2+}, Fe^{3+}, Ni^{2+}, Mn^{2+}, Co^{2+}) that iron addition promotes the reduction of the catalyst under oxygen-rich conditions while iron and copper promote the re-oxidation, under oxygen-poor conditions [70]. $Cs_{2.5}M^{n+}_{0.08}H_{1.5-0.08n}PVMo_{11}O_{40}$ (M = Ni^{2+}, Fe^{3+}) has also been found to catalyze the oxidation of propane and ethane [71, 72]. The state and role of V is quite important [73]. Light alkanes (C_1–C_3) have been observed to be transformed at reasonable yield to the corresponding carboxylic acid over $H_{3+x}PV_xMo_{11-x}O_{40}$ with x = 1–3 in presence of CO and in the $K_2S_2O_8$/CF_3COOH system [74].

The role of transition metals as counter-cations in polyoxometallates used as oxidation catalysts has been reviewed [75]. Transition metals have important and complex effects on textural, acid–base and redox properties of the heteropolyanions, as described in a number of studies. The interaction of the molybdophosphoric Keggin heteropolyanion with the iron counter-ion has been studied and

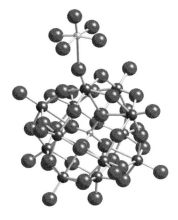

Figure 14.5 Schematic representation of the FeO(H$_2$O)$_5^{3+}$ counter-cation in interaction with the Keggin anion leading to a charge transfer between iron and molybdenum. Iron, yellow; molybdenum, black; oxygen, red.

the influence of the latter on the reducibility of iron-doped acid has been explained by an electron transfer between the heteropolyanion to the iron counter-ion as clearly demonstrated [76] and illustrated in Figure 14.5. Quantum-chemical calculations have further confirmed the existence of this transfer and explained why it was possible only under hydration of the solid [77]. The microscopic mechanism of this transfer is due to the modification of the relative position of the counter-ion, which enters into strong interaction with terminal oxygens of the heteropolyanion.

The effect of copper has also been studied in detail. Reduction experiments have shown that copper has a positive effect on the rate of reduction. Characterization of the samples after reduction by different techniques has clearly shown that Cu participates in the reduction of the heteropolycompound and that the reduction rate increases linearly with Cu content, with approximately 7 e$^-$/Cu [78]. This is shown in Figure 14.6 and Table 14.1.

Considering that one electron corresponds to the reduction of Cu^{2+} to Cu$^+$, the reduction of six Mo^{VI} species to six Mo^V can be attributed to one copper ion. The reduction proceeds in the vicinity of the copper cations via a concerted mechanism between the copper and one molybdenum cation of each of the six surrounding Keggin anions, the Cu^{2+} cations thus "catalyzing" the reduction of molybdenum cations:

$$Cu^{2+} + 2\,Mo^{6+} + O^{2-} + H_2 \rightarrow Cu^+ + Mo^{5+} + Mo^{6+} + H_2O$$

$$Cu^+ + Mo^{5+} + Mo^{6+} \rightarrow Cu^{2+} + 2\,Mo^{5+}$$

It is interesting to note that reduced heteropoly compounds show higher selectivity towards methacrylic acid than the non-reduced ones in the oxidation of isobutane. Mizuno and coworkers have also reported the oxidation of isobutane under oxygen-deficient conditions [67]. Ueda and coworkers have studied reduced 12-molybdophosphoric acid for the oxidation of propane [79]. This highly reduced

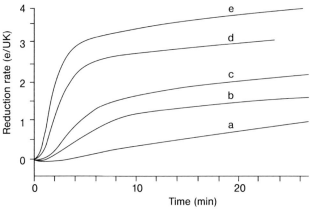

Figure 14.6 Extent of reduction at 613 K as a function of time for Cs_2Cu_x compounds. (a) Cs_2H_1, (b) $Cs_2Cu_{0.05}$, (c) $Cs_2Cu_{0.1}$, (d) $Cs_2Cu_{0.2}$ and (e) $Cs_2Cu_{0.3}$. (Taken from Ref. [77]).

Table 14.1 Extent and rate of the first rapid reduction period of the compounds Cs_2Cu_x by hydrogen at 613 K, calculated in electrons per KU and per copper ion. (Taken from Ref. [78]).

Compound	Reduction extent		Reduction rate	
	$(e^- KU^{-1})$	$(e^- Cu^{-1})$	$(e^- (KU\,min)^{-1})$	$(e^- (Cu\,min)^{-1})$
Cs_2	1.2	–	0.05	–
$Cs_2Cu_{0.05}$	1.5	6.5	0.19	3
$Cs_2Cu_{0.10}$	1.9	7.0	0.35	3.1
$Cs_2Cu_{0.20}$	2.7	7.5	0.81	3.8
$Cs_2Cu_{0.30}$	3.2	6.7	1.21	3.9

12-molybdophosphoric acid, formed by the heat treatment of the pyridinium or quinolinium salts, showed 50% selectivity to acrylic acid at 12% conversion.

Oxidative dehydrogenation at low temperatures and high pressures can result in the complete conversion of alkanes in comparison with simple dehydrogenation. Cavani and coworkers [80] have shown that Dawson-type mono-iron-substituted heteropolytungstates and Keggin-type heteropolymolybdates are active in the oxidative dehydrogenation of isobutane [80] and ethane [81], respectively. The rate per specific surface area of $K_7P_2W_{17}FeO_{61}$ for the oxidative dehydrogenation of isobutane was higher than those of active catalysts such as $Mg_3V_3O_8/MgO$ and Y_2O_3/CeF_3.

In a study employing negative differential resistance (NDR) features of $H_3PMo_{12}O_{40}$ HPA substituted with V, Barteau and coworkers have shown there is a relationship between their values and the redox potentials of the samples, and consequently with catalytic properties in partial oxidation reactions [82, 83]. For instance, this holds true for the oxidation of alkanes (propane, n-butane, isobutane) to the corresponding acids or alkenes, of acetaldehyde to acetic acid, of isobutyric acid to methacrylic acid [84]. The following order has been found: propane > butane, isobutane, isobutyric acid > acetaldehyde, corresponding to the weakest C–H bonds order. NDR values have been measured by scanning tunneling spectroscopy in a scanning tunneling microscope under a given atmosphere. This has been extended by the substitution of Mo with other elements such as Ag^+, Cu^{2+} and Pd^{2+}. A volcano curve of selectivity vs NDR or catalytic activity has been observed for the propane to acrylic acid reaction, with the optimum being at NDR ~0.8 V.

Interesting properties may also be obtained when using a mixed addenda system in the presence of a co-catalyst. The best known system [34d] is the V-substituted phosphomolybdate in conjunction with Pd^{2+} for the oxidation of olefins to carbonyl compounds. This is analogous to the Wacker oxidation process based on $CuCl_2$ and Pd^{2+}. Unlike the Wacker process, the HPA system works at very low chloride concentration, or even in its absence. In addition the HPA is more active and selective and less corrosive. Other examples of such two-component catalytic systems include Tl^{3+}/Tl^+, Pt^{4+}/Pt^{2+}, Ru^{4+}/Ru^{3+}, Ir^{4+}/Ir^{3+}, Br^{2-}/Br^- and I^-/I_2.

Although synergetic effects have been shown to be very important in heterogeneous catalysis, very few examples have been reported with polyoxometallates. Synergism is the overall improvement of performance obtained for a mixture of phases when it is greater than the sum of the performances of each individual phase. Such effects may have different origins. Synergy has been reported between polyoxometallate-based catalysts used for the partial oxidation of isobutane to methacrylic acid. These synergetic effects have led to the most efficient catalysts for that reaction. The first report on such synergies was based on the combination of the strong acid component corresponding to a sulfated tantalum oxide with a P–Mo–V Keggin-type phosphomolybdic acid. The strong acid site on the hydrated tantalum oxide treated with sulfuric acid would abstract an H^- ion from isobutane to form $i\text{-}C_4H_9^+$. It has been postulated that the $i\text{-}C_4H_9^+$ migrates to the polyoxometallate acid where it is oxidized to methacrolein and methacrylic acid [85]. The second report relates the combination of a lanthanum molybdate $La_2Mo_2O_9$ with a Keggin-type molybdophosphoric heteropolyacid with protons partially substituted by tellurium, vanadium and cesium cations, with the composition $Cs_2Te_{0.2}V_{0.1}H_{0.4}PMo_{12}O_{40}$. The synergy results primarily from a support effect, the lanthanum molybdate stabilizing the phosphomolybdic salt and preventing its sintering and degradation. The origin of this support effect has been related to the crystallographic fit between the two cubic structures of the phases [86].

14.4.2
Oxidation by Hydrogen Peroxide

Hydrogen peroxide is an important and widely used oxidant for organic substrates as it is cheap, easily available and yields water as a side-product, which is environmentally friendly, although H_2O_2 synthesis in not so eco-friendly. A wide range of oxidation processes such as epoxidation and hydroxylation have employed POMs with H_2O_2 as oxidant. Since POMs are generally insoluble in organic substrates they are rendered soluble by using alkylammonium groups as the counter-cations. W and Mo containing POMs have been shown to catalyze the oxidation of a wide range of organic substrates in either homogeneous or two-phase systems, with peroxo-type POMs being assumed to be the active intermediates. The most significant developments in this field have been reported by the groups of Venturello and Ishii. Venturello and coworkers observed that the tungstophosphate POM catalyzes the epoxidation of different alkenes with dilute H_2O_2 solution (15%) as oxidant [87]. Ishii and coworkers reported that $H_3PW_{12}O_{40}$ and cetylpyridinium chloride mixture catalyzes epoxidation of alkenes with commercially available H_2O_2 solution (35%) as oxidant [88]. More recently, the epoxidation mechanism on these catalysts was investigated by several groups [89–93]. It was demonstrated that $\{PO_4[WO(O_2)_2]_4\}^{3-}$ is the active species in the olefin epoxidation in the Venturello–Ishii system. Heteropolyacids with the Keggin structure, for example $H_3PW_{12}O_{40}$, are degraded in the presence of excess H_2O_2 to form peroxo species, for example $\{PO_4[WO(O_2)_2]_4\}^{3-}$ and $[W_2O_3(O_2)_4(H_2O)_2]^{2-}$, which are the true catalytic active intermediates.

Cyclohexene epoxidation by anhydrous urea–hydrogen peroxide adduct (UHP) has been studied over a series of Keggin-type heteropoly compounds using acetonitrile as an alternative solvent [94]. Among a series of Keggin-type POMs, tris(cetylpyridinium)12-tungstophosphate (($CPB)_3[PW_{12}O_{40}]$) gave 80% conversion of cyclohexene and 97% selectivity for cyclohexene oxide in the UHP/CH_3CN system. Epoxidation of 1-octene was achieved in a biphasic system [90a]. The oxidation of trimethoxybenzene to dimethoxy-p-benzoquinone in an acetic or formic acid medium was obtained at 303 K [95] over a mixture of molybdophosphoric, molybdosilicic and tungstophosphoric acids. Aromatic amines were oxidized with H_2O_2 catalyzed by cetylpyridinium salts of heteropolyoxometallates [96]. For instance, substituted anilines were oxidized to nitrosobenzenes at room temperature or nitrobenzene at elevated temperature under two-phase conditions in chloroform solvent and in azoxybenzene in aqueous medium. The oxidation of sulfides to sulfoxides and sulfones was observed in two-phase reaction conditions in chloroform solvent with 93–99% conversion [97]. Ballistreri and coworkers [98] were able to oxidize both internal and terminal alkynes by H_2O_2 in the presence of (cetylpyridinium)$_3$(PMo$_{12}$O$_{40}$), with an activity better than that for Na_2MO_4 (M = Mo^{VI} or W^{VI}). 1,2-Hexanediol and 1,2-octanediol were oxidized by H_2O_2 to 1-hydroxy-2-hexanone and 1-hydroxy-2-octanone respectively with yields above 90% on peroxotungstophosphates at reflux temperatures in chloroform [99]. Tris(cetylpyridinium)-12-tungstophosphate has been used to prepare epoxy acids from α,β-unsaturated acids, such as crotonic acid [100]. The epoxidation of allylic alcohols such as geraniol, 3-hydroxy-endotricyclo-deca-3,8-diene with H_2O_2 was performed over 12-molybdophosphoric acid and cetylpyridinium chloride at refluxing temperature in chloroform under two-phase conditions [101]. Epoxidation of cyclopentene was found to be more efficiently catalyzed by $H_3PMo_{12-n}W_nO_{40}$ with $n = 1$–11 than with $H_3PMo_{12}O_{40}$ and $H_3PW_{12}O_{40}$, when combined with cetylpyridinium bromide (CPB) as a phase transfer reagent with 50 equiv. H_2O_2 (30% solution) in acetonitrile [102]. It was then shown by UV-Vis, FTIR and ^{31}P NMR spectroscopies that these mixed Mo/W POMs are degraded during reaction into peroxo-type complexes $[(PO_4)(Mo_{4-x}W_xO_{20})]^{3-}$ with $x = 1–4$. These were not obtained from $H_3PW_{12}O_{40}$ although it was degraded during reaction. This explains the increased catalytic activity of the mixed Mo/W POMs. In the case of alcohol oxidation over mono-substituted $PM_{12}O_{40}$ (M = Mo or W) Keggin POMs, it was observed that they were degraded under reaction conditions into peroxo-phosphometallates $(PO_4[M(O_2)_2]_4^{3-})$, which are in fact the active catalysts.

Efficient H_2O_2-based oxidation has been observed with three types of polyoxometallate [103], $[\gamma\text{-}SiW_{10}O_{34}(H_2O)_2]^{4-}$, $[\gamma\text{-}1,2\text{-}H_2SiV_2W_{10}O_{40}]^{4-}$ and $[W_2O_3(O_2)_4(H_2O)_2]^{2-}$. The first POM catalyzed epoxidation of various olefins including non-activated terminal olefins such as propene and 1-octene with 99% selectivity to epoxide and 99% efficiency of H_2O_2 utilization. The second POM showed unique stereospecificity, regioselectivity and diastereoselectivity for the epoxidation of cis/trans olefins, non-conjugated dienes and 3-substituted cyclohexenes, respectively. The epoxidation of various allylic alcohols with only one equivalent H_2O_2 in water was catalyzed by the third POM and gave high yields of the corresponding epoxy alcohols.

The activation of the relatively inert C—H bonds as found in alkanes, alkenes and aromatic compounds via oxygen insertion was observed to be catalyzed by transition metal substituted heteropolyanions using H_2O_2. Di-iron substituted polyoxometallates were found to be highly efficient for the selective oxygenation of cyclohexane with H_2O_2. Other alkanes such as n-hexane, n-pentane and adamantane were oxidized employing such POMs. The efficiency and activity for the use of H_2O_2 greatly depends on the iron centers and the di-iron substituted complexes showed the highest efficiency for H_2O_2 conversion [104]. The catalytic properties of the transition metal substituted polyoxometallates $[PMW_{11}O_{39}]$ with (M = Fe^{3+}, Cr^{3+}, Ru^{4+}, Ti^{IV} and V^{IV}) were studied for substrates such as cyclohexene, benzene, alcohols and aldehydes with H_2O_2 and other oxidants [105]. Ti-substituted polyoxotungstate with the composition $[PTi_xW_{12-x}O_{40}]^{(3+2x)-}$ (where $x = 1, 2$) and peroxo titanium complexes were found to be efficient catalysts for alkene epoxidation reactions with H_2O_2 [106]. The epoxidation results from the synergistic interaction between a tungsten-peroxo site with an adjacent Ti-peroxo (µ) site which acts as an electrophilic centre for the alkene on the catalyst, involving OH radicals.

Catalytic properties of heteropoly complexes containing Fe^{3+} ions and the heteropolyanion $[PW_{11}O_{39}]^{7-}$, isolated from aqueous solution as tetrabutylammonium salts were studied for the oxidation of benzene by H_2O_2 in acetonitrile medium at 343 K [107]. The mechanism of H_2O_2 activation by one of the complexes, $[PW_{11}O_{39}Fe(OH)]^{5-}$, most likely involves the initial formation of a peroxo complex, which was observed spectroscopically. Chromium-containing derivatives of $[PW_{11}O_{39}]^{7-}$ were synthesized and used as catalysts for the oxidation of unsaturated hydrocarbons, such as benzene and cyclohexene, with H_2O_2. The surface location of Cr^{3+} was assumed to favor oxygen transfer from H_2O_2 to hydrocarbon. The resulting oxidized species PW$_{11}$Cr[O] are active oxidants in the reaction with unsaturated hydrocarbons [108]. The oxidation of cyclohexane by H$_2$O$_2$ was studied in acetonitrile medium, using tetrabutylammonium salts of Keggin-type polyoxotungstates. The polyanions [PW$_{11}$O$_{39}$]$^{7-}$ and [PW$_{11}$Fe(H$_2$O)O$_{39}$]$^{4-}$ showed higher catalytic activity and different selectivity for oxidation than the corresponding Cu, Co, Mn and Ni substituted complexes [109]. Oxidation of alkenes such as cyclooctene, 2-octene, 1-octene, cyclohexene, styrene and *trans*-stilbene was found to be affected by catalytic amounts of di-iron-substituted silicotungstate with high and efficient utilization of W$_{11}$O$_{39}$ [110]. The catalytic activity of di-iron-substituted silicotungstate was observed to be approximately 100 times higher than for non-, mono- and tri-iron-substituted silicotungstates [111].

Characteristic features of vanadium-containing heteropoly catalysts for the selective oxidation of hydrocarbons by H$_2$O$_2$ were described by Misono and coworkers [112] Conversion was 93% for oxidation of benzene to phenol with 100% selectivity. Over selectively V^{5+}-substituted Keggin heteropolytungstates, the catalytic activities for the hydroxylation of benzene in the presence of H$_2$O$_2$ was studied in a two-liquid aqueous and organic phase [113]. The activities and stabilities of catalysts were compared with those of V^V-substituted Dawson POMs, V^V-containing isopolyanions, Milas reagent and the picolinato–vanadium(V) oxo peroxo complex. It was observed that Keggin-type mixed addenda heteropolyanions containing vanadium such as [PMo$_{10}$V$_2$O$_{40}$]$^{5-}$ are effective as catalysts for the oxidation of alkyl aromatics to their respective acetates or alcohols and aldehydes or ketones using 30% H$_2$O$_2$ as oxidant in acetic acid [114]. The catalyst was not degraded during the catalytic cycle. The reaction proceeds by homolytic cleavage of the [PMo$_{10}$V$_2$O$_{40}$]$^{5-}$– peroxo intermediate, resulting in hydroperoxy and hydroxy radicals, which initiate the formation of benzyl radicals, and then to the products.

14.5
Characterization: Redox and Acid–Base Properties

Since the early 1970s, UV-Vis, infrared, Raman, MAS-NMR and inelastic neutron scattering spectroscopies and other techniques such as thermal desorption and thermogravimetry have been extensively used for identifying POM structure/type and chemical properties such as redox and acid–base behavior. MAS-NMR spectroscopy has been widely used, in particular for ^{31}P in the PM$_{12}$O$_{40}$ Keggin-type structure and in lacunary compounds, for structure determination and, in the case of ^{1}H, to characterize Brønsted acidic features. Such acidity has been shown to be high and to be rather difficult to characterize. The protons are in fact quite mobile and can move around the big POM anions, leading to strong acidity. It follows that hydroxyl groups do not exist on the IR timescale and IR bands are quite broad, which is a difference compared to other solid acid catalysts such as zeolites. Protonic sites have been identified in H$_3$PW$_{12}$O$_{40}$ and its Cs salts Cs$_x$H$_{3-x}$PW$_{12}$O$_{40}$, by *in situ* IR spectroscopy as a function of dehydration extent [115]. Similarly, protonic sites have been identified for H$_3$PW$_{12}$O$_{40}$ and Cs$_{1.9}$H$_{1.1}$PW$_{12}$O$_{40}$ by ^{1}H, ^{2}H, ^{31}P MAS-

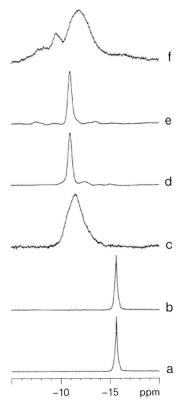

Figure 14.7 Room temperature MAS-NMR spectra of $H_3PW_{12}O_{40} \cdot nH_2O$ as a function of dehydration under dry N_2 flow for 2 h at: (a) 323 K, (b) 373 K, (c) 473 K, (d) 573 K, (e) 673 K, (f) 873 K. (Taken from Ref. [116]).

NMR spectroscopies, thermal analysis and inelastic neutron scattering as a function of the hydration state of the samples, as illustrated in Figure 14.7 and described in Section 14.5.4 [116].

An alternative way to determine acidity was to outgas the samples at increasing temperatures and to follow the weight losses by thermogravimetric analysis (TGA). Physisorbed water desorbs initially followed by water release from acid OH groups, according to the equations

$$H_3PW_{12}O_{40} \cdot n\,H_2O \rightarrow H_3PW_{12}O_{40} + n\,H_2O$$

followed by

$$H_3PW_{12}O_{40} \rightarrow PW_{12}O_{38.5} + 1.5\,H_2O$$

Another way to characterize acidity is to study the differential heat of adsorption of a basic probe compound, such as ammonia or pyridine, by microcalorimetry as a function of uptake. This technique yields the distribution of acid strength relative to coverage, but unfortunately does not differentiate between Brønsted and Lewis

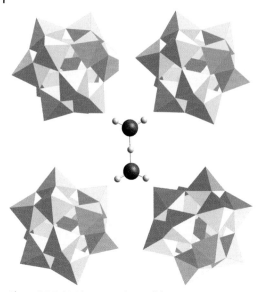

Figure 14.8 $H_5O_2^+$ species located between Keggin anions.

acid sites. It should thus be coupled with IR spectroscopic study of adsorption of the same basic probe molecule (Section 14.5.1).

X-ray and neutron diffraction and MAS-NMR techniques were successfully applied to identify the structure of hydrated and/or partially dehydrated HPAs [117–119]. The structure of 12-tungstophosphoric acid hexahydrate was solved using X-ray and neutron diffraction. The proton was shown to be coordinated to two water molecules in $H_5O_2^+$ species hydrogen bonded to four terminal W=O_t oxide ions as shown in Figure 14.8. The occupancy factor of the water O site was found to be equal to 0.5, supporting the suggestion of formation of $H_5O_2^+$, in the inter-Keggin anion space.

^{17}O NMR has been applied to differentiate the types of oxide ions in the Keggin anion [120], except for the PO_4 tetrahedron, which did not give rise to O exchange. The up-field NMR shift of the O_t resonance, upon dehydration of polycrystalline $H_3PW_{12}O_{40} \cdot xH_2O$ was interpreted as an indication that protonation sites in the anhydrous form of $H_3PW_{12}O_{40}$ are the terminal oxide ions. However, based on IR spectroscopy data, the identification of protonation sites in anhydrous POMs with the Keggin structure resulted in a different interpretation. Both terminal oxygens (M=O_t) and/or bridged oxygens (M–O–M) were proposed [121, 122]. Additionally, IR and Raman spectroscopies were used extensively to monitor the structural variations against the nature of the heteroatom (X) and/or of the transition metal element (M). Valuable information has been collected regarding both the assignment of the absorption frequency of the M=O, M–O–M and X–O vibrations [123] and the sensitivity of these vibrations to structural changes and/or hydration state or partial replacement of protons by alkaline cations.

14.5.1
IR Spectroscopy

The IR spectra of polyoxometallates have already been comprehensively discussed in the literature. The IR bands have been assigned previously [124–127]. The four distinct oxygen sites in a Keggin unit are represented in Figure 14.2(b) and correspond to the following description:

- Four O_a belong to the central tetrahedra PO_4.
- Twelve O are terminal oxygens, to a lone addendum M atom.
- Twelve O_b are involved in M–O_b–M bridges, between two different M_3O_{13} groups.
- Twelve O_c are involved in M–O_c–M bridge, in the same M_3O_{13} groups.

The most relevant assignments are as follows: v_{as}(P–O_a), (1080–1060 cm^{-1}), v_{as}(M–O_t) (990–960 cm^{-1}), v_{as}(M–O_b–M) (900–870 cm^{-1}) and v_{as}(M–O_c–M) (810–760 cm^{-1}). Some differences have been observed for lacunary Keggin-related compounds α-[$XM_{11}O_{39}$]$^{n-}$, which have a defect structure in which one metal atom and its terminal oxygen atoms are missing. These anions have a hole surrounded by five oxygen atoms and behave as pentadentate ligands. A general splitting of P–O stretching frequencies was observed and interpreted as a weakening of anion cohesion. In particular, the decrease in frequency of asymmetric bridge stretching is consistent with the lowering of M–Oc–M angles. The P–O stretching band for [$PM_{11}O_{39}$]$^{7-}$ is split into 1085 and 1040 cm^{-1}. This splitting is due to change of symmetry from T_d (XM_{12}) to C_s (XM_{11}) and may cause broadening of the band.

The protons were suggested to be localized on the most highly negatively charged O atoms, namely the O_b atoms. The v_{as}(M–O_b–M) mode is thus expected to be sensitive to the degree of hydration owing to hydrogen bonding, which decreases the M–O bond strength, thus increasing the bond length and decreasing the vibration frequency. Consequently, it is also sensitive to the nature of the cations exchanging protons. An increase in the v_{as}(M–O_b–M) frequency is thus expected upon dehydration as clearly shown in an *in situ* study of $Cs_{2.5}H_{1.5}W_{12}O_{40}$. Moreover, when the M–O bond strength increases, the bond becomes more covalent, its vibration frequency increases and the protons more free and thus more acidic. Thus an increase in v_{as}(M–O_b–M) frequency may correspond to an increase in Brønsted acidity strength and/or to a dehydrated state.

The acidic properties can also be estimated by NH_3 thermal desorption, although NH_3 can also act as a reductant at high temperature [128] and does not differentiate between Brønsted and Lewis acid sites. The acid subjected to evacuation at room temperature for 30 min showed a strong v_{OH} absorption at 3209 cm^{-1} with a prominent shoulder at 3355 cm^{-1}. In addition, the δ_{H2O} absorption band appeared at 1708 cm^{-1} as previously reported [129–134]. TGA performed under the same conditions, vacuum at ambient temperature for 30 min, showed that the HPA sample corresponded to $H_3PW_{12}O_{40} \cdot 5H_2O$, a state close to the stable hexahydrate form. The former absorption, at 1708 cm^{-1}, is typical of the presence of the protonated water clusters, probably the di-aqua-hydrogen ion $H_5O_2^+$, as demonstrated to be present

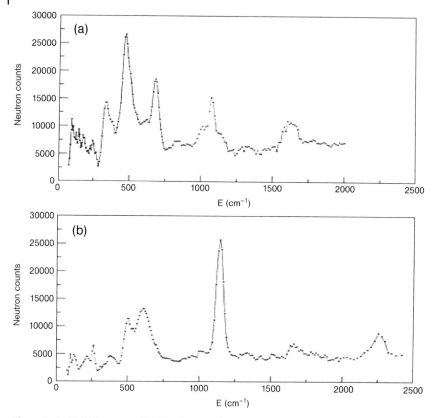

Figure 14.9 4 K INS spectra of $H_3PW_{12}O_{40} \cdot nH_2O$ as a function of dehydration under dry N_2 flow for 2 h at: (a) 473 K (H_3O^+ species) and (b) 573 K (lone protons). (Taken from Ref. [116]).

in $H_3PW_{12}O_{40} \cdot 6H_2O$ by X-ray and neutron diffraction studies [135], and/or hydroxonium ion which has been recently identified by inelastic neutron scattering (INS) in partially dehydrated 12-tungstophosphoric acid [116, 136]. INS data as a function of hydration state of $H_3PW_{12}O_{40}$ are illustrated in Figure 14.9. At 4K $H_5O_2^+$, H_3O^+ and H^+ (lone protons) were identified. $Cs_{1.9}H_{1.1}PW_{12}O_{40}$ was also characterized in the same way by INS.

14.5.2
Photoacoustic Spectroscopy

This technique is complementary to IR spectroscopy and well suited for dark colored samples. As for IR, it has been employed for structural determination using five or six bands below 1100 cm^{-1} that are characteristic of POMs and for water species such as H$_3$O$^+$ (~1710 cm^{-1}) and water (broad bands at ~3200 cm^{-1}, at ~2240 cm^{-1} due to H-bonded crystal water) [137].

14.5.3
UV-Visible Spectroscopy

Electronic spectroscopy of Keggin-type heteropoly compounds has been used for structural and quantitative analyses. POMs mainly absorb in the 180–270 nm region. The incorporation of a transition metal in the framework gives rise to additional bands in the UV-Vis region, depending on its nature and oxidation state. Postulation of reaction mechanisms, especially for oxidation reactions by performing *in situ* studies of the reaction mixture, has been based upon application of the technique. The changes observed in the spectra can be correlated to the formation of active intermediates formed due to the interaction between the active center in the catalyst and the oxidant, and the change in oxidation state of the transition metal elements.

14.5.4
Nuclear Magnetic Resonance Spectroscopy

NMR spectroscopy is a valuable tool in the study of the electron density distribution which results in large chemical shifts induced by both paramagnetic atoms and electron transfer between atoms in diamagnetic HPAs. Some paramagnetic contributions may exist depending on the presence of paramagnetic elements, such as V^{IV}, Mo^{V}, Cr$^{3+/V}$, Fe^{3+}, Ni^{2+}, Co^{2+}, and so on (d^n ions, $n \neq 0$ or 10) and leads to variations of the observed chemical shift. Several important elements can be studied easily. The most widely investigated nuclei include ^{31}P, ^{183}W, ^{51}V, ^{1}H, ^{13}C and ^{17}O. Studies of the electron density distribution in both diamagnetic and paramagnetic species are important for understanding the nature of chemical bonding in POMs and their role in chemical reactions.

^{17}O NMR spectroscopy provides information about the bonding nature of the oxygen atoms. There is a correlation between the downfield shift and the decreasing number of metal atoms to which the oxygen atom is bonded. Unfortunately, ^{17}O NMR spectroscopy is not often used, owing to the low natural abundance of ^{17}O. In contrast, ^{1}H NMR is an important and widely used tool for the detection of the different types of protons present in the heteropoly compound, especially because there is a change in the spectra with the number of hydration water molecules. ^{31}P NMR is widely used for structure description, especially in mixed addenda heteropoly compounds owing to the presence of a number of structural isomers, which become more numerous as the number of addenda atoms increases. It is greatly dependent on the degree of hydration in 12-phosphotungstic acid, containing n H$_2$O, values being −15.1 to −15.6 ppm for $n = 6$ and −11.1 to −10.5 ppm for $n = 0$ as shown in Figure 14.7. The difference is explained as follows: the former band is assigned to protonated water, $H(H_2O)_2^+$, connected with the heteropolyanion by hydrogen bonding at terminal oxygen and the latter band is assigned to the protons directly attached to the oxygen atoms of the polyanion. ^1H MAS-NMR spectra of $H_3PW_{12}O_{40}$ have been recorded at room temperature as a function of dehydration state [116] as for ^{31}P spectra in Figure 14.7. Samples a to c showed a single peak at about 7 ppm, samples d and e showed a broad peak at 7.5 ppm while sample f showed a peak at 6.6 ppm. The assignment of the different peaks was as follows: the peak at 10.5 ppm for bare protons, at 9.7 ppm for H_3O^+, at 7.5 ppm for $H_5O_2^+$ and at 7 ppm for $H(H_2O)_n^+$. A similar study of ^{31}P and ^1H on a $Cs_{1.9}H_{1.1}PW_{12}O_{40} \cdot nH_2O$ sample showed that the dehydration state was reached at a temperature about 100 K lower [116].

^{51}V NMR has also been widely used for structural elucidation of vanadium-containing mixed addenda heteropoly compounds, owing to the large natural abundance of ^{51}V nuclei. The number of structural isomers increases with the increase in the number of addenda atoms and their position in the Keggin structure is confirmed by ^{51}V NMR spectroscopy [138].

14.5.5
Electron Spin Resonance (ESR) Spectroscopy

Electron Spin Resonance (ESR) spectroscopy is well suited to the study of electron delocalization problems and spin density distribution, but it is limited to systems containing unpaired electrons. It gives information about the site symmetry and electronic structure of paramagnetic metal ions. ESR spectra may give information about mixed valence structure of reduced heteropoly compounds. The presence of unpaired spin in reduced HPA species gives rise to an ESR spectrum whose pattern depends on the number of atoms acquiring the unpaired electrons and also the temperature. For axial MO_6 complexes, to which the reduced form of HPA containing V^{IV}, Mo^V or W^V belong, an anisotropic ESR spectrum is observed [139]. The spectrum can become symmetrical at high temperature owing to molecular tumbling and electron delocalization over several atoms. The presence of a non-degenerate d_{xy} orbital for each octahedron allows these HPAs to be reversibly reduced by one or more electrons with retention of the original structure. Reduction of the anion results in the blue color arising from the intervalence band. The one-electron-reduced species gives rise to ESR spectra that have been interpreted to show complete delocalization of the unpaired spin over the twelve metal atoms. The ESR spectra at different temperatures have shown that one electron is trapped at quite low temperatures with only a partial delocalization, whereas it is completely delocalized at higher temperatures. This suggests that the spin is partially localized in the ground state but is involved in rapid thermal hopping from one site to another at elevated temperatures. This unpaired electron resides on a more easily reduced metallic atom. The degree of electron delocalization in the ground state is determined by the extent of interaction in the bridges [138].

14.5.6
Electrochemistry of Keggin Heteropoly Compounds

Heteropoly compounds can act as redox active materials when changing the heteroatom transition metal, without affecting the basic Keggin structure. The vast chemistry of heteropolyanion involves oxidation/reduction of addenda or the active transition metal. The electrochemical analysis of such species forms the basis of identification of redox active species. Thus electrochemical techniques such as cyclic voltammetry are means to understand the reactivity and mechanistic behavior of heteropolyanions as redox agents and redox catalysts in the liquid phase. The wide range of counter-cations assists in the study of the redox behavior of these anions, inducing solubility in a variety of solvents from water to a wide range of organic solvents. The heteropolyanions undergo rapid one- and two-electron reversible reduction to produce the heteropolyblues and further irreversible reduction can lead to their concomitant decomposition. Reduction is accompanied by an increase in the charge density and basicity. Reduction can also be accompanied by protonation depending on the pK_a of the produced oxometallate. The reduction potentials of the Keggin-type heteropoly tungstates and heteropoly molybdates are controlled by factors such as the nature of isomers. The reducibility increases in the sequence α-, β-, γ-. The reduction potential of one-electron waves is found to decrease linearly with a decrease in the valence of the central metal or an increase in the negative charge of the heteropolyanion. Transition metals incorporated into heteropolyanions reside in an octahedral environment, with one coordination site occupied by a solvent molecule, most commonly water. They can be oxidized to the corresponding oxometal, hydroxometal or peroxometal derivative depending on the nature of the metal and the oxidant. These species play an important role in oxidation catalysis.

The formal redox potential depends on a number of factors such as the nature of the heteroatom and addenda atom. The redox potential of the metal increases with increasing formal charge on the central atom. For a given oxidation state, the redox potential increases with the size and decreasing electronegativity of the central metal atom. The electrolytic conditions such as pH, counter-cation and the nature of the solvent also play an important role in determining the formal redox potential. By varying the addenda atom, the electrochemical character of the polyoxometallates can be changed. The addenda atoms can be arranged in decreasing oxidizing ability in the following order $V^V > Mo^{VI} > W^{VI}$. In case of one-electron-reduced mixed addenda heteropolyanions, the added electron is localized on the more reducible atom at room temperature [140]. For HPA-n = $H_{3+n}PMo_{12-n}V_nO_{40}$, $E_{Red} < E_{HPA-n} < E_{O2}$, where Red indicates reduced The reduction potential E_{HPA-n} equals about 0.7 V versus standard hydrogen electrode (SHE) for HPA-n with n = 1–4 at pH = 1 and the standard reduction potential of oxygen E_{O2} equals 1.23 V at 298 K.

14.5.7
Thermal Analysis

TGA, Differential Thermal Analysis (DTA) and Differential Thermogravimetric Analysis (DTG) measurements have been employed to determine the number and nature of water molecules present in the POMs. The results of TGA show the presence of two types of water in POM compounds, namely "water of crystallization" and "constitutional water molecules." The loss of the former usually occurs at temperatures below 443–473 K [141]. At temperatures exceeding 543 K for $H_3PMo_{12}O_{40}$ and 623 K for $H_3PW_{12}O_{40,}$ the constitutional water molecules (the acidic protons bound to the oxygens of the polyanion) are lost, according to the literature and for the acid forms [140]. For example the thermolysis of $H_3PMo_{12}O_{40}$ proceeds in two steps as schematized below:

$$H_3PMo_{12}O_{40} \cdot n\, H_2O \rightarrow H_3PMo_{12}O_{40} + n\, H_2O \text{ below } 543\,K$$

$$H_3PMo_{12}O_{40} \rightarrow H_{3-x}PMo_{12}O_{40-x/2} + (x/2)\, H_2O \text{ above } 623\,K$$

When protons are exchanged with metal cations, some water molecules physisorbed on these cations may be released at temperatures higher than 623 K, which unfortunately makes the assignment of water release above 270 °C to constitutional water, and therefore to protonic sites, questionable [142].

Thermal desorption of basic molecules such as ammonia and pyridine has been used to characterize acidic properties. For instance, ammonia has been observed to desorb at about 573 K on SiO_2–Al_2O_3 against >773 K on acid zeolites and >873 K on $H_3PW_{12}O_{40}$ and its $(NH_4)_xH_{1-x}$ and $Cs_{2.5}$ salts. Moreover it has also been shown that decomposition of ammonia to N_2, H_2, H_2O, as detected by MS analysis on line, has occurred during desorption at high temperature, which precludes the assignment to acid strength.

14.5.8
Microcalorimetry of Acid or Basic Probe Adsorption

The technique has been fruitfully used to characterize acid and basic sites in many catalysts, in particular for zeolites and metal oxides [143]. It has also been applied for POMs [144]. It consists of measuring the differential heats of adsorption when adsorbing successive increments of a basic probe molecule such as ammonia or pyridine for acidity characterization or of an acid probe molecule such as CO_2 or SO_2 to characterize basicity. The technique produces a histogram of the acid–base strength as a function of coverage, in particular when heterogeneity in strength exists. The data should then be compared with ammonia or pyridine desorption data from IR and thermal desorption experiments (see above).

14.6
Conclusions and Perspectives in Polyoxometallate Application in Heterogeneous Oxidation Catalysis

Owing to their multifunctional properties and easy preparation methods, wheel- or ball-shaped nanostructured POMs appear promising for applications in many fields, especially in heterogeneous catalysis for acid- and oxidation-type reactions. Industrial applications have already been established as mentioned at the end of Section 14.3. The problem of relatively weak stability under catalytic conditions has been partly circumvented by using alkali metal salts such as those of Cs^+. However, further progress is required and the stabilization of POMs in reaction conditions for long periods of time remains a key issue in their industrial development, and a real challenge for the future. The tertiary structure is modified under oxidation reaction conditions with a systematic decrease of the surface area, but more crippling is the observed change in the primary and consequently the secondary structures. For example, it was shown using Raman spectroscopy that the Keggin anion structure of $H_4PVMo_{11}O_{40}$ was relatively unstable upon heat treatment [145]. Vanadyl and molybdenyl species are expelled from the anions of the first layers of the compounds during oxidation reactions and defective Keggin structures are formed. These defective structures further disintegrate, presumably to form Mo_3O_{13} triads. In time, these fragments oligomerize to molybdenum oxygen clusters comparable to hepta- or octamolybdates, and finally to MoO_3-type oxides. The presence of water in the gas feeds, thought to be positive for the stabilization of the proton-containing phases, turned out to promote degradation of the anion. As stated above, only the presence of alkali metal cations, such as cesium, as counter-cations led to a partial stabilization of the structure. It has been shown that pure molybdophosphoric Keggin anions with Te or V cations capping these anions, as schematized in Figure 14.10, were very stable in the reaction conditions for isobutane partial oxidation up to 633 K and no degradation species could be detected [146]. Similar cappings have been observed with vanadium [147].

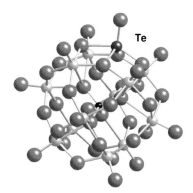

Figure 14.10 Example of Te capping a Keggin unit and helping to stabilize the anion. The Te cation is five coordinated, involving four oxygens from the same Keggin unit. (Taken from Ref. [146]).

In this chapter, we have reviewed the acid and redox properties of such materials which can be controlled and tuned by changing the chemical composition, whatever the addenda elements or the counter-cations, and the structure of the POMs.

The porosity of such materials has also been shown to vary depending on their chemical composition, and some examples have been given above based on "shape selectivity" properties. Moreover new POMs have been discovered (e.g. Figure 14.10) for instance the giant wheel- and ball-shaped anions, able to model metal oxide surfaces or to trap large entities such as metal oxide nanoparticles, metalloporphyrins, proteins, enzymes and so on, which opens up tremendous possibilities in many fields of application. Spherical clusters such as $\{Mo_{132}\}$ have pores about 0.5 nm in diameter which is comparable to zeolitic materials such as MFI or mordenite-type zeolites. However, unlike Si/Al zeolites, the POMs have acid and redox ($M^{VI/V}$) centers and are electron-rich, which are key parameters to promote catalytic reactions. The development of functionalized polyoxometallate [148] and polyoxometallate-based polymers [149] with creation of giant two-dimensional networks has been reported and provides new perspectives in the field of organic–inorganic frameworks.

The $\{Mo_{154}\}$ clusters are linked to form layered frameworks with nanosized channels [150] and could be considered as a model for bulk layered materials. The problem is how to remove cations between the layers, which prevent reactant access to the sites. These giant clusters can be considered as models for bulk layered metal oxides, including defects [151], as observed on large surfaces. The effect of such defects on catalytic properties, mainly for oxidation reactions, is quite important although still only partly known. For instance, local defects such as anionic or cationic vacancies leading to different coordination spheres of the surface O^{2-} anion in MgO have been characterized by DFT calculation, photoluminescence, IR and ^1H NMR [152] and have been shown to influence their basic properties as illustrated by the 2-methyl-but-3-yn-2-ol (MBOH) decomposition reaction, Such a reaction has been shown [153] to give 3-methyl-but-3-en-1-yne (Mbyne) and 3-methyl-but-2-enal (Prenal) on acid sites, 3-hydroxy-3-methyl-butanone (HMB) and 3-methyl-but-3ene-2-one (MIPK) on amphoteric sites and acetone and acetylene on basic sites. In a study of alkaline earth oxides involving Mg, Ca and Sr, insertion of Nd^{3+} cations and the use of different preparation and activation procedures (sol–gel against co-precipitation) have been shown in both cases by detailed XRD analysis to lead to local defects, which have been demonstrated to be of major importance for the propane oxidation reaction mechanism (radical-type) and selectivity to propene and ethene [154].

Great effort is now devoted to the design of functionalized POM catalysts to aid the development of green catalytic processes. Many examples have been presented in this chapter using either oxygen (air) or hydrogen peroxide as green oxidants for the up-grading of many organic chemicals to valuable compounds.

References

1. (a) Souchay, P. (1963) *Polyanions et Polycations*, Gauthier-Villars, Paris.
   (b) Souchay, P. (1969) *Ions Minéraux Condensés*, Masson, Paris.
   (c) Pope, M.T. (1983) *Heteropoly and Isopoly Oxometalates*, Springer, New York.
   (d) Pope, M.T. and Müller, A. (1991) *Angewandte Chemie – International Edition*, **30**, 34.
2. (a) Tsigdinos, G.A. (1978) *Topics in Current Chemistry*, **1**, 76.
   (b) Tsigdinos, G.A. and Hallada, C.J. (1968) *Inorganic Chemistry*, **7**, 437.
3. Misono, M. (2001) *Chemical Communications*, 1141.
4. Moffat, J.B. (2001) *Metal-oxygen Clusters. The Surface and Catalytic properties of Heteropoly Oxometallates*, Kluwer, New York, pp. 97–142.
5. Bonardet, J.L., Fraissard, J., Mc Garvey, G.B. and Moffat, J.B. (1995) *Journal of Catalysis*, **151**, 147.
6. Höchler, M., Englert, U., Zibrowius, B. and Hölderich, W.F. (1994) *Angewandte Chemie – International Edition*, **33**, 2991.
7. Inumaru, K., Nakajima, H., Itoh, T. and Misono, M. (1996) *Chemistry Letters*, 559.
8. Na, K., Lizaki, T., Okuhara, T. and Misono, M. (1995) *Chemistry Letters*, 155.
9. Langpape, M., Millet, J.-M.M., Ozkan, U.S. and Boudeulle, M. (1999) *Journal of Catalysis*, **181**, 80.
10. Berzelius, J.J. (1826) *Poggend. Ann. Phys. Chem.*, **6**, 369.
11. Galissard de Marignac, J.-C. (1864) *Ann. Chim. & Phys., 4è série*, **3**, 1.
12. Werner, A. (1907) *Berichte der Deutschen Chemischen Gesellschaft*, **40**, 40.
13. Miolati, A. and Pizzighelli, R. (1908) *Journal für Praktische Chemie*, **77**, 417;
14. Rosenheim, A. (1921) *Handbuch Der Anorganischen Chemie* (eds E. Abegg and F. Auerbach), Vol. 4, Part 1, ii, Hizel Verlag, Leipzig, pp. 997–1064.
15. Pauling, L. (1929) *Journal of the American Chemical Society*, **51**, 2868.
16. Keggin, J.F. (1934) *Proceedings of the Royal Society of London*, **A144**, 75.
17. Evans, H.T. (1948) *Journal of the American Chemical Society*, **70**, 1291.
18. Dawson, B. (1953) *Acta Crystallographica*, **6**, 113.
19. (a) Souchay, P. (1943) *Annales de Chimie France*, **19**, 102.
    (b) Souchay, P. (1945) *Contribution à l'étude des Hétéropolyacides Tungstiques*, Masson, Paris.
    (c) Souchay, P. (1965) *Talanta*, **12**, 1187.
    (d) Souchay, P., Massart, R. and Hervé, G. (1967) *Review of Polarography*, **14**, 270.
    (e) Jolivet, J.-P. (1994) *De la Solution à l'oxyde. Condensation des Cations en Solution Aqueuse. Chimie de Surface des Oxides*, EDP Sciences, Paris, CNRS Editions.
20. (a) Müller, A., Plass, W., Krickemeyer, E., Dillinger, S., Bögge, H., Armatage, A., Proust, A. and Beugholt, C. (1994) *Angewandte Chemie – International Edition*, **33**, 349.
    (b) Delgado, O., Dress, A. and Müller, A. (2001) *Polyoxometallate Chemistry from Topology via Self-Assembly to Applications* (eds M.T. Pope and A. Müller), Kluwer, Dordrecht, p. 69.
21. (a) Baker, L.C.W. and Glick, D.D. (1998) *Chemical Reviews*, **98**, 3.
    (b) Gouzerh, P. and Che, M. (2006) *Actualité en Chimie*, **298**, 9.
22. (a) Mialane, P., Dolbecq, A. and Secheresse, F. (2006) *Chemical Communications*, **33**, 3477.
    (b) Dolbecq, A., Mellot-Dranzieks, C., Mialane, P., Marrot, J., Ferey, G. and Secheresse, F. (2005) *European Journal of Inorganic Chemistry*, **15**, 3009.
23. Kong, X.J., Ren, Y.P., Zheng, P.Q., Long, Y.X., Long, L.S., Huang, R.B. and Zheng, L.S. (2006) *Inorganic Chemistry*, **45**, 10702.
24. Mioc, U.B., Todorovic, M.R., Davidovic, M., Colomban, P. and Holcljatner-Antunovic, I. (2005) *Solid State Ionics*, **176**, 3005.
25. Keita, B. and Nadjo, L. (1990) *Journal of Electroanalytical Chemistry*, **287**, 149.
26. Rhule, J.T., Hill, C.L. and Judd, D.A. (1998) *Chemical Reviews*, **98**, 327.

27 (a) Okuhara, T., Mizuno, N. and Misono, M. (1996) *Advanced Catalysis*, **41**, 113.
(b) Okuhara, T., Mizuno, N. and Misono, M. (2001) *Applied Catalysis A: General*, **222**, 63.
28 Kozhevnikov, I.V. (1998) *Chemical Reviews*, **98**, 171.
29 Chu, W., Yang, X., Shan, Y., Ye, X. and Wu, Y. (1996) *Catalysis Letters*, **42**, 201.
30 Kozhevnikov, I.V., Kloetstra, K.R., Sinnema, A., Zandbergen, H.W. and van Bekkum, H. (1996) *Journal of Molecular Catalysis A–Chemical*, **114**, 287.
31 Marme, F., Coudurier, G. and Védrine, J.C. (1998) *Microporous and Mesoporous Materials*, **22**, 151.
32 (a) Massart, R. (1969) *Annales de Chimie France*, **4**, 365 and 431.
(b) Souchay, P. and Hervé, G. (1965) *Comptes rendus de l'Académie des Sciences, Paris*, **261**, 2486.
(c) Hervé, G. (1966) *Comptes rendus de l'Académie des Sciences, Paris*, **263**, 1297.
(d) Hervé, G. (1967) *Comptes rendus de l'Académie des Sciences, Paris*, **265**, 805.
(e) Hervé, G. (1971) *Annales de Chimie France*, **6**, 219.
(f) Tézé, A. and Hervé, G. (1977) *Journal of Inorganic and Nuclear Chemistry.*, **39**, 999.
33 Otake, M. and Onoda, T. (1976) *Shokubai*, **18**, 169.
34 (a) Matveev, K. (1977) *Kinetika i Kataliz*, **18**, 862.
(b) Matveev, K.I. and Kozhevnikov, I.V. (1980) *Kinetika i Kataliz*, **21**, 1189.
(c) Kozhevnikov, I.V. (1995) *Catalysis Reviews–Science and Engineering*, **37**, 311.
(d) Kozhevnikov, I.V. (1998) *Chemical Reviews*, **98**, 171.
(e) Kozhevnikov, I. (2002) *Catalysis for Fine Chemical Synthesis. Catalysis by Polyoxometalates*, John Wiley & Sons, Ltd, Chichester, UK.
(f) Kozhevnikov, I.V. (2007) *Journal of Molecular Catalysis A–Chemical*, **262**, 86.
35 (a) Misono, M. (1987) *Catalysis Reviews–Science and Engineering*, **29**, 268.
(b) Misono, M. (1988) *Catalysis Reviews–Science and Engineering*, **30**, 339.
(c) Okuhara, T., Mizuno, N. and Misono, M. (1996) *Advanced Catalysis*, **41**, 113.
(d) Okuhara, T., Mizuno, N. and Misono, M. (2001) *Applied Catalysis A: General*, **222**, 63.
(e) Mizuno, N. and Misono, M. (1998) *Chemical Reviews*, **98**, 199.
(f) Misono, M. and Acad, C.R. (2000) *Comptes rendus de l'Académie des Sciences, Paris, Sér IIc Chimie/Chemistry*, **3**, 471.
(g) Min, J.S. and Mizuno, N. (2001) *Catalysis Today*, **71**, 89.
(h) Mizuno, N., Hikichi, S., Yamaguchi, K., Uchida, S., Nakagawa, Y., Uehara, K. and Kamata, K. (2006) *Catalysis Today*, **117**, 32.
36 (a) Moffat, J.B. (1987) *Reviews of Chemical Intermediates*, **8**, 1.
(b) Moffat, J.B. (1989) *Journal of Molecular Catalysis*, **52**, 169.
(c) Moffat, J.B. (2001) *Metal-oxygen clusters. The Surface and Catalytic properties of Heteropoly Oxometallates*, Kluwer, New York.
37 Corma, A. (1995) *Chemical Reviews*, **95**, 559.
38 Hill, C.L. (1998) *Chemical Reviews*, **98**, 1.
39 Alekar, N. (2000) PhD thesis, Pune, India.
40 Guo, Y. and Hu, C. (2007) *Journal of Molecular Catalysis A–Chemical*, **262**, 136.
41 (a) Védrine, J.C. (2002) *Topics in Catalysis*, **21**, 97.
(b) Novakova, E.K. and Védrine, J.C. (2006) *Metal Oxides. Chemistry and Applications* (ed. J.L.G. Fierro), CRC Taylor & Francis, Boca Raton, Fl., pp. 413–61.
42 Mars, P. and van Krevelen, D.W. (1953) *Chemical Engineering Science*, **3**, 41.
43 (a) Dimitratos, N. and Védrine, J.C. (2003) *Catalysis Today*, **81**, 561.
(b) Dimitratos, N. and Védrine, J.C. (2003) *Catalysis in Application* (eds S.D. Jackson, J.S.J. Hargreaves and D. Lennon), Royal Society of Chemistry, Cambridge, pp. 145–52.
(c) Dimitratos, N. and Védrine, J.C. (2003) *Applied Catalysis A: General*, **256**, 251–63.
(d) Dimitratos, N. and Védrine, J.C. (2006) *Journal of Molecular Catalysis A–Chemical*, **255**, 184–92.

44 Centi, G., Cavani, F. and Trifiro, F. (2001) *Selective Oxidation by Heterogeneous Catalysis*, Kluwer, New York.
45 Hill, C.L. and Proster-McCartha, C.M. (1995) *Coordination Chemistry Reviews*, **143**, 407.
46 Mizuno, N., Hirose, T., Tateishi, M. and Iwamoto, M. (1993) *Chemistry Letters*, 1839.
47 Mizuno, N., Hirose, T. and Iwamoto, M. (1994) *Studies in Surface Science and Catalysis*, **82**, 593.
48 Kholdeeva, O.A., Grigoriev, V.A., Maksimov, G.M., Fedotov, M.A., Golovin, A.V. and Zamaraev, K.I. (1990) *Journal of Molecular Catalysis*, **114**, 123.
49 Hamamoto, M., Nakayama, K., Nishitama, Y. and Ishii, Y. (1993) *Journal of Organic Chemistry*, **58**, 6421.
50 Kuznetova, N.I. and Yurchenko, E.N. (1989) *Reaction Kinetics and Catalysis Letters*, **39**, 399.
51 Harrup, M.K. and Hill, C.L. (1994) *Inorganic Chemistry*, **331**, 5448.
52 Khenkin, A.M. and Hill, C.L. (1993) *Journal of the American Chemical Society*, **115**, 8178.
53 Seo, Y.-J., Makai, Y., Taganawa, T. and Goto, S. (1997) *Journal of Molecular Catalysis*, **120**, 149.
54 Neumann, R. and Levin, M. (1992) *Journal of the American Chemical Society*, **114**, 7278.
55 Kholdeeva, O.A., Golovin, A.V. and Kozhevnikov, I.V. (1992) *Reaction Kinetics and Catalysis Letters*, **46**, 107
56 Lissel, M., In de Val, H.J. and Neumann, R. (1992) *Tetrahedron Letters*, **33**, 1795.
57 Neumann, R. and Assael, I. (1998) *Journal of the Chemical Society D – Chemical Communications*, 1285.
58 Atlamsani, A., Brégeault, J.-M. and Ziyad, M. (1993) *Journal of Organic Chemistry*, **58**, 5663.
59 Nakayama, K., Hamamoto, M., Nishiyama, Y. and Ishii, Y. (1993) *Chemistry Letters*, 1699.
60 Ali, B.E., Brégeault, J.-M., Mercier, J., Martin, J., Martin, C. and Convert, O. (1989) *Journal of the Chemical Society D – Chemical Communications*, 825.
61 Ai, M. (1982) *Applied Catalysis*, **4**, 245.
62 Akimoto, M., Tsuchida, Y., Sato, K. and Echigoya, E. (1981) *Journal of Catalysis*, **72**, 83.
63 Eguchi, K., Aso, I., Yamazoe, N. and Seiyama, T. (1979) *Chemistry Letters*, 1345.
64 Dimitratos, N. and Védrine, J.C. (2006) *Catalysis Communications*, **7**, 811.
65 Centi, G., Lopez Nieto, J.M. and Iapalucci, C. (1989) *Applied Catalysis*, **46**, 197.
66 (a) Cavani, F., Trifiro, F. and C. (1994) *Catalysis Today*, **11**, 247.
(b) Cavani, F., Mezzogori, R., Pigamo, A. and Trifirò, F. (2001) *Catalysis Today*, **71**, 97.
67 (a) Mizuno, N., Tateishi, M. and Iwamoto, M. (1994) *Journal of the Chemical Society D – Chemical Communications*, 1411.
(b) Mizuno, N., Tateishi, M. and Iwamoto, M. (1994) *Applied Catalysis A: General*, **118**, L1.
(c) Mizuno, N., Han, W., Kudo, T. and Iwamoto, M. (1996) *Studies in Surface Science and Catalysis*, **101**, 1001.
(d) Mizuno, N., Tateishi, M. and Iwamoto, M. (1997) *Journal of Catalysis*, **163**, 87.
68 Cavani, F., Etienne, E., Favaro, M., Gall, A., Trifiro, F. and Hecquet, G. (1995) *Catalysis Letters*, **32**, 215.
69 Cavani, F., Mezzogori, R., Pigamo, A. and Paris, F. Trifirò (2000) *Comptes rendus de l'Académie des Sciences, Paris, Série IIc, Chimie/Chemistry*, **3**, 523.
70 Min, J.-S. and Mizuno, N. (2001) *Catalysis Today*, **71**, 89.
71 Mizuno, N., Tateishi, M. and Iwamoto, M. (1995) *Applied Catalysis A: General*, **128**, L165.
72 Mizuno, N., Suh, D.J., Han, W. and Kudo, T. (1996) *Journal of Molecular Catalysis A – Chemical*, **114**, 309.
73 Bayer, R., Marchal, C., Liu, F.X., Tézé, A. and Hervé, G. (1996) *Journal of Molecular Catalysis A – Chemical*, **110**, 65.
74 Kirillova, M.V., da Silva, J.A.L., Fraústo da Silva, J.R. and Plombeiro, A.J.L. (2007) *Applied Catalysis A: General*, **332**, 159.
75 Marchal-Roch, C. and Millet, J.-M.M. (2001) *Comptes rendus de l'Académie des*

Sciences, Paris, Sér. IIc, Chimie/ Chemistry, **4**, 321.
76 (a) Langpape, M. and Millet, J.M.M. (2000) *Applied Catalysis A: General*, **200**, 89. (b) Huynh, Q. and Millet, J.M.M. (2005) *Journal of Physics and Chemistry of Solids*, **66**, 887.
77 Borshch, S.A., Duclusaud, H. and Millet, J.M.M. (2000) *Applied Catalysis A: General*, **200**, 103.
78 (a) Langpape, M., Millet, J.M.M., Ozkan, U.S. and Deliche̍re, P. (1999) *Journal of Catalysis*, **182**, 148. (b) Langpape, M., Millet, J.M.M., Ozkan, U.S. and Boudeulle, M. (1999) *Journal of Catalysis*, **181**, 80.
79 (a) Ueda, W., Suzuki, Y., Lee, W. and Imanoka, S. (1996) *Studies in Surface Science and Catalysis*, **101**, 1065. (b) Ueda, W. and Suzuki, Y. (1995) *Chemistry Letters*, 541.
80 (a) Cavani, F., Comuzzi, C., Dolcetti, G., Etienne, E., Finke, R.G., Selleri, G., Trifirò, F. and Trovarelli, A. (1996) *Journal of Catalysis*, **160**, 317. (b) Comuzzi, C., Dolcetti, G., Trovarelli, A., Cavani, F., Trifirò, F., Llorca, J. and Finke, R.G. (1996) *Catalysis Letters*, **36**, 75.
81 Albonetti, S., Cavani, F., Trifirò, F. and Koutyrev, M. (1995) *Catalysis Letters*, **30**, 253.
82 Barteau, M.A., Lyons, J.E. and Song, I.K. (2003) *Journal of Catalysis*, **216**, 236.
83 Song, I.K. and Barteau, M.A. (2002) *Journal of Molecular Catalysis A–Chemical*, **185**, 182–3.
84 Ballarini, N., Candiracci, F., Cavani, F., Degrand, H., Dubois, J.-L., Lucarelli, G., Margotti, M., Patinet, A., Pigamo, A. and Trifirò, F. (2007) *Applied Catalysis A: General*, **325**, 263.
85 Ushikubo, T. (2003) *Catalysis Today*, **78**, 79.
86 Huynh, Q., Selmi, A., Lacorre, L. P. and Millet, J.-M.M. *Applied Catalysis A: General* (in press).
87 (a) Venturello, C., Alneri, E. and Ricci, M. (1983) *Journal of Organic Chemistry*, **48**, 3831. (b) Venturello, C., Aloisio, R.D., Bart, J.C.J. and Ricci, M. (1985) *Journal of Molecular Catalysis*, **32**, 107.
(c) Venturello, C. and Aloisio, R.D. (1988) *Journal of Organic Chemistry*, **53**, 1553.
88 Ishii, Y., Yamawaki, K., Ura, T., Yamada, H., Yoshida, T. and Ogawa, M. (1988) *Journal of Organic Chemistry*, **53**, 3587.
89 Csanyi, L.J. and Jaky, K. (1990) *Journal of Molecular Catalysis*, **61**, 75.
90 (a) Aubry, C., Chottard, G., Platzer, N., Brégeault, J.-M., Thouvenot, R., Chauveau, F., Huet, C. and Ledon, H. (1991) *Inorganic Chemistry*, **30**, 4409; (b) Salle, L., Aubry, C., Thouvenot, R., Robert, F., Doremieux-Morin, C., Chottard, G., Ledon, H., Jeanin, Y. and Brégeault, J.-M. (1994) *Inorganic Chemistry*, **33**, 871. (c) Salle, L., Piquemal, J.Y., Thouvenot, R., Minot, C. and Brégeault, J.-M. (1997) *Journal of Molecular Catalysis A–Chemical*, **117**, 375.
91 Gresley, N.M., Griffith, W.P., Laemmel, A.C., Nogueira, H.I.C. and Parkin, B.C. (1997) *Journal of Molecular Catalysis A–Chemical*, **117**, 185.
92 Duncan, D.C., Chambers, R.C., Hecht, E. and Hill, C.L. (1995) *Journal of the American Chemical Society*, **117**, 681.
93 Gao, J., Chen, Y., Han, B., Feng, Z., Zhou, N., Li, C., Gao, S. and Xi, Z. (2004) *Journal of Molecular Catalysis A–Chemical*, **210**, 197.
94 (a) Ding, Y., Gao, Q., Li, G., Zhang, H., Wang, J., Yan, L. and Suo, J. (2004) *Journal of Molecular Catalysis A–Chemical*, **218**, 161. (b) Gao, Q., Ding, Y., Liu, H. and Suo, J. (2005) *Journal of Chemical Research*, 716.
95 Orita, H., Shimizu, H., Hayakawa, T. and Takehiro, K. (1991) *Reaction Kinetics and Catalysis Letters*, **44**, 3633.
96 Sakaue, S., Tsubakino, T., Nishiyama, Y. and Ishii, Y. (1993) *Journal of Organic Chemistry*, **58**, 3633.
97 Ishii, Y., Tanaka, H. and Nishiyama, Y. (1994) *Chemistry Letters*, 1.
98 Ballistreri, F.P., Failla, S., Spina, E. and Tomaselli, G.A. (1989) *Journal of Organic Chemistry*, **54**, 947.
99 Sakata, Y. and Ishii, Y. (1991) *Journal of Organic Chemistry*, **56**, 6233.
100 Oguchi, T., Sakata, Y., Takeuchi, N., Kaneda, K., Ishii, Y. and Ogawa, M. (1989) *Chemistry Letters*, 865.

References

101 Matoba, Y., Inoue, H., Akagi, J., Okabayashi, T., Ishii, Y. and Ogawa, M. (1984) *Synthetic Communications*, **14**, 865.
102 Li, G., Ding, Y., Wang, J., Wang, X. and Suo, J. (2007) *Journal of Molecular Catalysis A – Chemical*, **262**, 67.
103 Mizuno, N., Hikichi, S., Yamaguchi, K., Uchida, S., Nakagawa, Y., Uchara, K. and Kamata, K. (2006) *Catalysis Today*, **117**, 32.
104 Mizuno, N. (1998) *Journal of the American Chemical Society*, **120**, 9267.
105 Kuznetsova, L.I., Detusheva, L.G., Kuzenesova, N.I., Fedotov, M.A. and Likholobov, V.A (1997) *Journal of Molecular Catalysis*, **117**, 389.
106 Toshihiro, Y., Eri, I., Yasushi, A. and Kanai, S. (1996) *Journal of Molecular Catalysis*, **114**, 237.
107 Kuznetsova, N.I., Detusheva, L.G., Fedotov, M.A. and Likholobov, V.A. (1996) *Journal of Molecular Catalysis*, **111**, 81.
108 Kuznetsova, N.I., Kuznetsova, N.L.I. and Likholobov, V.A. (1996) *Journal of Molecular Catalysis*, **108**, 135.
109 Simoes, M.M.Q., Conceicao, C.M.M., Gamelas, J.A.F., Domingues, P.M.D.N., Cavaleiro, A.M.V., Ferre-Correia, A.J.V. and Johnstone, R.A.W. (1999) *Journal of Molecular Catalysis*, **144**, 461.
110 Mizuno, N., Nozaki, C., Kiyoto, I. and Misono, M. (1999) *Journal of Catalysis*, **182**, 285.
111 Mizuno, N., Kiyoto, I., Nozaki, C. and Misono, M. (1999) *Journal of Catalysis*, **181**, 171.
112 Misono, M., Mizuno, N., Inumaru, K., Koyano, G. and Lu, X.-H. (1997) *Studies in Surface Science and Catalysis*, **110**, 35.
113 Nomiya, K., Yanagibayashi, H., Nozaki, C., Kondoh, K., Hiramatsu, E. and Shimizu, Y. (1996) *Journal of Molecular Catalysis*, **114**, 181.
114 Neumann, R. and Delavega, M. (1993) *Journal of Molecular Catalysis*, **84**, 93.
115 Essayem, N., Holmqvist, A., Gayraud, P.-Y., Védrine, J.C. and Ben Taârit, Y. (2001) *Journal of Catalysis*, **197**, 273.
116 Essayem, N., Yong, Y.Y., Jobic, H. and Védrine, J.C. (2000) *Applied Catalysis A: General*, **194–195**, 109.
117 Brown, G.M., Noe-Spirlet, M.-R., Busing, W.R. and Levy, H.A. (1977) *Acta Crystallographica*, **B33**, 1038.
118 Fournier, M., Feumi-Jantou, C., Rabia, C., Hervé, G. and Launay, S. (1992) *Journal of Materials Chemistry*, **2**, 971.
119 Massart, R., Constant, R., Fruchart, J.-M., Ciabrini, J.-P. and Fournier, M. (1977) *Inorganic Chemistry*, **16**, 2916.
120 Kozhevnikov, I.V., Sinnema, A., Jansen, R.J.J. and van Bekkum, H. (1994) *Catalysis Letters*, **27**, 187.
121 Lee, K.Y., Mizuno, N., Okuhara, T. and Misono, M. (1989) *Bulletin of the Chemical Society of Japan*, **62**, 1731.
122 Bielanski, A., Maleka, A. and Kubelkova, L. (1989) *Journal of the Chemical Society – Faraday Transactions I*, **85**, 2847.
123 Rocchiccioli-Deltcheff, C.R., Thouvenot, R. and Franck, R. (1976) *Spectrochimica Acta*, **A32**, 587.
124 Okuhara, T., Mizuno, N. and Misono, M. (1996) *Advanced Catalysis*, **41**, 113.
125 Orita, H., Hayakawa, T., Shimizu, M. and Takehira, K. (1991) *Applied Catalysis*, **77**, 133.
126 Rocchiccioli-Deltcheff, C.R., Fournier, M., Franck, R. and Thouvenot, R. (1983) *Inorganic Chemistry*, **22**, 207.
127 Rocchiccioli-Deltcheff, C.R. and Fournier, M. (1991) *Journal of the Chemical Society – Faraday Transactions*, **87**, 3913.
128 Essayem, N., Fréty, R., Coudurier, G. and Védrine, J.C. (1997) *Journal of the Chemical Society – Faraday Transactions*, **93**, 3243.
129 Mizuno, N., Katamura, K., Yoneda, K. and Misono, M. (1983) *Journal of Catalysis*, **83**, 384.
130 Highfield, J.G. and Moffat, J.B. (1984) *Journal of Catalysis*, **88**, 177.
131 Bielanski, A., Maleka, A. and Kubelkova, L. (1989) *Journal of the Chemical Society – Faraday Transactions I*, **85**, 2847.
132 Bielanski, A., Datka, J., Gil, B. and Malecka-Lubanska, A. (1999) *Catalysis Letters*, **57**, 61.
133 Essayem, N., Kieger, S., Coudurier, G. and Védrine, J.C. (1996) *Studies in Surface Science and Catalysis*, **101**, 591.
134 Essayem, N., Coudurier, G., Védrine, J.C., Habermacher, D. and Sommer, J. (1999) *Journal of Catalysis*, **183**, 292.

135 Brown, G.M., Noe-Spirlet, M.-R., Busing, W.R. and Levy, H.A. (1977) *Acta Crystallographica*, **B33**, 1038.
136 Moic, U.B., Colomban, P., Davidovic, M. and Tomkinson, T. (1994) *Journal of Molecular Structure*, **326**, 99.
137 (a) Highfield, J.G. and Moffat, J.B. (1984) *Journal of Catalysis*, **88**, 177.
(b) Highfield, J.G. and Moffat, J.B. (1984) *Journal of Catalysis*, **89**, 185.
(c) Highfield, J.G. and Moffat, J.B. 1985) *Journal of Catalysis*, **95**, 108.
138 Kazansky, L.P. and Mc Garvey, B.R. (1999) *Coordination Chemistry Reviews*, **188**, 157.
139 (a) Che, M., Védrine, J. and Naccache, C. (1969) *Journal de Chimie Physique*, **65**, 579.
(b) Che, M., Védrine, J. and Naccache, C. (1970) *Bulletin de la Société chimique de France*, 3307.
(c) Derouane, E.G. and Védrine, J.C. (1973) *Industrie Chimique Belge-Belgische Chemische Industrie*, **38**, 375.
140 Sadakane, M. and Steckhan, E. (1998) *Chemical Reviews*, **98**, 219.
141 Bielanski, A., Malecka, A. and Pozniczek, J. (1989) *Journal of Thermal Analysis*, **35**, 1699.
142 Dimitratos, N. and Védrine, J.C. (2003) *Catalysis in Application* (eds S.D. Jackson, J.S.J. Hargreaves and D. Lennon), Royal Society of Chemistry, Cambridge, p. 145.
143 (a) Gravelle, P.C. (1972) *Advanced Catalysis*, **22**, 191.
(b) Gravelle, P.C. (1985) *Thermochimica Acta*, **96**, 365.
(c) Gravelle, P.C. (1977) *Catalysis Reviews–Science and Engineering*, **16**, 37.
(d) Auroux, A. (1994) *Catalyst Characterization: Physical Techniques for Solid Materials* (eds B. Imelik and J.C. Védrine), Plenum Press, New York, p. 611.
(e) Auroux, A. (1997) *Topics in Catalysis*, **4**, 71.
(f) Damjanovic, L. and Auroux, A. (2007) *Handbook of Thermal Analysis and Calorimetry*, Chapter 12 (in press).
144 (a) Lefebvre, F., Dupont, P. and Auroux, A. (1995) *Reaction Kinetics and Catalysis Letters*, **3**, 55.
(b) Liu-Cai, F.X., Sahut, B., Faydi, E., Auroux, A. and Hervé, G. (1999) *Applied Catalysis A: General*, **185**, 75.
145 Mestl, G., Ilkenhans, T., Spielbauer, D., Dieterle, M., Timpe, O., Kröhnert, J., Jentoft, F., Knözinger, H. and Schlögl, R. (2001) *Applied Catalysis A: General*, **210**, 13.
146 Huynh, Q. (2005) PhD thesis, University of Lyon, No. 168.
147 Rodriguez-Fortea, A., de Graaf, C. and Poblet, J.M. (2006) *Chemical Physics Letters*, **428**, 88.
148 (a) Mialane, P., Dolbecq, A. and Secheresse, F. (2006) *Chemical Communications*, **33**, 3477.
(b) Dolbecq, A., Mellot-Draznieks, C., Mialane, P., Marrot, J., Ferey, G. and Secheresse, F. (2005) *European Journal of Inorganic Chemistry*, **15**, 3009.
149 Kong, X.J., Ren, Y.P., Zheng, P.Q., Long, Y.X., Long, L.S., Huang, R.B. and Zheng, L.S. (2006) *Inorganic Chemistry*, **45**, 10702.
150 (a) Müller, A., Krickemeyer, E., Schmidtmann, H., Bögge, M., Beugholt, C., Das, S.K. and Peters, F. (1999) *Chemistry–A European Journal*, **5**, 1496.(b) Müller, A., Das, S.K., Bögge, H., Beugholt, C. and Schmidtmann, M. (1999) *Chemical Communications*, 1935.
151 Müller, A., Rabindranath, M., Schmidtmann, M., Bögge, H., Das, S.K. and Zhang, W. (2001) *Chemical Communications*, 2126.
152 (a) Chizallet, C., Costentin, G., Lauron-Pernot, H., Krafft, J.-M., Che, M., Saussey, J., Delbecq, F. and Sautet, P. (2006) *Journal of Physical Chemistry B*, **110**, 15878.
(b) Chizallet, C., Bailly, M.L., Costentin, G., Lauron-Pernot, H., Krafft, J.-M., Bazin, P., Saussey, J. and Che, M. (2006) *Catalysis Today*, **116**, 196.
153 Lauron-Pernot, H. (2006) *Catalysis Reviews–Science and Engineering*, **48**, 315.
154 Savova, B., Filkova, D., Petrov, L., Crisan, D., Crisan, M., Răileanu, M., Drăgan, N., Galtayries, A. and Védrine, J.C. (to be submitted).

15
Alkane Dehydrogenation over Vanadium and Chromium Oxides

S. David Jackson, Peter C. Stair, Lynn F. Gladden, and James McGregor

15.1
Introduction

The dehydrogenation of light alkanes over chromia systems has been known as a catalytic process since the early 1930s [1]. A comprehensive review of the pre-1954 literature can be found in Emmett's book on catalysis [2]. The dehydrogenation of light alkanes is highly endothermic (for example, propane to propene, $C_3H_8 \Leftrightarrow C_3H_6 + H_2$, $\Delta H^{\circ}_{298} = +110\,kJ\,mol^{-1}$) and is equilibrium limited at temperatures below 1000 K. Commercial processes based on chromia catalysts have been operational for over 60 years. During the Second World War chromia/alumina catalysts were used for the dehydrogenation of butane to give butenes, which were then dimerized and hydrogenated to produce high-octane aviation fuel. As was so often the case in the early years of the twentieth century, the dehydrogenation process using a chromia/alumina catalyst was first developed and commercialized in Germany at Leuna. There was also independent development by UOP (then Universal Oil Products) in the United States, together with ICI (Imperial Chemical Industries) in the UK. The first UOP-designed plant came on stream in Billingham, in the North-East of England, in 1940 [3]. Houdry made a significant process development by using a less than atmospheric pressure system, a highly innovative move in an industry where high pressure is the norm. In this way it was possible to achieve a higher per-pass conversion and, indeed, typical operating conditions for current commercial reactors are 0.3–0.5 atm and 823–923 K [4]. A very good review concerning the process chemistry is that of Bhasin [5]. In this chapter we will examine both chromia and vanadia, which although not used commercially, has been thoroughly examined for these reactions. The oxide surfaces have been subject to analysis by a wide range of techniques pre-, during and post-reaction.

Metal Oxide Catalysis. Edited by S. David Jackson and Justin S. J. Hargreaves
Copyright © 2009 WILEY-VCH Verlag GmbH & Co. KGaA, Weinheim
ISBN: 978-3-527-31815-5

15.2
Commercial LPG Dehydrogenation Process

The deactivation of chromia catalysts during the dehydrogenation of light alkanes, due to the deposition of carbonaceous species onto the catalyst surface, is well known. Regeneration of the catalyst is periodic and by oxidation. The designs of commercial reactors for this process take different approaches both to managing the deactivation and to the regeneration. Of specific interest in the present context is the following system that uses a chromium oxide catalyst.

15.3
Lummus/Houdry CATOFIN® Process [6]

The CATOFIN process technology is currently owned by Süd-Chemie and is offered for license by ABB Lummus. The CATOFIN process converts propane to propene over a fixed-bed chromia–alumina catalyst [7]. The process takes place in a series of fixed-bed reactors that operate on a cyclic basis. In one complete cycle, hydrocarbon is dehydrogenated and the reactor is purged with steam then subjected to air to reheat the catalyst and burn off coke, which is deposited during the reaction cycle. These steps are followed by an evacuation and reduction and then another cycle is begun. The cycle time is typically 15–25 minutes. In the classic process, the regenerating (heating) gas is fed in the same direction as the process gas, which leads initially to a higher temperature at the feed than the exit. With time on stream, a temperature front will move through the bed as the front of the bed is cooled by the endothermic reaction. Operation is generally at 823–873 K and ~0.3 atm. CATOFIN is also used commercially for isobutane to isobutene. The reaction conditions are usually 923 K and 0.5 atm. The overall operation is similar to that described for propane to propene, with conversion in region of ~50% per pass and selectivity for isobutane to isobutene of greater than 90%; the selectivity of propane to propene is >86% [8]. A catalyst formulation taken from a recent patent [9] has the following make-up, ~20% Cr, ~1% Zr as ZrO_2, ~0.75% Mg as MgO, supported on alumina with an alkali metal oxide promoter such as Na_2O at ~0.75%.

15.4
Chromia

A wide range of techniques has been used on both fresh and used catalysts to characterize the nature of the oxide surface, for example X-ray powder diffraction (XRD) [10–12], UV-Visible diffuse reflectance spectroscopy (UV-Vis DRS) [11, 12], Raman spectroscopy [10, 11, 14–17], X-ray photoelectron spectroscopy (XPS) [11, 12, 18], electron paramagnetic resonance (EPR) [12, 19], infrared spectroscopy [10, 20] and temperature programmed reduction (TPR) [16, 21]. Given the number of publications we will not review them all but will summarize the main conclusions from this body of work.

Chromium can exist in the (II), (III), (V) and (VI) oxidation states; however, in an as-prepared catalyst the (III) and (VI) oxidation states are the predominant species. At low loadings the +6 state is favored on alumina. On zirconia however Cr(V) was found to be ~50% of the chromium present [22]. As loading increases the surface species change from monochromate to polychromate to α-Cr_2O_3. Typically catalysts are prepared by impregnation of alumina, although zirconia is also used, from a variety of chromium-containing solutions such as ammonium dichromate, chromic acid and chromium nitrate. After calcination in air at 823 K, UV-visible analysis of a 6% w/w chrome oxide/γ-alumina revealed the presence of Cr(III) with bands at 280 nm ($^4A_{2g} \rightarrow \,^4T_{1g}[P]$), 354 nm ($^4A_{2g} \rightarrow \,^4T_{1g}[F]$), 454 nm ($^4A_{2g} \rightarrow \,^2T_{2g}$), 590 nm ($^4A_{2g} \rightarrow \,^4T_{2g}$) and 706 nm ($^4A_{2g} \rightarrow \,^2E_g, \,^2T_{1g}$). XPS analysis indicated that the predominant chromium species was Cr(III) with a small amount of Cr(VI) also detected. This was confirmed by wet chemical analysis of a typical chromium oxide supported on alumina catalyst which gave <10% Cr(VI) [23]. XRD analysis revealed that the only detectable chromium species was Cr_2O_3. Usually no XRD detectable phase is present until the loading is greater than monolayer coverage, typically ~7 wt% for alumina supported catalysts. Raman spectroscopic analysis [13], however, reveals the complexity of the chromia species on the surface (Figure 15.1). Analysis of the spectra gave approximate values for the various components as 14% monomer/dimer, 50% polymer, 7% crystalline, with the balance found in mixed bands.

To obtain the active catalyst the chromia can be reduced with hydrogen, alkane or carbon monoxide [12, 24, 25]; as reported above the commercial process uses a hydrogen reduction. TPR analyses reveal that reduction starts at 543–553 K (T_{int}) increases up to 693 K (T_{max}) and then decreases, with the reduction finishing by

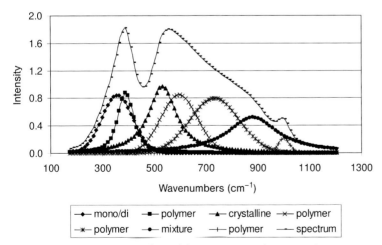

Figure 15.1 Raman spectrum obtained from a 6% CrO_x/alumina catalyst.

723–743 K (T_{fin}) [9, 26]. Burwell and coworkers [27] studied the exchange reaction between alkanes and deuterium over chromia and found that no activity was present if the catalyst had not been activated above 573 K, and maximum activity was observed with samples that had been activated at 743 K, confirming the need for reduction to activate chromia samples.

The active species was initially suggested to be Cr(II) by Sachtler and coworkers [28]. In more recent studies, Cr(III) as the active species has gained prominence [12, 24, 25, 29]. However, it has not been possible to definitively state that Cr(III) is the active site. Other groups have suggested a Cr(II)/Cr(III) pair [18]. In general spectroscopic analyses favor Cr(III), but quantification of the reduction and re-adsorption/absorption of oxygen suggests that hydrogen reduction reduces Cr_2O_3 to CrO [26].

The presence of residual hydrogen retained by a chromia catalyst after reduction was shown by pulsing aliquots of deutero-propane (C_3D_8 or [^2H]C_3H_8) over a hydrogen-reduced catalyst and aliquots of propane over a deuterium-reduced ($^2H_2$ or D_2) catalyst. In both cases isotope exchange occurred such that deuterium incorporation was observed in the propane for a deuterium-reduced catalyst and hydrogen incorporation was observed in the deutero-propane for a hydrogen-reduced catalyst [30].

The rate determining step for propane dehydrogenation over chromia was found to be the loss of the second hydrogen:

$$C_3H_7 \text{ (ads)} \rightarrow C_3H_6 \text{ (ads)} + H \text{ (ads)}$$

with an activation energy for propene formation of 72 ± 7 kJ mol^{-1}.

At 823 K and 16,800 GHSV (gas hourly space velocity) a chromia/alumina catalyst gives the following product distribution for propane dehydrogenation; 1% methane, 1% ethane, 1% ethene, 39% propene and 54% unreacted propane (S.D. Jackson, unpublished results), showing nearly equilibrium conversion and high selectivity. For butane dehydrogenation at 873 K and 14,500 GHSV the product distribution is 70% unreacted butane, 5% trans-2-butene, 6% 1-butene, 4% cis-2-butene, 1.5% butadiene and ~1% C_1–C_3. For butane dehydrogenation, 1-butene is the primary product [31].

Both systems deposit carbon, and when run for extended cycles under propane dehydrogenation polynuclear aromatics form. A Raman spectrum from a 1% chromia/alumina coked catalyst is shown in Figure 15.2. The main peak appears at 1592 cm^{-1} with a shoulder at 1550 cm^{-1}, also broad features are observable at 1170 cm^{-1} and 1380 cm^{-1}. With the exception of the shoulder at 1550 cm^{-1} the Raman spectrum is characteristic of polynuclear, aromatic carbon. Espinat and coworkers [32] characterized a wide variety of carbon compounds including coke on a series of mono- and bi-metallic catalysts and assigned the Raman bands found from 1300 to 1700 cm^{-1}. The D (1350 cm^{-1}) and G (1600 cm^{-1}) bands can be used to identify the type of coke deposited on the catalyst. For the spectrum shown in Figure 15.2, the D band is half the intensity of the G band. This ratio indicates pre-graphitic coke, which is composed of graphite with crystal defects

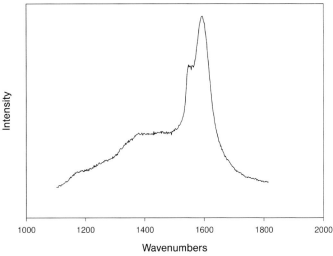

Figure 15.2 UV Raman spectrum of coke deposited on 1% Cr/Al_2O_3 catalyst used for propane dehydrogenation.

or pre-graphitic paracrystals [32] The shoulder at 1550 cm^{-1} is characteristic of conjugated olefins or polyenes.

Confirmation of this interpretation of the Raman spectrum came from analysis of yellow oil produced during extended use of a catalyst. The analysis revealed the presence of pyrene ($C_{16}H_{10}$), fluoranthene ($C_{16}H_{10}$), benzanthracene ($C_{18}H_{12}$) and perylene ($C_{20}H_{12}$) among other compounds. The alumina support principally produced perylene in the absence of chromia but with chromia a wider range of polynuclear aromatics were detected [26].

Studies using isotopically labeled propane [30] have shown that the route to the coke deposits during propane dehydrogenation goes through a C_1 species, probably CH(a). However, before these species convert to coke there is a period when they are labile and can be hydrogenated off the surface as methane [26, 30]. This was shown using isotopically labeled propane [26] and by using a mixed continuous/pulse reaction system where a chromia/alumina catalyst was reduced and subjected to a continuous flow of propane for 10 min at 873 K followed by two pulses of propane in helium [20]. With both pulses the H:C ratio was ~3.3 much higher than the inlet 2.7; also the mass balance was >100%. However the propene : propane ratio was what would have been expected from a continuous flow experiment after 10 min. Clearly carbonaceous material retained by the catalyst was being desorbed into the pulse as methane. Therefore under continuous flow conditions there is a reservoir of C_1 fragments on the surface of the catalyst that is continually in flux between removal as methane and consolidation on the surface to generate higher molecular weight species. A study by Krause and coworkers [33] showed that if the chromia is not pre-reduced the reduction of chromates by propane may proceed through the formation of isopropoxide species, which react

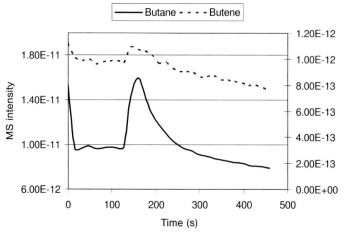

Figure 15.3 Desorption of butane and butene at room temperature in a flow of 2% O_2/Ar from a 6% chromia/alumina catalyst after butane dehydrogenation at 873 K.

further first to adsorbed acetone and then to formates and acetates. They also found that dehydrogenation of propane resulted in sequential formation of aliphatic, unsaturated/aromatic and graphite-like hydrocarbon deposits, which deactivated the catalyst.

The situation with butane is similar but with significant differences. The formation of coke on the surface is very similar to that found with propane; however, Edussuriya and coworkers (unpublished results) found that although the majority of the catalyst deactivation was due to polynuclear aromatics or other forms of coke strongly adsorbed reaction intermediates were also detected. These species can be desorbed at room temperature by treating with a low concentration O_2 in Ar mix. Figure 15.3 shows the desorption of butane. The catalyst had been catalyzing the dehydrogenation reaction for 2 h at 873 K when the gas stream was switched from butane to argon. Online mass spectrometry indicated that all hydrocarbon components were rapidly swept out of the reactor at 873 K and no other gases apart from the argon carrier were detected during the cool-down period. However, when 2% O_2/Ar was passed over the sample at room temperature, butane, butenes and butadiene were evolved. This desorption process can be viewed as an oxidative displacement, resulting in recombination of reaction intermediates (hydrogen and alkyl, alkenyl and alkadienyl) from the surface of reduced CrO_x species. Removal of these species did not regenerate the catalyst. The polynuclear aromatics were only removed when the catalyst was subjected to a full regeneration involving high-temperature oxidation. Only after this treatment did the catalyst regain its activity.

15.5
Vanadia

Although vanadia catalysts have not been used commercially there is a significant literature concerning the dehydrogenation reaction over supported vanadia systems. The structure of supported vanadia catalysts under redox reaction conditions is directly related to the catalytic performance. Vanadia catalysts are usually reduced to some extent during redox reactions, and the reduced vanadia species have been proposed as the active sites [20, 34, 35]. Therefore, information on the valence state and molecular structure of the reduced vanadia catalysts is of great interest. A number of techniques have been applied to investigate the reduction of supported vanadia catalysts, such as TPR [36–38], X-ray photoelectron spectroscopy (XPS) [35], electron spin resonance (ESR) [39], UV-Vis DRS [40–45], X-ray absorption fine structure spectroscopy (XAFS) [46] and Raman spectroscopy [37, 47–50]. Most of these techniques give information only on the oxidation state of vanadia species. Although Raman spectroscopy is a powerful tool for characterization of the molecular structure of supported vanadia [34, 42, 51], it has been very difficult to detect reduced supported vanadia species with conventional (visible) Raman measurements [37, 47–52]. A widely accepted explanation for this phenomenon is that the Raman cross-section of reduced vanadium oxide species is very small or near zero [53]. Thus, it remains challenging to use Raman spectroscopy for obtaining information on the molecular state of reduced vanadia species.

It is notable that most Raman studies of supported VO_x catalysts were carried out using a single excitation wavelength in the visible region (488, 514 or 532 nm) [34, 46, 47, 54]. However, several recent investigations on supported transition metal oxides [55–58], including vanadium oxides [55, 56] under ambient conditions, using both UV and visible wavelength Raman excitations suggest that more complete and sometimes new structural information of supported metal oxides can be achieved by using multiple excitation wavelengths. The reason lies in the strong electronic absorptions in the UV and visible wavelength regions exhibited by most transition metal oxides, which make it possible to measure resonance-enhanced Raman spectra. Under circumstances where supported VO_x species are present in a distribution of cluster sizes or coordination geometries, it is likely that these species also possess a corresponding distribution of electronic absorption wavelengths. Excitation of Raman spectra within the absorption region will produce resonance-enhanced spectra from the subset of VO_x species with absorptions at the excitation wavelength. By measuring the Raman spectra at several wavelengths, more information can be obtained about the various VO_x species in the distribution. Moreover, when UV excitation is employed, even Raman spectra from supported VO_x at low loadings (<1 wt%) on oxides having strong fluorescence are possible because of the avoidance of fluorescence and enhanced sensitivity [59, 60]. In addition, the decreased self-absorption effects in the UV region indicated by *in situ* UV-Vis DRS studies of reduced VO_x and CrO_x [40–44, 61, 62] suggests

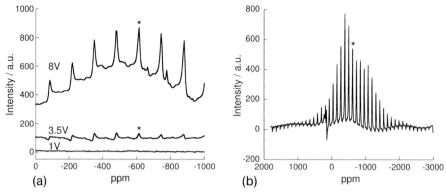

Figure 15.4 51-V MAS NMR spectra. (a) As-prepared catalysts, from top 8, 3.5 and 1 V; (b) 8 V after regeneration at 873 K. * indicates an isotropic resonance. In spectrum (b) the peaks to negative intensity correspond to Al atoms in the support.

that UV Raman spectroscopy may be capable of detecting reduced supported metal oxides.

Studies on alumina supported catalysts with a range of vanadia weight loadings revealed [63, 64] that isolated VO_x species dominated at surface densities below 1 V/nm^2; polyvanadates coexisted with monovanadates at surface densities between 1.2–4.4 V/nm^2; and V_2O_5 formed at a surface density higher than 4.4 V/nm^2. The vanadia densities of the catalysts were 1.1 V/nm^2 for the sample with the 1% loading, 3.7 V/nm^2 for the 3.5% loading and 10.4 V/nm^2 for the 8% loading. Only the XRD pattern for the catalyst with the 8% loading exhibited additional characteristic peaks of crystalline V_2O_5 at 2θ values 26.3 and 34.6 [65, 66]. However, analysis by ^{51}V NMR revealed V_2O_5 on the 3.5% vanadium as well as on the 8% (Gladden and McGregor, unpublished results).

Figure 15.4a shows simple pulse-acquire ^{51}V MAS NMR spectra for catalysts with vanadium loadings of 1, 3.5 and 8 wt%. This simple 1-D spectroscopy identifies a ^{51}V resonance only in the 3.5 and 8 wt% catalysts. The spectra are dominated by spinning sidebands, with the isotropic chemical shift (denoted by *) at −612 ppm being indicative of V^{5+} in a V_2O_5 environment. The broad background upon which these peaks are superimposed in the case of the 8 wt% sample is characteristic of V^{5+} in more amorphous environments. The spectra clearly suggest that the population of V_2O_5-like species in the 8 wt% catalyst is larger than in the 3.5 wt% catalyst, consistent with the Raman data. Alongside characterizing the ^{51}V environment, NMR spectroscopy of the ^{27}Al species is also of interest. Figure 15.5a shows a simple pulse-acquire ^{27}Al spectrum of an as-prepared 8 wt% catalyst. Two broad resonances are observed; the peak at ~6 ppm is characteristic of sixfold, octahedrally coordinated Al, while Al in four-fold and five-fold coordinated sites contribute to the broader resonance in the range 25–70 ppm. Interpretation of the broader

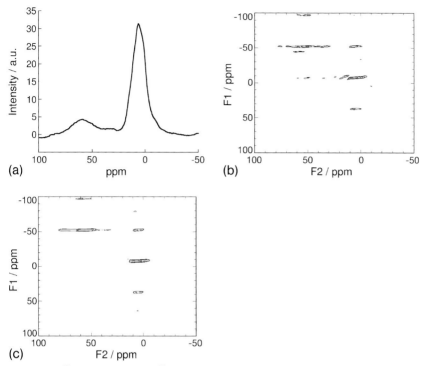

Figure 15.5 ^{27}Al NMR spectra. (a) ^{27}Al MAS NMR spectrum of as-prepared 8%V/alumina catalyst; (b) ^{27}Al 3Q MAS NMR spectrum of as-prepared 8%V/alumina; (c) ^{27}Al 3Q-MAS NMR spectrum of 8%V/alumina after regeneration at 873 K.

resonance is made significantly easier – although the NMR technique itself is more challenging – by using multiple quantum NMR spectroscopy. A triple-quantum spectrum from the 8 wt% catalyst is shown in Figure 15.5b. The projection of such spectra in the F1 dimension yields the isotropic peaks, while the anisotropic lineshapes (corresponding to a standard MAS spectrum) are displayed by the projection in the F2 dimension. The resolution of the isotropic resonance provided by the triple-quantum MAS experiment unambiguously confirms the presence of three distinct ^{27}Al environments, assigned to 4-, 5- and 6-coordinate Al.

Studies reveal that reduction takes place over a broad range of temperatures and the onset of reduction follows the weight loading with 8% < 3.5% < 1% but reduction is complete for each sample by ~873 K. This difference in onset of reduction may be attributed to different reducibilities of vanadia species, monovanadates, polyvanadates and V_2O_5 coexisting on the catalyst's surface. The weight loss at the reduction step increases with vanadia loading: when these weights are translated into oxygen loss we see that there is a 1:1 relationship between vanadium atoms and oxygen atom loss (Table 15.1). The same trend in mass loss is observed during

Table 15.1 Extent of oxygen removal during reduction of vanadia/alumina catalysts.

Catalyst (V/nm^2)	V atoms in 1 g of catalyst ($\times 10^{20}$)	O atoms removed from 1 g of catalyst ($\times 10^{20}$)	V:O ratio
1.1	1.18	1.13	1.0:1
3.7	4.13	4.14	1.0:1
10.4	9.45	7.90	1.2:1

Tapered Element Oscillating Microbalance (TEOM) studies of catalyst reduction (Gladden and McGregor, unpublished results). These studies also yield an average 1:1 relationship between vanadium atoms and oxygen atom loss.

This would suggest a change of 2 in the vanadium oxidation state, so if the vanadium were in a +5 oxidation state before reduction it would be converted to +3 after reduction. This was confirmed by *in situ* UV-Vis DRS. Before reduction the catalysts were in a V^{5+} oxidation state with typical charge transfer bands at 267 nm for a 1% loading, 290 nm for a 3.5% loading and 357 nm for an 8% loading. This change in position of the charge transfer band is indicative of the increasing polymeric nature of the vanadia [67]. Over the course of the reduction the spectrum changes to give spectra typical for a V^{3+} state with d–d transitions around 575 nm and 625 nm for 3.5%V and 8%V. However, the spectrum of the 1% sample is more indicative of V^{4+}, with a single band at 585 nm [31]. ^{51}V NMR spectroscopy revealed a total loss in signal after reduction, indicating the presence of V(III) or V(IV) species (Gladden and McGregor, unpublished results).

Using oxygen chemisorption at 293 K and 873 K the average oxidation state of the 1% VO$_x$/alumina was calculated as 3.8, whereas the average oxidation state for the 3.5% VO$_x$/alumina was calculated as 2.6 [31]. Both of these figures are in excellent agreement with the conclusions from the UV-Vis DRS.

Bulk re-oxidation data in conjunction with thermogravimetric analysis (TGA) and TEOM results indicated that the ease of replacing oxygen that had been removed during reduction was 8% > 3.5% > 1% [31], (Gladden and McGregor, unpublished results). This difference was also seen in the UV-Vis DRS and in TEOM data. TEOM data provide information on the quantity of oxygen taken up by the catalysts and on their rate of re-oxidation. The rate of catalyst re-oxidation was seen to be greater for the 8 wt% catalyst than for materials of lower vanadium loading (Gladden and McGregor, unpublished results). Once reduced, the 3.5%V sample did not regain its original state even after high temperature oxidation; however, the reduced 8%V sample did convert to a species that was similar to the original oxidation state.

Supported VO$_x$ catalysts show good catalytic performance in both oxidation and reduction reactions; for example, they are among the most active and selective simple metal oxides for dehydrogenation and oxidative dehydrogenation of alkanes. The possibility of using oxide systems other than chromia is being explored in an effort to make the alkane–alkene conversion process more selective and stable. Supported VO_x catalysts have been extensively studied for the oxidative dehydrogenation of light alkanes [40, 54, 68–72]. They have also been explored in the dehydrogenation of light alkanes including propane and butane [35, 39, 73, 74] and have shown promise to be both active and selective for the dehydrogenation reactions. However deactivation of VO_x catalysts by coke species is also observed with time-on-stream in butane dehydrogenation reactions. Owing to the limited number of studies of supported VO_x for alkane dehydrogenation, the mechanism of deactivation/coke deposition and also the structure of supported VO_x species under dehydrogenation conditions are unclear. Conventional visible Raman spectroscopy usually cannot provide meaningful spectra for a catalytic system that produces carbonaceous deposits because of strong fluorescence interference. These two issues were addressed in an investigation of butane dehydrogenation reaction on $V/\theta\text{-}Al_2O_3$ catalysts with various surface VO_x densities (0.03–14.2 V/nm^2) via *in situ* UV Raman spectroscopy [75].

$V/\theta\text{-}Al_2O_3$ catalysts with surface VO_x density from 0.03 to 14.2 V/nm^2 were tested for butane dehydrogenation under low partial pressures of butane at 873 K, and the activity and selectivity data on these different catalysts were compared [75]. In essence, the initial dehydrogenation activity increases as a function of surface VO_x density. However, the activity decreases significantly with prolonged reaction time for samples with VO_x density higher than 1.2 V/nm^2, apparently due to the formation of surface coke deposits and blocking of surface sites during the reaction. Interestingly, the dehydrogenation activity remains nearly unchanged on the catalysts with surface VO_x density no higher than 1.2 V/nm^2. However, at higher butane pressures, although the behavior of the 1.2 V/nm^2 sample is unchanged, catalysts with higher V/nm^2 densities show enhanced activity (Table 15.2).

In the *in situ* UV Raman study [75], $V/\theta\text{-}Al_2O_3$ samples (0.03 V, 1.2 V, 4.4 V and 14.2 V) that contain surface VO_x species with different structures were treated in dilute butane flow at different temperatures and examined by Raman spectra measurements at each temperature. Taking the 1.2 V sample as an example, the Raman spectra collected during butane dehydrogenation at different temperatures are shown in Figure 15.6.

Table 15.2 Activities and turnover frequencies for butane dehydrogenation at 1 bar pressure and 873 K with time-on-stream.

Time-on-stream (min)	Rate ($\mu mole\,g^{-1}\,s^{-1}$)			Turnover frequency (s^{-1})		
	1.1 V/nm^2	3.7 V/nm^2	10.4 V/nm^2	1.1 V/nm^2	3.7 V/nm^2	10.4 V/nm^2
15	1.1	20.3	5.8	0.051	0.269	0.037
30	1.0	5.8	3.7	0.047	0.077	0.023
45	1.0	4.7	2.9	0.044	0.063	0.018
60	1.0	4.2	2.5	0.044	0.056	0.016
105	0.9	3.5		0.041	0.046	–

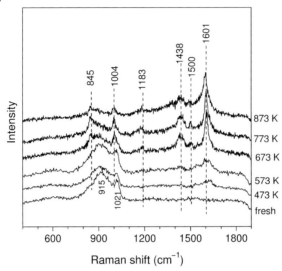

Figure 15.6 UV Raman spectra of butane dehydrogenation over oxidized 1.2 V at different temperatures.

At temperatures below 673 K, a weak Raman band at 1620 cm^{-1}, assigned to C=C stretching in polyalkenes, is observed together with the two bands at 1021 and 915 cm^{-1} due to the V=O and V—O—Al modes of surface VO$_x$ species, respectively. After butane dehydrogenation at 673 and 773 K, an intense band at 1601 cm^{-1} due to polyaromatic hydrocarbons develops. Simultaneously, bands at 1500, 1438, 1183, 1004 and 845 cm^{-1} are also observed. Meanwhile, Raman bands at 1021 and 915 cm^{-1} from surface VO$_x$ species are no longer observable, owing to a combination of the reduction of VO$_x$ species [43, 45, 47, 65, 71] and strong optical absorption by surface coke species. The band at 1500 cm^{-1} is usually assigned to conjugated polyalkenes or cyclopentadienyl species [76, 77]. The band at 1438 cm^{-1} is due to the bending mode of CH$_3$/CH$_2$ or C—H in aromatic rings. C—H bending in aromatics usually appears near 1180 cm^{-1}. The two sharp bands at 1004 and 845 cm^{-1} are rarely reported for coke species, and thus one might ascribe them to surface VO$_x$ species. However, the absence of an isotope shift on an ^{18}O-exchanged V/θ-Al$_2$O$_3$ sample confirms that these bands are due to surface coke deposits. As the dehydrogenation temperature is increased to 873 K, the intensity of the Raman bands below 1500 cm^{-1} decreases significantly. Raman spectra obtained after the reaction of butane with pre-reduced 1.2%V (not shown) are very similar to those on pre-oxidized 1.2%V. This suggests that the initial valence state of surface VO$_x$ does not affect the nature of coke species and most likely the VO$_x$ species is at least partially reduced under butane dehydrogenation conditions. This conclusion can also be inferred from the similar activity/selectivity results from butane dehydrogenation tests on either oxidized or pre-reduced 1.2%V and also the similar UV-Visible spectra from V/θ-Al$_2$O$_3$ treated in either hydrogen or butane.

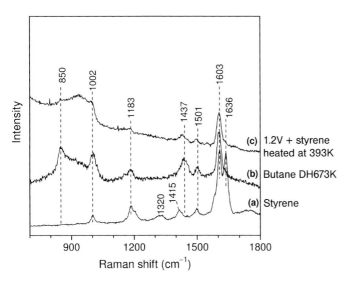

Figure 15.7 Comparison of Raman spectra from (a) styrene, (b) butane dehydrogenation over 1.2 V at 673 K and (c) a mixture of 1.2 V and styrene heated at 393 K.

The set of Raman bands near 850, 1002, 1183, 1437, 1501 and 1603 cm^{-1} is also observed to grow simultaneously on all other V/θ-Al$_2$O$_3$ samples when treated in butane at high temperatures. To aid the assignment of these bands, the adsorption and reaction of different C$_4$ olefins were studied on the V/θ-Al$_2$O$_3$ samples via Raman spectroscopy. It turned out that they all form similar coke species, indicating that they are precursors to coke species in butane dehydrogenation on V/θ-Al$_2$O$_3$ catalysts. It is interesting to find out that 1,3-butadiene adsorption at room temperature results in similar spectra to that from butane dehydrogenation on 1.2 V at 673 K. Considering the low temperature (298 K), a possible chemical change for 1,3-butadiene is its cyclization to form styrene on the surface. A comparison of the spectrum from styrene with that after butane dehydrogenation at 673 K on 1.2%V (Figure 15.7) reveals a very similar set of bands. Also included is the spectrum from a mixture of 1.2%V with styrene heated at 393 K (Figure 15.7c).

Styrene is known to polymerize easily upon heating. Comparison of spectrum c to that reported for polystyrene [78] confirms the identity of spectrum c: the set of Raman bands near 850, 1002, 1183, 1437, 1501 and 1603 cm^{-1} is due to polystyrene. Thus, it appears that polystyrene is a key intermediate in the coke formation process during butane dehydrogenation on the 1.2 V catalyst. Further Raman spectroscopy study of styrene on V/θ-Al$_2$O$_3$ samples heated to 873 K showed that the characteristic Raman bands of polystyrene disappear, owing to its decomposition. This is consistent with catalytic studies of polystyrene degradation over solid acid catalysts [79]. An interesting new observation is that although the Raman spectrum seems to indicate that the catalyst surface is free of coke species after polystyrene decomposition, the follow-up TPO (Temperature Programmed Oxidation) shows the evolution of a considerable amount of CO_2. This suggests that the absence of typical Raman bands (1000–1650 cm^{-1} range) due to coke species does not necessarily mean the catalyst is coke free. Investigation of the catalytic surface with multi-wavelength Raman spectroscopy should be helpful in determining the chemical nature of this surface carbon. Microanalysis of a used 1.2 V catalyst for carbon and hydrogen gave a C:H value of 0.9 as per polystyrene monomer in agreement with the Raman spectra [31].

One important general characteristic of coke is its topology, which can be assessed by the intensity ratio of the band at around 1600 cm^{-1} (G band) to the band at around 1400 cm^{-1} (D band) [76, 80]. UV Raman spectra from a series of polyaromatic compounds [76, 81] show that the intensity in the spectral range 1600–1650 cm^{-1} is significantly higher than that in the region 1300 to 1450 cm^{-1} for coke species with a 2-D, sheet-like topology. By contrast, the intensity is more nearly equal in these two spectral regions for coke species with chain-like topologies. The topology of coke species formed from butane dehydrogenation on the various V/θ-Al_2O_3 catalysts at 873 K is compared by plotting the intensity ratio of I_G to I_D as a function of surface VO_x density. It is shown that the coke species are more 2-D, sheet-like from butane dehydrogenation on V/θ-Al_2O_3 with high surface VO_x density (>1.2 V/nm^2) while more 1-D-like on V/θ-Al_2O_3 with lower surface VO_x density (≤1.2 V/nm^2). The 2-D coke species can, presumably, reorganize into pre-graphitic entities that have been thought to be the kind of coke that causes the deactivation of dehydrogenation catalysts [33, 82, 83]. A study of coke deposited on a 1 wt% V/Al_2O_3 catalyst, in which the deactivation was studied as a function of time-on-stream (TOS) using ^{13}C CP MAS (cross-polarization magic angle spinning) NMR reveals a significant increase in the aromatic peak at 130 ppm up to 3 h, then a gradual decrease in this peak [84]. The initial increase in the peak intensity was explained by an initial coke build-up, whereas the decrease in intensity was due to formation of very slowly relaxing, NMR "invisible", highly polyaromatic coke, which formed under prolonged exposure to the reaction conditions. This interpretation is consistent with the constant decrease in T_2 of adsorbed pentane observed, which also suggests an increasing aromaticity of coke with increasing TOS. Further, a monotonic decrease in pentane self-diffusion coefficient as a function of TOS was observed; these data are consistent with restriction in molecular motion resulting from pore blockage due to coke deposition as deactivation proceeds. However, in another study it was found that the majority of the catalyst deactivation in the early stages of reaction was not due to polynuclear aromatics or other forms of coke but to strongly adsorbed reaction intermediates [31]. These species can be desorbed at room temperature by treating with a low concentration O_2 in Ar mix. Figure 15.8 shows desorption of butane from a series of VO_x/alumina catalysts.

The catalyst had been running under butane for 2 h at 873 K when the gas stream was switched from butane to argon. Online mass spectrometry indicated that all hydrocarbon components were rapidly swept out of the reactor at 873 K and no

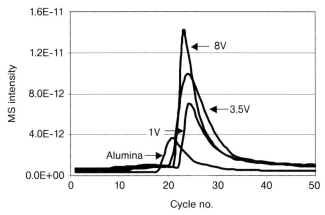

Figure 15.8 Desorption of butane at room temperature in a flow of 2% O_2/Ar from vanadia/alumina catalysts after butane dehydrogenation at 873 K and 1 bar.

other gases apart from the argon carrier were detected during the cool-down period. However when 2% O_2/Ar was passed over the sample at room temperature butane, butenes and butadiene were evolved. This desorption process can be viewed as an oxidative displacement, resulting in recombination of reaction intermediates (hydrogen and alkyl, alkenyl and alkadienyl) from the surface of reduced VO_x species. The polynuclear aromatics were only removed when the catalyst was subjected to a full regeneration involving high temperature oxidation [31].

The regeneration of samples of $V/\theta\text{-}Al_2O_3$ coked from low partial pressure butane dehydrogenation via oxidation was also followed by UV Raman spectroscopy as well as gas chromatography (GC). The Raman spectroscopy indicates that the structure of the VO_x species can be restored by oxidation of the coke deposits up to 873 K. UV-Vis DRS, however, suggests that only high loaded samples are fully restored. ^{51}V MAS NMR confirms the re-oxidation of reduced vanadium species and, as such, spectra of the regenerated catalyst can be acquired (Gladden and McGregor, unpublished results). Figure 15.4b shows the spectrum of the 8 wt% catalyst after reaction at 973 K and regeneration at 873 K (note the peaks to negative intensity correspond to Al atoms in the support). By comparison with Figure 15.4a it can be seen that the spectrum of the regenerated catalysts indicates the presence of a greater quantity of ordered V_2O_5-like species than were present in the as-prepared catalyst. However, one must take care as each technique has its "blind spots" and it is possible that it is the inability of each of the techniques to "see" all of the vanadia species that leads to apparent differences. At low butane partial pressure, TPO quantification results showed that the amount of coke formed in butane dehydrogenation follows the sequence: polymeric VO_x > monomeric VO_x > V_2O_5, Al_2O_3. However, at 1 bar butane pressure the order is changed slightly with V_2O_5 > polymeric VO_x > monomeric VO_x > Al_2O_3. After regeneration the surface area of an 8%V catalyst was reduced by 50%, a change assigned to transformations in the nature of the support. A similar effect was observed after propane dehydrogenation over a vanadia/alumina catalyst [73], where a conversion of the γ-alumina support to α-alumina was confirmed by XRD. ^{27}Al triple-quantum (3Q) MAS NMR studies have also demonstrated that the nature of the support undergoes a transformation. Figure 15.5c shows the ^{27}Al 3Q MAS NMR spectrum of 8V after reaction and regeneration. Comparison with Figure 15.5b shows the loss of an isotropic resonance; this resonance was assigned to 5-coordinate Al and the loss of this species suggests a transformation towards a more crystalline state in agreement with the reduced surface area.

It is rare for a structure–function relationship to be shown unambiguously in heterogeneous catalysis. However, a structure–coke relationship has been established in the dehydrogenation of butane on $VO_x/\theta\text{-}Al_2O_3$ catalysts by combined reactivity, UV-Vis DRS, NMR, TEOM and UV Raman investigations. In this case, both the nature and amount of coke show a dependence on the structure of the $VO_x/\theta\text{-}Al_2O_3$ catalysts.

15.6
Conclusions

In this chapter we have reviewed the use of chromia and vanadia in alkane dehydrogenation. In both systems there is a rich and complex chemistry that is still not yet fully understood, although considerable advances have been made over the last few years. Both patent and academic literature are active and it can be expected to be so for many years to come.

References

1 Frey, F.E. and Huppke, W.F. (1933) *Industrial and Engineering Chemistry*, **25**, 54.
2 Kearby, K.K. (1955) *Catalysis*, Vol. 3 (ed. P.H. Emmet), Reinhold, New York, p. 453.
3 Hornaday, G.F., Ferrell, F.M. and Mills, G.A. (1961) *Advances in Petroleum Chemistry and Refining*, Vol. 4, Interscience, Paris, p. 451.
4 Waddams, A.L. (1978) *Chemicals From Petroleum*, 4th edn, Gulf Publishing Company, Houston, p. 1980.
5 Bhasin, M.M., McCain, J.H., Vora, B.V., Imai, T. and Pujado, P.R. (2001) *Applied Catalysis A: General*, **221**, 397.
6 http://www.sud-chemie.com/scmcms/web/page_en_4504.htm (May 2008)
7 http://www.sud-chemie.com/scmcms/web/page_en_4506.htm (May 2008)
8 http://www.cbi.com/lummus/process-technology/pdfs/catofindehydrogenation.pdf (May 2008)
9 Fridman, V. and Rokicki, A. (2005) WO2,005,040,075, assigned to Sud Chemie.
10 Zaki, M.I., Fouad, N.E., Leyrer, J. and Knozinger, H. (1986) *Applied Catalysis*, **21**, 359.
11 Cavani, F., Koutyrev, M., Trifirò, F., Bartolini, A., Ghisletti, D., Iezzi, R., Santucci, A. and Del Piero, G. (1996) *Journal of Catalysis*, **158**, 236.
12 Puurunen, R.L. and Weckhuysen, B.M. (2002) *Journal of Catalysis*, **210**, 418.
13 Jackson, S.D., Matheson, I.M., Naeye, M.-L., Stair, P.C., Sullivan, V.S., Watson, S.R. and Webb, G. (2000) *Studies in Surface Science and Catalysis*, Vol. 130 (eds A. Corma, F.V. Melo, S. Mendioroz, J.L.G. Fierro), Elsevier, Amsterdam, p. 2213.

14 Mentasty, L.R., Gorriz, O.F. and Cadùs, L.E. (2001) *Industrial and Engineering Chemistry Research*, **40**, 136.
15 Vuurman, M.A., Wachs, I.E., Stufkens, D.J. and Oskam, A. (1993) *Journal of Molecular Catalysis*, **80**, 209.
16 Grzybowska, B., Sloczyñski, J., Grabowski, R., Wcislo, K., Kozlowska, A., Stoch, J. and Zieliñski, J. (1998) *Journal of Catalysis*, **178**, 687.
17 Sullivan, V.S., Jackson, S.D. and Stair, P.C. (2005) *Journal of Physical Chemistry B*, **109**, 352.
18 Rahman, A., Mohamed, M.H., Ahmed, M. and Aitani, A.M. (1995) *Applied Catalysis A: General*, **121**, 203.
19 Ashmawy, F.M. and McAuliffe, C.A. (1984) *Journal of the Chemical Society – Faraday Transactions I*, **80**, 1985.
20 De Rossi, S., Ferraris, G., Fremiotti, S., Garrone, E., Ghiotti, G., Campa, M.C. and Indovina, V. (1994) *Journal of Catalysis*, **148**, 36.
21 Kanervo, J.M. and Krause, A.O.I. (2002) *Journal of Catalysis*, **207**, 57.
22 De Rossi, S., Casaletto, M.P., Ferraris, G., Cimino, A. and Minelli, G. (1998) *Applied Catalysis A: General*, **167**, 257.
23 Cutrufello, M.G., De Rossi, S., Ferino, I., Monaci, R., Rombia, E. and Solinas, V. (2005) *Thermochimica Acta*, **434**, 62.
24 Weckhuysen, B.M. and Schoonheydt, R.A. (1999) *Catalysis Today*, **51**, 223.
25 Hakuli, A., Kytokivi, A. and Krause, A.O.I. (2000) *Applied Catalysis A: General*, **190**, 219.
26 Jackson, S.D. and Stitt, E.H. (2003) *Current Topics in Catalysis*, **3**, 245.
27 Pass, G., Littlewood, A.B. and Burwell, R.L., Jr (1960) *Journal of the American Chemical Society*, **82**, 6281.
28 van Reijin, L.L., Sachtler, W.M.H., Cossee, P. and Brouwer, D.M. (1965) *Proceedings of the 3rd International Congress on Catalysis*, Vol. 2 (eds W.M.H. Sachtler, G.C.A. Schuit and P. Zwietering), North Holland, Amsterdam, p. 829.
29 Weckhuysen, B.M., Wachs, I.E. and Schoonheydt, R.A. (1996) *Chemical Reviews*, **96**, 3327.
30 Jackson, S.D., Grenfell, J., Matheson, I.M. and Webb, G. (1999) *Reaction Kinetics and the Development of Catalytic Processes*, Studies in Surface Science and Catalysis, **122** (eds G.F. Froment and K.C. Waugh), Elsevier, Amsterdam, pp. 149–55.
31 Jackson, S.D. and Rugmini, S. (2007) *Journal of Catalysis*, **251**, 59.
32 Espinat, D., Dexpert, H., Freund, E., Martino, G., Couzi, M., Lespade, P. and Cruege, F. (1985) *Applied Catalysis*, **16**, 343
33 Airaksinen, S.M.K., Bañares, M.A. and Krause, A.O.I. (2005) *Journal of Catalysis*, **230**, 507.
34 Wachs, I.E. and Weckhuysen, B.M. (1997) *Applied Catalysis A: General*, **157**, 67.
35 Harlin, M.E., Niemi, V.M. and Krause, A.O.I. (2000) *Journal of Catalysis*, **195**, 67.
36 Gao, X., Banares, M.A. and Wachs, I.E. (1999) *Journal of Catalysis*, **188**, 325.
37 Ruitenbeek, M., Van Dillen, A.J., de Groot, F.M.F., Wachs, I.E., Geus, J.W. and Koningsberger, D.C. (2000) *Topics in Catalysis*, **10**, 241.
38 Wachs, I.E. and Chan, S.S. (1984) *Applied Surface Science*, **20**, 181.
39 Harlin, M.E., Niemi, V.M., Krause, A.O.I. and Weckhuysen, B.M. (2001) *Journal of Catalysis*, **203**, 242.
40 Wachs, I.E., Jehng, J.-M., Deo, G., Weckhuysen, B.M., Guliants, V.V., Benziger, J.B. and Sundaresan, S. (1997) *Journal of Catalysis*, **170**, 75.
41 Wachs, I.E., Jehng, J.-M., Deo, G., Weckhuysen, B.M., Guliants, V.V. and Benziger, J.B. (1995) Proceedings of the 5th European workshop meeting on selective oxidation by heterogeneous catalysis. *Catalysis Today*, **32**, 47.
42 Sun, Q., Jehng, J.-M., Hu, H., Herman, R.G., Wachs, I.E. and Klier, K. (1997) *Journal of Catalysis*, **165**, 91.
43 Mul, G., Banares, M.A., Garcia Cortez, G., van der Linden, B., Khatib, S.J. and Moulijn, J.A. (2003) *Physical Chemistry Chemical Physics*, **5**, 4378.
44 Banares, M.A., Cardoso, J.H., Agullo-Rueda, F., Correa-Bueno, J.M. and Fierro, J.L.G. (2000) *Catalysis Letters*, **64**, 191.
45 Christodoulakis, A., Machli, M., Lemonidou, A.A. and Boghosian, S. (2004) *Journal of Catalysis*, **222**, 293.
46 Olthof, B., Khodakov, A., Bell, A.T. and Iglesia, E. (2000) *Journal of Physical Chemistry B*, **104**, 1516.
47 Banares, M.A. and Wachs, I.E. (2002) *Journal of Raman Spectroscopy*, **33**, 359.
48 Li, C. (2003) *Journal of Catalysis*, **216**, 203.

49 Gao, X. and Wachs, I.E. (2000) *Journal of Physical Chemistry B*, **104**, 1261.
50 Xie, S., Iglesia, E. and Bell, A.T. (2000) *Langmuir*, **16**, 7162.
51 Xie, S., Iglesia, E. and Bell, A.T. (2001) *Journal of Physical Chemistry B*, **105**, 5144.
52 Argyle, M.D., Chen, K.D., Iglesia, E. and Bell, A.T. (2005) *Journal of Physical Chemistry B*, **109**, 2414.
53 Dupuis, A.C., Abu Haija, M., Richter, B., Kuhlenbeck, H. and Freund, H.J. (2003) *Surface Science*, **539**, 99.
54 Argyle, M.D., Chen, K.D., Bell, A.T. and Iglesia, E. (2002) *Journal of Catalysis*, **208**, 139.
55 Xiong, G., Li, C., Li, H.Y., Xin, Q. and Feng, Z.C. (2000) *Chemical Communications*, 677.
56 Chua, Y.T., Stair, P.C. and Wachs, I.E. (2001) *Journal of Physical Chemistry B*, **105**, 8600.
57 Xiong, G., Li, C., Feng, Z.C., Ying, P.L., Xin, Q. and Liu, J.K. (1999) *Journal of Catalysis*, **186**, 234.
58 Xiong, G., Feng, Z., Li, J., Yang, Q., Ying, P., Xin, Q. and Li, C. (2000) *Journal of Physical Chemistry B*, **104**, 3581.
59 Li, C. and Stair, P.C. (1996) *11th International Congress on Catalysis – 40th Anniversary*, Studies in Surface Science and Catalysis, 101 (Pt. B) (eds J.W. Hightower, W.N. Delgass, E. Iglesia and A.T. Bell), Elsevier, Amsterdam, p. 881.
60 Stair, P.C. and Li, C. (1997) *Journal of Vacuum Science & Technology. A, Vacuum, Surfaces, and Films*, **15**, 1679.
61 Tian, H., Ross, E.I. and Wachs, I.E. (2006) *Journal of Physical Chemistry B*, **110**, 9593.
62 Catana, G., Rao, R.R., Weckhuysen, B.M., Van Der Voort, P., Vansant, E. and Schoonheydt, R.A. (1998) *Journal of Physical Chemistry B*, **102**, 8005.
63 Wu, Z., Kim, H.S., Stair, P.C., Rugmini, S. and Jackson, S.D. (2005) *Journal of Physical Chemistry B*, **109**, 2793.
64 Jackson, S.D., Rugmini, S., Stair, P.C. and Wu, Z. (2006) *Chemistry-Engineering Journal*, **120**, 127.
65 Kanervo, J.M., Harlin, M.E., Krause, A.O.I. and Banares, M.A. (2003) *Catalysis Today*, **78**, 171.
66 Khodakov, A., Olthof, B., Bell, A.T. and Iglesia, E. (1999) *Journal of Catalysis*, **181**, 205.
67 Arena, F., Frusteri, F., Martra, G., Coluccia, S. and Parmaliana, A. (1997) *Journal of the Chemical Society – Faraday Transactions*, **93**, 3849.
68 Albonetti, S., Cavani, F. and Trifiro, F. (1996) *Catalysis Reviews – Science and Engineering*, **38**, 413.
69 Mamedov, E.A. and Corberan Cortes, V. (1995) *Applied Catalysis, A: General*, **127**, 1.
70 Blasco, T. and Lopez Nieto, J.M. (1997) *Applied Catalysis, A: General*, **157**, 117.
71 Cavani, F. and Trifiro, F. (1995) *Catalysis Today*, **24**, 307.
72 Madeira, L.M. and Portela, M.F. (2002) *Catalysis Reviews – Science and Engineering*, **44**, 247.
73 Jackson, S.D., Lennon, D., Webb, G. and Willis, J. (2001) *Catalyst Deactivation*, Studies in Surface Science and Catalysis, 139, Elsevier, Amsterdam, p. 271.
74 Volpe, M., Tonetto, G. and de Lasa, H., (2004) *Applied Catalysis, A: General*, **272**, 69.
75 Wu, Z. and Stair, P.C. (2006) *Journal of Catalysis*, **237**, 220.
76 Chua, Y.T. and Stair, P.C. (2003) *Journal of Catalysis*, **213**, 39.
77 Baruya, A., Gerrard, D.L. and Maddams, W.F. (1983) *Macromolecules*, **16**, 578
78 Noda, L.K. and Sala, O., (2000) *Spectrochimica Acta Part A – Molecular and Biomolecular Spectroscopy*, **56**, 145.
79 Ukei, H., Hirose, T., Horikawa, S., Takai, Y., Taka, M., Azuma, N. and Ueno, A. (2000) *Catalysis Today*, **62**, 67.
80 Kuba, S. and Knozinger, H. (2002) *Journal of Raman Spectroscopy*, **33**, 325.
81 Asher, S.A. (1993) *Analytical Chemistry*, **65**, 201A.
82 Davis, S.M., Zaera, F. and Somorjai, G.A. (1982) *Journal of Catalysis*, **77**, 439.
83 Tinnemans, S.J., Kox, M.H.F., Nijhuis, T.A., Visser, T. and Weckhuysen, B.M. (2005) *Physical Chemistry Chemical Physics*, **7**, 211.
84 Huang, Z., McGregor, J., Steiner, P., Gladden, L.F., Rugmini, S. and Jackson, S. D. (2007) *Magnetic Resonance Imaging*, **25**, 562.

16
Properties, Synthesis and Applications of Highly Dispersed Metal Oxide Catalysts

Juncheng Hu, Lifang Chen, and Ryan Richards

16.1
Introduction

Metal oxides can be found in the form of single crystals (pure or defective), powder (with large numbers of crystals), polycrystalline samples (crystals with various orientations) or thin films and constitute an important class of materials covering the entire range from metals to insulators. The insulating oxides are made up of the metals from the far left and right sides of the periodic table. Typical examples of insulating oxides include MgO, CaO, Al_2O_3 and SiO_2. The oxides of the metals in the middle of the periodic table (Sc to Zn) make up semiconducting and metallic oxides. Typical examples include ZnO, TiO_2, NiO, Fe_2O_3 and Cr_2O_3. Additionally, transition metal oxides, which include the oxides of Ru, Mo, W, Pt, V and so forth, exhibit a wide-ranging array of properties and phenomena. The diverse structures adopted by metal oxides demonstrate the relationship between structure and properties. Correlation of structure and physical properties of transition metal oxides requires an understanding of the valence electrons that bind the atoms in the solid state [1]. Band theory and ligand field theory have been invoked to explain the electronic properties of transition metal oxides. The important properties of metal oxides that are of interest are magnetic, electrical, dielectric, optical, Lewis acid–base and redox behaviour. The transition metal oxides are of particular interest for applications in catalysis, sensor materials and other potential applications. Indeed, metal oxides are the key components for a variety of catalytic reactions, functioning directly as reactive components or as supports for dispersed active metal species, or as additives or promoters to enhance the rate of catalytic reactions. The importance of metal oxides in the field of catalysis is thus profound. Hitherto, the main interest in heterogeneous catalysis had centered on the kinetics of the reactions rather than on their mechanisms; now it is the catalyst itself and, in particular, its surface, that has become the focus of attention. In considering the action of a catalyst and its surface, two particular factors have been explored. One is the geometrical factor, for which experiments with chemisorption and catalysis on the various faces of single crystals have helped to define the role. The second factor is the role of electrons in the catalytic process, and the surface science of metal oxides has received tremendous attention during recent years. The advent of several advanced techniques to probe surfaces has led to a new understanding of their surface properties and structures. This has helped catalyst scientists tailor the properties of the oxide materials for appropriate applications.

This chapter presents an overview of structures, properties, synthetic methods and applications in catalysis for highly dispersed metal oxides, with special attention given to recent developments in nano-scale metal oxides.

16.2
Properties

In the case of nano-scale materials one of the most influential factors towards their properties is the large number of atoms at the surface. For example, spherical nanoparticles with a diameter of 3 nm have nearly 50% of the atoms or ions on the surface, allowing the possibility of manipulation of bulk properties by surface effects and allowing near-stoichiometric chemical reaction [2]. Delocalization can vary with size when strong chemical bonding is present, and can lead to different chemical and physical properties [3]. The formation of ultra-small particles is facilitated by the refractory nature of most of the metal oxides [2]. For some materials, especially MgO, Al_2O_3, ZrO_2 and TiO_2, their highly ionic nature allows the formation of many stable defect sites, including edges, corners and anion/cation vacancies. MgO and CaO were found to be attractive materials to study because they are highly ionic and have high melting points and it would be expected that samples of very small particle size might be stable and isolable. Furthermore, reactive surface sites on these oxides have been extensively studied for macroscopic and conventionally prepared samples.

Metal oxides prepared via aerogel methods have found various applications, including as detectors for radiation, super insulators, solar concentrators, coatings, glass precursors, catalysts, insecticides and destructive adsorbents. Nanoparticles of crystalline substances have been found to possess about 10^{19} interfaces cm^{-3} and surface areas up to $800 m^2 g^{-1}$. Upon compaction, but without growing the nanocrystals, solids with multitudinous grain boundaries are formed. It has been proposed that further work in the area of consolidated nano-phase materials may lead to ceramics with increased flexibility, less brittleness and perhaps greater strength [2]. It may also be possible to form materials with a large fraction of atoms at grain boundaries, perhaps in unique arrangements.

16.2.1
Structure and Bonding

Characterization of the exact nature of structure and bonding in nanomaterials is particularly difficult because they generally consist of very small crystallites or are

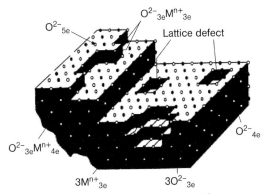

Figure 16.1 A representation of the various defects present on metal oxides (from [4]).

amorphous. Advances in crystallography for powders and crystals employing X-ray, electron and neutron diffraction have provided insight into the structures of metal oxides. Metal oxides crystallize in a variety of structures and the bonding in these materials can range from ionic (MgO, $Fe_{1-x}O$) to metallic (TiO, ReO_3) [1].

An understanding of both crystal structure and bonding, as well as the local microstructures that are due to defects, is necessary to understand the structure of complex transition metal oxides. Of course, on the nanometer scale the number of defects due to edges, corners, "F" centers and other surface imperfections is greatly enhanced by the large surface area (Figure 16.1) [4]. The pursuit of an understanding of the structure/property relationships at an atomic or molecular level is integral to the understanding of the unique properties observed on the nano-scale.

Five types of crystals found in bulk structures can be defined based on bonding considerations: covalent, ionic, metallic, molecular (van der Waals) and hydrogen bonded. These are also present on the nano-scale. However, the number of atoms at the surface must also be considered when examining the structure of nano-scale materials. When highly electronegative and highly electropositive elements are combined in a lattice, ionic crystals are formed. It has been found that the ionic model is a poor approximation for crystals containing large anions and small cations (e.g. oxides and sulfides) where the covalent contribution to bonding becomes significant [5]. Van der Waals interactions play a crucial role in many transition metal oxides, especially those with layered structures. In many oxide hydrates or hydroxy oxides, hydrogen bonding also contributes to the cohesive energy. In most transition metal oxides the bonding is only partly ionic; in other words, there is a considerable overlap between the orbitals of the cations and anions. However, there are several examples of metal oxides that are primarily ionic, such as MgO and CaO. Many transition metal oxides also exhibit metallic properties. Inorganic compounds of the formula AB can have the rock salt (B1),

CsCl (B2), zinc blende (B3), Wurtzite (B4), or NiAs (B8) structure [6]. Alkaline earth metal oxides such as MgO and monoxides of 3d transition metals as well as of lanthanides and actinides such as TiO, NiO, EuO and NpO, exhibit the rock salt structure with 6:6 octahedral coordination.

16.2.2
Defects

The physical and chemical properties of metal oxides are dominated by defects. Thermodynamic considerations imply that all crystals must contain a certain number of defects at temperatures above absolute zero. Structural defects can arise from a variety of causes: they may be thermally generated or they may arise in the course of fabrication of the solid, incorporated either unintentionally or deliberately. Defects are important because they are much more abundant at surfaces than in bulk, and in oxides they are usually responsible for many of the catalytic and chemical properties. Because of the number of atoms at the surface and the limited number of atoms within the lattice, the chemistry and bonding of oxide nanoparticles is greatly affected by the defect sites present. Point defects in crystals, such as vacancies and interstitials described by Schottky and Frenkel, account for the transport properties of ionic solids [6]. However, it appears that the point defect model is valid only when the defect concentration (or the deviation from stoichiometry) is extremely small. The defects that occur in ionic solids are grouped into the following classes: point, linear, planar and volumetric defects. Point defects are a result of the absence of one of the constituent atoms (or ions) on the lattice sites, or their presence in interstitial positions. Foreign atoms or ions present in the lattice represent another type of point defect. Point defects cause displacements on neighboring atoms or ions because of polarization in the surrounding region. A cationic vacancy in an ionic solid will have an electronegative charge, causing displacements of neighboring anions [7]. The energy of formation of a point defect depends primarily on the atomic arrangement in the immediate environment of the corresponding rows of atoms that do not possess the proper coordination. Boundaries between small crystallites (grain boundaries), stacking faults, crystallographic shear planes, twin boundaries and anti-phase boundaries are planar defects. Three-dimensional volumetric defects are a result of segregating point defects. The common point defects in ionic solids are Schottky pairs (pairs of cation and anion vacancies) and Frenkel defects (cation or anion interstitial plus a vacancy) [8]. When there is a large concentration of Schottky pairs, the measured pyknometric density of the solid is considerably lower than the density calculated from the X-ray unit cell dimensions (e.g. VO_x) [5]. Creation of defects is generally an endothermic process with the formation energies of vacancies in ionic solids generally 2 eV or more. Therefore, the intrinsic defect concentration in these solids is extremely low even at high temperatures [7]. The surface of a crystal constitutes a planar, 2-D defect. The environment of atoms or ions on the surface of a crystal is considerably different from that in the bulk. In polycrystalline materials, there are grain boundaries between the particles. The interface between two solid phases is an important factor in determining the course of reactions, crystal growth, and so on. An interface may be coherent, incoherent or semicoherent. It is coherent when the interface matches perfectly between the contact planes of two solid planes. Epitaxial growth occurs even when there is considerable mismatch (semicoherent interface). In close packed solids, one also often encounters stacking faults. For example, in a solid with cubic close packing, ABC ABC ABC, there can be a fault such as ABC AB ABC. Other types of planar defects include tilt boundary (array of periodically spaced edge dislocations), twist boundary (array of screw dislocations), twin boundary (a layer with mirror plane symmetry with respect to the rotation of one part of the crystal, on a specific plane, with respect to another) and antiphase boundary (boundary across which the sublattice occupation becomes interchanged).

The radius of the oxide anion (1.44 Å) given by Shannon and Prewitt is larger than most cations. However, in crystals, the anions are subjected to a positive Madelung potential, which gives rise to a contraction of the charge cloud, while cations are subjected to negative potential causing an opposite effect [7]. Generally, phase purity is hard to achieve, especially for phases containing more than one cation [9]. Often, phase segregation occurs and so the particles are a mixture of other possible phases derivable form the precursors.

The high surface areas of nano-scale particles yield a number of defect sites. There have been numerous studies of their surfaces in an attempt to clarify the type of defect sites that can exist [10–13]. The most common defects are coordinatively unsaturated ions arising from planes, edges, corners, anion/cation vacancies and electron excess centers. Such sites are often attributed as the active sites for many useful and interesting reactions, including methane activation [14], D_2-CH exchange [15], methanol oxidation [16, 17], CO oligomerization [18], oxygen exchange in CO_2 [7] and H_2O [8].

16.2.3
Acid–Base Properties of Metal Oxides

Metal oxides are often hard acids or bases and thus possess sites capable of catalyzing acid–base chemistry. Acidic metal oxides include Al_2O_3, TiO_2, CeO_2, V_2O_5, WO_3, etc. Basic metal oxides include MgO, CaO, SrO, BaO, Li_2O, etc. The interactions of the oxide surfaces with gases or liquids, which are relevant in the fields of adsorption and catalysis, are mainly governed by acid–base interactions. Acidic properties of metal oxides sometimes play an important role in the catalytic oxidations of hydrocarbons, where reactions, especially the selectivities to the target products, are strongly affected by the functions of acidic sites.

Acid–base characterizations of metal oxide surfaces are often used to explain their catalytic behavior. Extensive studies have been performed on the interaction of acid–base probe molecules with powders or supported metal oxides. The adsorption of NH_3 at cation sites has been used to characterize the Lewis acidity of metal oxides. The systematic use of CO adsorption at room temperature as a probe for (strong) Lewis acidity at the surface of a number of metal oxides (TiO_2, ZrO_2, HfO_2 and Al_2O_3) of interest in catalysis was also illustrated. The adsorption of CO_2, a standard acidic probe molecule, has also been used to characterize the Lewis basicity of the oxygen anions on metal oxides.

Several insulating oxides and oxide composites were found to be potential catalysts for a variety of important reactions as a result of their surface basicity or acidity [14, 19–22]. Some selected reactions typical to metal oxides include dehydration of alcohols, cracking of hydrocarbons, isomerization of olefins and paraffins, dehydrohalogenation, alkylation and esterification. Several metal oxides, such as MgO, CaO and SrO, exhibit surface basic behavior while others, such as Al_2O_3 and WO_3, are considered to be acidic solids that possess more and stronger acidic sites on their surfaces. Acid–base behavior and the presence of several types of deficiencies in the lattice and on the surface are two major driving forces for surface reactivity of metal oxides. When metal oxides are prepared on the nanoscale, the percentage of coordinatively unsaturated ions, especially on edges and corners, increases significantly. Consequently, surface chemistry effects, which are barely noticeable in macroscopic systems, become prevalent in nanoparticle systems. These effects are demonstrated by enhanced surface reactivities and catalytic potentials possessed by many nanoparticle systems of metal oxides [23–27]. As indicated earlier, two of the most intensively studied nanoparticulate systems of the metal oxides are MgO and CaO. In particular, two types of nanocrystalline oxides have been prepared and thoroughly studied: a "conventional preparation" (CP) and an "aerogel preparation" (AP). Nanocrystalline MgO prepared by a modified AP yields a fine, white powder of 400–500 $m^2 g^{-1}$ and 4 nm average crystallite size. High-resolution transmission electron microscope (TEM) imaging of a single crystallite indicated a polyhedral structure suggesting the presence of high surface concentrations of edge/corner sites and various exposed crystal planes (such as (002), (001), (111)) [28]. Conversely, CP yields particles with surface areas of 150–200 $m^2 g^{-1}$ and 8 nm average crystallites. If intrinsic surface chemistry differences due to size are to be uncovered, it is important to consider that in bulk MgO the effective ionic charges are close to +2, whereas the MgO molecule is much more covalent with effective charges close to +1. Lower coordination surface ions such as Mg^{2+}_{3c}, Mg^{2+}_{4c}, O^{2-}_{3c} and O^{2-}_{4c} are expected to have effective charges between +1 and +2. Surface sites on crystalline and powdered MgO have been probed by theoretical as well as experimental efforts. *Ab initio* calculations with H_2 have been used to probe perfect crystal surfaces and various defect sites. On the perfect (100) MgO surface, H_2 has a small adsorption energy and does not dissociate. However, temperature programmed desorption methods have shown that polycrystalline samples do dissociate H_2, probably on O_{3c}–Mg_{3c} sites. These sites are apparently very active for heterolytic H_2 dissociation. The microfaceted (111) surface of MgO is particularly reactive, and steps, kinks and point defects (ion vacancies and substitutions) are also important. Indeed, the unique catalytic properties of defective MgO surfaces also depend on a plethora of unusual coordination sites.

16.2.4
Redox Property of Metal Oxides

In some catalytic reactions, metal oxides often undergo reduction and re-oxidation simultaneously by loss and gain of surface lattice oxygen to and from the gas phase. This phenomenon is called redox catalysis. The redox property as well as the acidic and basic nature are the most important properties of metal oxide catalysis. It is well known that some simple metal oxides such as V_2O_5, CoO_2, NiO, MnO_2, CeO_2, MgO and some mixed metal oxides have redox properties. Mars and van Krevelen first proposed a redox mechanism to describe oxidation of compounds over oxide catalysts [29]. The Mars–van Krevelen mechanism (MvK) is now commonly accepted. When an adsorbate is oxidized at the surface, the oxidant is often a surface lattice oxygen atom, thus creating a surface oxygen vacancy. Surface oxygen vacancies are proposed to participate in many chemical reactions catalyzed by metal oxides. Vacancies also bind adsorbates more strongly than normal oxide sites and assist in their dissociation. The redox property of a metal oxide catalyst can be characterized by using the techniques of temperature-programmed reduction/temperature-programmed oxidation (TPR/TPO).

16.3
Synthesis

The development of synthetic methods is one of the fundamental aspects to the understanding and development of nano-scale materials. The novel properties and numerous applications of nano-scale materials have encouraged many researchers to invent and explore preparation methods that allow control over such parameters as particle size, shape, size distributions and composition. While considerable progress has taken place, one of the major challenges is the development of a "synthetic toolbox" which would afford access to size and shape control of structures on the nano-scale and conversely allow scientists to study the effects these parameters impart to the chemical and physical properties of the nanoparticles.

The two principal approaches to the preparation of nano-scale materials are "bottom-up" and "top-down". While "top-down" preparations involve approaching the nano-scale by breaking down larger starting materials, the "bottom-up" preparation methods are of primary interest to chemists and materials scientists because the fundamental building blocks are atoms or molecules, and these methods will be the focus of this chapter. Among the most sought after goals of synthetic chemists is gaining control over the way these fundamental building blocks come together and form particles. Interest in "bottom-up" approaches to nano-scale oxides and other materials is clearly indicated by the number of reports and reviews on this subject [30–46]. There are, of course, numerous "bottom-up" approaches to the preparation of nano-scale materials, and metal oxides are no exception. Gas–solid (wet chemical) and liquid–solid (physical) transformations are two different approaches to synthesizing nanomaterials by "bottom-up" preparation methods. Several physical aerosol methods have been reported for the synthesis of nano-size particles of oxide materials including gas condensation techniques [47–53], spray pyrolysis [51, 54–60], thermochemical decomposition of metal–organic precursors in flame reactors [53, 61–63] and other aerosol processes named after the energy sources applied to provide the high temperatures during gas–particle conversion. The sol–gel process is by far the most common and widely used "bottom-up" wet chemical method for the preparation of nano-scale oxides. Other wet chemistry methods including novel micro-emulsion techniques, oxidation of metal colloids and precipitation from solutions have also been used.

The methods of sample preparation are critical as they determine the morphology of the resulting material [64]. For example, burning Mg in O_2 (MgO smoke) yields 40–80 nm cubes and hexagonal plates, while thermal decomposition of $Mg(OH)_2$, $MgCO_3$ and especially $Mg(NO_3)_2$ yields irregular shapes often exhibiting hexagonal platelets. Surface areas can range from $10\,m^2\,g^{-1}$ (MgO smoke) to $250\,m^2\,g^{-1}$ for $Mg(OH)_2$ thermal decomposition, but surface areas of about $150\,m^2\,g^{-1}$ are typical. In the case of calcium oxide, surface areas can range from 1 to $100\,m^2\,g^{-1}$ when prepared by analogous methods, but about $50\,m^2\,g^{-1}$ is typical. In the following discussion, the most common methods for the synthesis of metal oxides will be the focus.

16.3.1
Sol–Gel Technique

Sol–gel processing describes a type of solid materials synthesis procedure, performed in a liquid and at low temperature (typically $T < 100\,°C$). The development of sol–gel techniques has long been known for preparations of metal oxides and has been described many times [30–38, 40–46, 65]. The process is typically used to prepare metal oxides via the hydrolysis of reactive precursors, usually alkoxides in an alcoholic solution, resulting in the corresponding hydroxide. It is usually easy to maintain such hydroxide in a dispersed state in the solvent. Condensation of the hydroxide molecules with loss of water leads to the formation of a network. When hydroxide species undergo polymerization by condensation of the hydroxy

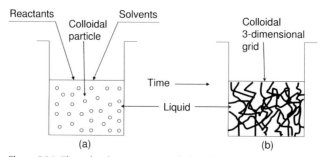

Figure 16.2 The sol–gel process: (a) sol, (b) gel.

network, gelation is achieved and a dense porous gel is obtained. The gel is a polymer with a three-dimensional skeleton surrounding interconnected pores, and the gels that are obtained are termed colloidal gels (Figure 16.2b). Removal of the solvents and appropriate drying of the gel are important steps that result in an ultra-fine powder of the metal hydroxide. Heat treatment of the hydroxide is a final step that leads to the corresponding ultra-fine powder of the metal oxide. Depending upon the heat treatment procedure, the final product may be in the form of a nano-scale powder, bulk material or oxygen-deficient metal oxide. A flow chart of typical sol–gel processing of nano-scale metal oxides is shown in Figure 16.3.

The chemical and physical properties of the final product are primarily determined by the hydrolysis and drying steps.

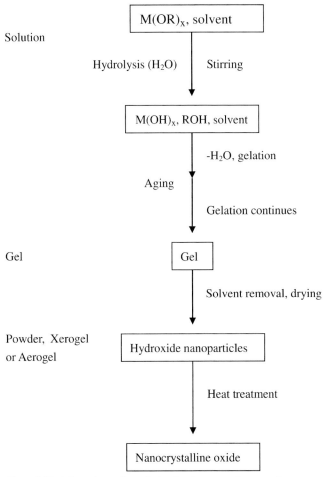

Figure 16.3 A flow chart of a typical sol–gel process for preparing nano-scale metal oxide powder.

16.3.1.1 Hydrolysis and Condensation of Metal Alkoxides

The initial species used in sol–gel processing are metal alkoxides ($M(OR)_y$). Hydrolysis of metal alkoxides involves nucleophilic reactions with water as follows:

$$M(OR)_y + xH_2O \leftrightarrow M(OR)_{y-x}(OH) + xROH$$

The mechanism of this reaction involves the addition of a negatively charged $HO^{\delta-}$ group to the positively charged metal center ($M^{\delta+}$). The positively charged proton is then transferred to an alkoxy group followed by the removal of ROH.

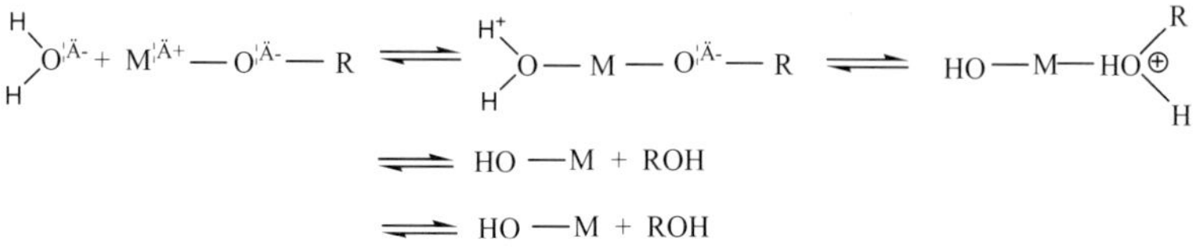

$$\begin{array}{c}
H \\
\diagdown \\
O^{\ddot{A}-} + M^{\ddot{A}+} - O^{\ddot{A}-} - R \\
\diagup \\
H
\end{array}
\rightleftharpoons
\begin{array}{c}
H^+ \\
\diagdown \\
O - M - O^{\ddot{A}-} - R \\
\diagup \\
H
\end{array}
\rightleftharpoons
HO - M - HO \begin{array}{c} R \\ \diagup \\ \diagdown \\ H \end{array}$$

$$\rightleftharpoons HO - M + ROH$$

$$\rightleftharpoons HO - M + ROH$$

Condensation occurs when the hydroxide molecules bind together as they release water molecules and a gel/network of the hydroxide is obtained as shown below:

$$\diagup\!\!\!M-OH + HO-M\!\!\!\diagdown \longleftarrow \diagup\!\!\!M-O-M\!\!\!\diagdown + H_2O$$

The rates at which hydrolysis and condensation take place are important parameters that affect the properties of the final product, as slower and more controlled hydrolysis typically leads to smaller particle sizes and more unique properties. Hydrolysis and condensation rates have been found to depend on the electronegativity of the metal atom, the alkoxy group, the solvent system and the molecular structure of the metal alkoxide. Those metals with higher electronegativities undergo hydrolysis more slowly than those with lower electronegativities. For example, the hydrolysis rate of $Ti(OEt)_4$ is about five orders of magnitude greater than that of $Si(OEt)_4$. Hence, the gelation times of silicon alkoxides are much longer (on the order of days) than those of titanium alkoxides (few seconds or minutes) [55]. The sensitivity of metal alkoxides toward hydrolysis decreases as the OR group size increases, with smaller OR groups leading to higher reactivity of the corresponding alkoxide toward water, in some cases resulting in uncontrolled precipitation of the hydroxide.

The choice of solvents in sol–gel processes is very important because alcohol interchange reactions are possible. As an example, when silica gel was prepared from $Si(OMe)_4$ and heated to 600 °C the surface area was $300\,m^2\,g^{-1}$ with a mean pore diameter of 29 Å when ethanol was used as a solvent. However, when methanol was used, the surface area dropped to $170\,m^2\,g^{-1}$ and the mean pore diameter increased to 36 Å [32]. The rate of hydrolysis also becomes slower as the coordination number around the metal center in the alkoxide increases. Therefore, alkoxides that tend to form oligomers usually show slower rates of hydrolysis and, hence, are easier to control and handle. n-Butoxide (O-n-Bu) is often preferred as a precursor to different oxides including TiO_2 and Al_2O_3, because it is the largest alkoxy group that does not prevent oligomerization [33].

Because most metal alkoxides are highly reactive toward water, careful handling in dry atmospheres is required to avoid rapid hydrolysis and uncontrolled precipitation. For alkoxides that have low rates of hydrolysis, acid or base catalysts can be used to enhance the process. The relatively negative alkoxides are protonated by acids, creating a better leaving group and eliminating the need for proton transfer in the transition state. Alternatively, bases provide better nucleophiles (OH^-) for hydrolysis; however, deprotonation of metal hydroxide groups enhances their condensation rates. In the case of highly reactive compounds, controlling the hydrolysis ratio may necessitate the use of non-aqueous solvents, where hydrolysis is controlled by strict control of water in the system rather than by acids or bases. Klabunde and coworkers have demonstrated the effectiveness of this approach in the preparation of gels from $Mg(OEt)_2$ in methanol and methanol–toluene solvents [66].

16.3.1.2 Solvent Removal and Drying

Developments in the areas of solvent removal and drying have further facilitated the production of nano-scale metal oxides with novel properties. When drying is achieved by solvent evaporation at ambient pressure with moderate shrinkage, the gel network shrinks as a result of capillary pressure, and the hydroxide product obtained is referred to as xerogel. However, if supercritical drying is applied using a high-pressure autoclave reactor at temperatures higher than the critical temperatures of solvents, less shrinkage of the gel network occurs, as there is no capillary pressure and no liquid–vapor interface, which allows the pore structure to remain largely intact by avoiding the pore collapse phenomenon. In practice, supercritical drying consists of heating the wet gel in a closed container, so that the pressure and temperature exceeds the critical temperature, T_c, and critical pressure, P_c, of the liquid entrapped in the pores inside the gel. The critical conditions are very different depending on the fluid which impregnates the wet gel. A few values are given in Table 16.1 [67]. The hydroxide product obtained in this manner, which is the traditional drying technique, is referred to as an aerogel and is the origin of the label "aerogel". Aerogel powders usually demonstrate higher porosities and larger surface areas than analogous xerogel powders. Aerogel processing has been very useful in producing highly divided powders of different metal oxides [28, 68, 69] (Figures 16.4–16.6).

Sol–gel processes have several advantages over other techniques for the synthesis of nano-scale metal oxides. Because the process begins with a relatively homogeneous mixture, the resulting product is generally a uniform ultra-fine porous powder. Sol–gel processing also has the advantage that it can be scaled up to accommodate industrial-scale production.

Numerous metal oxide nanoparticles have been produced by making some modifications to the traditional AP method. One modification involved the addition of large amounts of aromatic hydrocarbons to the alcohol–methoxide

Table 16.1 Critical point parameters of common fluids.

Fluid	Formula	T_c (°C)	P_c (MPa)
water	H_2O	374.1	22.04
carbon dioxide	CO_2	31.0	7.37
Freon 116	$(CF_3)_2$	19.7	2.97
acetone	$(CH_3)_2O$	235.0	4.66
nitrous oxide	N_2O	36.4	7.24
methanol	CH_3OH	239.4	8.09
ethanol	C_2H_5OH	243.0	6.3

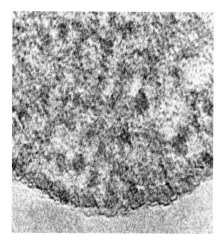

Figure 16.4 TEM micrograph of the nanostructure of CP-MgO. Note the absence of porosity and that all of the nanocrystals have agglomerated (from [68]).

Figure 16.5 TEM micrograph of the nanostructure of AP-MgO (supercritical solvent removal). Porosity is formed by the interconnected cubic nanocrystals of MgO (from [28]).

Figure 16.6 TEM micrograph of MgO(111) (supercritical solvent removal). Here the MgO nanoplates possess the (111) surface on the edges (from [69]).

solutions before hydrolysis and alcogel formation. This was done in order to further reduce the surface tension of the solvent mix and to facilitate solvent removal during the alcogel–aerogel transformation [30, 64, 70]. The resulting nanoparticles exhibited higher surface areas, smaller crystallite sizes and more porosity for samples of MgO, CaO, TiO$_2$ and ZrO$_2$ [71, 72].

Nano-structured MgO and CaO have been reported to be extremely effective for the destructive adsorption of numerous environmental toxins and several chemical warfare agents. Figure 16.7 outlines briefly how nano-crystalline and micro-crystalline MgO and CaO were prepared. For nano-crystalline MgO (AP-MgO) surface areas ranged from 250 to 500 m^2 g^{-1} and for AP-CaO 100–160 m^2 g^{-1}. For CP micro-crystalline MgO and CaO, the surface areas are 130–250 and 50–100 m^2 g^{-1}, respectively, whereas commercially available MgO and CaO had the lowest surface areas. Richards and coworkers have reported a simple method for preparing a sheet-like MgO with a thickness of less than 10 nm and, more interestingly, exhibiting the highly ionic (111) facet as the major surface of the "nanosheet". In this system, a mixture of water and methanol was added slowly to hydrolyze a magnesium methoxide/benzyl alcohol solution to produce the nanosheets. Theoretical studies suggest that water plays an important role as chemisorption of water forms a hydroxyl surface which stabilizes the otherwise unstable (111) surface of MgO [69].

Mesoporous and spherical TiO$_2$ materials are interesting candidates for applications in catalysis, biomaterials, microelectronics, optoelectronics and photonics. Zhang and coworkers have put forward a new method to synthesize mesoporous TiO$_2$ as well as hollow spheres, using titanium butoxide, ethanol, citric acid, water and ammonia [73]. The mesoporous solid and hollow spheres were formed with the same reactants but adding them in a different order. The TiO$_2$ possessed a mesoporous structure with the particle diameters of 200–300 nm for solid spheres and 200–500 nm for hollow spheres. The average pore sizes and BET surface areas of the mesoporous TiO$_2$ solid and hollow spheres are 6.8 and 7.0 nm and 162

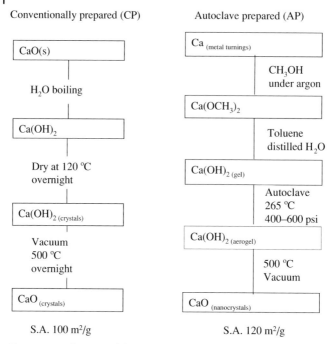

Figure 16.7 Schematic of the preparative scheme for nanocrystalline MgO (CaO) labeled AP-MgO (CaO) and microcrystalline MgO (CaO) labeled CP-MgO (CaO).

and 90 m^2 g^{-1}, respectively. Optical adsorption studies showed that the TiO$_2$ solid and hollow spheres possess a direct band gap structure with the optical band gap of 3.68 and 3.75 eV, respectively. The advantage of this route is that it does not use surfactants or templates, as is more usual for synthesizing mesoporous materials. In the proposed preparation mechanism (Figure 16.8), ammonium citrate forms and plays a key role in the formation of the mesoporous solid [73]. The formation of solid mesoporous TiO$_2$ or of hollow spheres was highly dependent on the extent of TiO$_2$ condensation at the onset of ammonium citrate crystal growth. That is, mesoporous solid spheres form when the TiO$_2$ condensation process takes place simultaneously with the formation of ammonium citrate crystals; mesoporous hollow spheres are produced when the nucleation and growth of ammonium citrate crystals occurs before the TiO$_2$ condensation process.

The chemistry of transition metals differs from those systems previously discussed, and only a few transition metals exhibit metal–alkoxide chemistry amenable to sol–gel synthesis. Cerda and coworkers prepared BaSnO$_3$ [74] by calcining a gel formed between Ba(OH)$_2$ and K$_2$SnO$_3$ at pH ≈ 11. The material possessed a particle size of 200–500 nm and contained BaCO$_3$ as an impurity, which can be eliminated by high temperature calcination. O'Brien and coworkers [75] synthe-

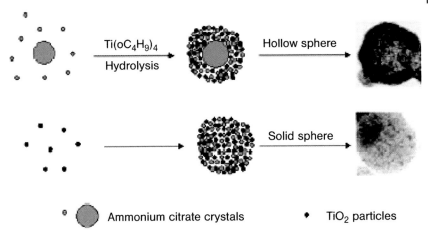

Figure 16.8 Schematic illustration of the formation mechanisms for mesoporous TiO_2 hollow and solid spheres (from [73]).

sized samples of monodisperse nanoparticles of barium titanate with diameters ranging from 6 to 12 nm by the sol–gel method. The technique was extended by Meron and coworkers [76] to the synthesis of colloidal cobalt ferrite nanocrystals. This synthesis involved the single-stage high-temperature hydrolysis of metal–alkoxide precursors to obtain crystalline, uniform, organically coated nanoparticles, which were well dispersed in an organic solvent. They were also able to form Langmuir–Blodgett films consisting of a close packed nanocrystal monolayer. The structural and magnetic properties of these nanocrystals were similar to those of bulk Fe_3O_4 with a very high coercivity at low temperatures. Other aerogel oxides, such as V_2O_5 [77, 78] MoO_3 [78] and MnO_2 [79, 80], have also been prepared. Nanocomposites of RuO_2–TiO_2 [81] are of particular interest because of their supercapacitor properties.

16.3.2
Co-precipitation Methods

One of the conventional syntheses to prepare nanoparticles is the precipitation of sparingly soluble products from aqueous solutions, followed by thermal decomposition of those products to oxides [82–84]. This process involves dissolving a salt precursor, usually a chloride, oxychloride or nitrate, such as $AlCl_3$ to make Al_2O_3, $Y(NO_3)_3$ to make Y_2O_3 and $ZrOCl_2$ to make ZrO_2. The corresponding metal hydroxides are usually formed and precipitated in water by the addition of a basic solution such as sodium hydroxide or ammonia solution. The resulting chloride or nitrate salts, such as $NaCl$ or NH_4Cl, are then washed away and the hydroxide is calcined after filtration and washing to obtain the final oxide powder. This method is useful in preparing composites of different oxides by co-precipitation of the corresponding hydroxides in the same solution. One of the disadvantages of this method is the difficulty in controlling the particle sizes and size distribution. Very often, fast (uncontrolled) precipitation takes place, resulting in large particles. In order to overcome this shortcoming, some new co-precipitation methods have developed, such as sonochemical co-precipitation and microwave-assisted co-precipitation.

16.3.2.1 Co-precipitation from Aqueous Solution at Low Temperature

The products of co-precipitation reactions are usually amorphous at or near room temperature. It is difficult to determine experimentally whether the as-prepared precursor is a single-phase solid solution or a multi-phase, nearly homogeneous mixture of the constituent metal hydroxides, carbonates and oxides that react to form a single phase mixed metal oxide when heated.

Many nano-particulate metal oxides are prepared by calcining hydroxide co-precipitation products. As an example, the zirconia system is highlighted here because of its interesting properties and applications. Zirconia has very interesting properties as an acid–base catalyst, as a promoter for other catalysts and as an inert support material [85, 86]. Sulfated ZrO_2 is of great interest on account of its high activity as a solid acid catalyst in alkylation, while ZrO_2 has also been found to act as a photocatalyst, owing to its n-type semiconductor nature.

The typical preparation consists of the calcination of a hydroxylated gel prepared by hydrolysis of zirconium salts in various media [72, 87–94]. Amorphous zirconia undergoes crystallization at around 450 °C and hence its surface area decreases dramatically at that temperature. At room temperature the stable crystalline phase of zirconia is monoclinic while the tetragonal phase forms upon heating to 1100–1200 °C For several applications, it is desirable to have the tetragonal phase with a high surface area. However, preparation protocols are needed to obtain the tetragonal phase at lower temperatures. Many researchers have tried to maintain the high surface area (HSA) of zirconia by several means. Usually the ZrO_2 is mixed with CaO, MgO, Y_2O_3, Cr_2O_3, or La_2O_3 for stabilization of the tetragonal phase at low temperature [95]. Bedilo and Klabunde [92, 93] have prepared HSA sulfated zirconia by a supercritical drying technique and found that the resulting material was active towards alkane isomerization reactions [92, 94]. Recently, Chane-Ching and coworkers [96] reported a general method to prepare HSA materials through the self-assembly of functionalized nanoparticles. This process involves functionalizing the oxide nanoparticles with bifunctional organic anchors such as aminocaproic acid and taurine. After the addition of a copolymer surfactant, the functionalized nanoparticles will slowly self-assemble on the copolymer chain through a second anchor site. Using this approach the authors prepared several metal oxides, including CeO_2, ZrO_2 and CeO_2–$Al(OH)_3$ composites. The method yielded ZrO_2 of surface area $180\,m^2\,g^{-1}$ after calcining at 500 °C, $125\,m^2\,g^{-1}$ for CeO_2 and $180\,m^2\,g^{-1}$ for CeO_2–$Al(OH)_3$ composites.

Table 16.2 [97–100] shows the literature data for zirconia obtained by different processes and the resulting surface area obtained at different calcination temperatures. Richards and coworkers [101] obtained stable ZrO_2 by the cetyltrimethylammonium chloride stabilization route and the data indicate that digesting the

Table 16.2 Survey from the literature of ZrO_2 prepared using precipitation methods.

Starting material	Ppt agent	Surfactant[a]	Conditions	Surface area ($m^2 g^{-1}$)	Ref.
$ZrOCl_2$	NH_3 (aq)	none	calcined 450 °C	247	[97]
$ZrOCl_2$	NH_3 (aq)	$C_{12}TACl$	calcined 450 °C	274	[97]
$ZrOCl_2$	NH_3 (aq)	$C_{14}TACl$	calcined 450 °C	300	[97]
$ZrOCl_2$	NH_3 (aq)	$C_{16}TACl$	calcined 450 °C	312	[97]
$ZrOCl_2$	NH_3 (aq)	$C_{18}TACl$	calcined 450 °C	313	[97]
$ZrOCl_2$	NaOH	P_{123}	calcined 450 °C	103	[98]
$ZrOCl_2$	NaOH	P_{123}	calcined 600 °C	51	[98]
$Zr(NO_3)_4$	TEA	CTAB	calcined 500 °C	–	[99]
$ZrOCl_2$	NH_3 (aq)	CTAB	calcined 600 °C	168	[100]
$ZrOCl_2$	NH_3 (aq)	CTAB	calcined 800 °C	105	[100]

a $C_{12}TACl$, dodecyltrimethylammonium chloride; P_{123}, poly(ethylene oxide)-b-poly(propylene oxide)-b-poly(ethylene oxide) triblock copolymer; CTAB, cetyltrimethylammonium bromide.

material in the presence of cetyltrimethylammonium chloride and ammonia provided higher surface areas. Moreover, surface area and thermal stability were found to depend on digestion time. In their study, digestion at 110 °C for 100 hours provided high thermal stability and the highest surface area of $370 m^2 g^{-1}$; the same sample when calcined at 700 °C possessed a surface area of $160 m^2 g^{-1}$, which was the best value obtained to date. Zirconia samples that were digested more than 100 hours had high thermal stability but no increases in surface area were observed. The same samples that were dried by applying supercritical techniques did not yield better results.

As previously stated, a common method for producing crystalline nanoparticle oxides is the co-precipitation of metal cations as carbonates, dicarbonates or oxalates, followed by their subsequent calcination and decomposition. Unfortunately, the calcination invariably leads to agglomeration or, at high temperatures, aggregation and sintering. As an example, CeO_2 nanopowders have been prepared by calcining the product of the precipitation between $Ce(NO_3)_3$ and $(NH_4)_2CO_3$, resulting in crystalline 6 nm particles of CeO_2 at calcination temperatures as low as 300 °C [106]. NiO with 10–15 nm particles has been similarly prepared by precipitating aqueous Ni^{2+} solutions with $(NH_4)_2CO_3$ and calcining the products at 400 °C [107]. In some rare instances, crystalline oxides can be precipitated from aqueous solution, eliminating the need for a calcination step and greatly reducing the risk of agglomeration. For example, TiO_2 can be prepared by precipitation using aqueous $TiCl_3$ with NH_3(aq) under ambient conditions. The products were 50–60 nm aggregates and stabilized with poly(methyl methacrylate) [108].

The direct co-precipitation of complex ternary oxides is somewhat uncommon, but is nonetheless possible, particularly when the product assumes a very thermodynamically favorable structure such as spinel. In such cases, the precipitation reactions are normally carried out at elevated temperatures (50–100 °C). Condensation of the two hydroxide intermediates into oxides and the induction of co-precipitation were carried out in the same reaction vessel. For instance, Pr^{3+}-doped ceria has been precipitated to yield monodispersed 13 nm particles by aging aqueous solutions of $Ce(NO_3)_3$ and $PrCl_3$ at 100 °C in the presence of a hexamethylenetetramine stabilizer [109]. In such a case, the stabilizer indirectly serves as the precipitating agent by raising the pH sufficiently to induce precipitation of the metal hydroxides. Likewise, $MnFe_2O_4$ was prepared from aqueous Mn^{2+} and Fe^{2+} at temperatures up to 100 °C to yield 5–25 nm particles [110].

Chinnasamy and coworkers have reported an extensive series of experiments for the preparation of spinel-structured $CoFe_2O_4$ [111]. The sizes of the products were determined by the influence of reaction temperature, reactant concentration and reactant addition rate. In each case, aqueous solutions of Fe^{3+} and Co^{2+} were precipitated with dilute NaOH. The results were predominantly in line with expectations based on the considerations outlined in reaction temperature, reactant concentration and reactant addition rate. Increasing the temperature from 70 °C to 98 °C increased the average particle size from 14 to 18 nm. Increasing the NaOH concentration from 0.73 to 1.13 M increased particle size from 16 to 19 nm. NaOH concentrations of 1.5 M or greater resulted in the formation of a secondary FeOOH phase and slowing the NaOH addition rate appeared to broaden the particle size distribution. Li and coworkers also prepared 12 nm $CoFe_2O_4$ by a similar route but stabilized the product by acidification with dilute nitric acid [112].

A summary of oxides precipitated from aqueous solutions, including the relevant reaction conditions, is given in Table 16.3 [113–119].

16.3.2.2 Sonochemical Co-precipitation

The principle of sonochemistry is breaking the chemical bond with the application of high-power ultrasound waves, usually between 10 and 20 MHz. The physical

Table 16.3 Summary of reactions for the precipitation of oxides from aqueous solution.

Oxide	Starting material	Ppt agent	Stabilizer	Conditions	Product size (nm)	Ref.
Fe_2O_3	$FeCl_2$	NH_3 (aq)	H^+	calcined 550 °C	53.5	[113]
TiO_2	$Ti(SO_4)_2$	NH_3 (aq)	H^+	calcined 550 °C	9.2	[113]
Al_2O_3	$Al(NO_3)_3$	NH_3 (aq)	H^+	calcined 550 °C	13.2	[113]
CeO_2	$Ce(NO_3)_3$	NH_3 (aq)	none	calcined 650 °C	12–15	[114]
TiO_2	$TiCl_3$	Na_2O_2	NaCl	calcined 700 °C	30	[115]
VO_2	NH_4VO_3	$N_2H_4 \cdot H_2O$	none	calcined 300 °C	35	[116]
Cr_2O_3	$K_2Cr_2O_7$	$N_2H_4 \cdot H_2O$	none	calcined 500 °C	30	[116]
γ-Mn_2O_3	$KMnO_4$	$N_2H_4 \cdot H_2O$	none	–	8	[116]
Fe_3O_4	$FeCl_2$	NH_3 (aq)	H^+	N_2 atm	8–50	[117]
NiO	$NiCl_2$	NH_3 (aq)	CTAB	annealed 500 °C	22–28	[118]
ZnO	$ZnCl_2$	NH_3 (aq)	CTAB	annealed 500 °C	40–60	[118]
SnO_2	$SnCl_4$	NH_3 (aq)	CTAB	annealed 500 °C	11–18	[118]
Sb_2O_3	$SbCl_3$	NaOH	PVA	annealed 350 °C	10–80	[119]

phenomenon responsible for the sonochemical process is acoustic cavitation. According to published theories for the formation of nanoparticles by sonochemistry, the main events that occur during the preparation are creation, growth and collapse of the solvent bubbles that are formed in the liquid. These bubbles are in the nanometer size range. Solute vapors diffuse into the solvent bubble and when the bubble reaches a certain size, its collapse takes place. During the collapse very high temperatures of 5000–25 000 K [127] are obtained, which is enough to break chemical bonds in the solute. The collapse of the bubble takes place in less than a nanosecond [128, 129] hence a high cooling rate (10^{11} K s^{-1}) is also obtained. This high cooling rate hinders the organization and crystallization of the products. Since the breaking of bonds in the precursor occurs in the gas phase, amorphous nanoparticles are obtained. Though the reason for the formation of amorphous products is well understood, the formation of nanostructures is not. One possible explanation is that in each collapsing bubble a few nucleation centers are formed and while the fast kinetics does not stop the growth of nuclei that growth is limited by the collapse. Another possibility is that the precursor is a non-volatile compound and the reaction occurs in a 200 nm ring surrounding the collapsing bubble [130]. In the latter case, the sonochemical reaction occurs in a liquid phase and the products could be either amorphous or crystalline depending on the temperature in the ring region of the bubble. Suslick has estimated the temperature of the ring region as 1900 °C. The sonochemical method has been found useful in many areas of material science, from the preparation of amorphous products [131, 132] through the insertion of nanomaterials into mesoporous materials [102, 103] to the deposition of nanoparticles on ceramic and polymeric surfaces [104, 105].

Sonochemical methods for the preparation of nanoparticles were pioneered by Suslick in 1991 [127]. If a reaction is carried out in the presence of oxygen by similar methods, the product formed will be an oxide. $NiFe_2O_4$ has been synthesized by sonicating the mixture of $Fe(CO)_5$ and $Ni(CO)_4$ in decalin under 1–1.5 atm pressure of oxygen [133]. ZrO_2 was synthesized using $Zr(NO)_3$ and $NH_3(aq)$ in aqueous phase under ultrasonic irradiation, followed by calcination at 300–1200 °C to improve the crystallinity or, for temperatures of 800 °C and above, to induce the tetragonal to monoclinic phase transformation [100, 134]. Nanocrystalline $La_{1-x}Sr_xMnO_3$ was prepared in a similar manner [135]. The method has also been successfully applied to synthesize $Ni(OH)_2$ and $Co(OH)_2$ nanoparticles [136] and iron(III) oxide [137]. Polycrystalline CeO_2 nanorods 5–10 nm in diameter and 50–150 nm in length were synthesized via ultrasonication using polyethylene glycol (PEG) as a structure-directing agent at room temperature. The content of PEG, the molecular weight of PEG and the sonication time were confirmed to be the crucial factors determining the formation of one-dimensional CeO_2 nanorods. A possible ultrasonic formation mechanism is suggested in Figure 16.9 [138].

16.3.2.3 Microwave-Assisted Co-precipitation

The microwave processing of nanoparticles results in rapid heating of the reaction mixtures, particularly those containing water. As a consequence, the precipitation

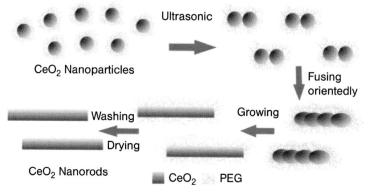

Figure 16.9 Possible formation mechanism of CeO$_2$ nanorods (from [139]).

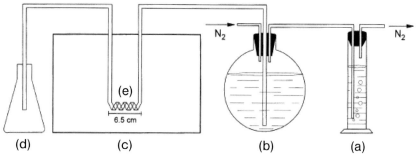

Figure 16.10 Schematic of a continuous-flow microwave reactor consisting of (a) a liquid column as a pressure regulator, (b) a metal salt solution container, (c) a microwave oven cavity, (d) a metal cluster dispersion receiver and (e) a spiral tube reactor (from [139]).

of particles from such solutions tends to be rapid and nearly instantaneous, leading to very small particle sizes and narrow size distributions within the products. The method offers the additional benefit of requiring very short reaction times and has similarly been used for the synthesis of Ag and Au nanoparticles and transition metal oxides. Tu and coworkers [139] adapted a continuous flow reactor which is depicted in Figure 16.10. For example, two nano-structured cerium(IV) oxide powders were synthesized by different synthetic routes: two samples were obtained by precipitation from a basic solution of cerium nitrate and treated at 523 and 923 K, respectively. A broad particle size distribution is observed for CeO$_2$ obtained by precipitation (8.0–15.0 nm). Smaller particles (sizes around 3.3–4.0 nm) with a narrow particle size distribution characterize the ceria obtained by microwave irradiation. Uniform α-Fe$_2$O$_3$ nanoparticles were prepared by forced hydrolysis of ferric salts under microwave irradiation. Gedanken's group has also published extensively on various oxides and chalcogenides prepared by microwave-assisted irradiation [120–122, 140].

16.3.3
Solvothermal Technique

Hydrothermal processes [123–126] involve using water at elevated temperatures and pressures in a closed system, often in the vicinity of its critical point. A more general term, "solvothermal," refers to a similar reaction in which a non-aqueous solvent (organic or inorganic) is used. Under solvo(hydro)thermal conditions, certain properties of the solvent, such as density, viscosity and diffusion coefficient, change dramatically and the solvent behaves much differently from what is expected under ambient conditions [126]. Consequently, the solubility, the diffusion process and the chemical reactivity of the reactants (usually solids) are greatly increased or enhanced, enabling the reaction to take place at a much lower temperature than normal. The method has been widely applied and adopted for crystal growth of many inorganic materials, such as zeolites, quartz, metal carbonates, phosphates and other oxides and halides [141–148].

Solvothermal techniques have been extensively developed for the synthesis of metal oxides [149–152]. Unlike many other synthetic techniques, solvothermal synthesis concerns a much milder and softer chemistry conducted at low temperatures. The mild and soft conditions make it possible to leave polychalcogen building-blocks intact while they reorganize themselves to form various new structures, many of which might be promising for applications in catalysis, electronic, magnetic, optical and thermoelectronic devices [153–155]. They also allow the formation and isolation of phases that may not be accessible at higher temperatures because of their metastable nature [156, 157].

Although some solvothermal processes involve supercritical solvents, most simply take advantage of the increased solubility and reactivity of metal salts and complexes at elevated temperatures and pressures without bringing the solvent to its critical point. The metal complexes are decomposed thermally either by boiling the contents in an inert atmosphere or by using an autoclave. A suitable capping agent or stabilizer such as a long-chain amine, thiol, trioctylphosphine oxide (TOPO), etc. is added to the reaction contents at a suitable point to hinder the growth of the particles and hence stabilize them against agglomeration. The stabilizers also help in dissolution of the particles in different solvents. Unlike the cases of co-precipitation and sol–gel methods, solvothermal processes also allow substantially reduced reaction temperatures, and the products of solvothermal reactions are usually crystalline and do not require post-annealing treatments.

The synthesis of nanocrystalline TiO_2, which is an important photocatalyst for the decomposition of toxic chemicals, is one of the more thoroughly investigated solvothermal/hydrothermal reactions. Approaches to this preparation have involved the decomposition of metal alkoxides, [158] a TOPO-capped autoclave synthesis of TiO_2 by metathetic reaction [159] and decomposition of a metal N-nitroso-N-phenyl hydroxylamine complex [160, 161]. In 1988, Oguri and coworkers reported the preparation of anatase by hydrothermally processing hydrous titania prepared by the controlled hydrolysis of $Ti(OEt)_4$ in ethanol [162]. The reaction conditions leading to monodispersed anatase nanoparticles by this approach were elucidated by others [163]. Extension of the method to the preparation of lanthanide-doped titania particles has also been reported [164]. Cheng and coworkers developed a method for preparing nano-scale TiO_2 by hydrothermal synthesis using an aqueous $TiCl_4$ solution [165]. They found that acidic conditions favored rutile while basic conditions favored anatase. It was also found that higher temperature favored the highly dispersed product and that grain size could be controlled by the addition of minerals such as $SnCl_4$ or NaCl, although the presence of NH_4Cl led to agglomeration of particles. The approach was extended, and revealed that phase purity of the products depends primarily on concentration, with higher concentrations of $TiCl_4$ favoring the rutile phase, while particle size depends primarily on reaction time [166]. Yin and coworkers produced 2–10 nm crystallites of monodispersed, phase-pure anatase by using citric acid to stabilize the TiO_2 nanoparticles and treating the precursors hydrothermally in the presence of KCl or NaCl mineralizers [167].

Niederberger and coworkers [150] reported a widely applicable solvothermal route to nanocrystalline iron, indium, gallium and zinc oxides based on the reaction between the corresponding metal acetylacetonate as metal oxide precursor and benzylamine as solvent. They proved that with the exception of the iron oxide system, in which a mixture of the two phases magnetite and maghemite is formed, only phase-pure materials are obtained, γ-Ga_2O_3, zincite ZnO and cubic In_2O_3. The particle sizes lie in the ranges 15–20 nm for the iron, 10–15 nm for the indium, 2.5–3.5 nm for the gallium and around 20 nm for the zinc oxide (Figure 16.11). Moreover, the same group developed mixed nanocrystalline $BaTiO_3$, $SrTiO_3$ and (Ba,Sr)-TiO_3. $BaTiO_3$ nanoparticles are nearly spherical in shape, with diameters ranging from 4 to 5 nm, while $SrTiO_3$ particles display less-uniform particle shapes, the size vaying between 5 and 10 nm [168].

Masui and coworkers [169] reported the hydrothermal synthesis of nanocrystalline, monodispersed CeO_2 with a very narrow size distribution. They combined $CeCl_3 \cdot 6H_2O$ and aqueous ammonia with a citric acid stabilizer and heated the solution in a sealed Teflon container at 80 °C. The CeO_2 nanoparticles exhibited a 3.1 nm average diameter. The ceria particles were subsequently coated with turbostratic boron nitride by combining them with a mixture of boric acid and 2,2′-iminodiethanol, evaporating the solvent and heating at 800 °C under flowing ammonia. Inoue and coworkers were able to reduce the particle size of hydrothermally prepared colloidal CeO_2 to 2 nm using a similar approach by autoclaving a mixture of cerium metal and 2-methoxyethanol at 250 °C [170]. In this case, the autoclave was purged with nitrogen prior to heating and the colloidal particles were coagulated after heating with methanol and ammonia. The higher reaction temperatures resulted in increased particle size.

In many cases, anhydrous metal oxides have been prepared by solvothermal treatments of sol–gel or micro-emulsion-based precursors. Wu and coworkers prepared anatase and rutile TiO_2 by a micro-emulsion-mediated method, in which the micro-emulsion medium was further treated by hydrothermal reaction [171]. This micro-emulsion-mediated hydrothermal (MMH) method could lead to the formation of crystalline titania powders under much milder reaction conditions.

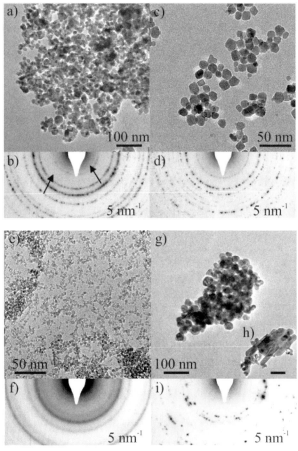

Figure 16.11 TEM overview images of (a) iron oxide, (c) indium oxide, (e) gallium oxide and (g) and (h) zinc oxide nanoparticles (scale bar 100 nm) and their respective selected-area electron diffraction patterns (b, d, f and i); (i) corresponds to (g) (from [150]).

The micro-emulsion medium was heated to 120–200 °C in a stainless steel autoclave. Micro-emulsions acidified with HNO_3 produced monodispersed anatase nanoparticles while those acidified with HCl produced rutile nanorods. Titanium dioxide (TiO_2) nanoparticles prepared this way have been shown to be active toward the photocatalytic oxidation of phenol [172].

Metal oxides can also be synthesized by the decomposition of metal–cupferron complexes, M^xCup_x (Cup = $C_6H_5N(NO)O^-$) [160]. Seshadri and coworkers were able to replace amine-based solvents with toluene, and prepared ≈10 nm diameter γ-Fe_2O_3 and ≈7 nm $CoFe_2O_4$ by hydrothermal processes [161]. They synthesized maghemite γ-Fe_2O_3 nanoparticles from a Fe^{III}–cupferron complex and spinel $CoFe_2O_4$ nanoparticles starting from Co^{II}–cupferron complex and Fe^{III}–cupferron complex. However, the authors found that the presence of at least trace quantities of a strongly coordinating amine was necessary to act as a capping agent and prevent aggregation. They were also able to synthesize nanoparticulate $ZnFe_2O_4$ by a similar approach [173].

16.3.4
Micro-Emulsion Technique

Micro-emulsions or micelles (including reverse micelles) represent an approach based on the formation of micro/nano reaction vessels for the preparation of nanoparticles, and has received considerable interest in recent years [174–180]. A literature survey indicates that ultra-fine nanoparticles in the size range 2–50 nm can be easily prepared by this method. This technique uses an inorganic phase in water-in-oil micro-emulsions, which are isotropic liquid media with nano-sized water droplets dispersed in a continuous oil phase. In general, micro-emulsions consist of, at least, a ternary mixture of water, a surfactant (or a mixture of surface-active agents) and oil. The dispersion of the aqueous phase is shown in Figure 16.12 [68]. The classical examples of emulsifiers are sodium dodecyl sulfate (SDC) and aerosol bis(2-ethylhexyl) sulfosuccinate (AOT). The surfactant (emulsifier) molecule stabilizes the water droplet because they have polar head groups and non-polar organic tails. The organic (hydrophobic) portion faces towards the oil phase and the polar (hydrophilic) group towards water. In diluted water (or oil) solutions, the emulsifier dissolves and exists as a monomer, but when its concentration exceeds a certain limit called the critical micelle concentration (CMC), the molecules of emulsifier associate spontaneously to form aggregates called micelles. These micro-water droplets then form nano-reactors for the formation of nanoparticles. The nanoparticles formed usually have monodisperse properties. One method of formation consists of mixing two micro-emulsions or macro-emulsions and aqueous solutions carrying the appropriate reactants in order to obtain the

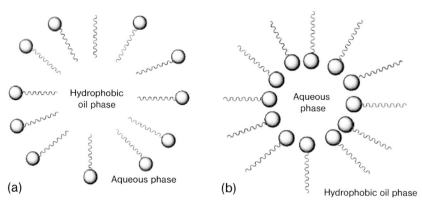

Figure 16.12 Schematic representation of (a) a micelle and (b) an inverse micelle (from [68]).

desired particles. The interchange of the reactants takes place during the collision of the water droplets in the micro-emulsions. The interchange of the reactant is very fast so that, for the most commonly used micro-emulsions, it occurs simply during the mixing process. The reduction, nucleation and growth occur inside the droplets, which controls the final particle size. The chemical reaction within the droplet is very fast, so the rate determining step will be the initial communication step of the micro-droplets with different droplets. The rate of communication has been defined by a second-order communication-controlled rate constant and represents the fastest possible rate constant for the system. The reactant concentration has a major influence on the reaction rate. The rate of both nucleation and growth are determined by the probabilities of the collisions between several atoms, between one atom and a nucleus and between two or more nuclei. Once a nucleus forms with the minimum number of atoms, the growth process starts. For the formation of monodisperse particles, all of the nuclei must form at the same time and grow simultaneously and with the same rate.

The preparation of metal oxide nanoparticles within micelles involves forming two micro-emulsions, one with the metal salt of interest and the other with the reducing or oxide-containing agent and mixing them together. When the two different reactants mix, the interchange of the reactants takes place through the collision of water micro-droplets. The reaction (reduction, nucleation and growth) takes place inside the droplet, which controls the final size of the particles. The interchange of nuclei between two micro-droplets does not take place owing to the special restrictions from the emulsifier. Once the particle inside the droplets attains its full size, the surfactant molecules attach to the metal surface, thus stabilizing and preventing further growth.

Reverse micelles are used to prepare nanoparticles by using an aqueous solution of reactive precursors that can be converted to insoluble nanoparticles. Nanoparticle synthesis inside the micelles can be achieved by a variety of methods, including hydrolysis of reactive precursors, such as alkoxides, and precipitation reactions of metal salts [181, 182]. Solvent removal and subsequent calcination leads to the final product. A variety of surfactants can be used in these processes such as pentadecaoxyethylene nonylphenylether (TNP-35) [182], decaoxyethylene nonylphenyl ether (TNT-10) [182], poly(oxyethylene)$_5$ nonylphenylether (NP5) [183] and many others that are commercially available. Several parameters, such as the concentration of the reactive precursor in the micelle and the weight percentage of the aqueous phase in the micro-emulsion, affect the properties, including particle size, particle size distribution, agglomerate size and phases of the final oxide powders. There are several advantages to using this method, including the preparation of very small particles and the ability to control the particle size. Disadvantages include low production yields and the need to use large amounts of solvents and surfactants.

This method has been successfully applied for the synthesis of metals, metal oxides, alloys and core–shell nanoparticles. The synthesis of metal oxides from reverse micelles is similar in most aspects to their synthesis in aqueous phase by a precipitation process. For example, precipitation of hydroxides is obtained by addition of a base such as NH_3(aq) or NaOH to a reverse micelle solution containing aqueous metal ions at the micellar cores. This method has been also used to synthesize mixed metal iron oxides [162],

$$M^{2+} + 2Fe^{2+} + OH^- \text{(excess)} \xrightarrow{\Delta, O_2} MFe_2O_4 + xH_2O \uparrow$$

where M = Fe, Mn or Co. The resultant particles obtained are of the order of 10–20 nm in diameter. The cation distribution in the case of spinels depends on the temperature used in the reaction [184]. If the transition metal cation is unstable or insoluble in aqueous media, nanoparticles of those metals can be prepared by hydrolysis of suitable precursors; for example, TiO_2 has been prepared from tetraisopropyl titanate [185] in the absence of water. The hydrolysis occurs when a second water-containing solution of reverse micelles is added to the solution of the first micelle containing the metal precursor. Adjusting the concentration of reactants can change the final crystal form, that is between amorphous and crystalline. The reaction can be described as follows,

$$Ti(O^iPr)_4 + 2H_2O \rightarrow TiO_2 + 4(^iPr\text{-}OH)$$

Micro-emulsions have been also employed to prepare precursors that decompose during calcination, resulting in desired mixed metal oxides. A series of mixed-metal ferrites have been prepared by precipitating the metal precursor in a H_2O-AOT-isooctane system and calcining the products at 300–600 °C [186]. A superconductor material $YBa_2Cu_3O_{7-\delta}$ with 10 nm particles have been prepared by combining a micellar solution of Y^{3+}, Ba^{2+} and Cu^{2+} prepared in an Igepal CO-430–cyclohexene system with a second micellar solution containing oxalic acid in the aqueous cores (Note: Igepal CO-430 is a surfactant with chemical name "Nonyl phenol 4 mole ethoxylate"). The precipitate was subsequently calcined to 800 °C to remove oxalic precursors [187]. Other metals and metal oxides that are prepared by a similar method include tungsten and tungsten oxide nanoparticles [188] and high-surface-area Al_2O_3 [189].

16.3.5
Combustion Methods

The combustion synthesis technique consists of bringing a saturated aqueous solution of the desired metal salts and a suitable organic fuel to the boil, until the mixture ignites a self-sustaining and rather fast combustion reaction, resulting in a dry, usually crystalline, fine oxide powder. By simple calcination, the metal nitrates can, of course, be decomposed into melt oxides upon heating to or above the phase transformation temperature.

Flame processes have been widely used to synthesize nanosize powders of oxide materials. Chemical precursors are vaporized and then oxidized in a combustion process using a fuel/oxidant mixture such as propane/oxygen or methane/air [190]. They combine the rapid thermal decomposition of a precursor/carrier gas stream in a reduced pressure environment with thermophoretically driven deposition of the rapidly condensed product particles on a cold substrate [63]. The flame usually provides a high temperature (1200–3000 K), which promotes rapid gas-phase chemical reactions [191]. Several types of flame reactors have been used in research settings and have produced numerous types of nano-scale metal oxides [63, 190–196].

Chromium oxides have been synthesized by Lima and coworkers using solution combustion [197]. The results suggest that glycine is a better complexing/combustible agent for ammonium dichromate than urea. The addition of extra ammonium nitrate to stoichiometric compositions improved the specific surface area and reduced the crystallite size. The smallest crystallite size (≈ 20 nm) of Cr_2O_3 was obtained with glycine as fuel/complexant agent in fuel-lean mixtures. The highest specific surface area (63 $m^2 g^{-1}$) was observed with urea in fuel-rich mixtures, forming amorphous CrO_3. Pratsinis and coworkers synthesized well structured nanocrystals of ceria–zirconia mixed oxides with high temperature stability and surface area. They sprayed cerium and zirconium precursors dissolved in carboxylic acid into a methane–oxygen flame [198]. The mixed products had surface areas from 60–90 $m^2 g^{-1}$. ZrO_2 powder was prepared by the gel combustion technique using citric acid as a fuel and nitrate as an oxidant. Calcination at 600 °C of the dried powder, obtained after sluggish combustion of the citrate–nitrate gel, produced spherical nanocrystalline ZrO_2 with diameter about 10 nm [199]. Nanocrystalline $LiMn_2O_4$ with cubic spinel structure has been prepared by combustion of reaction mixtures containing Li(I) and Mn(II) nitrates that operate as oxidizers and sucrose that acts as fuel [200]. Annealing at 700 °C led to single-phase cubic spinels. In these phases, the smallest average particle size (ca. 30 nm) corresponded to the sample obtained with a hyperstoichiometric amount of fuel.

16.3.6
Others

There are numerous approaches to the synthesis of highly dispersed metal oxides in addition to those discussed. These include some methods that have been less commonly used to prepare metal oxides, such as vapor condensation methods, spray pyrolysis and templated techniques.

16.3.6.1 Vapor Condensation Methods

Among the earliest methods for producing nanoparticles were gas condensation techniques, which produce nanoparticles directly from a supersaturated vapor of metals. These methods generally involve two steps: first, a metallic nano-phase powder is condensed under inert convection gas after a supersaturated vapor of the metal is obtained inside a chamber. A high pressure of inert gas is usually needed to achieve supersaturation, then the powder is oxidized by allowing oxygen into the chamber. This post-oxidation is a critical step and it is very often necessary for the process to be performed slowly. Because of the highly exothermic reaction, particles heat up for short times (usually less than one second) to temperatures as high as 1000 °C, resulting in their agglomeration into large particles by rapid diffusion processes. A subsequent annealing process at higher temperature is often required to complete the oxidation.

Gas condensation methods to prepare nanoparticles directly from supersaturated vapor have many advantages over other techniques including versatility, the ease of performance and analysis and the high purity of the products. These methods can also be employed to produce films and coatings. However, in spite of their success, there are drawbacks in the high production cost because of low yields and the difficulty in scaling-up. Heating techniques have other disadvantages, which include the possibility of reactions between the metal vapors and the heating source materials. Furthermore, the operating temperature is limited by the choice of the source material and, because of that, they cannot be used to make a wide variety of materials.

16.3.6.2 Spray Pyrolysis

Another useful method for the synthesis of high-purity homogeneous oxide powders is spray pyrolysis [201–206]. This technique has been known by several other names, including solution aerosol thermolysis [203], evaporative decomposition of solutions [204], plasma vaporization of solutions [205] and aerosol decomposition [206]. The starting materials in this process are chemical precursors, usually appropriate salts, in solution, sol or suspension form. The process involves the generation of aerosol droplets by nebulizing or atomizing the starting solution, sol or suspension. The generated droplets undergo evaporation and solute condensation within the droplet, drying, thermolysis of the precipitate particle at higher temperature to form microporous particles and, finally, sintering to form dense particles.

Aqueous solutions are usually used because of their low cost, safety and the availability of a wide range of water-soluble salts. Metal chloride and nitrate salts are commonly used as precursors because of their high solubility. Precursors that have low solubility or those that may induce impurities, such as acetates that lead to carbon in the products, are less suitable [202, 204]. During the transformation of the aerosol droplets into particles, different processes are involved, including solvent evaporation, precipitation of dissolved precursor and thermolysis of precipitated particles. One advantage to this process is that all of these processes take place in one step. Other advantages include the production of high-purity nano-size particles, the homogeneity of the particles as a result of the homogeneity of the original solution, the fact that each droplet/particle undergoes the same reaction conditions and that no subsequent milling is necessary. Disadvantages of spray pyrolysis include the large amounts of solvents necessary and the difficulty of scaling-up the production. The use of large amounts of non-aqueous solvents increases production expenses because of the high cost of pure solvents and the need for proper disposal.

16.3.6.3 Templated/Surface Derivatized Nanoparticles

In recent years, varieties of porous materials have been obtained by templated techniques. Generally two types of templates have been reported in the literature, soft templates (surfactants [207]) and hard templates (porous solids such as carbon or silica). In the case of hard templates, the formation of porous material takes place in a confined space formed by the porosity of the template. Two types of hard templates have been employed in template synthesis, active carbon [208] and mesoporous silica materials [209–212]. Commercial active carbons were used as templates to prepare different types of HSMO [213] and monodisperse and porous spheres of oxides and phosphates [214]. However, the use of active carbons as templates has certain limitations since, at the high treatment temperature employed during the synthesis, infiltrated salts and the carbon may react with each other, destroying the intended material. Moreover, if the heat treatment is performed in air, the carbon may be rapidly oxidized (ignition) even at relatively low temperature owing to the catalytic effect of the infiltrated salts. On the other hand, some metallic salts may end up as metal instead of metal oxide as carbon is a good reducing agent at high temperature under inert atmosphere.

16.4
Applications in Catalysis

As discussed in a previous section, metal oxides represent an important class of materials exhibiting a broad range of properties from insulators to semiconductors and conductors and have found applications as diverse as electronics, cosmetics and catalysts. Metal oxides have been widely used in many valuable heterogeneous catalytic reactions. Typical metal oxide-catalyzed reactions, including alkane oxidation, biodiesel production, methanol adsorption and decomposition, destructive adsorption of chlorocarbons and warfare agents, olefin metathesis and the Claisen–Schmidt condensation will be briefly discussed as examples of metal oxide-catalyzed reactions.

16.4.1
Oxygenation of Alkanes

Oxidation remains one of the principal paradigms for the activation of alkanes, which are well known for their low reactivity, and is of interest both academically and commercially. Among hydrocarbons, the oxidation of n-alkanes has attracted much attention because they are abundant as feedstocks [215–218]. Catalytic oxidation of alkanes has been explored using several oxidants and, for economic and environmental reasons, oxidation processes used in bulk chemical industries predominantly involve the use of atmospheric oxygen as the primary oxidant. Their success depends largely on the use of metal catalysts to promote both the rate of reaction and the selectivity to partial oxidation products of organic substrates for successful commercial processes [219, 220]. There have been a number of publications concerning alkane oxidation with molecular oxygen under mild conditions, using transition metal oxide catalysts systems (summarized in Table 16.4) [221–232].

Table 16.4 Aerobic partial oxidation of alkanes by metal oxide catalysts.

Entry	Alkane	Catalyst	T (°C)	Oxidant/others	Conversion (%)	Selectivity (%)	Products	Ref.
1	ethane	Mo-V-Nb-O	400	O_2/He	22.7	47.8	ethene	[221]
2	propane	V_2O_5/SBA-15	400	O_2	8	84	acrylic acid	[222]
3	propane	Co_3O_4	100	O_2/Ar	0.8	100	propene	[223]
4	propane	V-P-O	400	air/H_2O	23.0	48.1	acrylic acid	[224]
5	propane	V-P-O/TiO_2-SiO_2	300	O_2/H_2O	21.8	61.2	acrylic acid	[225]
6	propane	V-P-O	300	O_2/H_2O	15	47.2	acrylic acid	[225]
7	propane	Mo-V-Sb-O	380	O_2	46	27	acrylic acid	[226]
8	propane	$MoV_{0.3}Te_{0.23}P_{0.15}O_n$	500	O_2	46.6	8.1	acrylic acid	[227]
9	n-butane	V_2O_5/MgO	500	O_2	9.9	66.6	butene	[228]
10	isobutane	Mo-V-O	300	O_2	34.9	54.7	ethanoic acid	[229]
11	n-pentane	Mo-V-O	300	O_2	41.4	70.8	maleic anhydride	[229]
12	cyclohexane	V-P-O	65	H_2O_2/acetonitrile	84	50/50	cyclohexanol/cyclohexanone	[230]
13	n-heptane	Co_3O_4	450	O_2/H_2	20.4	14.2	CO_2	[231]
14	n-hexane	Co_3O_4	450	O_2/H_2	66.5	46.2	CO_2	[231]
15	n-heptane	CeO_2	450	O_2/H_2	7.5	9.0	CO_2	[231]
16	n-hexane	CeO_2	450	O_2/H_2	24.4	29.2	CO_2	[231]
17	n-hexane	Ti-Si-O	50	H_2O_2	7.0 (TON)	56	alcohols	[232]
18	n-heptane	Ti-Si-O	50	H_2O_2	45 (TON)	80	alcohols	[232]

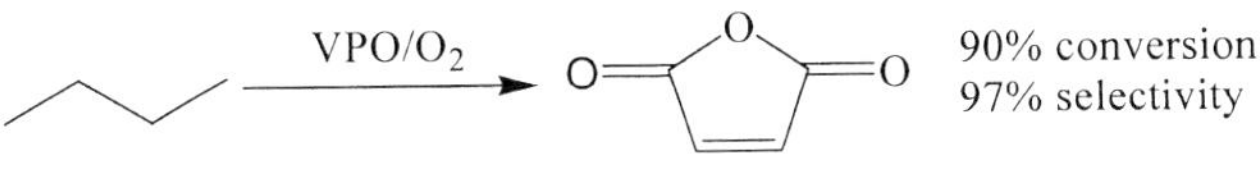

Scheme 16.1

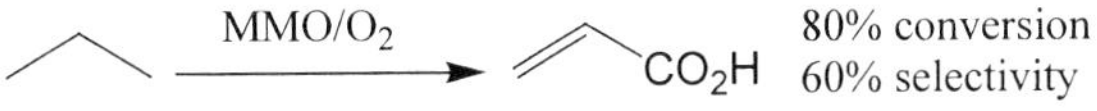

n-butane $\xrightarrow{\text{VPO/O}_2}$ maleic anhydride, 90% conversion, 97% selectivity

Scheme 16.2

propane $\xrightarrow{\text{MMO/O}_2}$ $CH_2=CHCO_2H$, 80% conversion, 60% selectivity

The conversion of butane to maleic anhydride over a VPO catalyst remains the only selective alkane oxidation technology that has achieved commercial operation. The yields of maleic anhydride vary from 45 to 67%, with n-butane conversion at ~90% and selectivity to maleic anhydride ranging from 65% to as high as 97% (Scheme 16.1) [232, 245].

The success of VPO catalysts in n-butane oxidation has stimulated great interest in the selective oxidation of propane to acrylic acid [236, 237]. On the basis of the excellent performance of VPO for converting n-butane to maleic anhydride, it is reasonable to believe that they also have potential for converting propane to acrylic acid effectively. Many studies are focused on the supported vanadium-catalyzed partial oxidation of hydrocarbons and the supports can be SiO_2 [246, 247], TiO_2 [248, 249], Al_2O_3 [250] and CeO_2 [251]. Most supported vanadium catalysts and vanadium-based mixed-metal oxides MMO [252] were been studied for the oxidation of propane (Scheme 16.2). Although the history of the MMOs in propane to acrylic acid oxidation is relatively short, they have shown excellent conversion (80%) and selectivity (60%).

16.4.2
Biodiesel Production

The increasing concern over greenhouse gas emissions and the search for alternative renewable fuels has led to increased interest in fuels developed from renewable feedstocks. One such alternative fuel is biodiesel, which is recognized as a "green fuel" produced from vegetable oils (VO) and other feedstocks that primarily contain triglycerides (TG) and free fatty acids (FFA) [253]. Biodiesel as a non-petroleum-based fuel is normally synthesized through a chemical process called transesterification, whereby TGs react with a low molecular weight alcohol in the presence of a catalyst to produce a complex mixture of fatty acid alkyl esters (biodiesel) and glycerol. Transesterification of TGs with low molecular weight alcohols proceeds via three successive and reversible reactions. Figure 16.13 shows the consecutive chemical reactions involved in the transesterification of triacetin with methanol. There are different processes that can be applied to the synthesis of biodiesel: (1) base-catalyzed transesterification [254], (2) acid-catalyzed transesterification [255], (3) integrated acid-catalyzed pre-esterification of FFAs and base-catalyzed transesterification [256], (4) enzyme-catalyzed transesterification

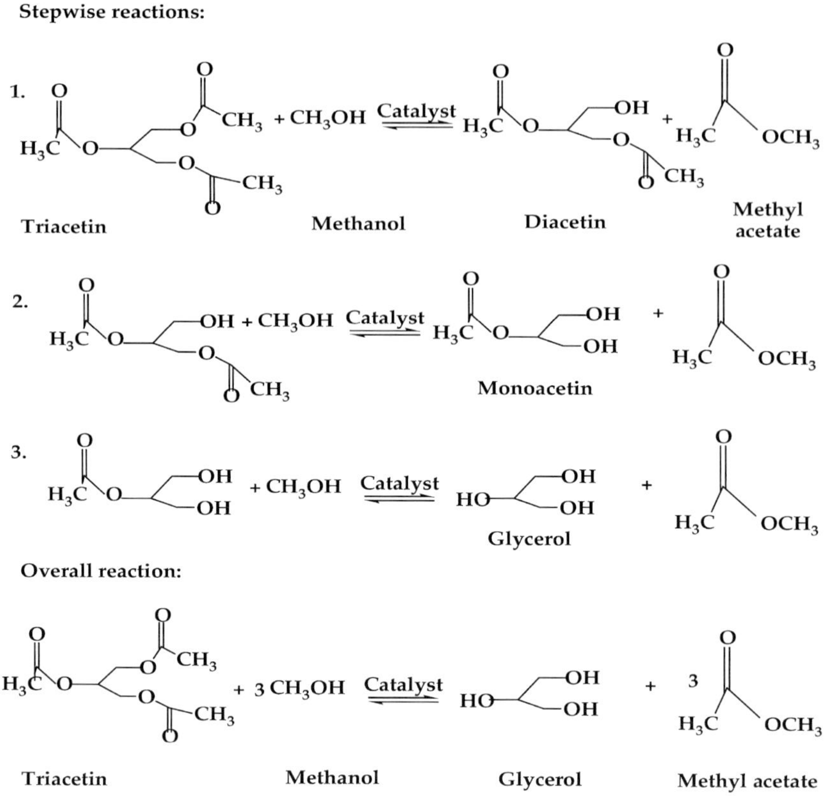

Stepwise reactions:

1. Triacetin + CH₃OH ⇌ (Catalyst) Diacetin + Methyl acetate

2. Diacetin + CH₃OH ⇌ (Catalyst) Monoacetin + Methyl acetate

3. Monoacetin + CH₃OH ⇌ (Catalyst) Glycerol + Methyl acetate

Overall reaction:

Triacetin + 3 CH₃OH ⇌ (Catalyst) Glycerol + 3 Methyl acetate

Figure 16.13 The transesterification reactions of triacetin and methanol.

[257], (5) hydrolysis and acid-catalyzed esterification, (6) pyrolysis [258] and (7) supercritical alcohol transesterification [259].

Industrially, homogeneous base catalysts (such as NaOH, KOH and NaOCH₃) are used. Base catalysis is preferred to the use of acid catalysts such as sulfuric or sulfonic acids, given the corrosivity and lower activity of the latter. However, the application of an alkaline catalyst in the transesterification of waste cooking oil is somewhat limited, because the FFA in waste cooking oil reacts with the alkaline catalyst (KOH, NaOH) and forms soap. The soap formed during the reaction prevents the glycerol separation, which drastically reduces the ester yield. The water in waste cooking oil also affects the methyl ester yield by favoring a saponification reaction. Further, removal of the base catalyst after reaction is problematic.

In order to circumvent these problems, the use of heterogeneous catalysts has been explored. This approach eliminates the need for an aqueous quench and largely eliminates the formation of metal salts, thereby simplifying downstream separation steps; consequently, biodiesel production can be more readily performed as a continuous process. Based on their ready availability, solid acid catalysts such as zeolites, clays and ion exchange resins are attractive for this purpose. Supported tungsten oxide catalysts have also received much attention because of their acid properties and their ability to catalyze both esterification and transesterification reactions, which play major roles in biodiesel production. Furuta and coworkers [260] reported that soybean oil can be efficiently converted to methyl esters at a reaction temperature of 250 °C using tungstated zirconia alumina (WZA) calcined at 800 °C. However, reaction rates are generally found to be unacceptably low. Consequently, solid base catalysts have attracted attention. Catalysts of this type include simple metal oxides such as MgO and CaO in supported or unsupported form, Zn–Al mixed oxides, cesium-exchanged zeolite X, anion exchange resins, polymer-supported guanidines, Na/NaOH/Al_2O_3 and K- and Li-promoted oxides, prepared by impregnating the corresponding nitrate or halide salt onto an oxidic carrier such as Al_2O_3, ZnO or CaO. Nano-crystalline calcium oxide is an efficient catalyst for the production of environmentally compatible biodiesel fuel in high yields at room temperature using soybean oil (SBO) and poultry fat as raw materials. An SBO : MeOH ratio of 1 : 27 is suitable for obtaining high product yields: the conversion can reach 99%. Under the same conditions, laboratory-grade CaO gave only 2% conversion in the case of SBO and there was no observable reaction with poultry fat [261].

16.4.3
Methanol Adsorption and Decomposition

One possible way to extend the application of metal oxides as catalysts is to tailor their surface chemistry, as the catalytic process occurs only on solid surfaces and surface structure typically controls activity and selectivity kinetically. Therefore, it is desirable to obtain a high surface area of inexpensive metal oxides with a preferentially grown outer surface to provide catalytically active sites, which requires a combination of both surface science and catalysis. One case in point is the typical rock salt structure of MgO. The traditional method of preparing MgO is the thermal decomposition of either magnesium salts or magnesium hydroxides, which results in inhomogeneity of morphology and crystallite size, with low surface area. Many attempts were therefore made to tailor the texture and morphology of MgO [262–267]. Most catalytic reactions reported to date are related to the well known and conventional (100) surface [233, 268]; however, the (111) surface is more interesting as it is composed of alternating monolayers of anions and cations so that a strong electrostatic field perpendicular to the (111) surface is created [234]. Such a surface has provided a prototype for the study of surface structure and surface reactions, and has received great attention for theoretical and experimental studies [235, 236]. Figure 16.14 shows the relation between the exposed face and the coordination of MgO [237].

Identification of the active sites of the MgO surface is of crucial importance for the understanding of the reaction mechanisms and the properties of MgO(111) nanosheets. Methanol is a "smart" molecular probe that can provide fundamental information about the number and nature of active surface sites [238]. Methanol

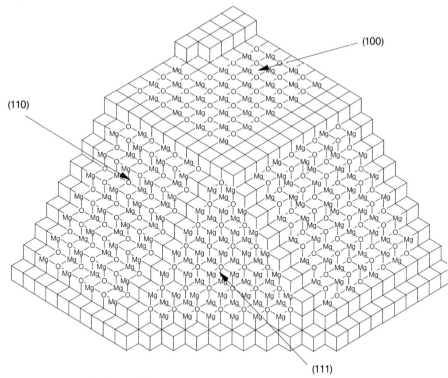

Figure 16.14 The disposal of MgO on (001), (110) and (111) facets (from [240]).

decomposition has been found to be structure-sensitive, in that the selectivity depends on the arrangement of the surface atoms. To satisfy ever-increasing energy needs, all feasible alternative energy sources to our available oil and gas resources must be considered. Methanol is an excellent fuel in its own right and it can also be blended with gasoline, although it has half the volumetric energy density relative to gasoline or diesel [239]. It is also used in the direct methanol fuel cell (DMFC). The performance of liquid-feed methanol fuel cells is already attractive for some applications and is approaching the levels required for electric vehicle propulsion [240]. In this electrochemical cell, methanol is directly oxidized with air to carbon dioxide and water to produce electricity, without the need to first generate hydrogen [241–244]. This greatly simplifies the fuel-cell technology and makes it accessible to a wide range of applications. For low-temperature fuel cells, restricted by the extreme sensitivity of fuel-cell catalysts to CO poisoning, it is very important to control the selectivity in this process [269]. Low-temperature fuel cells require an external processor to produce H_2 and remove CO, which not only reduces the efficiency of the system but also increases its complexity and cost [270]. The conventional Cu/ZnO-based methanol synthesis catalysts performed

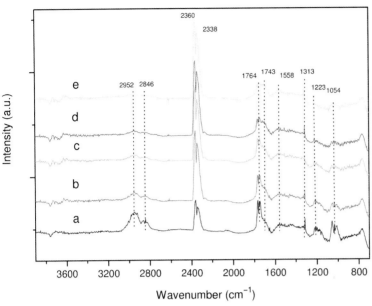

Figure 16.15 DRIFT spectra of methanol adsorption and reaction on MgO(111) nanosheets at 70 °C after different times: (a) 3 minutes, (b) 10 minutes, (c) 15 minutes, (d) 20 minutes, (e) 25 minutes (from [273]).

poorly for methanol decomposition and the catalysts suffered from rapid deactivation [271]. The activity and stability of the catalyst systems have been two major challenges in the decomposition of methanol. MgO(111) nanosheets can decompose methanol at low temperature and oxidize the surface C=O species formed during methanol decomposition. The preparation of MgO(111) nanosheets is simple and can be readily scaled-up. The large-scale application of MgO(111) nanosheet catalysts without the necessity for transition metals for low-temperature methanol decomposition may therefore be feasible.

The decomposition of methanol over MgO(111) nanosheets has been studied via *in situ* IR spectroscopy, and both undissociated and dissociated methanol were observed when MgO(111) nanosheets were exposed to methanol at 70 °C (Figure 16.15) [273]. A large amount of CO_2 formed upon exposure to methanol at 70 °C (peaks at 2360 and 2341 cm^{-1}) and increased with time. The bands centered at 1558, 1313 and 1223 cm^{-1} can be attributed to chemisorbed CO_2. A weak pair of peaks at 1764 and 1743 cm^{-1} can be attributed to the C=O asymmetric stretching of CO and formic acid [274, 275], free CO stretching (2036, 2014 cm^{-1}) peaks were also visible at 3 min. The intensity of the pair of peaks at 1764 and 1743 cm^{-1} decreased and the peaks at 2036 and 2014 cm^{-1} disappeared with time. This may indicate that the C=O species formed were oxidized by the oxygen anions on the surface of MgO(111) nanosheets. From the DRIFTS results we can see that methanol was oxidized and decomposed to near completion on the surface of MgO(111)

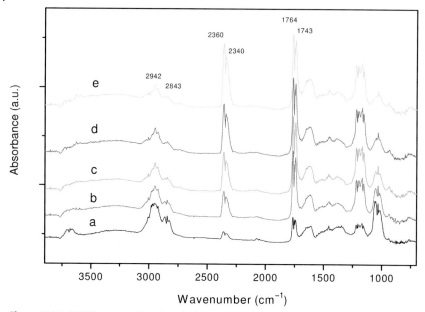

Figure 16.16 DRIFT spectra of methanol adsorption and reaction on CM-MgO at 70°C after different times: (a) 3 minutes, (b) 10 minutes, (c) 15 minutes, (d) 20 minutes and (e) 25 minutes (from [273]).

nanosheets by 25 min, indicating that MgO(111) nanosheets are highly active for methanol decomposition. In comparison with the spectra at room temperature, the region of O–H stretching has no distinct change; indicating that methanol interacts primarily with surface oxygen anions and oxygen defects but not hydroxyl groups at 70°C. This suggests that the main active sites for methanol decomposition are oxygen defects and oxygen anions. First, methanol interacts with oxygen defect sites to form methoxides, then the formed methoxides are oxidized by surface oxygen anions. In order to better understand the surface structure of MgO(111) nanosheets, methanol decomposition behavior and the relationship between them, methanol adsorption on the surface of Commerial MgO (CM-MgO) at 70°C under high vacuum was studied for comparison. The spectra are shown in Figure 16.16. The bands of CO_2 at 2360 and 2341 cm^{-1} increase and the bands of C–H stretching between 2800 and 3100 cm^{-1} decrease in intensity with time, indicating that methanol decomposition continues over time. Compared to the spectra in Figure 16.15e, the intensity of the bands between 2800 and 3100 cm^{-1} in Figure 16.16e are much stronger. This suggests that the methanol conversion is significantly lower and the activity of CM-MgO is lower than that of MgO(111) nanosheets. Comparison of the peaks at 1764 and 1743 cm^{-1} in Figures 16.15 and 16.16, which have been attributed to surface C=O species, shows that the pair of peaks in Figure 16.15 are weak and decrease in intensity with time, indicating that the rate of formation of the surface C=O species is lower than the rate of further oxidation; but in Figure 16.16 the pair of peaks are strong and increase in intensity with time, indicating that the rate of formation is higher than the further oxidation rate. This shows that the surface C=O species formed can be oxidized quickly by the high concentration of oxygen anions on the surface of MgO(111) nanosheets.

16.4.4
Destructive Adsorption of Chlorocarbons

The carcinogenic and toxic properties of chlorocarbons have encouraged scientists to find efficient ways to destroy them. The desired reaction is the complete oxidation of chlorocarbons to water, CO_2 and HCl, without the formation of any toxic by-products. High surface area and high surface reactivity are desired properties of such destructive adsorbents and nanoparticles fall into this category. Nanoparticles will react to a greater extent than normal metal oxides because there are more molecules of metal oxide available for reaction with chemicals adsorbed on the surface of the particles. Also, nanocrystals exhibit intrinsically higher chemical reactivities owing to the presence of a large number of defect sites such as edges/corners and high concentrations of coordinatively unsaturated ions. The mechanism to describe the oxidation of chlorocarbons has been developed and involves chemisorption followed by surface-to-adsorbate oxygen transfer and adsorbate-to-surface chlorine transfer, that is a chloride–oxide exchange [272].

CaO and MgO destroy chlorocarbons such as CCl_4, $CHCl_3$ and C_2Cl_4 at temperatures around 400 to 500 °C in the absence of an oxidant, stochiometrically yielding mainly CO_2 and the corresponding metal chlorides [276–278]. Nanoparticles of MgO and CaO react with CCl_4 to yield $MgCl_2$, $CaCl_2$, and CO_2. The reaction with $CHCl_3$ yields $CaCl_2$, CO and H_2O, while C_2Cl_4 yields $CaCl_2$, carbon and $CaCO_3$. All of these reactions are thermodynamically favorable; however, kinetic parameters demand that high-surface-area metal oxides are used. The mobility of oxygen and chlorine atoms in the bulk of the material becomes important and kinetic parameters involving the migration of these atoms are rate-limiting. The reaction efficiencies can be improved by the presence of small amounts of transition metal oxide as catalyst; for example, Fe_2O_3 on CaO. Morphological studies indicate that iron chloride intermediates help shuttle chloride and oxide anions in and out of the base metal oxide (CaO or MgO). Nano-crystalline MgO and CaO also allow the destruction of chlorinated benzenes (mono-, di- and trichlorobenzenes) at lower temperatures (700 to 900 °C) than incineration [279]. The presence of hydrogen as a carrier gas allows still lower temperatures to be used (e.g. 500 °C). MgO was found to be more reactive than CaO, as the latter induces the formation of more carbon.

In 1993, some environmental groups proposed the need for a chlorine-free economy. The cost of complete elimination of chlorinated compounds is quite staggering, with the latest estimate as high as $160 billion/year [280]. The most common method of destroying chlorocarbons is by high-temperature thermal oxidation (incineration). The toxic chlorinated compounds seem to be completely destroyed at high temperatures; however, there is concern about the formation of toxic by-products such as dioxins and furans. The catalytic oxidation of chlorinated compounds over metal oxides yields mainly HCl and CO_2. Catalysts such as V_2O_3 and Cr_2O_3 have been used with some success [281, 282]. Nano-scale MgO and CaO prepared by a modified aerogel/hypercritical drying procedure (AP-MgO and AP-CaO) were found to be superior to conventionally prepared (CP) CP-CaO, CP-MgO and commercial CaO/MgO catalysts for the dehydrochlorination of several toxic chlorinated substances [283, 284]. The interaction of 1-chlorobutane with nanocrystalline MgO at 200 to 350 °C results in both stoichiometric and catalytic dehydrochlorination of 1-chlorobutane to isomers of butene and simultaneous topochemical conversion of MgO to $MgCl_2$ [285, 286]. The particle sizes in these nano-scale materials are of the order of nanometers (~4 nm). These oxides are efficient owing to the presence of a high concentration of low-coordinated sites, structural defects on their surfaces and high specific surface area.

16.4.5
Alkene Metathesis

Maximizing the value of the steam cracker- or refinery-based C_4 stream is a major objective for most petrochemical companies. One approach towards this objective, olefin metathesis, has become an important reaction for producing a wide variety of more valuable products from available olefin substrates [287]. Figure 16.17 shows the possible reaction pathways of 1-butene. Tungsten oxide supported on silica is a well known metathesis catalyst [288, 289]. The activity of 1-butene metathesis over WO_3/SiO_2 was dependent on the states of tungsten oxide species on the surface of the catalyst, and the tetrahedrally coordinated tungsten oxide species on the surface of silica were suggested to be the active sites of metathesis [290]. Owing to the weak interaction of silica and tungsten oxide, WO_3/SiO_2 catalysts prepared by the traditional impregnation method have low amounts of surface tetrahedrally coordinated tungsten oxide species.

SBA-15 is a mesoporous silica molecular sieve with high surface area, tunable uniform hexagonal channels ranging from 5 to 30 nm and thick framework walls (3–6 nm), which is therefore thermally and hydrothermally robust [291, 292]. The active sites in the molecular sieves are often from heteroatoms such as Al and Ti species and it is therefore greatly significant to introduce heteroatoms into the

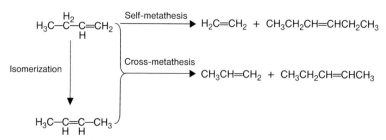

Figure 16.17 Possible reaction pathways of 1-butene (from [294]).

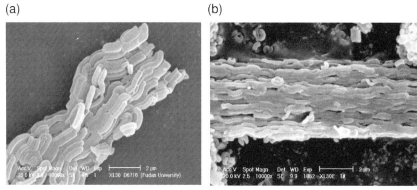

Figure 16.18 SEM images of (a) SBA-15 and (b) MWS (Si/W = 30) (from [294]).

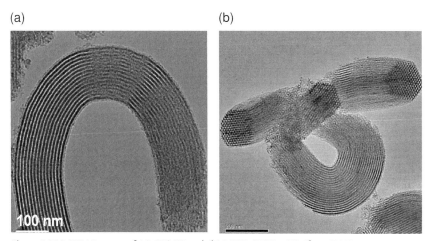

Figure 16.19 TEM images of (a) SBA-15 and (b) MWS (Si/W = 30) (from [294]).

framework of SBA-15 [293]. A one-step co-condensation sol–gel method improves dispersion of the tungsten species, alleviates aggregation of WO_3 and strengthens interactions between tungsten species and silica [294]. Figure 16.18 shows scanning electron microscopy (SEM) images of (a) SBA-15 and (b) MWS (Si/W = 30). The SEM image (Figure 16.18a) reveals that the SBA-15 sample consists of well defined wheat-like macrostructures aggregated together with rope-like domains with relatively uniform sizes of 1 μm. After tungsten oxide incorporation, a significant degradation in macroscopic structure can be observed; however, the rope-like domains with average sizes of 1 μm were still largely maintained (Figure 16.18b). Transmission electron microscopy (TEM) images (Figure 16.19) show the well ordered hexagonal arrays of one-dimensional mesoporous channels and confirm that the MWS samples have a two-dimensional P_{6mm} hexagonal structure, as does SBA-15, and unambiguously confirm that the hexagonal pore structure of the SBA-15 is robust enough to survive the tungsten-incorporation process and thus offers a good matrix to support highly dispersed tungsten oxide. The distance between two consecutive centers of the hexagonal pores estimated from the TEM image is ca. 10 nm. The average thickness of the wall is ca. 3 nm, which is much larger than for MCM-41 and the pore diameter is around 7 nm, in agreement with N_2 adsorption measurements. TEM analysis also reveals that no obvious extra-framework phases of the WO_3 species are present outside the mesoporous structure. The resulting tungsten-substituted mesoporous SBA-15 materials exhibit the best catalytic performance at present for the metathesis of 1-butene using tungsten-based catalysts.

16.4.6
Claisen–Schmidt Condensation

MgO nanoparticles have proven to be very effective chemical reagents in the Claisen–Schmidt condensation, which is a very valuable C–C bond-formation reaction commonly employed in the pharmaceutical and fine chemical industries [295]. When MgO(111) nanosheets were employed for the Claisen–Schmidt condensation of benzaldehyde and acetophenone, they were found to exhibit activity superior to other systems, such as $AlCl_3$, BF_3, $POCl_3$, alumina and other reported nano-crystalline MgO samples [296]. This development is particularly noteworthy as it represents a potential heterogenization of the Claisen–Schmidt catalytic process, which offers numerous advantages including easier product recovery and catalyst recycling.

It has been established that for the base-catalyzed Claisen–Schmidt condensation reaction, shape and exposed crystal facets are crucial parameters for the utilization of magnesium oxide as a solid base catalyst. Polar anionic surfaces are of particular interest for the reaction, and we have therefore evaluated the catalytic activity of MgO(111) nanosheets. The catalytic activities for the Claisen–Schmidt condensation of benzaldehyde with acetophenone at 110 °C over different crystallites of magnesium oxide samples are shown in Figure 16.20. CM-MgO samples are generally large cubic crystals, the NA-MgO samples are thin hexagonal platelets about 150 nm in diameter and 10 nm thick having large exposed areas of the (100) crystal face while the NAP-MgO samples are very small, irregular stacks of square plates exhibiting numerous crystal faces, edges and corners. As shown in Figure 16.20, NAP-MgO displays higher activity than that of NA-MgO and CM-MgO, which has been attributed to the NAP-MgO possessing a single-crystallite polyhedral structure, which has high surface concentrations of edges/corners and various exposed crystal planes (such as (001) and (111)), hence leading to an inherently high surface reactivity per unit area. It is notable that the MgO(111) nanosheets display a much higher catalytic activity for the Claisen–Schmidt condensation than the best of the reported NAP-MgO catalysts. Previous studies indicate that the Claisen–Schmidt condensation is largely driven by Lewis basic O^{2-} sites while the Brønsted basic –OH surface groups are less important in the reaction. The pronounced activity for the Claisen–Schmidt condensation taken together

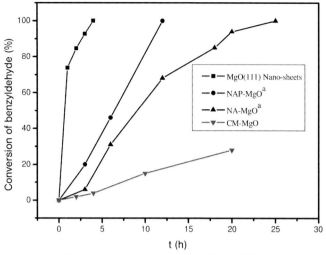

Figure 16.20 Claisen–Schmidt condensation of benzaldehyde with acetophenone using different crystallites of magnesium oxide at 110 °C. (NAP-MgO: aerogel prepared MgO, SA = 590 m² g⁻¹; NA-MgO: conventionally prepared MgO, SA = 250 m² g⁻¹; CM-MgO: commercial MgO, SA = 30 m² g⁻¹.) (from [69]).

with CO_2 temperature programmed desorption studies demonstrates clearly that the MgO(111) nanosheets are Lewis basic and the MgO(111) face which has O^{2-} anions on one surface is particularly important for the reaction. The high concentration of oxygen anions on the surface of MgO(111) thus provides a possible explanation for the excellent catalytic activity observed for the Claisen–Schmidt condensation.

16.5 Conclusions

Nano-scale metal oxides represent an important class of material, owing to their chemical and physical properties as well as their potential widespread applications. While a great deal of knowledge about these materials has developed during recent years, an atomic/molecular level understanding of the materials, their chemistry and their preparation will greatly impact their future. Ultimately, breakthroughs in the areas of synthesis, instrumentation and modeling will aid scientists in their quest to understand relationships between physical, electronic and chemical properties.

Metal oxides display a wide range of properties, from metallic to semiconductor to insulator. Owing to the compositional variability and their more localized electronic structures compared to metals, the presence of defects (such as corners, kinks, steps and coordinatively unsaturated sites) play a very important role in oxide surface chemistry and hence in catalysis. As described, the catalytic reactions also depend on the surface crystallographic structure and the example of MgO(111) has been highlighted. The catalytic properties of the oxide surfaces can often be explained in terms of Lewis acidity and basicity. The electronegative oxygen atoms accumulate electrons and act as Lewis bases while the metal cations act as Lewis acids.

As is hopefully obvious from this chapter and the entire book, the science of metal oxides is vast and requires synergistic contributions from multiple disciplines if the potential of nano-scale metal oxides is to be reached. As the general field of nano-scale materials develops, metal oxides are likely to be at the forefront owing to their stability and the intensive studies that have already been reported for their bulk counterparts.

References

1 Rao, C.N.R. and Gopalakrishnan, J. (1986) *New Directions in Solid State Chemistry*, Cambridge University Press, Cambridge, UK.
2 Klabunde, K.J., Stark, J.V., Koper, O., Mohs, C., Park, D.G., Decker, S., Jiang, Y., Lagadic, I. and Zhang, D. (1996) *Journal of Physical Chemistry*, **100**, 12142–53.
3 Klabunde, K.J. and Mohs, C. (1998) Nanoparticles and nanostructural materials, in *Chemistry of Advanced Materials: An Overview* (eds L.V. Interrante and M.J. Hampden-Smith), Wiley-VCH Verlag GmbH, New York, p. 317.
4 Dyrek, K. and Che, M. (1997) *Chemical Reviews*, **97**, 305–31.
5 Rao, C.N.R. and Raveau, B. (eds) (1998) *Transition Metal Oxides, Structure, Properties and Synthesis of Ceramic Oxides*, 2nd edn, Wiley-VCH Verlag GmbH, New York.
6 Shriver, D.F., Atkins, P.W. and Langford, C.H. (1990) *Inorganic Chemistry*, W.H. Freeman and Co., New York.
7 Zhang, X.G. and Armentrout, P.B. (2003) *Journal of Physical Chemistry A*, **107**, 8915–22.
8 Li, Y.X. and Klabunde, K.J. (1992) *Chemistry of Materials*, **4**, 611–15.
9 Klabunde, K.J., Stark, J.V., Koper, O., Mohs, C., Khaleel, A., Glavee, G.N., Zhang, D., Sorensen, C.M. and Hadjipanayis, G.C. (1994) *Nanophase Materials, Synthesis, Properties and Applications* (eds G.C. Hadjipanayis and R.W. Siegel), Kluwer Academic Publishers, Dordrecht, The Netherlands, p. 73.
10 Liu, H., Feng, L., Zhang, X. and Xue, O. (1995) *Journal of Physical Chemistry*, **99**, 332–4.
11 Radha, A.V., Thomas, G.S., Kamath, P.V. and Shivakumara, C. (2007) *Journal of Physical Chemistry*, **111**, 3384–90.
12 Utiyama, M., Hattori, H. and Tanabe, K. (1978) *Journal of Catalysis*, **53**, 237–42.
13 Morris, R.M. and Klabunde, K.J. (1983) *Inorganic Chemistry*, **22**, 682–7.
14 Driscoll, D.J., Martin, W., Wang, J.X. and Lunsford, J.H. (1985) *Journal of the American Chemical Society*, **107**, 58–63.
15 Hoq, M.F. and Klabunde, K.J. (1986) *Journal of the American Chemical Society*, **108**, 2114–16.
16 Zhang, Z., Bondarchuk, O., White, J.M., Kay, B.D. and Dohnalek, Z. (2006) *Journal of the American Chemical Society*, **128**, 4198–9.
17 Sanchez de Armas, R., Oviedo, J., San Mihuel, M.A. and Sanz, J.F. (2007) *Journal of Physical Chemistry C*, **111**, 10023–8.

18 Wyrwas, R.B. and Jarrold, C.C. (2006) *Journal of the American Chemical Society*, **128**, 13688–9.

19 Philipp, R., Omata, K., Aoki, A. and Fujimoto, K. (1992) *Journal of Catalysis*, **134**, 422–33.

20 Ito, T., Wang, J.X., Liu, C.H. and Lunsford, J.H. (1985) *Journal of the American Chemical Society*, **107**, 5062–8.

21 Baily, M.L., Chizallet, C., Costentin, G., Krafft, J.M., Lauron-Pernot, H. and Che, M. (2005) *Journal of Catalysis*, **235**, 413–22.

22 Stakheev, A.Yu., Shapiro, E.S. and Apijok, J. (1993) *Journal of Physical Chemistry*, **97**, 5668–72.

23 Vishwanathan, B., Tanka, B. and Toyoshima, L. (1986) *Langmuir*, **2**, 113–16.

24 Chizallet, C., Bailly, M.L., Costentin, G., Lauron-Pernot, H., Krafft, J.M., Bazin, P., Saussey, J. and Che, M. (2006) *Catalysis Today*, **116**, 196–205.

25 Sun, N. and Klabunde, K.J. (1999) *Journal of the American Chemical Society*, **121**, 5587–8.

26 Stankic, S., Sterrer, M., Hofmann, P., Diwald, O. and Knoziger, E. (2005) *Nano Letterss*, **5**, 1889–93.

27 Napoli, F., Chiesa, M., Giamello, E., Finazzi, E., Di Valentin, C. and Pacchioni, G. (2007) *Journal of the American Chemical Society*, **129**, 10575–81.

28 Richards, R., Li, W., Decker, S., Davidson, C., Koper, O., Zaikovski, V., Volodin, A., Rieker, T. and Klabunde, K.J. (2000) *Journal of the American Chemical Society*, **122** (20), 4921–5.

29 Mars, P. and van Kravelen, D.W. (1954) *Chemical Engineering Science*, **3**, 41–59.

30 Itoh, H., Utamapanya, S., Stark, J.V., Klabunde, K.J. and Schlup, J.R. (1993) *Chemistry of Materials*, **5** (1), 71–7.

31 Palkar, V.R. (1999) *Nanostructured Materials*, **11** (3), 369–74.

32 Klabunde, K.J. and Mohs, C. (1998) *Chemistry of Advanced Materials: An Overview* (eds L.V. Interrante and M.J. Hampden-Smith), Wiley-VCH Verlag GmbH, New York, p. 317.

33 Feldmann, C., Matschulo, S. and Ahlert, S. (2007) *Journal of Materials Science*, **42**, 7076–80.

34 Gesser, H.D. and Goswami, P.C. (1989) *Chemical Reviews*, **89** (4), 765–88.

35 Agarwala, M.K., Bourell, D.L. and Persad, C. (1992) *Journal of the American Ceramic Society*, **75**, 1975–7.

36 Chatry, M., Henry, M. and Livage, J. (1994) *Materials Research Bulletin*, **29**, 517–22.

37 Kumazawa, H., Inoue, T. and Sada, E. (1994) *Chemical Engineering Journal*, **55**, 93–6.

38 Laberty-Robert, C., Long, J.W., Lucas, E.M., Pettigrew, K.A., Strous, R.M., Doescher, M.S. and Rolison, D.R. (2006) *Chemistry of Materials*, **18**, 50–8.

39 Hench, L.L. and West, J.K. (1990) *Chemical Reviews*, **90** (1), 33–72.

40 Malenfant, P.R.L., Wan, J.L., Taylor, S.T. and Manoharan, M. (2007) *Nature Nanotechnology*, **2**, 43–6.

41 Polarz, S., Roy, A., Lehmann, M., Driess, M., Kruis, F.E., Hoffmann, A. and Zimmer, P. (2007) *Advanced Functional Materials*, **17**, 1385–91.

42 Chandler, C.D., Roger, C. and Hampden-Smith, M.J. (1993) *Chemical Reviews*, **93** (3), 1205–41.

43 Brinker, C.J. and Scherer, G.W. (1990) *Sol–Gel Science*, Academic Press, San Diego, CA.

44 Segal, D. (1989) *Chemical Synthesis of Advanced Ceramic Materials* (eds A.R. West and E.H. Baxter), Cambridge University Press, Cambridge, UK, p. 58.

45 Hench, L.L. and Nogues, J.L. (1993) *Sol–Gel Optics: Processing and Applications* (ed. L.C. Klein), Kluwer Academic Publishers, Boston, MA, p. 39.

46 Narula, C.K. (1995) *Ceramic Precursor Technology and Its Applications*, Marcel Dekker, New York.

47 Siegel, R.W., Ramasamy, S., Hahn, H., Li, Z.Q., Lu, T. and Gronsky, R. (1988) *Journal of Materials Research*, **3** (6), 1367–72.

48 El-Shall, M.S., Slack, W., Vann, W., Kane, D. and Hanley, D. (1994) *Journal of Physical Chemistry*, **98** (12), 3067–70.

49 Epifani, M., Diaz, R., Arbiol, J., Comini, E., Sergent, N., Pagnier, T., Siciliano, P., Taglia, G. and Morante, J.R. (2006)

Advanced Functional Materials, **16**, 1488–98.
50 Skarman, B., Nakayama, T., Grandjean, D., Benfiels, R.E., Niihara, K. and Wallenberg, L.R. (2002) *Chemistry of Materials*, **14**, 3686–99.
51 Hadjipanayis, G.C. and Siegel, R.W. (eds) (1994) *Nanophase Materials: Synthesis, Properties and Application*, Kluwer Academic Publishers, Dordrecht, The Netherlands.
52 Xu, C.H., Shi, S.Q., Surya, C. and Woo, C.H. (2007) *Journal of Materials Science*, **42**, 9855–8.
53 Cow, G.M. and Gonsalves, K.E. (eds) (1996) *Nanotechnology, Molecularly Designed Materials*, American Chemical Society, Washington, DC.
54 Majumdar, D., Kodas, T.T. and Glicksman, H.D. (1996) *Advanced Materials*, **8** (12), 1020–2.
55 Janackovic, D., Jokanovic, V., Kostic-Gvozdenovic, L. and Uskokovic, D. (1998) *Nanostructured Materials*, **10** (3), 341–8.
56 Jossen, R., Pratsinis, S.E., Stark, W.J. and Masler, L. (2005) *Journal of the American Chemical Society*, **88**, 1388–93.
57 Parguay-Delgado, F., Antunez-Flores, W., Miki-Yoshida, M., Aguilar-Elguezaba, A., Santiago, P., Siaz, R. and Ascencio, J.A. (2005) *NanoTechnology*, **16**, 688–94.
58 Witanachchi, S., Dedigamuwa, G. and Mukherjee, P. (2007) *Journal of Materials Research*, **22**, 649–54.
59 Jayanthi, G.V., Zhang, S.C. and Messing, G.L. (1993) *Aerosol Science and Technology*, **19** (4), 478–90.
60 Kavitha, R., Meghani, S. and Jayaram, V. (2007) *Materials Science and Engineering B*, **2007** (139), 134–40.
61 Ulrich, G.D. and Riehl, J.W. (1982) *Journal of Colloid and Interface Science*, **87** (1), 257–65.
62 Lindackers, D., Janzen, C., Rellinghaus, B., Wassermann, E.F. and Roth, P. (1998) *Nanostructured Materials*, **10** (8), 1247–70.
63 Skandan, G., Chen, Y.-J., Glumac, N. and Kear, B.H. (1999) *Nanostructured Materials*, **11** (2), 149–58.
64 Klabunde, K.J., Stark, J.V., Koper, O., Mohs, C., Park, D.G., Decker, S., Jiang, Y., Lagadic, I. and Zhang, D.J. (1996) *Journal of Physical Chemistry*, **100** (30), 12142–53.
65 Klabunde, K.J. (ed.) (2001) *Nanoscale Materials in Chemistry*, Wiley Interscience, New York, pp. 85–114.
66 Diao, Y.L., Walawender, W.P., Sorensen, C.M., Klabunde, K.J. and Ricker, T. (2002) *Chemistry of Materials*, **14** (1), 362–8.
67 Matson, D.W. and Smith, R.D. (1989) *Journal of the American Ceramic Society*, **72** (6), 871–81.
68 Schwarz, J.A., Contescu, C. and Putyera, K. (eds) (2004) *Encyclopedia of Nanoscience and Nanotechnology*, Marcel Dekker, New York, pp. 1905–9.
69 Zhu, K.K., Hu, J.C., Kübel, C. and Richards, R. (2006) *Angewandte Chemie – International Edition*, **45** (43), 7277–81.
70 Utamapanya, S., Klabunde, K.J. and Schlup, J.R. (1991) *Chemistry of Materials*, **3** (1), 175–81.
71 Klabunde, K.J., Stark, J.V., Koper, O., Mohs, C., Khaleel, A., Glavee, G.N., Zhang, D., Sorensen, C.M. and Hadjipanayis, G.C. (1994) *Nanophase Materials, Synthesis, Properties and Applications* (eds G.C. Hadjipanayis and R.W. Siegel), Kluwer Academic Publishers, Dordrecht, The Netherlands, p. 71.
72 Bedilo, A.F. and Klabunde, K.J. (1997) *Nanostructured Materials*, **8** (2), 119–35.
73 Zhang, Y.X., Li, G.H., Wu, Y.C., Luo, Y.Y. and Zhang, L.D. (2005) *Journal of Physical Chemistry B*, **109** (12), 5478–81.
74 Cerda, J., Arbiol, J., Diaz, R., Dezanneau, G. and Morante, J.R. (2002) *Material Letters*, **56** (3), 131–6.
75 O'Brien, S., Brus, L. and Murray, C.B. (2001) *Journal of the American Chemical Society*, **123** (48), 12085–6.
76 Meron, T., Rosenberg, Y., Lereah, Y. and Markovich, G. (2005) *Journal of Magnetism and Magnetic Materials*, **292**, 11–16.
77 Harreld, J.H., Dong, W. and Dunn, B. (1998) *Materials Research Bulletin*, **33** (4), 561–7.
78 Dong, W., Rolison, D.R. and Dunn, B. (2000) *Electrochemical and Solid State Letters*, **3** (10), 457–9.

References

79 Long, J.W., Swider-Lyons, K.E., Stroud, R.M. and Rolison, D.R. (2000) *Electrochemical and Solid State Letters*, **3** (10), 453–6.
80 Long, J.W., Young, A.L. and Rolison, D.R. (2003) *Journal of the Electrochemical Society*, **150** (9), A1161–5.
81 Swider, K.E., Merzbacher, C.I., Hagans, P.L. and Rolison, D.R. (1997) *Chemistry of Materials*, **9** (5), 1248–51.
82 Gao, L., Wang, H.Z., Hong, J.S., Miyamoto, H., Miyamoto, K., Nishikawa, Y. and Torre, S.D.D.L. (1999) *Nanostructured Materials*, **11** (1), 43–9.
83 Qian, Z. and Shi, J.L. (1998) *Nanostructured Materials*, **10** (2), 235–44.
84 Rao, K.R.M., Rao, A.V.P. and Komarneni, S. (1996) *Materials Letters*, **28** (4–6), 463–7.
85 Corma, A. (1995) *Chemical Reviews*, **95**, 559–614.
86 Srinivasan, R., Watkins, T.R., Hubbard, C.R. and Davis, B.H. (1995) *Chemistry of Materials*, **7** (4), 725–30.
87 Pârvulescu, V., Coman, S., Grange, P. and Pârvulescu, V.I. (1999) *Applied Catalysis A*, **176** (1), 27–43.
88 Pârvulescu, V.I., Pârvulescu, V., Endruschat, U., Lehmann, C.W., Grange, P., Poncelet, G. and Bönnemann, H. (2001) *Microporous and Mesoporous Materials*, **44–45**, 221–6.
89 Pârvulescu, V.I., Bönnemann, H., Pârvulescu, V., Endruschat, U., Rufinska, A., Lehmann, C.W., Tesche, B. and Poncelet, G. (2001) *Applied Catalysis A*, **214** (2), 273–87.
90 Stichert, W. and Schüth, F. (1998) *Journal of Catalysis*, **174** (2), 242–5.
91 Chuah, G.K., Jaenicke, S. and Pong, B.K. (1998) *Journal of Catalysis*, **175** (1), 80–92.
92 Bedilo, A.F. and Klabunde, K.J. (1998) *Journal of Catalysis*, **176** (2), 448–58.
93 Bedilo, A.F. and Klabunde, K.J. (1997) *Nanostructured Materials*, **8** (2), 119–35.
94 Bedilo, A.F., Timoshok, A.V. and Volodin, A.M. (2000) *Catalysis Letters*, **68** (3–40), 209–24.
95 (a) Heuer, A.H. and Hobbs, L.W. (eds) (1981) *Science and Technology of Zirconia I*, American Ceramic Society, Westerville, OH.
(b) Claussen, N. Rühle, M. and Heuer, A.H. (eds) (1984) *Science and Technology of Zirconia II*, American Ceramic Society, Westerville, OH.
(c) Claussen, N., Rühle, M. and Heuer, A. H. (eds) (1988) *Science and Technology of Zirconia III*, American Ceramic Society, Westerville, OH.
96 Chane-Ching, J.Y., Cobo, F., Aubert, D., Harvey, H.G., Airiau, M. and Corma, A. (2005) *Chemistry – A European Journal*, **11** (3), 979–87.
97 Hudson, M.J. and Knowles, J.A. (1996) *Journal of Materials Chemistry*, **6** (1), 89–95.
98 Liu, X.M., Lu, G.Q. and Yan, Z.F. (2004) *Journal of Physical Chemistry B*, **108** (40), 15523–8.
99 Zhou, L.P., Xu, J., Li, X.Q. and Wang, F. (2006) *Materials Chemistry and Physics*, **97** (1), 137–42.
100 Rezaei, M., Alavi, S.M., Sahebdelfar, S., Yan, Z.F., Teunissen, H., Jacobsen, J.H. and Sehested, J. (2007) *Journal of Materials Science*, **42** (4), 1228–37.
101 D'Souza, L., Saleh-Subaie, J. and Richards, R. (2005) *Journal of Colloid and Interface Science*, **292**, 476–85.
102 Landau, M.V., Vradman, L., Herskowitz, M., Koltypin, Y. and Gedanken, A. (2001) *Journal of Catalysis*, **201** (1), 22–36.
103 Perkas, N., Wang, Y.Q., Koltypin, Y., Gedanken, A. and Chandrasekhar, S. (2001) *Chemical Communications*, 988–9.
104 Ramesh, S., Koltypin, Y., Prozorov, R. and Gedanken, A. (1997) *Chemistry of Materials*, **9** (2), 546–51.
105 Pol, V.G., Reisfeld, R. and Gedanken, A. (2002) *Chemistry of Materials*, **14**, 3920–4.
106 Li, J.G., Ikegami, T., Wang, Y.R. and Mori, T. (2002) *Journal of Solid State Chemistry*, **168** (1), 52–9.
107 Xiang, L., Deng, X.Y. and Jin, Y. (2002) *Scripta Materialia*, **47** (4), 219–24.
108 Borse, P.H., Kankate, L.S., Dassenoy, F., Vogel, W., Urban, J. and Kulkarni, S.K. (2002) *Journal of Materials Science, Materials in Electronics*, **13** (9), 553–9.
109 Rojas, T.C. and Ocana, M. (2002) *Scripta Materialia*, **46** (9), 655–60.
110 Tang, Z.X., Sorensen, C.M., Klabunde, K.J. and Hadjipanayis, G.C. (1991) *Journal of Colloid and Interface Science*, **146** (1), 38–52.

111 Jeyadevan, B., Chinnasamy, C.N., Perales-Perez, O., Iwasaki, Y., Hobo, A., Shinoda, K., Tohji, K. and Kasuya, A. (2002) *IEEE Transactions on Magnetics*, **38** (5), 2595–7.

112 Li, J., Dai, D.L., Zhao, B.G., Lin, Y.Q. and Liu, C.Y. (2002) *Journal of Nanoparticle Research*, **4** (3), 261–4.

113 Wang, M.L., Wang, C.H. and Wang, W. (2007) *Journal of Materials Chemistry*, **17** (20), 2133–8.

114 Natile, M.M., Boccaletti, G. and Glisenti, A. (2005) *Chemistry of Materials*, **17** (25), 6272–86.

115 Perera, S., Zelenski, N. and Gillan, E.G. (2006) *Chemistry of Materials*, **18** (9), 2381–8.

116 Gui, Z., Fan, R., Mo, W.Q., Chen, X.H., Yang, L. and Hu, Y. (2003) *Materials Research Bulletin*, **38** (1), 169–76.

117 Liu, Z.L., Liu, Y.J., Yao, K.L., Ding, Z.H., Tao, J. and Wang, X. (2002) *Journal of Materials Synthesis and Processing*, **10** (2), 83–7.

118 Sun, X.D., Ma, C.L., Wang, Y.D. and Li, H.D. (2002) *Inorganic Chemistry Communications*, **5** (10), 747–50.

119 Zhang, Z.L., Guo, L. and Wang, W.D. (2001) *Journal of Materials Research*, **169** (3), 803–5.

120 Palchik, O., Kerner, R., Gedanken, A., Weiss, A.M., Slifkin, M.A. and Palchik, V. (2001) *Journal of Materials Chemistry*, **11**, 874–8.

121 Grisaru, H., Palchik, O., Gedanken, A., Palchik, V., Slifkin, M.A. and Weiss, A.M. (2002) *Journal of Materials Chemistry*, **12**, 339–44.

122 Kerner, R., Palchik, O. and Gedanken, A. (2001) *Chemistry of Materials*, **13**, 1413–19.

123 Laudise, R.A. (1962) *Progress in Inorganic Chemistry*, Vol III (ed. F.A. Cotton), Interscience, New York, p. 1.

124 Barrer, R.M. (1982) *Hydrothermal Chemistry of Zeolites*, Academic Press, London.

125 Rabenau, A. (1985) *Angewandte Chemie – International Edition in English*, **24**, 1026–40.

126 Laudise, R.A. (1987) Hydrothermal synthesis of crystals, C&EN News, September 28, 30–43.

127 Suslick, K.S., Choe, S.B., Cichowlas, A.A. and Grinstaff, M.W. (1991) *Nature*, **353**, 414–16.

128 Hiller, R., Putterman, S.J. and Barber, B.P. (1992) *Physical Review Letters*, **69** (8), 1182–4.

129 Barber, B.P. and Putterman, S.J. (1991) *Nature*, **352**, 318–20.

130 Suslick, K.S., Hammerton, D.A. and Cline, R.E. (1986) *Journal of the American Chemical Society*, **108** (18), 5641–2.

131 Livage, J. (1981) *Journal of Physics*, **42** (NC4), 981–92.

132 Sugimoto, M. (1994) *Journal of Magnetism and Magnetic Materials*, **133** (1–3), 460–2.

133 Koltypin, K.V.P.M., Shafi, Y., Gedanken, A., Prozorov, R., Balogh, J., Lendavi, J. and Felner, I. (1997) *Journal of Physical Chemistry B*, **101** (33), 6409–14.

134 Liang, J.H., Jiang, X., Liu, G., Deng, Z.X., Zhuang, J., Li, F. and Li, Y.D. (2003) *Materials Research Bulletin*, **38** (1), 161–8.

135 Pang, G., Xu, X., Markovich, V., Avivi, S., Palchik, O., Koltypin, Y., Gorodetsky, G., Yeshurun, Y., Buchkremer, H.P. and Gedanken, A. (2003) *Materials Research Bulletin*, **38**, 11–16.

136 de Torresi, S.I.C., Vidotti, M., Ponzio, E.A. and Greco, C.V. (2006) *ElectroChemical Communications*, **8** (4), 554–60.

137 Fulton, J.L., Matson, D.W., Pecher, K.H., Amonette, J.E. and Linehan, J.C. (2006) *Journal of Nanoscience and Nanotechnology*, **6** (2), 562–7.

138 Zhang, D., Fu, H., Shi, L., Pan, C., Li, Q., Chu, Y. and Yu, W. (2007) *Inorganic Chemistry*, **46** (7), 2446–51.

139 Tu, W.X. and Liu, H.F. (2000) *Chemistry of Materials*, **12** (2), 564–7.

140 Palchik, O., Zhu, J.J. and Gedanken, A. (2000) *Journal of Materials Chemistry*, **10**, 1251–4.

141 Breck, D.W. (1974) *Zeolite Molecular Sieves, Structure, Chemistry and Use*, John Wiley & Sons, Inc., New York.

142 Corbett, J.D. (1985) *Chemical Reviews*, **85**, 383–97.

143 Lii, K.H. and Haushalter, R.C. (1987) *Journal of Solid State Chemistry*, **69**, 320.

144 Haushalter, R.C. and Lai, F.W. (1988) *Journal of Solid State Chemistry*, **76**, 218.

145 Haushalter, R.C., Strohmaier, K.G. and Lai, F.W. (1989) *Science*, **246**, 1289–91.

146 Mundi, L.A., Strohmaier, K.G., Goshorn, D.P. and Haushalter, R.C. (1990) *Journal of the American Chemical Society*, **112** (22), 8182–3.

147 Mundi, L.A. and Haushalter, R.C. (1990) *Inorganic Chemistry*, **29** (16), 2879–81.

148 Haushalter, R.C. and Mundi, L.A. (1992) *Chemistry of Materials*, **4** (1), 31–48.

149 Cao, Y., Hu, J., Hong, Z., Deng, J. and Fan, K. (2002) *Catalysis Letters*, **81** (1), 107–12.

150 Pinna, N., Garnweitner, G., Antonietti, M. and Niederberger, M. (2005) *Journal of the American Chemical Society*, **127** (15), 5608–12.

151 Niederberger, M., Pinna, N., Polleux, J. and Antonietti, A. (2004) *Angewandte Chemie – International Edition*, **43** (17), 2270–3.

152 Su, C.Y., Goforth, A.M., Smith, M.D., Pellechia, P.J. and zur Loye, H.C. (2004) *Journal of the American Chemical Society*, **126** (11), 3576–86.

153 Parise, J.B., Ko, Y.H., Rijssenbeck, J., Nellis, D.M., Tan, K.M. and Koch, S. (1994) *Journal of the Chemical Society – Chemical Communications*, 527.

154 Tan, K.M., Ko, Y.H., Parise, J.B. and Darovsky, A. (1996) *Chemistry of Materials*, **8** (2), 448–53.

155 Ramirez, A.P., Cava, R.J. and Krajewski, J. (1997) *Nature*, **386**, 156–9.

156 Stein, A., Keller, S.W. and Mallouk, T.E. (1993) *Science*, **259**, 1558–64.

157 Sheldrick, W.S. and Wachhold, M. (1997) *Angewandte Chemie – International Edition in English*, **36** (3), 207–24.

158 Chemseddine, A. and Moritz, T. (1999) *European Journal of Inorganic Chemistry*, 235–45.

159 Trentler, T.J., Denler, T.E., Bertone, J.F., Agrawal, A. and Colvin, V.L. (1999) *Journal of the American Chemical Society*, **121** (7), 1613–14.

160 Rockenberger, J., Scher, E.C. and Alivisatos, A.P. (1999) *Journal of the American Chemical Society*, **121** (49), 11595–6.

161 Thimmaiah, S., Rajamathi, M., Singh, N., Bera, P., Meldrum, F., Chandrasekhar, N. and Seshadri, R. (2001) *Journal of Materials Chemistry*, **11** (12), 3215–21.

162 Oguri, Y., Riman, R.E. and Bowen, H.K. (1988) *Journal of Materials Science*, **23**, 2897–904.

163 Kondo, M., Shinozaki, K., Ooki, R. and Mizutani, N. (1994) *Journal of the Ceramic Society of Japan*, **102** (2), 742–846.

164 Jeon, S. and Braun, P.V. (2003) *Chemistry of Materials*, **15** (6), 1256–63.

165 Cheng, H.M., Ma, J.M., Zhao, Z.G. and Qi, L.M. (1995) *Chemistry of Materials*, **7** (4), 663–71.

166 Zheng, Y.Q., Shi, E.R., Chen, Z.Z., Li, W.J. and Hu, X.F. (2001) *Journal of Materials Chemistry*, **11** (5), 1547–51.

167 Yin, H.B., Wada, Y., Kitamura, T., Kambe, S., Murasawa, S., Mori, H., Sakata, T. and Yanagida, S. (2001) *Journal of Materials Chemistry*, **11** (6), 1694–703.

168 Niederberger, M., Garnweitner, G., Pinna, N. and Antonietti, M. (2004) *Journal of the American Chemical Society*, **126** (29), 9120–6.

169 Masui, T., Hirai, H., Hamada, R., Imanaka, N., Adachi, G., Sakata, T. and Mori, H. (2003) *Journal of Materials Chemistry*, **13** (3), 622–7.

170 Inoue, M., Kimura, M. and Inui, T. (1999) *Chemical Communications*, 957–8.

171 Wu, M.M., Long, J.B., Huang, A.H. and Luo, Y.J. (1999) *Langmuir*, **15** (26), 8822–5.

172 Andersson, M., Österlund, L., Ljungström, S. and Palmqvist, A. (2002) *Journal of Physical Chemistry B*, **106**, 10674–10679.

173 Gautam, U.K., Ghosh, M., Rajamathi, M. and Seshadri, R. (2002) *Pure and Applied Chemistry*, **74** (9), 1643–9.

174 He, P., Shen, X.H. and Gao, H.C. (2005) *Journal of Colloid and Interface Science*, **284** (2), 510–15.

175 Wallin, M., Cruise, N., Klement, U., Palmqvist, A., Skoglundh, M. and Klement, U. (2004) *Colloids and Surfaces A*, **238** (1–3), 27–35.

176 Lu, C.H. and Wang, H.C. (2004) *Journal of the European Ceramic Society*, **24** (5), 717–23.

177 Koetz, J., Bahnemann, J., Lucas, G., Tiersch, B. and Kosmella, S. (2004) *Colloids and Surfaces A*, **250** (1–3), 423–30.

178 Chen, D.L. and Gao, L. (2004) *Journal of Colloid and Interface Science*, **279** (1), 137–42.

179 Zhang, X. and Chan, K.Y. (2003) *Chemistry of Materials*, **15** (2), 451–9.
180 Holzinger, D. and Kickelbick, G. (2003) *Chemistry of Materials*, **15** (26), 4944–8.
181 Bruch, C., Kruger, J.K., Unruh, H.G., Krauss, W., Zimmermeier, B., Beck, C. and Hempelmann, R. (1997) *Berichte Der Bunsen-Gesellschaft-Physical Chemistry Chemical Physics*, **101** (11), 1761–4.
182 Hartl, W., Beck, C., Roth, M., Meyer, F. and Hempelmann, R. (1997) *Berichte Der Bunsen-Gesellschaft-Physical Chemistry Chemical Physics*, **101** (11), 1714–17.
183 Fang, J.Y., Wang, J., Ng, S.C., Chew, C.H. and Gan, L.M. (1997) *Nanostructured Materials*, **8** (4), 499–505.
184 Zhang, Z.J., Wang, Z.L., Chakoumakos, B.C. and Yin, J.S. (1998) *Journal of the American Chemical Society*, **120** (8), 1800–4.
185 Moran, P.D., Bartlett, J.R., Bowmaker, G.A., Woolfrey, J.L. and Cooney, R.P. (1999) *Journal of Sol-Gel Science and Technology*, **15** (3), 251–62.
186 Yener, D.O. and Giesche, H. (1999) *Journal of the American Chemical Society*, **84** (9), 1987–95.
187 Kumar, P., Pillai, V., Bates, S.R. and Shah, D.O. (1993) *Materials Letters*, **16** (2–3), 68–74.
188 Xiong, L.F. and He, T. (2006) *Chemistry of Materials*, **18** (9), 2211–18.
189 Pang, Y.X. and Bao, X.J. (2002) *Journal of Materials Chemistry*, **12** (12), 3699–704.
190 Ulrich, G.D. and Riehl, J.W. (1982) *Journal of Colloid and Interface Science and Technology*, **87** (1), 257–65.
191 Cow, G.M. and Gonsalves, K.E. (eds) (1996) *Nanotechnology, Molecularly Designed Materials* American Chemical Society, Washington, DC, pp. 64–78, 79–99.
192 Singhal, A., Skandan, G., Wang, A., Glumac, N., Kear, B.H. and Hunt, R.D. (1999) *Nanostructured Materials*, **11** (4), 545–52.
193 Wu, M.K., Windeler, R.S., Steiner, C.K.R., Bors, T. and Friedlander, S.K. (1993) *Aerosol Science and Technology*, **19** (4), 527–48.
194 Lindackers, D., Strecker, M.G.D., Roth, P., Janzen, C. and Pratsinis, S.E. (1997) *Combustion Science and Technology*, **123** (1–6), 287–315.
195 Glumac, N.G., Chen, Y.J., Skandan, G. and Kear, B. (1998) *Materials Letters*, **34** (3–6), 148–53.
196 Zachariah, M.R., Chin, D., Semerjian, H.G. and Katz, J.L. (1989) *Combustion and Flame*, **78** (3–4), 287–98.
197 Lima, M.D., Bonadimann, R., de Andrade, M.J., Toniolo, J.C. and Bergmann, C.P. (2006) *Journal of the European Ceramic Society*, **26** (7), 1213–20.
198 Stark, W.J., Mädler, L., Maciejewski, M., Pratsinis, S.E. and Baiker, A. (2003) *Chemical Communications*, 588–9.
199 Purohit, R.D., Saha, S. and Tyagi, A.K. (2006) *Materials Science and Engineering B*, **130** (1–3), 57–60.
200 Kovacheva, D., Gadjov, H., Petrov, K., Mandal, S., Lazarraga, M.G., Pascual, L., Amarilla, J.M., Rojas, R.M., Herrero, P. and Rojo, J.M. (2002) *Journal of Materials Chemistry*, **12** (4), 1184–8.
201 Kodas, T.T. (1989) *Advanced Materials*, **6**, 180–92.
202 Janackovic, D., Jokanovic, V., Kostic-Gvozdenovic, L. and Uskokovic, D. (1998) *Nanostructured Materials*, **10** (3), 341–8.
203 Messing, G.L. and Gardner, T. (1984) *American Ceramic Society Bulletin*, **64**, 1498–501.
204 Pollinger, J.P. and Messing, G.L. (1993) *Journal of Aerosol Science and Technology*, **19** (4), 217–27.
205 Jayanthi, G.V., Zhang, S.C. and Messing, G.L. (1993) *Journal of Aerosol Science and Technology*, **19** (4), 478–90.
206 Kodas, T.T., Datye, A., Lee, V. and Engler, E. (1989) *Journal of Applied Physics*, **65**, 2149–51.
207 Zhang, H.F. and Cooper, A.I. (2005) *Soft Matter*, **1** (2), 107–13.
208 Li, W.-C., Lu, A.-H., Weidenthaler, C. and Schuth, F. (2004) *Chemistry of Materials*, **16** (26), 5676–81.
209 Tian, B.Z., Liu, X.Y., Yang, H.F., Xie, S.H., Yu, C.Z., Tu, B. and Zhao, D.Y. (2003) *Advanced Materials*, **15** (15), 1370–4.
210 Tian, B.Z., Liu, X.Y., Solovyov, L.A., Liu, Z., Yang, H.F., Zhang, Z.D., Xie, S.H., Zhang, F.Q., Tu, B., Yu, C.Z., Terasaki,

O. and Zhao, D.Y. (2004) *Journal of the American Chemical Society*, **126** (3), 865–75.
211 Fuertes, A.B. (2005) *Journal of Physics and Chemistry of Solids*, **66**, 741–7.
212 Laha, S.C. and Ryoo, R. (2003) *Chemical Communications*, 2138–9.
213 Schwickardi, M., Johann, T., Schmidt, W. and Schuth, F. (2002) *Chemistry of Materials*, **14**, 3913–19.
214 Dong, A.G., Ren, N., Tang, Y., Wang, Y.J., Zhang, Y.H., Hua, W.M. and Gao, Z. (2003) *Journal of the American Chemical Society*, **125** (17), 4976–7.
215 Shilov, A.E. and Shul'pin, G.B. (1997) *Chemical Reviews*, **97**, 2879–932.
216 Punniyamurthy, T., Velusamy, S. and Iqbal, J. (2005) *Chemical Reviews*, **105**, 2329–63.
217 Barton, D.H.R., Csuhai, E. and Ozbalik, N. (1990) *Tetrahedron*, **46**, 3743–52.
218 Collman, J.P., Tanaka, H., Hembre, R.T. and Brauman, J.I. (1990) *Journal of the American Chemical Society*, **112**, 3689–90.
219 Grasselli, R.K. (1999) *Catalysis Today*, **49**, 141–53.
220 Sigman, M.S. and Schultz, M.J. (2004) *Organic and Biomolecular Chemistry*, **2**, 2551–4.
221 Botalla, P., Nieto, J.M.L., Dejoz, A., Vazquez, M.I. and Martinez-Arias, A. (2003) *Catalysis Today*, **78**, 507–12.
222 Hess, C., Looi, M.H., Hamid, S.B.A. and Schlögl, R. (2006) *Chemical Communications*, 451–3.
223 Davies, T.E., García, T., Solsona, B. and Taylor, S.H. (2006) *Chemical Communications*, 3417–19.
224 Batis, N.H., Batis, H., Ghorbel, A., Vedrine, J.C. and Volta, J.C. (1991) *Journal of Catalysis*, **128**, 248–63.
225 Han, Y.-F., Wang, H.-M., Cheng, H. and Deng, J.-F. (1999) *Chemical Communications*, 521–2.
226 Ueda, W., Endo, Y. and Watanabe, N. (2006) *Topics in Catalysis*, **38** (4), 261–8.
227 Ushikubo, T., Nakamura, H., Koyasu, Y. and Wajiki, S. (1995) U.S. Patent 5,380,933.
228 Pillai, U.R. and Sahle-Demessie, E. (2002) *Chemical Communications*, 2142–3.
229 Pedrosa, A.M.G., Souza, M.J.B., Melo, D.M.A., Araujo, A.S., Zinner, L.B., Fernandes, J.D.G. and Martinelli, A.E. (2003) *Solid State Sciences*, **5**, 725–8.
230 Raja, R., Thomas, J.M., Xu, M., Harris, K.D.M., Greenhill-Hooper, M.L and Quill, K. (2006) *Chemical Communications*, 448–50.
231 Chen, C.X., Tang, J., Li, W.S., Au, C.T. and Zhou, X.P. (2006) *Catalysis Letters*, **111**, 103–9.
232 Tatsumi, T., Nakamura, M., Negishi, S. and Tominaga, H. (1990) *Journal of the Chemical Society–Chemical Communications*, 476–7.
233 Rudberg, J. and Foster, M. (2004) *Journal of Physical Chemistry B*, **108**, 18311–17.
234 Tasker, P.W. (1979) *Journal of Physics C: Solid State Physics*, **12**, 4977–84.
235 Zuo, J.M., O'Keeffe, M., Rez, P. and Spence, J.C.H. (1997) *Physical Review Letters*, **78**, 4777–80.
236 Plass, R., Egan, K., Collazo-Davila, C., Grozea, D., Landree, E., Marks, L.D. and Gajdardziska-Josifovska, M. (1998) *Physical Review Letters*, **81**, 4891–4.
237 Verziu, M., Cojocaru, B., Hu, J., Richards, R., Ciuculescu, C., Filip, P. and Parvulescu, V.I. (2007) *Green Chemistry*, **4**, 373–81.
238 Badlani, M. and Wachs, I.E. (2001) *Catalysis Letters*, **75**, 137–49.
239 Olah, G.A. (2005) *Angewandte Chemie–International Edition*, **44**, 2636–9.
240 Surampudi, S., Narayanan, S.R., Vamos, E., Frank, H., Halpert, G., LaConti, A., Kosek, J., Surya, G.K. and Olah, G.A. (1994) *Journal of Power Sources*, **47**, 377–85.
241 Surumpudi, S., Narayanan, S.R., Vamos, E., Frank, H., Halpert, G., Prakash, G.K.S. and Olah, G.A. (2001) US Patent 6,248,460.
242 Surumpudi, S., Narayanan, S.R., Vamos, E., Frank, H., Halpert, G., Laconti, A., Kosek, J., Prakash, G.K.S. and Olah, G.A. (1997) US Patent 5,599,638.
243 Surumpudi, S., Narayanan, S.R., Vamos, E., Frank, H., Halpert, G., Laconti, A., Kosek, J., Prakash, G.K.S. and Olah, G.A. (2002) US Patent 6,444,343.

244 Prakash, G.K.S., Smart, M.C., Wang, Q.J., Atti, A., Pleynet, V., Yang, B., McGrath, K., Olah, G.A., Narayanan, S.R., Chun, W., Valdez, T. and Surumpudi, S. (2004) *Journal of Fluorine Chemistry*, **125**, 1217–30.

245 Cavani, F. and Trifiro, F. (1992) *Applied Catalysis A: General*, **88**, 115–35.

246 Abon, M., Bere, K.E., Tuel, A. and Delichere, P. (1995) *Journal of Catalysis*, **156**, 28–36.

247 Zhao, Z., Yamada, Y., Teng, Y., Ueda, A., Sakurai, H. and Kobayashi, T. (2000) *Journal of Catalysis*, **190**, 215–27.

248 Zhao, Z., Yamada, Y., Ueda, A., Sakurai, H. and Kobayashi, T. (2004) *Catalysis Today*, **93–95**, 163–71.

249 Santacesaria, E., Cozzolino, M., Serio, M.D., Venezia, A.M. and Tesser, R. (2004) *Applied Catalysis A: General*, **270**, 177–92.

250 Jiménez-Jiménez, J., Mérida-Robles, J., Rodríguez-Castellón, E., Jiménez-López, A., Granados, M.L., del Val, S., Cabrera, I.M. and Fierro, J.L.G. (2005) *Catalysis Today*, **99**, 179–86.S.

251 Yang, S., Iglesia, E. and Bell, A.T. (2005) *Journal of Physical Chemistry B*, **109**, 8987–9000.

252 Daniell, W., Ponchel, A., Kuba, S., Anderla, F., Weingand, T., Gregory, D.H. and Knözinger, H. (2002) *Topics in Catalysis*, **20**, 65–74.

253 Lopez, D., Goodwin, J., Bruce, D. and Lotero, E. (2005) *Applied Catalysis*, **295**, 97–105.

254 Dorado, M.P., Ballesteros, E., Almeida, J.A., Schellert, C., Lohrlein, H.P. and Krause, R. (2002) *Transactions of the American Society of Agricultural Engineers*, **45**, 525.

255 Lotero, E., Liu, Y., Lopez, D.E., Suwannakarn, K., Bruce, D.A. and Goodwin, J.G., Jr (2005) *Industrial and Engineering Chemistry Research*, **44**, 5353–63.

256 Ramadhas, A.S., Jayaraj, S. and Muraleedharan, C. (2005) *Fuel*, **84**, 335–40.

257 Du, W., Xu, Y.Y., Liu, D.H. and Zeng, J. (2004) *Journal of Molecular Catalysis B Enzyme*, **30**, 125–9.

258 Demirbas, A. (2003) *Energy Conversion and Management*, **44**, 2093–109.

259 Saka, S. and Kusdiana, D. (2001) *Fuel*, **80**, 225–31.

260 Furuta, S., Matsuhashi, H. and Arata, K. (2004) *Catalysis Communications*, **5**, 721–3.

261 Venkat Reddy, C.R., Oshel, R. and Verkade, J.G. (2006) *Energy and Fuels*, **20**, 1310–14.

262 Ranjit, K.T. and Klabunde, K.J. (2005) *Chemistry of Materials*, **17**, 65–73.

263 Utamapanya, S., Klabunde, K.J. and Schlup, J.R. (1991) *Chemistry of Materials*, **3**, 175–81.

264 Diao, Y., Walawander, W.P., Sorensen, C.M., Klabunde, K.J. and Rieker, T. (2002) *Chemistry of Materials*, **14**, 362–8.

265 Li, W.C., Lu, A.H., Weidenthaler, C. and Schueth, F. (2004) *Chemistry of Materials*, **16**, 5676–81.

266 Roggenbuck, J. and Tiemann, M. (2005) *Journal of the American Chemical Society*, **127**, 1096–7.

267 Stankic, S., Mueller, M., Dewald, O., Sterrer, M., Knoezinger, E. and Bernardi, J. (2005) *Angewandte Chemie – International Edition*, **44**, 4917–20.

268 Jensen, M.B., Pettersson, L., Swang, O. and Olsbye, U. (2005) *Journal of Physical Chemistry B*, **109**, 16774–81.

269 Rozovskii, A.Y. and Lin, G.I. (2003) *Topics in Catalysis*, **22**, 137–43.

270 Steele, B.C.H. (1999) *Nature*, **400**, 619–21.

271 Cheng, W. (1999) *Accounts of Chemical Research*, **32**, 685–91.

272 Richards, R. (2006) *Surface and Nanomolecular Catalysis*, CRC Press, Taylor & Francis Group, Boca Raton, FL, p. 56.

273 Hu, J., Zhu, K., Chen, L., Kubel, C. and Richards, R. (2007) *Journal of Physical Chemistry C*, **111**, 12038–44.

274 Millikan, R.C. and Pitzer, K.S. (1958) *Journal of the American Chemical Society*, **80**, 3515–21.

275 Kustov, L.M., Ostgard, D. and Sachtler, W.M.H. (1991) *Catalysis Letters*, **9**, 121–6.

276 Koper, O., Li, Y.X. and Klabunde, K.J. (1993) *Chemistry of Materials*, **5**, 500–5.

277 Koper, O., Lagadic, I. and Klabunde, K.J. (1997) *Chemistry of Materials*, **9**, 838–48.

278 Jiang, Y., Decker, S., Mohs, C. and Klabunde, K.J. (1998) *Journal of Catalysis*, **180**, 24–35.

279 Li, Y.X., Li, H. and Klabunde, K.J. (1994) *Environmental Science and Technology*, **28**, 1248–53.
280 Amato, I. (1993) *Science*, **261**, 152–4.
281 Mochida, I. and Yoneda, Y. (1968) *Journal of Organic Chemistry*, **33**, 2163–5.
282 Pistarino, C., Brichese, F., Finocchio, E., Romezzano, G., Di Felice, R., Balde, M. and Busca, G. (2000) *Studies in Surface Science and Catalysis*, **130**, 1613–18.
283 Wagner, G.W., Koper, O., Lucas, E., Decker, S. and Klabunde, K.J. (2000) *Journal of Physical Chemistry B*, **104**, 5118–23.
284 Wagner, G.W., Bartram, P.W., Koper, O. and Klabunde, K.J. (1999) *Journal of Physical Chemistry B*, **103**, 3225–8.
285 Fenelonov, V.B., Mel'gunov, M.S., Mishakov, I.V., Richards, R.M., Chesnokov, V.V., Volodin, A.M. and Klabunde, K.J. (2001) *Journal of Physical Chemistry B*, **105**, 3937–41.
286 Mishakov, I.V., Bedilo, A.F., Richards, R.M., Chesnokov, V.V., Volodin, A.M., Zaikovskii, V.I., Buyanov, R.A. and Klabunde, K.J. (2002) *Journal of Catalysis*, **206**, 40–8.
287 Mol, J.C. (1999) *Catalysis Today*, **51**, 289–99.
288 Pennella, F. and Banks, R.L. (1973) *Journal of Catalysis*, **31**, 304–8.
289 Luckner, R.C., Mcconchie, G.E. and Wills, G.B. (1973) *Journal of Catalysis*, **28**, 63–8.
290 Wang, Y.D., Chen, Q.L., Yang, W.M., Xie, Z.X., Xu, W. and Huang, D.Y. (2003) *Applied Catalysis A: General*, **250**, 25–37.
291 Zhao, D.Y., Feng, J.L., Huo, Q.S., Melosh, N., Fredrickson, G.H., Chmelka, B.F. and Stucky, G.D. (1998) *Science*, **279**, 548–52.
292 Cao, Y., Hu, J.C., Yang, P., Dai, W.L. and Fan, K.N. (2003) *Chemical Communications*, 908–9.
293 Han, Y., Xiao, F.S., Wu, S., Sun, Y., Meng, X., Li, D., Lin, S., Deng, F. and Ai, X. (2001) *Journal of Physical Chemistry B*, **105**, 7963–6.
294 Hu, J., Wang, Y., Chen, L., Richards, R., Yang, W., Liu, Z. and Xu, W. (2006) *Microporous and Mesoporous Materials*, **93**, 158–63.
295 Liu, Z., Cortés-Concepción, J.A., Mustian, M. and Amiridis, M.D. (2006) *Applied Catalysis A: General*, **302**, 232–6.
296 Choudary, B.M., Kantam, M.L., Ranganath, K.V.S., Mahendar, K. and Sreedhar, B. (2004) *Journal of the American Chemical Society*, **126**, 3396–7.

17
Preparation of Superacidic Metal Oxides and Their Catalytic Action
Kazushi Arata

17.1
Introduction

When an acid solution is highly diluted in water, the pH acidity scale is used, which becomes problematic when the acid concentration increases, and more so, when non-aqueous media are employed. Considering the limited application of the pH scale, a quantitative scale was provided by Hammett and Deyrup to express the acidity of more concentrated or non-aqueous solutions: that is, Hammett acidity function, H_0 [1].

The equilibrium between an acid, BH^+, and its conjugate base, B, can be written

$$BH^+ \rightleftharpoons B + H^+ \qquad (17.1)$$

The corresponding thermodynamic equilibrium constant is K_{BH^+}, which is expressed as

$$K_{BH^+} = \frac{a_{H^+} \cdot a_B}{a_{BH^+}} = \frac{a_{H^+} \cdot C_B}{C_{BH^+}} \cdot \frac{f_B}{f_{BH^+}} \qquad (17.2)$$

in which a is the activity, C the concentration, and f the activity coefficient. This equation can be written in the more usual form:

$$H_0 = pK_{BH^+} + \log\frac{[B]}{[BH^+]} \qquad (17.3)$$

H_0 is also expressed by

$$H_0 = -\log\frac{a_{H^+} \cdot f_B}{f_{BH^+}} = -\log\frac{[H^+] \cdot f_{H^+} \cdot f_B}{f_{BH^+}} \qquad (17.4)$$

Metal Oxide Catalysis. Edited by S. David Jackson and Justin S. J. Hargreaves
Copyright © 2009 WILEY-VCH Verlag GmbH & Co. KGaA, Weinheim
ISBN: 978-3-527-31815-5

Table 17.1 Base indicators used for the measurement of superacid strength.

Indicator[a]	pK_{BH^+}
p-Nitrotoluene	−11.35
m-Nitrotoluene	−11.99
p-Nitrofluorobenzene	−12.44
p-Nitrochlorobenzene	−12.70
m-Nitrochlorobenzene	−13.16
2,4-Dinitrotoluene	−13.75
2,4-Dinitrofluorobenzene	−14.52
1,3,5-Trinitrobenzene	−16.04
1,3,5-Trichlorobenzene	−16.12

a Color of base form, colorless: color of acid form, yellow.

In dilute aqueous solution, as the activity coefficients tend to unity, the Hammett acidity function, H_0, becomes identical to pH.

The Hammett acidity function relates to the ability of a solution to convert a neutral base (B) into its conjugate acid (BH⁺). Use is made of this in an experimental procedure which requires the determination of the BH⁺ concentration derived from a very small amount of B (a Hammett indicator) diluted in the solution. This concept allows quantitative determination of the acidity of concentrated and non-aqueous acidic solutions. Gillespie and co-workers determined the acid strength of 100% H_2SO_4 to be $H_0 = -11.93$ [2]. After 1 mg of p-nitrotoluene ($pK_{BH^+} = -11.35$) (B) was dissolved in 1000 ml of H_2SO_4 and the equilibrium was attained, the concentration of p-nitrotoluene-H⁺, BH⁺, was measured at λ_{max} 376 nm with a spectrophotometer. The value of log[B]/[BH⁺] obtained was −0.67 which corresponded to an H_0 value of −12.02. Further measurements using other indicators such as m-nitrotoluene ($pK_{BH^+} = -11.99$) and nitrobenzene (−12.14) gave the average H_0 value of 100% H_2SO_4 to be −11.93. Lower values of H_0 correspond to greater acid strength. Table 17.1 shows the Hammett indicators used for the measurement of solid-superacidic strength, which we have usually used. The indicators with higher pK_{BH^+} values are (2,4-dinitrofluorobenzene)H⁺ ($pK_{BH^+} = -17.35$), (1,3,5-trinitrobenzene)H⁺ (−18.93), and (p-methoxybenzaldehyde)H⁺ (−19.50), but we have not applied these yet.

Any acid is termed a *superacid* when its acidity is stronger than that of 100% H_2SO_4, that is, $H_0 \leq -11.93$ [3–5], and a rapid development of superacid chemistry was seen in the 1960s and 1970s. Superacids can be both Brønsted and Lewis type and their conjugate combinations. Table 17.2 shows several superacids, which are generally made up by mixing a fluorine-containing Brønsted acid (e.g. HF, FSO_3H, CF_3SO_3H, etc.) and a fluorinated Lewis acid (BF_3, SbF_5, TaF_5, etc.) [6–8]. The Hammett acidity function is a logarithmic scale on which 100% H_2SO_4 has an H_0 of ∼−12 and FSO_3H-SbF_5 in the table has an H_0 of ∼−19. Thus, the acid strength of the latter is estimated to be seven orders of magnitude stronger than 100% sulfuric acid. This combination is termed *magic acid* [9].

Table 17.2 Liquid superacids and their acid strength.

Superacid	H_0
HClO$_4$	−13.0
ClSO$_3$H	−13.8
CF$_3$SO$_3$H	−14.1
FSO$_3$H	−15.1
H$_2$SO$_4$-SO$_3$ (1:1)	−14.5
FSO$_3$H-TaF$_5$ (1:0.2)	−16.7
FSO$_3$H-SbF$_5$ (1:0.1)	−18.9

Solid acids have been extensively studied and used as catalysts or catalyst carriers in the chemical industry, and in particular in the petroleum field, for many years. Aluminosilicates such as zeolites and silica-aluminas, together with mixed metal oxides such as TiO$_2$-ZrO$_2$, are typical examples. Replacement of homogeneous liquid acids by heterogeneous solid acids as catalysts is anticipated to bring about ease of continuous operation. Furthermore, the use of heterogeneous solid catalysts can lead to additional advantages, for example fewer problems in terms of reactor corrosion and environmental benefits in terms of the disposal of the used catalyst. However, their acidity is generally below the superacidity range. They are inactive for alkane isomerization and cracking at temperatures below 100 °C.

Novel organic syntheses that are possible in usual acidic media can be accomplished in superacids, including syntheses of economically important hydrocarbons. The remarkable ability of superacids to bring about hydrocarbon transformations can open up new fields in chemistry. In consideration of the exceptionally high activity of liquid superacids, research was extended to prepare solid superacids. As for chemical applications of liquid superacids, efforts were made to attach them to solid materials, and the results are found in extensive patent literature [10–13].

Primary studies to obtain the acidic materials of superacids attached to solid supports included the individual components of BF$_3$, SbF$_5$, TaF$_5$, FSO$_3$H, CF$_3$SO$_3$H, and their mixtures, supported on metals, alloys, resins, polymers, graphites, and various metal oxides such as zeolites and TiO$_2$-ZrO$_2$ [14–17]. The next general effort made to immobilize liquid superacids was to intercalate them in the lattice of graphite. An very rapid expansion occurred in this area of research, since graphite-intercalated compounds are also useful materials with applications as electronic conductors and battery components [18]. Graphite possesses a layered structure and consists of sheets of sp^2 carbon atoms, whose sheets are held together by weak van der Waals forces. SbF$_5$ can be intercalated easily in the lattice simply by heating a mixture of SbF$_5$ and graphite [19]. In addition to SbF$_5$, other superacids, AlCl$_3$, AlBr$_3$, NbF$_5$, and HF-SbF$_5$ were also synthesized as graphite intercalates and used for various superacid-catalyzed reactions [20–25]. Mixtures of aluminum halides and metal salts such as AlCl$_3$-CuSO$_4$ and AlCl$_3$-CuCl$_2$ are active for the isomerization of paraffins at room temperature, and the acidity of the former was estimated to be below $H_0 = -13.75$ [26–28].

Although much effort was devoted to the immobilization of superacids, the major drawback in these catalysts was their short lifetime and ease of deactivation. Although initial conversions were high, the catalysts totally deactivated after a long period on stream. The immobilized superacids were not stable enough for the catalytic process, and metal halides were leached out by the feed from the graphite layers; reduction of metals in the superacids was also easily brought about [23, 29].

Superacidic metal oxides prepared by calcination at a high temperature can be used at elevated temperatures and, thus, provide new trends for developing environmentally benign processes. Superacidity is generated on the oxides of Fe, Ti, Zr, Hf, Sn, Si, and Al by treatment with sulfate, tungstate, and molybdate. Sulfated and tungstated zirconias have attracted much attention as potential catalysts; the latter are thermally stable superacids and can be calcined at temperatures above 1000 °C.

Table 17.3 summarizes the solid superacids we have prepared, together with their acid strengths. The acid strengths were estimated by the color change of Hammett indicators, TPD of pyridine, TPR of furan, and catalytic activities for various reactions. The temperatures shown in parentheses are the calcination temperatures to generate the highest acidity or activity. Although the optimum calcination temperature used varies with the type of vessel employed, the data shown in Table 17.3 were obtained for the catalysts calcined in a Pyrex glass tube below 700 °C and the catalysts calcined in a ceramic crucible above 700 °C for 3 h. The optimum calcination may vary more than 50 °C depending on the type of cal-

Table 17.3 Solid superacids and their acid strength.

Catalyst (calcination temperature, °C)[a]	Highest acid strength (H_0 value)
SO_4/SnO_2 (550)	−18.0
SO_4/ZrO_2 (650)	−16.1
SO_4/HfO_2 (700)	−16.0
SO_4/TiO_2 (525)	−14.6
SO_4/Al_2O_3 (650)	−14.6
SO_4/Fe_2O_3 (500)	−13.0
SO_4/SiO_2 (400)	−12.2
WO_3/ZrO_2 (800)	−14.6
MoO_3/ZrO_2 (800)	−13.3
WO_3/SnO_2 (1000)	−13.3
WO_3/TiO_2 (700)	−13.1
WO_3/Fe_2O_3 (700)	−12.5
B_2O_3/ZrO_2 (650)	−12.5

a Named sulfated stannia, zirconia, hafnia, titania, alumina, ferria, and silica; tungstated zirconia; molybdated zirconia; tungstated stannia, titania, ferria; borated zirconia, respectively.

cination vessel, time, and so on. The notation SO_4/MeO_x means a sulfated metal oxide, such as sulfated zirconia and sulfated alumina. The valence of S in the sulfated metal oxides is +6, and this expression is not correct in the strict sense, since the SO_4 component does not carry a neutral charge.

Among the superacids listed in Table 17.3, sulfated zirconia (SO_4/ZrO_2) has been most frequently investigated, modified, and applied to various reactions. This may be because the sulfated zirconia possesses strong acid sites, is relatively easy to prepare, and was found to be a superacid early in the history of solid superacid development. Sulfated zirconia is now commercially available and used as a catalyst for organic synthesis in industry.

17.2 Preparation

17.2.1 Sulfated Metal Oxides of Zr, Sn, Ti, Fe, Hf, Si, and Al

17.2.1.1 Preparation of Zirconia Gel [30]

A sample of $ZrOCl_2 \cdot 8H_2O$ (200 g) is dissolved in water (3 l), and aqueous ammonia (28%) is added dropwise into the solution with stirring until the solution has a pH value of 8. The effect of pH of the solution can be dramatic in terms of the final catalytic activity, with the maximum activity being observed at pH 8 [31]. The aqueous portion is decanted from the precipitates, and fresh water is added followed by stirring and again decanting the aqueous portion; washing of the precipitates by decantation is repeated until the total amount of water used is 60 l, with almost no chloride ion being detected in the washings. The precipitates are dried at 100 °C for 24 h.

The catalysts prepared using the 'heating method' show higher activity [31]. For the 'heating method' aqueous ammonia is added dropwise to $ZrOCl_2 \cdot 8H_2O$ (25 g) dissolved in hot distilled water (0.5 l; 60–70 °C); the precipitated solution is kept in a water bath warmed at 60–70 °C for 3 h followed by washing the precipitates twice with hot water (0.25 l each time) and drying at 100 °C.

The gel can also be prepared in the manner described above from $ZrO(NO_3)_2 \cdot 2H_2O$ as a starting material [32]. Since residual nitrate ions are thermally decomposed, thorough washing of the precipitates is not needed; a few times would be enough for the decantation washing; in this respect the preparation is easier.

17.2.1.2 Preparation of Stannia Gel [33, 34]

A sample of $SnCl_4 \cdot nH_2O$ (100 g) is dissolved in distilled water (3 l) followed by addition of a 28% NH_3 solution dropwise with stirring, and the final pH of the solution is adjusted to 8. The precipitated product is collected by filtration, suspended in a solution containing 2–4% CH_3COONH_4, and washed three to five times. The precipitates are filtered by suction, dried at 100 °C for 24 h, and finally ground.

The gel is obtained as fine particles when it is washed with distilled water, and a large part of the precipitate passes through a conventional filter paper, resulting in diminished yields. This difficulty can be avoided by washing the gel with aqueous ammonium acetate solution, which provides a quantitative yield.

17.2.1.3 Preparation of H_4TiO_4 [30]

A volume of $Ti[OCH(CH_3)_2]_4$ (290 ml) is added to distilled water (2 l) with stirring, and the white precipitates formed are dissolved by gradually adding conc. HNO_3 (250 ml) with stirring. Ammonia solution (28%, ~300 ml), is added into the aqueous solution with stirring until pH 8 is attained. The solution is then allowed to stand for a day. Washing is then undertaken by decantation of a 5 l beaker of deionized water twice. Finally the resultant material is dried at 100 °C for 24 h.

Another method of preparing H_4TiO_4 is by hydrolysis of $TiCl_4$ as follows. A volume of $TiCl_4$ (80 ml) is gradually added to distilled water (2 l) in a 5 l beaker cooled by ice water, with large amounts of HCl gas being formed. Ammonia solution (28%) is added at room temperature until a pH of 8 is attained. The resultant precipitates are washed thoroughly by decantation using 60 l of water until no chloride ions are detected in the filtrate. The aqueous portion might become cloudy during washing, but the white washings can be decanted.

17.2.1.4 Preparation of $Fe(OH)_3$ [30]

To a solution of $Fe(NO_3)_3 \cdot 9H_2O$ (500 g) dissolved in 2 l of water in a 5 l beaker, ammonia solution (28%, ~300 ml used, pH 8) is added with stirring to precipitate $Fe(OH)_3$. The aqueous portion is decanted from the precipitate after allowing the solution to stand. The precipitates are washed by decantation until the liquid portion becomes cloudy (7–8 times), and dried.

17.2.1.5 Preparation of $Hf(OH)_4$ [35]

$HfCl_4$ is gradually dissolved in distilled water with care, and the hydroxide is prepared in the manner described above for $Zr(OH)_4$.

17.2.1.6 Sulfation, Calcination, and Catalytic Action [30]

The above prepared materials are powdered below 100 mesh and treated with sulfate ions by exposing 2 g of the hydroxides (gel) in 30 ml of aqueous sulfuric acid for 1 h, filtering, drying in a desiccator at room temperature, and finally calcining [36]. The iron materials are again powdered because of solidification after drying [36, 37]. The concentration of H_2SO_4 is 0.5 M for the hydroxides of Zr and Ti [38, 39], 3 M for Sn [40, 41], 0.25 M for Fe [37], and 1 M for Hf [35]. A recent study shows that the optimum concentration for Zr is 0.25 N [42].

After calcination of the sulfate-adsorbed materials in air, the substances are catalytically active for the skeletal isomerization of butane to isobutane at room temperature. The activities are dependent on the calcination temperature. The maximum activity is observed with calcination at 575–650 °C for the Zr catalyst [32], 500–550 °C for Sn [34, 41], 525 °C for Ti [43], 500 °C for Fe [36, 37], and 700 °C for Hf [35].

All the catalysts are calcined in Pyrex glass tubes in air for 3 h and sealed in ampoules while being hot, to minimize exposure to humidity until use.

17.2.1.7 Preparation of Sulfated Silica [44]

Silica gel is obtained by hydrolyzing $Si(OC_2H_5)_4$ (100 ml) with water (100 ml) and a few drops of HNO_3. The mixture is stirred until gel formation. The precipitates are obtained by evaporation of excess water and ethanol, formed by hydrolysis of $Si(OC_2H_5)_4$, followed by drying at 100 °C, and powdering. The silica (3 g) is exposed to SO_2Cl_2 for 1 h followed by evacuating HCl evolved by the reaction of surface OH group with SO_2Cl_2 and excess SO_2Cl_2 in vacuum, and calcining in air at 400 °C.

17.2.1.8 Preparation of Sulfated Alumina

In the case of the sulfate-treated superacids of Zr, Sn, Ti, Fe, Hf, and Si, superacid sites are not created by the treatment of sulfate ion on the crystallized oxides but rather on the amorphous forms, followed by calcination to crystallization. The superacid of Al_2O_3 is prepared from the crystallized oxide, γ-Al_2O_3 [45, 46].

A highly active catalyst is obtained by hydrolysis of aluminum isopropoxide [47]. Distilled water is added with stirring to a solution of $Al[OCH(CH_3)_2]_3$ (10 g) dissolved in isopropanol (300 ml) to precipitate $Al(OH)_3$, followed by washing the precipitates, drying, and calcining for crystallization at 500 °C for 3 h. The crystallized materials are then hydrated before sulfation. The sample (5 g) is suspended in water (5 l) at 80 °C for 3 h, filtered, and dried at 100 °C. SO_4/Al_2O_3 is obtained by treatment with 2.5 M H_2SO_4 and calcination at 600–650 °C for 3 h.

17.2.1.9 Property and Characterization

Properties of the sulfated materials thus prepared are summarized as follows [43, 48].

1. Superacidity is generally created by adsorbing sulfate ions onto amorphous metal oxides followed by calcination in air to convert to the crystalline forms. However in the case of Al_2O_3, a superacid is prepared from the crystallized oxide.

2. Specific surface areas of the catalysts are much larger than those of the oxides without the sulfate treatment except for the Al_2O_3. A particularly large increase in the area is observed on the highly active and acidic catalysts. The main reason for the increase in surface area is the retardation of crystallization by sulfate treatment. The areas of the Al_2O_3 catalysts are smaller than those of the oxides without the sulfate treatment.

3. By XRD analysis the degree of crystallization of the sulfated oxides is much lower than that of the oxides without the sulfate treatment. Temperatures of the crystallization or phase transformation for SO_4/ZrO_2 and SO_4/TiO_2 are circa 150 and 200 °C higher than those for pure ZrO_2 and TiO_2, respectively; the XRD pattern of SO_4/ZrO_2 (650 °C) and SO_4/TiO_2 (525 °C), whose superacidities are highest, are pure tetragonal and anatase forms, respectively.

4. The sulfate samples have IR spectra that are different from those of metal sulfates; the materials show absorption bands at 980–990, 1040, 1130–1150, and 1210–1230 cm^{-1}, which are assigned to the bidentate sulfate coordinated to metal ions.

5. XPS spectra of SO_4/ZrO_2 and SO_4/Fe_2O_3 show that the surface is not $Zr(SO_4)_2$ and $Fe_2(SO_4)_3$, but composed of ZrO_2 and Fe_2O_3 with SO_4^{2-}, respectively. On the other hand, the spectra of SO_4/TiO_2 and SO_4/Al_2O_3 are consistent with the presence of surface $Ti(SO_4)_2$ and $Al_2(SO_4)_3$, respectively.

6. The IR spectra of pyridine adsorbed on SO_4/ZrO_2 and SO_4/TiO_2 show the facile conversion of Lewis sites to Brønsted sites by water molecules. Lewis and Brønsted sites are easily interchangeable by adsorption or desorption of water molecules [49–55].

7. Upon dehydration in N_2 at 375 °C, a new IR peak centered at 1370 cm^{-1} appears; the band is assigned to an S=O stretching vibration and is observed for the catalytically active material [56, 57].

8. The sulfated materials also show oxidizing action at elevated temperatures. In particular the superacids of Fe_2O_3 and SnO_2 demonstrate strongly oxidizing potential at temperatures above 100 °C [58, 59].

9. Catalysts obtained by treatment with sulfuric acid are usually more active than those obtained with ammonium sulfate treatment. However, a recent study shows that the analogous activity is generated on SO_4/ZrO_2 by the kneading method with ammonium sulfate [60].

17.2.1.10 One-Step Method for Preparation of SO_4/ZrO_2 [61–65]

Ward and Ko investigated preparation of sulfate-zirconia aerogels in a one-step synthesis by the sol–gel method followed by supercritical drying [61]. Sulfuric acid is mixed with zirconium n-propoxide (16.2 ml of 70 wt% in propanol) in n-propanol (30 ml) and reacted with water (1.3 ml) and nitric acid (1.9 ml of 70% w/w) to form a zirconia-sulfate co-gel; supercritical drying with carbon dioxide removes the alcohol solvent forming a high surface area aerogel. The properties and catalytic activities are similar to those of the compounds we have prepared. A single-step sol–gel method has also been used for the preparation of SO_4/TiO_2 [66].

17.2.1.11 Commercial Gels for Preparation of SO_4/ZrO_2 and SO_4/SnO_2

Commercial zirconia-gels are now supplied by chemical companies such as MEL, Nakarai, and Daiichi Kigenso. For instance, XZO631 and 632 are the gels, and XZO682 is a sulfate-adsorbed gel from MEL. According to our experience, commercial gels have often led to catalysts superior to those prepared in the laboratory [67, 68]. Similar observations have been made with SO_4/SnO_2. A commercial gel prepared from *meta*-stannic acid gives satisfactory results for this type of catalysis [69]. SO_4/SnO_2 was prepared using *meta*-stannic acids ($SnO_2 \cdot H_2O$, commercial grade) supplied by Kojundo Kagaku, Ltd. and Yamanaka & Co. Ltd. The activity for the acid-catalyzed conversion of methanol into dimethyl ether was higher than for catalysts obtained by hydrolysis of $SnCl_4$ and much higher for the isomerization of butane.

17.2.1.12 Effect of Drying and Calcination Temperatures on the Catalytic Activity of SO_4/ZrO_2 [68]

The catalytic activity of SO_4/ZrO_2 varies with the type of zirconia gel and the drying and calcination conditions. The calcination temperature showing the maximum activity and acidity often varies with the type of prepared gel. For instance, the maximum activity for the conversion of butane to isobutane is observed with calcination at 575 and 650 °C, respectively, for the materials prepared from $ZrO(NO_3)_2$ and $ZrOCl_2$ as starting reagent [32].

The activity also depends on the drying temperature of the gel before sulfation. For the gels obtained by hydrolysis of $ZrO(NO_3)_2$ followed by drying at 100 °C and 300 °C, the difference in the calcination temperature showing the maximum activity for the butane conversion was 50 °C, and the difference in maximum activity was a factor of two.

Several sulfated zirconias were prepared by changing the drying temperature of gel in the range 100–400 °C and the final calcination temperature in the range 475–700 °C. It was found that the drying temperatures exhibiting the highest activity for the butane conversion are not always fixed, for instance 200 °C for one and 300 °C for another.

Figure 17.1 summarizes the effect of the calcination temperature on the catalytic activity for butane isomerization over sulfated zirconias prepared from different zirconia gels dried at the optimum temperatures. The figure indicates that the maximum activities are approximately the same for different catalysts even though

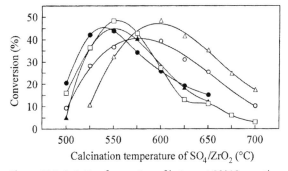

Figure 17.1 Activities for reaction of butane at 180 °C over the SO_4/ZrO_2 catalysts prepared from various Zr gels: MEL 631 dried at 300 °C (□), MEL 632 dried at 200 °C (▲), Nakarai dried at 300 °C (●), and the gels prepared by hydrolysis of $ZrO(NO_3)_2$ (△) and $ZrOCl_2$ (○) followed by drying at 300 and 200 °C, respectively.

the temperatures to give maximum activity are different. The calcination temperatures required to give maximum activity for different catalysts fall within a temperature range <50 °C. Residual species such as Cl^- and NO_3^- in the gel result in differences in the optimum drying and calcination temperatures owing to differences in the state of dehydration of the zirconia gel or to reactivity of the zirconia support with the sulfate species. However, selection of the optimum drying and calcination temperature generate the same ultimate catalytic activity. The present results point out that the optimum temperature for drying the Zr gel and the final calcination should be determined according to the type of zirconia gel.

17.2.2
Tungstated, Molybdated, and Borated Metal Oxides

17.2.2.1 Preparation of WO_3/ZrO_2 and MoO_3/ZrO_2 [30]

A sample of $Zr(OH)_4$ (10 g), obtained from $ZrOCl_2$, is heated at 300 °C, the gels are impregnated with aqueous ammonium metatungstate $[(NH_4)_6(H_2W_{12}O_{40})] \cdot nH_2O$, 50 wt% WO_3, 3.8 g] and water (15 ml) in a 100 ml beaker followed by evaporating water at room temperature, drying, and calcining in air at 800 °C for 3 h. The concentration is 15 wt% W based on the hydroxide, and 13 wt% W after calcination at 650–950 °C. The analogous material is also formed by the kneading method with tungstic acid (H_2WO_4) which is insoluble in water; a wet mixture of $Zr(OH)_4$ (10 g) and H_2WO_4 (2 g) with a little water is kneaded for 3 h [70, 71].

After heating $Zr(OH)_4$ (10 g) at 300 °C the sample is impregnated with molybdic acid (H_2MoO_4, 2.5 g) dissolved in ammonium hydroxide (28%, 2 ml) and water (15 ml) followed by evaporating water at room temperature, drying, and calcining at 800 °C in air for 3 h. The concentration is 5 wt% Mo metal based on the hydroxide [72].

17.2.2.2 Preparation of WO_3/SnO_2, WO_3/TiO_2, and WO_3/Fe_2O_3 [30]

After the hydroxides of Sn, Ti, and Fe, prepared by hydrolysis of $SnCl_4$, $TiCl_4$, and $Fe(NO_3)_3$, respectively, are dried at 300 °C, the gels are impregnated with aqueous ammonium metatungstates $[(NH_4)_6(H_2W_{12}O_{40})]$ followed by evaporating water, drying, calcining in air for 3 h at 1000, 700, and 700 °C for the Sn, Ti, and Fe materials, respectively. The concentration is 15 wt% W based on the hydroxides (11–13 wt% W after calcination) [73, 74].

WO_3/SnO_2 is also prepared from a commercial gel, used after drying metastannic acid of Kojundo Kagaku, Ltd. at 100 °C [69]. Calcination at 1000 °C after impregnation of the tungstate generates the highest activity for acid-catalyzed reactions. High temperatures such as 1000 °C for calcination do not discriminate precursor stannia gels, though the calcination temperature giving the highest activity for SO_4/SnO_2 differs according to the gel.

17.2.2.3 Preparation of B_2O_3/ZrO_2 [75, 76]

Zirconium hydroxide is impregnated with aqueous boric acid followed by evaporating water and calcining in air at 650 °C (3 wt% B). The same catalyst is obtained by suspending the hydroxide in 2-propanol solution of trimethyl borate followed by adding water to hydrolyze the borate.

17.2.2.4 Property and Characterization

Properties of the catalysts thus prepared are summarized as follows [43, 48].

1. Superacid sites are not created by impregnation on the crystallized oxides, but on the amorphous forms whose calcination then converts them to the crystalline forms. Recent work shows that a WO_3/ZrO_2 catalyst prepared by impregnation of the crystalline zirconia (65% tetragonal, 35% monoclinic) exhibits comparable behavior [77, 78].

2. $WO_3/ZrO_2(800\,°C)$ and $MoO_3/ZrO_2(800\,°C)$, whose activities are highest, are observed to be 100% tetragonal ZrO_2 by XRD, with the TiO_2 in $WO_3/TiO_2(700\,°C)$ being the anatase polymorph [74].

3. Specific surface areas of the WO_3 and MoO_3 catalysts are much larger than those of the oxides without tungsten and molybdenum oxides, pure metal oxides.

4. XPS spectra of the WO_3 and MoO_3 supported catalysts show their surface to be WO_3 or MoO_3 and ZrO_2, SnO_2, TiO_2 or Fe_2O_3 [73].

17.3
Determination of Acid Strength

Several methods have been used to determine the surface acidity of solid acids, but each method has its limitations. Common methods are titration with the Hammett indicators, temperature-programmed desorption (TPD), adsorption microcalorimetry, catalytic test reactions, and IR and NMR spectroscopies. These techniques exhibit several advantages and disadvantages.

Titration with a variety of Hammett indicators is one of the most widely used techniques to determine the distribution of acid strengths on the solid surface [79]. However, many arguments have been raised in the past against the use of Hammett indicators for evaluation of solid acidity [80–83].

A commonly used technique is TPD of adsorbed bases such as ammonia and pyridine [84–86]. However, there are critical questions in this method [87–89]. The NH_3 molecule is well known to interact with both the acidic OH group and the basic oxygen [90]. Thus, NH_3 gives information on dual acid–base sites.

Instead of TPD, microcalorimetry of adsorption shows the heat evolved during the adsorption of probe molecules, usually ammonia, on acid sites [91–94]. This measurement can determine the distribution of adsorption enthalpies but cannot differentiate between adsorption on Lewis and Brønsted acid sites.

Catalytic activity can be used to rank solid acidity, and activity for the skeletal isomerization of butane is often used to indicate very strong acidity, in particular superacidic strength [95]. A comparative study using the isomerizations of butane and pentane as test reactions gave good correlations between their rate constants and the Hammett acid strengths [96]. These test reactions do not differentiate between the acid strength and the number of acid sites. The mechanism of butane or pentane isomerization, however, has been shown to be bimolecular, so rates of alkane isomerization alone cannot be used to compare acidities [97–100]. One paper indicates that the isomerization of α-pinene enables solid acids to determine the Brønsted acid strength; superacidity promotes the formation of limonene over camphene [101].

A comparative study using IR spectroscopy can rank solid acidity by determining the frequency shift of the adsorption band of pyridine [102], the band shift of OH groups due to the adsorption of benzene or CD_3CN [103, 104], and the shift of the adsorbed CO on Lewis sites [105]. These IR techniques, however, are not widely used for evaluation of solid acidity.

Solid-state NMR spectroscopy enables site-specific characterization of solid acids such as protonated zeolites [106, 107]. Attempts have been made to relate the acid strength of solids to the ^1H chemical shift of surface OH groups, the shift brought by the adsorbed bases such as CD_3CN and CCl_3CN, and the ^{31}P shift of the adsorbed $^{31}P(CH_3)_3$ [108–113]. A disadvantage of NMR spectroscopy is the complicated chemical shift due to hydrogen bonding.

17.3.1
Hammett Indicators

The acid strength of a solid is defined as the ability of the surface to convert an adsorbed neutral base into its conjugate acid. The strength is expressed by the Hammett acidity function, H_0, as explained in the Introduction. The color of suitable indicators adsorbed on a surface can give a measure of acid strength. If the color is that of the acid form of the indicator, then the value of the H_0 function of the surface is equal to or lower than the pK_{BH^+} of the conjugate acid of the indicator. Lower values of H_0 correspond to greater acid strength.

The acid strengths shown in Table 17.3 were examined by the visual color change method using the Hammett indicators shown in Table 17.1 [43, 48]. The indicator dissolved in solvent was added to the sample in powder form in a nonpolar solvent, sulfuryl chloride [38] or cyclohexane [40]. The strength of colored materials such as SO_4/Fe_2O_3 and MoO_3/ZrO_2 was estimated from their catalytic activities in comparison with those of the catalysts determined by the Hammett-indicator method.

Determination of the acid strength of solid catalysts using Hammett indicators, however, has been criticized frequently because of the heterogeneity of the solid surface [81, 104, 110, 114–116]. The principle of the Hammett acidity function is based on the equilibrium equation in a homogeneous solution, and its application to the heterogeneous condition is subject to severe criticism. In addition, the color change of the adsorbed indicators on solids as determined by the naked eye is subjective. The effects of interactions between the solvent and the solid surface has also been raised [9].

Table 17.4 Acid-catalyzed reactions.

Dehydration of alcohol → MeOH, EtOH, i-PrOH
Decomposition of alkylbenzene → Ph-Me, Ph-Et, Ph-^{i}Pr
Friedel–Crafts acylation → acetylation, benzoylation, and so on
Isomerization of paraffin → open-chain C_1–C_7, cyclic C_6–C_{12}
Esterification → AcOH + MeOH, EtOH, and so on
C_8OH + phthalic acid, and so on
Cationic polymerization → Me, Et, ^{i}Bu vinyl ether
Oligomerization → β-pinene, 1-octene, 1-decene
Others → aldol condensation, and so on

17.3.2
Test Reactions

Owing to the heterogeneity of solid superacids, accurate acidity measurements are difficult to perform and interpret. The most simple and useful way to estimate the acidity of a solid catalyst is to test its catalytic activity in acid-catalyzed reactions. We usually compare activity with those of aluminosilicates such as silica-aluminas and zeolites. These materials have strong enough acidities to cause the Friedel–Crafts reactions, and their acidities are known to be in the range of superacidity [117]. The Hammett-indicator method indicates that the acid strength of the SiO_2-Al_2O_3 used is in the range of $-12.70 < H_0 \leq -11.35$, whose acidity is $H_0 = {\sim}{-}12$ and superacidic [37]. The acidity and catalytic activity of zeolites are generally higher than the silica-aluminas, with mordenites being highest among them ($H_0 = {\sim}{-}14$) (ZSM-5: $H_0 = {\sim}{-}13$) [67].

In order to confirm the acidity results measured using the indicators shown in Table 17.1, we have investigated as many acid-catalyzed reactions as possible. The reactions are summarized in Table 17.4 [43, 48, 118, 119]. Among them, the skeletal isomerization of light paraffins, in particular butane and pentane, has been the most widely applied. The isomerization of butane at room temperature was a well known test reaction for superacidity at the beginning of this work [43, 48, 118]. The activity for many of the reactions tested correspond to the acidities as determined by use of the Hammett indicators.

17.3.3
Temperature-Programmed Desorption (TPD)

TPD using base adsorbents such as ammonia and pyridine has been one of the common techniques to evaluate the amount and strength of acid sites on solid acids. However, the desorption temperature of ammonia and pyridine adsorbed on solid superacids is so high (500 °C and above) owing to the strong interaction that the adsorbed compounds are decomposed or oxidized before desorption. The surface structure can even be destroyed by reactions with probe molecules [120–122].

TPD using pyridine was attempted for estimating the acidity of superacidic materials, even though this method suffers a major drawback. An examination of the termination temperature of pyridine desorption was made using TG-DTA [123]. When the adsorbed compound is decomposed or oxidized to destruction before desorption, its temperature must be below that of the real desorption. The temperatures of pyridine desorption obtained were generally proportional to the H_0 values of the highest acid strength determined by the Hammett method or estimated by catalytic activity, except that of SO_4/SnO_2. All of the temperatures for the superacids were higher than those of Al_2O_3, ZrO_2-TiO_2, SiO_2-ZrO_2, and SiO_2-Al_2O_3, whose acidities are below superacidic, and whose oxidative action is quite weak. In the case of SO_4/SnO_2 whose temperature was unexpectedly low, oxidation of the adsorbed pyridine occurred on the catalyst surface at a temperature below that expected from the acid strength by the Hammett method.

17.3.4
Temperature-Programmed Reaction (TPRa)

Another method similar to TPD, is temperature-programmed reaction (TPRa) using furan as a probe molecule [118]. Furan is resinified on the surface of solid acids when heated at a temperature that depends on the acid strength. Solids with weak acidity give rise to the reaction at elevated temperatures, and the reaction temperature is dependent on the surface acidity. A relationship is observed between decrease of the temperature and increase of the acid strength. The resinification of furan adsorbed on the surface of catalyst is exothermic, and TPRa using DTA is a technique for estimating the acid strength of superacids.

The initial temperatures of resinification were compared with the H_0 values of superacids as determined by the Hammett method, and a linear relationship was observed. The TPRa results gave the H_0 value of SO_4/SnO_2 as -18; the value was not determined by the Hammett method because -16.12 relates to 1,3,5-trichlorobenzene indicator which has the lowest pK_{BH^+} value among the Hammett indicators used to date (Table 17.1).

17.3.5
Ar-TPD [124, 125]

A new method has involved the use of Ar as a probe atom. We tried Ar as a probe TPD and found that Ar-TPD could be applied to the evaluation of the acid strength of solid superacids.

Noble gases such as Ar and Xe interact only with the acidic sites, giving information on these sites. Hence these atoms are useful as probes of the intrinsic nature of the acidic sites. For noble gases, polarization and dispersion are dominant. Ar shows an acid–base-like interaction in a polarized state with acid sites at low temperature owing to its induced dipole.

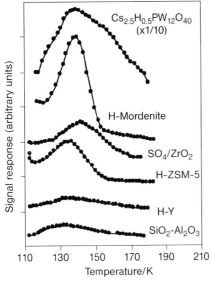

Figure 17.2 Ar-TPD profiles of solid acids; temperature programmed rate: 2 K min^{-1}.

The acid strength was first evaluated as an activation energy of Ar desorption. The activation energy was calculated by the following equation:

$$2\ln T_m - \ln b = E_d/RT_m + \text{const} \quad (17.5)$$

Here T_m is the peak temperature of desorption, b the rate of temperature increase, and E_d the activation energy. A linear plot is obtained between $2 \ln T_m - \ln b$ and $1/T_m$, and E_d is determined from the gradient.

The Ar-TPD experiments were performed in the temperature range 113–223 K at the programmed rate of 2–5 K min^{-1} using a cooling system with N_2 gas bubbled through liquid N_2 along with an electric heater regulated by a temperature controller. The profiles at the rate of 2 K min^{-1} are shown Figure 17.2. The activation energies of Ar desorption are calculated from the estimated values of T_m to be 5.5, 6.0, 6.6, 6.7, 7.6, and 9.3 kJ mol^{-1} for SiO_2-Al_2O_3, zeolites of H-Y, H-ZSM-5, and H-Mordenite, heteropoly acid ($Cs_{2.5}H_{0.5}PW_{12}O_{40}$), and SO_4/ZrO_2, respectively. The value for SO_4/SnO_2 is determined to be 10.6 kJ mol^{-1} [33, 34].

The heat of adsorption of ammonia correlates with the Hammett acidity function, H_0. Similarly, the activation-energies of SiO_2-Al_2O_3, three zeolites, and SO_4/ZrO_2 relate well with their H_0 values determined by the Hammett method and represented by their highest acid strengths, with a linear plot being obtained. By extrapolation of the linear plot to the activation energy 9.3 kJ mol^{-1} for SO_4/ZrO_2, the H_0 value is estimated to be −19 [68], which agrees with that estimated from the heat of adsorption of ammonia [93]. The value of 10.6 kJ mol^{-1} found for SO_4/SnO_2 is estimated to correspond to $H_0 = -21$.

17.3.6
Ar-Adsorption [126, 127]

The heat of adsorption of Ar was also measured for acidity evaluation. In the case of Ar-TPD, an effect of the probe molecule diffusion in micropores is observed with some samples, such as zeolites, at high temperature-programmed rates. The adsorption method is not influenced by diffusion of the adsorbed molecule because the Ar isotherm is measured at static equilibrium. It is also advantageous that the usual BET apparatus can be used to obtain the adsorption isotherm. In addition, the adsorption behavior of Ar is of the Henry type at temperatures around room temperature.

Langmuir's adsorption equation can be converted into the following equation by the introduction of an approximation at low coverage:

$$\ln(V/P) = -(\Delta H/RT) + \ln b_0 + \ln V_m \tag{17.6}$$

where V is the volume adsorbed, P is the equilibrium pressure, ΔH is the heat of adsorption, R is the gas constant, T is the adsorption temperature, b_0 is a constant, and V_m is the volume corresponding to a monolayer. The heat of adsorption (ΔH) is calculated from the gradient of a plot of $\ln(V/P)$ vs $1/T$. This method of calculation is known as the H-method.

Experimental results by means of the volumetric method using a conventional BET system were observed to be Henry type in the temperature range 233–313 K and the pressure range $P = 5$–$30\,kPa$, achieving the condition of small coverage. An example adsorption isotherm, for H-mordenite, is shown in Figure 17.3.

In addition to the above technique, the heat of adsorption can also be determined from Langmuir-type adsorption isotherms. Langmuir's adsorption equation pro-

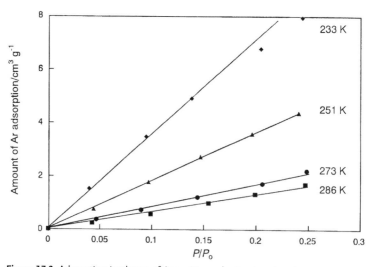

Figure 17.3 Adsorption isotherm of Ar on H-mordenite at various temperatures.

17.3 Determination of Acid Strength

Table 17.5 Adsorption heats of Ar and acid amounts of various solid acids.

Sample	$-\Delta H_H{}^{a)}$ (kJ mol^{-1})	$-\Delta H_L{}^{b)}$ (kJ mol^{-1})	$V_m{}^{b)}$ (mmol g^{-1})
SO$_4$/SnO$_2$	23.5	29.6	0.10
SO$_4$/ZrO$_2$	22.4	26.3	0.009
WO$_3$/ZrO$_2$	21.6		
SO$_4$/Fe$_2$O$_3$	18.9		
ZSM-5	17.3	17.9	0.39
Mordenite	17.3	17.7	1.44
Beta		17.1	0.10
HY	14.8	15.6	0.35
SiO$_2$-Al$_2$O$_3$	14.4	14.9	0.35
3 wt% Pt-SO$_4$/ZrO$_2$	22.8		
7.5 wt% Pt-SO$_4$/ZrO$_2$	24.6		

a Calculated by H-method.
b Calculated by L-method.

vides information regarding the saturated adsorption amount. It is expected that the number of adsorbed atoms translates into the number of acid sites on the solid surface. In this case, the heat of adsorption (ΔH) is related to the adsorption constant (b) as follows:

$$b = b_0 \exp(-\Delta H/RT) \tag{17.7}$$

and the adsorption heat is calculated from the gradient of a plot of $\ln b$ vs $1/T$. This method of calculation is known as the L-method.

On comparing the two methods of measurement, the advantage of the H-method is its ease of operation and its shorter experimental time. However, the adsorption isotherm must be measured with care, as the concentration of adsorbate is quite small in the low equilibrium pressure range.

The calculated heats of adsorption are summarized in Table 17.5 [125, 128]. The heats of adsorption presented are comparable with those obtained by applying the Henry and Langmuir equations. The relative order of SO$_4$/ZrO$_2$, mordenite, and ZSM-5 is consistent with that of the acid strength evaluated by the heat of ammonia adsorption [93]. The value of SO$_4$/SnO$_2$ is higher than that of SO$_4$/ZrO$_2$, indicating higher acidity for the former, the strongest acidity among the solid metal oxide-based superacids. For both materials, the heats of adsorption determined by the H-method are lower than those by the L-method. The heats of adsorption of Ar obtained from the L-method indicate the highest acid strength. On the other hand, the H-method provides an average value that includes the weak acid sites, owing to the wide pressure range used in the measurement. The saturated adsorption amount is the number of acid sites with high acidity. In general, ammonia cannot be used as a probe molecule for solid acids containing a component that actively decomposes ammonia, for example platinum. The table indicates that the present technique is applicable for metal-containing solid acids.

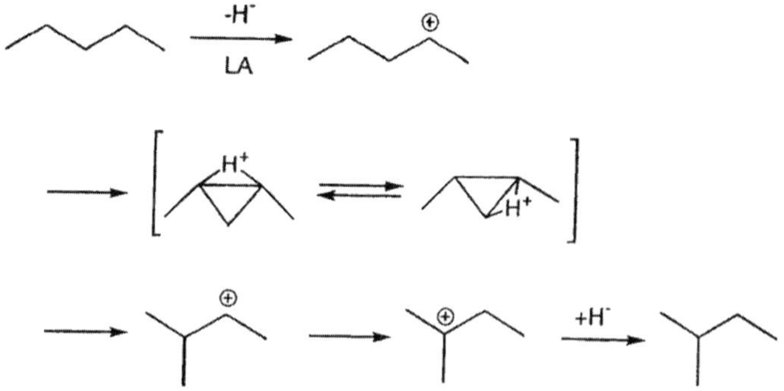

Scheme 17.1 Isomerization of pentane via monomolecular intermediate.

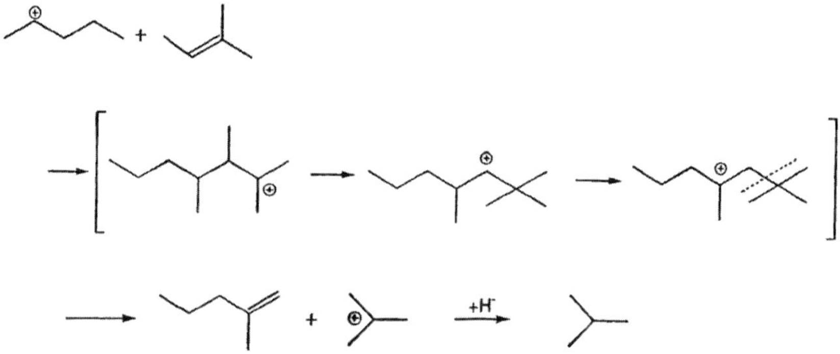

Scheme 17.2 Disproportionation of pentane via bimolecular intermediate.

17.4
Nature of Acid Sites

The skeletal isomerization of butane to isobutane is a typical reaction catalyzed by superacidity. Early in the history of this work, SO_4/Fe_2O_3, SO_4/TiO_2, and SO_4/ZrO_2, were termed superacids owing to their ability to isomerize butane at room temperature or below [32, 37, 39] The formation of isobutane from butane, however, does not necessarily require superacidic strength. A bimolecular reaction pathway based on the intermediacy of butane is energetically lower than a monomolecular mechanism [129–133]. The monomolecular and bimolecular mechanisms are shown in Schemes 17.1 and 17.2, respectively, using pentane as a model.

The processes are the monomolecular reaction through a protonated cyclopropane produced by the abstraction of H⁻ over Lewis acid sites and the bimolecular mechanism where an olefin takes part in the reaction. The olefin is produced over Brønsted acid sites. In the case of butane in the monomolecular mechanism, isobutane is formed through protonated methylcyclopropane with an activation energy of $8.4\,kcal\,mol^{-1}$ followed by the formation of the primary isobutyl cation with high energy [134].

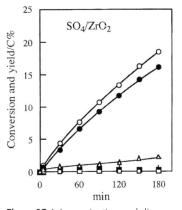
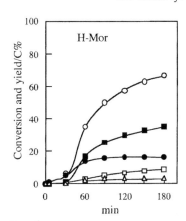

Figure 17.4 Isomerization and disproportionation of pentane on SO_4/ZrO_2 at 0 °C and H-mordenite at 200 °C; conversion (○), butane (□), isobutane (■), isopentane (●), hexanes (△).

On the Brønsted acid sites, H^+ is added to the C—H bond of the substrate, followed by elimination of H_2 and formation of the secondary butyl cation. The catalyst surface would be in a proton-deficient condition owing to the elimination of H^+ as H_2. As a result, the secondary butyl cation releases H^+ to convert into an alkene, which is an intermediate of the bimolecular reaction. The thermodynamic stability of the tertiary cation is higher than that of the secondary cation. The former cation is easily formed by methyl migration followed by a 1,2-hydrogen shift. The tertiary carbenium ion releases H^+ readily, converting into isopentene, and this would be the intermediate of the bimolecular reaction. The reaction proceeds via the bimolecular mechanism of oligomerization–cracking involving the formation of a C_{10} intermediate. This is formed from a C_5 alkene and C_5 cation followed by rearrangement and β-scission to yield isobutane as the final product, as shown in Scheme 17.2. Butane is also converted into isobutane in the same manner.

In order to compare the catalytic action of SO_4/ZrO_2 with that of H-mordenite (H-Mor), the reaction of pentane was performed in a closed recirculation system. The changes of product yields against reaction time are shown in Figure 17.4 [135]. The reaction rate with SO_4/ZrO_2 at 0 °C is almost constant during the reaction, and the main product is isopentane, with small amounts of isobutane and hexanes being detected.

In contrast, the reaction over H-Mor at 200 °C has different characteristics than that over SO_4/ZrO_2. H-Mor is inactive at 0 °C, and at 200 °C a short induction period is observed for the catalytic activity and product selectivity. Although the production of isopentane is predominant in the induction period, an increase in the activity is observed along with the formation of isobutane, butane, and hexanes. The main product is isobutane after the induction period, highlighting the effect of Brønsted acid sites. The surface alkenes for the bimolecular mechanism

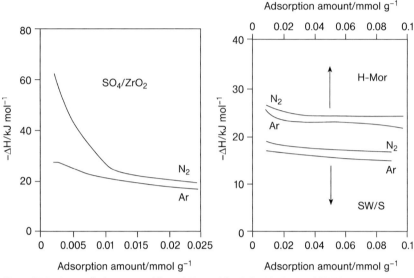

Figure 17.5 Heats of adsorption of Ar and N_2 on SO_4/ZrO_2, H-Mor, and SW/S.

are produced on the Brønsted acid sites, and alkenes accumulate on the surface during the induction period. The activity increases with increase in the amount of the accumulated alkenes. This leads to the conclusion that the monomolecular reaction proceeds on SO_4/ZrO_2 and the bimolecular mechanism proceeds on H-mordenite.

In order to examine the nature of acid sites on SO_4/ZrO_2, the heats of adsorption of N_2 and Ar were measured and compared with those of H-Mor and a heteropoly acid ($H_4SiW_{12}O_{40}$) supported on SiO_2 (SW/S). Both heats plotted against the quantity adsorbed are shown in Figure 17.5 [136]. The heat of adsorption of N_2 is larger than that of Ar, with the difference being 2–3 kJ mol^{-1} on H-Mor and SW/S. Both heats decrease gradually with increasing adsorption amounts, with an almost constant relationship. On the other hand, SO_4/ZrO_2 behaves differently. In particular, a very large heat of adsorption, more than 60 kJ mol^{-1}, is obtained for N_2 adsorption on SO_4/ZrO_2, which is much larger than that of H-Mor and the difference in the adsorption heats of the two gases is more than 30 kJ mol^{-1}.

The nitrogen molecule is strongly adsorbed on Lewis acid sites by interaction between the 5σ electron pair and the vacant molecular orbital of the Lewis site [137]. The stabilization of the N–N bond in addition to the stabilization by adsorption on the acid sites is responsible for the large heat of N_2 adsorption. The acid sites on SW/S are of Brønsted type [138], and the acidity on H-Mor is also regarded as Brønsted type, though a small number of Lewis sites are generated by treatment at high temperatures [139]. This is consistent with the formation of isobutane from pentane. Therefore it is concluded that the acidity of sulfated zirconia is of Lewis type.

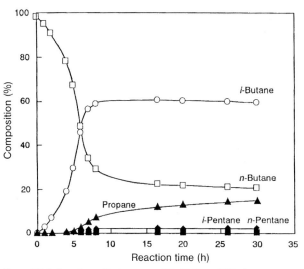

Figure 17.6 Reaction of butane over SO_4/ZrO_2 at 0 °C; the catalyst was pre-treated at 250 °C in vacuum for 3 h.

17.5
Isomerization of Butane Catalyzed by Sulfated Zirconia [140]

The skeletal isomerization of butane catalyzed by SO_4/ZrO_2 was carried out in a closed recirculation system under mild conditions, at 0 °C for a long reaction period; the results are shown Figure 17.6. A long induction period, ~4 h, is observed with the sole formation of isobutane. At longer times on stream, an increase in the activity is observed along with the formation of propane and pentanes. The ratio of butane to isobutane is close to the equilibrium value after 8 h. The apparent activation energy is 13.0 kcal mol^{-1} in the induction period and 8.8 kcal mol^{-1} after that period. The activation energy of the monomolecular mechanism is higher than that of the bimolecular mechanism. The isomerization of butane proceeds via a primary carbenium ion in the former mechanism and via a secondary carbenium ion in the latter. The monomolecular reaction is predominant on superacidic Lewis sites in the induction period, and the reaction changes to the bimolecular process on Brønsted sites by the formation of surface alkene intermediates, giving rise to the additional C_3 and C_5 products. The number of Brønsted sites over SO_4/ZrO_2 is reduced, and the proportion of Lewis sites is increased with increasing evacuation temperature. On the catalyst pretreated in vacuum at 250 °C, some Brønsted acidity remains and causes the bimolecular reaction. However, the activity increase disappears and only isobutane is obtained on catalysts treated at 450 °C or higher, which is indicative of the monomolecular reaction.

17.6
Isomerization of Cycloalkanes [141]

Cyclohexane is known to be isomerized to methylcyclopentane when catalyzed by strong acids. In fact, the SO_4/ZrO_2 catalyst converts cyclohexane into methylcyclopentane and methylcyclopentane into cyclohexane [119, 142, 143]. The reactions proceed by the monomolecular mechanism via the intermediacy of secondary and tertiary carbenium ions followed by protonated cyclopropanes.

The isomerization of open-chain alkanes with more than six carbon atoms gives isobutane as the main product, together with disproportionated materials, even though the reaction proceeds by the monomolecular pathway [144]. On the other hand, for cyclic alkanes the monomolecular process with preservation of the cyclic structure seems to be the most probable, judging from the results for cyclohexane. The absence of isobutane in the products indicates that the reaction path does not involve open-chain intermediate species. Therefore, it is of interest to try cycloalkanes larger than cyclohexane for clarification of the reaction mechanism along with the catalytic action of SO_4/ZrO_2.

The skeletal isomerization of cycloalkanes with more than six carbon atoms, that is cycloheptane, cyclooctane, cyclodecane, and cyclododecane, was performed over SO_4/ZrO_2 in the liquid phase at 50 °C.

The results for cycloheptane after 30 min reaction time are shown in Table 17.6. Methylcyclohexane is the major product, with a selectivity of 97%, in addition to small amounts of four dimethylcyclopentanes and ethylcyclopentane. The reaction of methylcyclohexane under the same conditions also produced the latter five compounds, and the system reached its equilibrium state after 90 min of reaction.

The reaction results of cyclooctane are shown in Table 17.7. The major product is ethylcyclohexane with a selectivity of 93%. Small amounts of five dimethylcyclohexanes and methylcycloheptane were formed in addition. The comparable reaction of ethylcyclohexane produced dimtheylcyclohexanes and methylcycloheptane.

The reaction results of cycloheptane and cyclooctane indicate that the monomolecular pathway is followed, preserving the cyclic structure, without the formation

Table 17.6 Product distribution for the reaction of cycloheptane at 50 °C after 30 min reaction time over SO_4/ZrO_2.

Product	Yield (%)	Selectivity (%)
Methylcyclohexane	53	97
trans-1,2-Dimethylcyclopentane	0.7	1.3
trans-1,3-Dimethylcyclopentane	0.4	0.7
1,1-Dimethylcyclopentane	0.3	0.4
cis-1,3-Dimethylcyclopentane	0.2	0.3
Ethylcyclopentane	0.2	0.3

Table 17.7 Product distribution in the reaction of cyclooctane at 50 °C after 30 min reaction time over SO_4/ZrO_2.

Product	Yield (%)	Selectivity (%)
Ethylcyclohexane	57	93
trans-1,2-Dimethylcyclohexane	2.1	3.3
Methylcycloheptane	1.1	1.7
cis-1,3-Dimethylcyclohexane	1.0	1.3
trans-1,4-Dimethylcyclohexane	0.2	0.3
1,1-Dimethylcyclohexane	0.1	0.2
trans-1,3-Dimethylcyclohexane	0.1	0.2

Protonated [4.1.0]bicycloheptane intermediate

Protonated [4.2.0]bicyclooctane intermediate

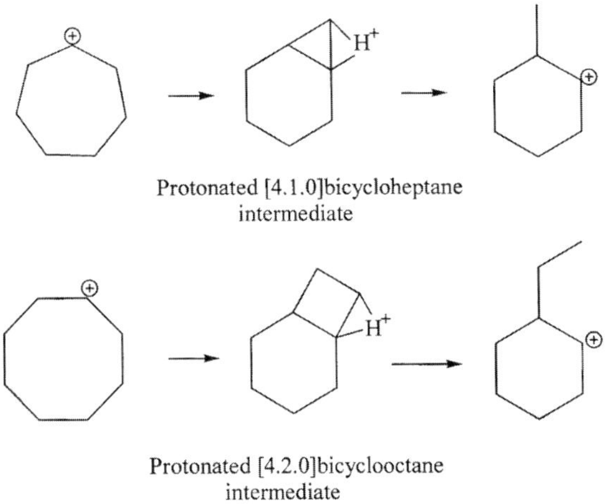

Scheme 17.3 Intermediates for the isomerizations of cycloheptane into methylcyclohexane and of cyclooctane into ethylcyclohexane.

of isobutane through protonated cyclopropane and cyclobutane intermediates, as shown in Scheme 17.3. The structure of the cyclic hydrocarbon has a large effect on reaction. No isomerization was observed with cyclodecane, with decalins being formed by dehydrogenation, and cyclododecane was converted into more than 30 product species by rearrangement, dehydrogenation, and cracking.

17.7
Structure of Sulfated Zirconia

In the very early stages of this work, we studied the catalytic surface using mainly XPS and IR spectroscopy [32], and proposed the surface structure to be SO_4 combined with two zirconium species in a bridging bidentate state (refer to Scheme 17.4). An analogous model was proposed by Segawa and coworkers [53], and Bolis and coworkers [145].

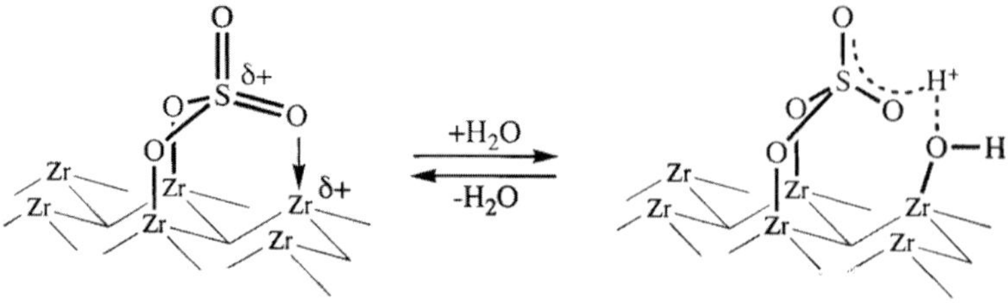

Scheme 17.4 A monosulfate structure.

Later, a number of studies attempted to determine the nature of acid sites in the catalyst. Also applying XPS and IR spectroscopy, Tanabe and others proposed a structure of chelating bidentate complexes, in which the sulfate species chelates to a single Zr atom [102, 146]. This model, a chelating bidentate species, was also proposed by Ward and Ko [61].

Morrow and coworkers proposed a structure, on the basis of ^{18}O exchange using $H_2^{18}O$ combined with IR analysis, in which three oxygens from the sulfate groups are bonded to Zr species in a tridentate form [147]. They also pointed out the possibility of the formation of a polysulfate structure at high sulfate loading [148]. This structure was supported by Morterra and coworkers using IR studies of adsorbed pyridine [149].

A monodentate structural model, which contains a bisulfate group, has been proposed by several workers [103, 104]. The bisulfate OH group is hydrogen-bonded to an oxygen on the surface of zirconia. A similar model has been proposed for the surface of sulfated alumina on the basis of NMR studies [150].

Another bisulfate structure was proposed by Riemer and coworkers using NMR and Raman spectroscopies, in which two sulfate oxygens are bonded to Zr atoms in a bridged bidentate state [106]. The same model was also proposed by Lunsford [151] and Clearfield [152].

Models in which SO_3 species are coordinated with zirconia have also been proposed. One of them, suggested by Vedrine and coworkers [52], is coordination of the SO_3 sulfur with lone pairs of the oxygen in ZrO_2 in addition to one of the SO_3 oxygens coordinating with a Zr. Another proposal by White and coworkers [153] is depicted in such a way that two of the SO_3 oxygens are coordinated with surface zirconium atoms, leaving a single S=O moiety. A thionyl tetraoxide species with four oxygens bonded to zirconia together with a single S=O has been proposed at low sulfate content [154].

Finally, sulfated zirconia was investigated by thermal analyses in addition to XPS in order to provide additional information on the surface. On the basis of the observations made, we have proposed possible surface structures of sulfated zirconia [155a]. The active species of sulfated zirconia are decomposed over a quite broad range of temperature, from 700 °C to >1000 °C. XPS data indicates that the S species comprise SO_4^{2-}. An example of the models is shown in Scheme 17.4, where two oxygens are bonded to Zr in addition to coordination of an S=O group with Zr, resulting in three grafting bonds in total. The addition of water causes the breakage

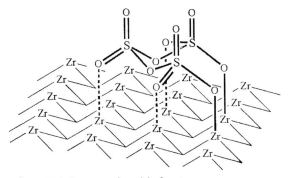

Scheme 17.5 A structural model of a trimer.

of this coordination, generating Brønsted acid sites. Another example, consisting of a cyclic trimer of SO_3, is shown in Scheme 17.5, where two terminal S=O anions are bonded to Zr cations including three coordinations of S=O with Zr. These coordination sites at Zr are also positions for water molecules giving Brønsted acid sites, as shown in Scheme 17.4 in the case of monosulfate species.

The surface is composed of a wide range of coordinated oligomeric species, predominantly species containing 3 or 4 S atoms, with two ionic bonds between S—O— and Zr being formed, which is identical to the models for monomer and trimer (Schemes 17.4 and 17.5). The active site is not on the metal species, but rather on the S atoms.

The addition of water causes the breakage of the coordination bonds to yield Brønsted acid sites strengthening Lewis acid sites, as shown in Scheme 17.4, for example. Many research groups report the simultaneous existence of Brønsted and Lewis acid sites or the reversible transformation between Brønsted and Lewis acidity upon hydration or dehydration [61, 106, 152]. Fraenkel suggests that in order to be an effective superacid, sulfated zirconia should contain a critical amount of moisture [155b]. Several workers propose that the strong acidity requires the presence of both Lewis and Brønsted sites.

17.8 Promoting Effect

17.8.1 Effect of Addition of Metals to Sulfated Zirconia on the Catalytic Activity

A large number of metal-promoted superacids, which are highly active for butane conversion, have been prepared by the addition of metal salts to SO_4/ZrO_2. Metal promoters include Fe-Mn [156–162]: Pt, Pd [163]: Pd [164]: Ga, In, Tl [165]: Pt-Ni [166]: Ni [167]: Mn [168]: Ce [169]: V, Cr, Mn, Fe, Co, Ni, Cu, Zn [170]: Pt, Pd, Ir [171]: Al, Ga [172]: Cu [173]: Al [174, 175a–175d]: Ni-Al [176]: Ga [177–181]: Pt, Nb

[182]: Pt, Cu, Zn, Cd [183]: Pt, Ni, Fe, Cr, Co, W [184]: Pt [185–194]. Among those metals, Fe, Pt, and Ga are promising additives and the simultaneous promotion with Fe and Mn has attracted much attention [195, 196], but their role is not clear. Catalysts containing small amounts of Pt were also developed and the role and state of Pt have been comprehensively summarized in a review by Song and Sayari [197].

A highly active catalyst for the conversion of butane is obtained by adding a large quantity (7–8 wt%) of Pt to SO_4/ZrO_2 [198], though an active component such as platinum is usually present in low quantity, 0.1–0.5%. The most active catalyst is prepared by impregnation of Zr gel with aqueous sulfuric acid followed by impregnation of the sulfated gel with a solution of H_2PtCl_6, followed by calcination [199]. The Pt-SO_4/ZrO_2 thus prepared gives the same butane conversion as SO_4/ZrO_2 but at much lower temperature, the temperature difference being 136 °C. The analogous enhancement of activity is also observed with other noble metals, Ir, Rh, Ru, Os, and Pd [198, 200].

The Pt-added SO_4/ZrO_2 catalyst is generally prepared by impregnation of the sulfated zirconia with $H_2[PtCl_6]$. The charge on the platinum compound has a large effect on the catalytic action [201]. Platinum particles in catalysts prepared using a cationic platinum compound, $[(NH_3)_4Pt]^{2+}$, showed much higher dispersion and more-even distribution than those in the catalysts prepared by an anionic precursor, $[PtCl_6]^{2-}$, because of strong interaction of the former with the support.

17.8.2
Effect of Mechanical Mixing of Pt-Added Zirconia on the Catalytic Activity

An active catalyst for the conversion of butane to isobutane was obtained by mechanically mixing WO_3/ZrO_2 and Pt/ZrO_2 (0.5 wt% Pt) as shown in Table 17.8. The activity of Pt-WO_3/ZrO_2 prepared by co-impregnation of zirconia with W and Pt materials was lower than that of the mechanical mixture of WO_3/ZrO_2 and Pt/ZrO_2 [202]. The reaction was carried out in a reactor in which two catalysts could

Table 17.8 The application of physical mixtures of WO_3/ZrO_2 and Pt/ZrO_2 for the reaction of butane into isobutane.

Catalyst	Products[a] (%)				
	$C_1 + C_2$	C_3	i-C_4[b]	i-C_4'[c]	C_5[d]
Pt-WO_3/ZrO_2	0	0.1	1.0	trace	0
WO_3/ZrO_2 + Pt/ZrO_2	0	2.9	7.8	0.6	0.2
$Pt/ZrO_2 \rightarrow WO_3/ZrO_2$	1.3	2.3	2.2	0.2	0
$WO_3/ZrO_2 \rightarrow Pt/ZrO_2$	1.4	2.6	1.5	1.2	0

a C_1: methane, C_2: ethane, C_3: propane.
b   i-C_4: isobutene.
c   i-C_4': isobutene.
d C_5: pentane.

Table 17.9 The reaction profile of butane over solid acids mixed with Pt/ZrO$_2$

Catalyst	Reaction temperature (°C)	Products (%)	
		C$_3$	i-C$_4$
SO$_4$/TiO$_2$	300	0.3	1
SO$_4$/TiO$_2$ + Pt/ZrO$_2$	200	6	13
SO$_4$/Al$_2$O$_3$	300	0	0.2
SO$_4$/Al$_2$O$_3$ + Pt/ZrO$_2$	200	4	19
SO$_4$/Fe$_2$O$_3$ + Pt/ZrO$_2$	220	3	7
Mordenite	300	1	2
Mordenite + Pt/ZrO$_2$	200	22	17
ZSM-5	300	0	0.1
ZSM-5 + Pt/ZrO$_2$	200	10	3
SiO$_2$-Al$_2$O$_3$ + Pt/ZrO$_2$	240	6	13

Table 17.10 Conversion of propane into butane.

Catalyst	Temperature (°C)	Conversion (%)	Products (%)			
			C$_1$	C$_2$	C$_4$	C$_5$
SO$_4$/ZrO$_2$	200	0.1			0.1	
Pt-SO$_4$/ZrO$_2$	200	0.2			0.2	
SO$_4$/ZrO$_2$ + Pt/ZrO$_2$	200	3.2	trace	0.2	3.0	trace
SO$_4$/ZrO$_2$ + Pt/ZrO$_2$	225	6.1	0.3	1.6	4.1	0.1

C$_4$: butanes.

be segregated and mixed. In the two arrangements where Pt/ZrO$_2$ was placed ahead of WO$_3$/ZrO$_2$ and WO$_3$/ZrO$_2$ was placed ahead of Pt/ZrO$_2$, the conversions of n-butane were less than that of the mechanical mixture, and methane and ethane were significant by-products.

The catalytic activities of several sulfated metal oxides, zeolites, and silica-aluminas were enhanced by addition of Pt/ZrO$_2$ as shown in Table 17.9. The order of activity of SO$_4$/MeO$_x$ was TiO$_2$ > Al$_2$O$_3$ > Fe$_2$O$_3$ [203]. A mixture of SO$_4$/Al$_2$O$_3$ and Pt/ZrO$_2$ gave the highest selectivity for the skeletal isomerization. The mechanical mixtures of Pt/ZrO$_2$ with zeolites and silica-aluminas showed satisfactory conversions of butane, but their selectivities for isobutane were low.

Although mixing Pt/ZrO$_2$ with SO$_4$/ZrO$_2$ showed no appreciable enhancement of the activity of SO$_4$/ZrO$_2$, a positive effect was observed for the conversion of propane into butanes as shown in Table 17.10 [204]. The catalysts that are effective for butane conversion are not always efficient for propane conversion. As the acidic sites on SO$_4$/ZrO$_2$ are stronger than those on SO$_4$/TiO$_2$, SO$_4$/Al$_2$O$_3$, and SO$_4$/Fe$_2$O$_3$, it is suggested that a high superacidity is required for activity for propane conversion to butanes by mixing with Pt/ZrO$_2$.

The effect of mixing was specific to Pt/ZrO$_2$, because no efficacy was observed with Pt/other metal oxides such as Pt/TiO$_2$ and Pt/SiO$_2$. XPS analysis of Pt/ZrO$_2$ showed the binding energy of Pt 4f to be 72.6 eV, close to that of Pt^{2+}. On the other hand, the binding energies, 70.5–71.2 eV, for Pt-WO$_3$/ZrO$_2$, Pt/TiO$_2$, and Pt/SiO$_2$ were close to 71.7 eV, the binding energy for Pt^0. The results indicate that the affinity of Pt to ZrO$_2$ is quite specific.

Aubke and co-workers reported the existence of a molecular [Pt(CO)$_4$]$^{2+}$ cation from the following synthesis:

$$2\,Pt(SO_3F)_4 + 5\,CO \text{ in } HSO_3F \rightarrow [Pt(SO_3F)_6]\cdot[Pt(CO)_4] \qquad (17.8)$$

where Pt^{2+} is a coordinating species [205]. Therefore, it is proposed that Pt^{2+} species on ZrO$_2$ act as coordinatiom sites for H$^-$ formed during reaction or reservoirs for the coordination of H$^-$. A stable Pt^{2+} complex with a hydride ion is also known [206].

The catalytic activity of Fe- and Mn-added SO$_4$/ZrO$_2$ for butane is three orders of magnitude greater than that of SO$_4$/ZrO$_2$ around room temperature [196]. The promotional effect of Fe and Mn is explained in terms of the dehydrogenative action of Fe species giving butene and subsequent conversion via a bimolecular reaction pathway involving C$_8$ intermediates. Thermodynamically, the dehydrogenation step requires a high temperature. Similar to the case of mixing Pt/ZrO$_2$, the Fe species might act partly as the coodination sites, or reservoirs, for transfer of H$^-$.

17.9
Friedel–Crafts Acylation of Aromatics

The Friedel–Crafts reaction has been used as a test reaction since this work started [43, 48]; with the benzylation of toluene with benzyl chloride being achieved with heat-treated iron sulfates. Of the two main reactions, alkylation and acylation, the latter has been predominantly investigated, since it is more difficult than the former. In addition, the catalytic acylation of aromatics via the Friedel–Crafts reaction is attractive for organic synthesis and a prerequisite challenge to green technology. Traditional methods depend on the use of at least stoichiometric quantities of Lewis acids, such as aluminum trichloride and boron trifluoride, because the Lewis catalysts are consumed by coordination with the formed aromatic ketones [207, 208]. A large quantity of acids, and their waste matter after work-up procedures, cause serious environmental problems. There have been several efforts to conduct the reaction using catalytic amounts of acidic promoters. However, a high acidic strength is required on the surface to catalyze the acylation and hence a superacid is desired.

An important test reaction has been the benzoylation of toluene because of the ease of procedure operating in a batch mode at temperatures around 100 °C [209], an example on a comparative study is shown in Table 17.11 [67]. Both sulfated and

Table 17.11 Friedel–Crafts benzoylation of toluene with benzoyl chloride or benzoic anhydride at 100°C.

Catalyst	Yield (%)	
	Ph-Cl	(Ph-CO)$_2$O
SO$_4$/ZrO$_2$	22	92
SO$_4$/SnO$_2$	52	48
WO$_3$/ZrO$_2$	19	55
WO$_3$/SnO$_2$	14	16

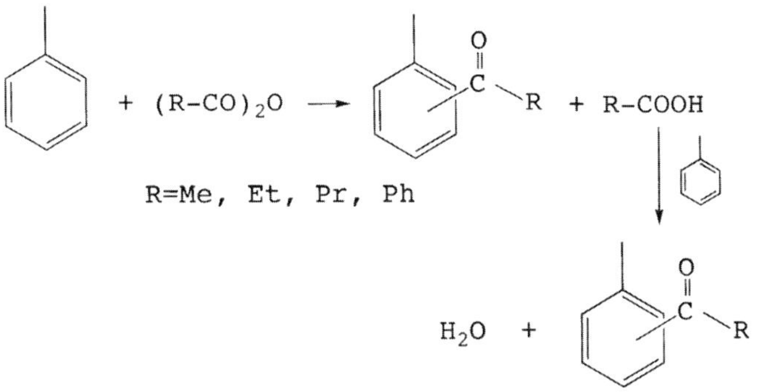

Scheme 17.6 Benzoylation of toluene with alkyl acid anhydride.

tungstated zirconias (Scheme 17.6) along with stannias show satisfactory activities. Yields in the reactions with benzoyl chloride are lower than those with the anhydride, though the reactivity with acylating reagents is generally PhCOCl > (PhCO)$_2$O, with strong acid sites being required for the formation of an acyl cation (PhCO$^+$) from the anhydride. However, the results are not consistent with the reactivity, in particular in the case of SO$_4$/ZrO$_2$. The reason is probably associated with strong interactions between oxygens of the anhydride and acid sites on the surface. Benzoylation by the benzoic acid produced does not occur when the reaction is performed with (PhCO)$_2$O; since the reaction occurs at temperatures above the reflux temperature of the mixture.

The catalysts were examined in the acetylation of toluene, one of the difficult acylations because of the difficulty in the formation of an intermediate acetyl cation (MeCO$^+$) from alkyl chain acid anhydrides and hydrides. The reactivities increase as the length of the hydrocarbon chain of the acylating agent is lengthened [210], which shows acetylation to be the most difficult acylation. The heterogeneous acetylation of toluene using solid acidic promoters has not been reported in detail. In fact, to complete acetylation with the anhydride is quite difficult under reflux conditions [211].

Acylations by more reactive acylating reagents with longer hydrocarbon chains, propionylation and butyrylation, have been examined [211]. The SO_4/ZrO_2 catalyst gave higher yields for propionylation and butyrylation than for acetylation, with buyrylation giving the highest yield. The reactivity with acylating reagents is $(PrCO)_2O > (EtCO)_2O > (MeCO)_2O$, which agrees with the stabilities of the intermediates $PrCO^+ > EtCO^+ > MeCO^+$.

The present catalyst was examined in a more reactive acetylation, the acetylation of anisole with acetic anhydride. A quantitative yield of methoxyacetophenone was obtained under the reaction conditions, showing how reactive anisole is in comparison with toluene – demonstrating the effect an oxygen has on reactivity. Several papers concerning the Friedel–Crafts reaction of anisole have been published [212–216], but studies of the reaction of toluene or even benzene are desired. The difference in reactivity between anisole and toluene is close to 100 °C in terms of the reaction temperature.

Although acid anhydride is predominantly used in place of the acid chloride for environmental considerations, the successive acylation with the carboxylic acid produced could be more advantageous, with the most efficient utilization of carbon. In this case, the reaction is equivalent to a dehydrated acylation, which requires strong acidity. In fact, the benzoylation of toluene with benzoic anhydride was carried out at 180 °C, a temperature above reflux, using an autoclave, and marked conversion with benzoic acid produced was observed, though the reaction with the anhydride took place even at 30 °C, as shown in Figure 17.7 [217]. In this point of view, a remarkable reaction is the acetylation of toluene, or benzene if possible, with acetic anhydride up to the consumption of acetic acid produced. A solid acid able to catalyze the reaction of benzene at temperatures below its boiling point, 80 °C, would be highly desirable, since the reactivity of benzene to toluene

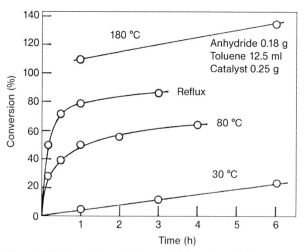

Figure 17.7 Benzoylation of toluene with benzoic anhydride catalyzed by SO_4/ZrO_2.

for acylation is 1/150. In addition, the ultimate catalyst would be one that facilitates the acetylation of deactivated aromatics, for instance nitrobenzene.

17.10
Ceramic Acid

A solid acid calcined at temperatures above 1000 °C is termed a "Ceramic Acid". Ceramics are made by calcination of inorganic materials at elevated temperatures and they possess favorable stability. Brick, usually produced by calcination at temperatures above 1000 °C, is a typical example of one of these materials. Solid acids are generally prepared by calcination at temperatures around 500 °C to generate the highest acidity on the surface, and it is known that the treatment with temperatures above 500 °C causes the surface acidity to reduce. A solid acid prepared by calcination at similar temperatures to a ceramic, would be highly desirable in terms of green chemistry, as a result of the corresponding stability of acid sites.

A pathway into this area arose in a study of the catalytic activity of molybdated zirconia (MO_3/ZrO_2) for the benzoylation of toluene with benzoic anhydride [72]. The catalysts were prepared by addition of different amounts of molybdic acid (H_2MoO_4) to zirconia gel and calcination at various temperatures. The activities are shown as a function of calcination temperature of the catalysts in Figure 17.8 [218]. The figure shows that the effect of modifying the proportion of Mo in the

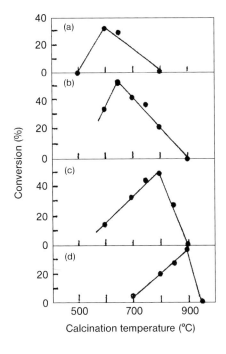

Figure 17.8 Catalytic activities of MoO_3/ZrO_2 for the benzoylation of toluene with benzoic anhydride. Quantity of molybdenum (wt% Mo): 20 (a), 10 (b), 5 (c), and 2 (d).

catalyst is very large. The maximum activity is a strong function of the Mo quantities and observed with 600, 650, 800, and 900 °C for 20%-, 10%-, 5%-, and 2%-Mo/ZrO$_2$ catalysts, respectively, indicating the lower the quantity, the higher the optimum calcination temperature. Analogous phenomena are also observed with tungstated zirconia [219]. The results indicate that the use of a small quantity of tungsten or molybdenum is desirable.

17.10.1
Tungstated Stannia [220]

A 13%W-added stannia generates remarkable acidity on the surface when calcined at 1000 °C (Table 17.3). The WO$_3$/SnO$_2$ catalysts were prepared by drying, at 100 °C, *meta*-stannic acid of Koujundo Kagaku, Ltd. (commercial grade, powdered below 100 mesh) followed by impregnation with aqueous ammonium metatungstate [(NH$_4$)$_6$(H$_2$W$_{12}$O$_{40}$), Nippon Inorganic Color & Chemical Co.], evaporating the water, drying, and calcining in air for 3 h. The concentrations were 2, 5, 10, 20, 40, and 80 wt% W based on the stannic acid.

The materials thus prepared were examined in the acid-catalyzed conversion of methanol into dimethyl ether. The conversions are shown as a function of calcination temperature of the catalysts in Figure 17.9. The highest activity for 40%- and 80%-WO$_3$/SnO$_2$ was observed at a calcination temperature of 800 °C, but materials with lower quantities of W required higher temperatures of calcination for optimum activity. This indicates the main tendency that the lower the quantity, the higher the temperature showing the optimum activity: namely, 900, 1000, 1100, and 1150 °C for the materials with 20, 10, 5, and 2% W, respectively.

The above results led to the synthesis of a ceramic acid, 5–10 wt% W-added SnO$_2$ materials calcined at 1000–1100 °C. The surface acidity determined by the heat

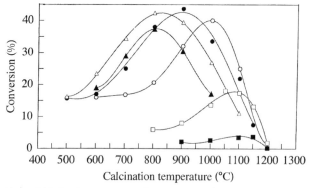

Figure 17.9 Catalytic activities of WO$_3$/SnO$_2$ with various concentrations of W for the dehydration of methanol to dimethyl ether at 210 °C. The concentration of W: 80% (▲), 40% (△), 20% (●), 10% (○), 5% (□), 2% (■).

of adsorption of Ar and the catalytic activity for decompositions of toluene and ethylbenzene were intermediate between the values for mordenite and for silica-alumina.

As stable materials, just like brick, ceramics have excellent durability against water. The ceramic catalysts were tested in the esterification of *n*-octanoic acid with methanol to methyl *n*-octanoate to determine their stability to the water liberated. The 5%- and 10%-WO_3/SnO_2 compounds heat-treated at 1100 °C showed activities much higher than mordenites and SiO_2-Al_2O_3 in the presence of water at a reaction temperature of 64 °C, though their activities for alkylbenzenes, together with their acidities, were lower than those of the mordenites.

17.10.2
Tungstated Alumina [221]

Tungstated aluminas (WO_3/Al_2O_3) were prepared by drying aluminum hydroxide [Koujundo Kagaku, Ltd., commercial grade (3N), powdered below 100 mesh] at 100 °C, followed by impregnation with the *meta*tungstate [$(NH_4)_6(H_2W_{12}O_{40})$], evaporation of the water, drying, and calcination in air for 3 h. The concentration applied was from 1 to 20 wt% W based on the alumina gel, denoted as 1%-WO_3/Al_2O_3 (1.18 wt% W after calcination at 1200 °C) and 5%-WO_3/Al_2O_3(4.43 wt% W after calcination at 1000 °C), for instance.

The materials thus prepared were examined in an acid-catalyzed reaction – the cracking of cumene into benzene and propylene. The conversions of cumene into benzene at 250 °C over the catalysts with 1, 3, 5, 10, and 20 wt% W for the impregnated concentration are shown in Figure 17.10. The highest activities for 5%-,

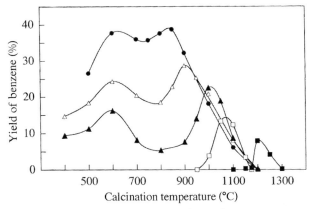

Figure 17.10 Catalytic activities of WO_3/Al_2O_3 with various concentrations of W for the cracking of cumene to benzene and propylene at 250 °C. The concentration of W: 20% (●), 10% (△), 5% (▲), 3% (□), 1% (■).

10%- and 20%-WO$_3$/Al$_2$O$_3$ were observed at two temperatures of calcination; the first temperature was identical among the three materials, 600 °C, and the second one was 850, 900, and 1000 °C for 20%-WO$_3$/Al$_2$O$_3$, 10%-WO$_3$/Al$_2$O$_3$, and 5%-WO$_3$/Al$_2$O$_3$, respectively. For the materials with 3 and 1% W the highest activity is produced by calcination at 1050 and 1200 °C, respectively, though their first peaks are not observed. The materials with lower quantities of W gave their second peaks in the higher temperature range of calcination, again illustrating the tendency: the lower the quantity, the higher the temperature.

The above studies resulted in novel ceramic acids, 1–5 wt%W-added aluminas calcined at 1000–1200 °C. The surface acidity determined by the heat of adsorption of Ar and the catalytic activity for decompositions of toluene, ethylbenzene, and cumene were higher than those of silica-alumina. The crystallographic phase was θ- or α-Al$_2$O$_3$.

17.11
Application to Sensors and Photocatalysis

Titanium and tin are important in terms of photocatalytic and sensor applications. There is insufficient space to continue this description here, but the literature on these topics can be found elsewhere [222–232]. Tungstated zirconia, for example, is a promising candidate for an ammonia exhaust gas sensor required to control a Urea-SCR (Selective Catalytic Reduction) system in diesel-engined cars [233, 234].

References

1 Hammett, L.P. and Deyrup, A.J. (1932) *Journal of the American Chemical Society*, **54**, 2721.
2 Gillespie, R.J., Peel, T.E. and Robinson, E.A. (1971) *Journal of the American Chemical Society*, **93**, 5083.
3 Hall, N.F. and Conant, J.B. (1927) *Journal of the American Chemical Society*, **49**, 3047.
4 Gillespie, R.J. (1968) *Accounts of Chemical Research*, **1**, 202.
5 Gillespie, R.J. and Peel, T.E. (1972) *Advances in Physical Organic Chemistry*, **9**, 1.
6 Gillespie, R.J. and Peel, T.E. (1973) *Journal of the American Chemical Society*, **95**, 5173.
7 Olah, G.A., Prakash, G.K.S. and Sommer, J. (1979) *Science*, **206**, 13.
8 Gillespie, R.J. and Liang, J. (1988) *Journal of the American Chemical Society*, **110**, 6053.
9 Olah, G.A., Prakash, G.K.S. and Sommer, J. (1985) *Superacids*, John Wiley & Sons, Inc., New York.
10 Oelderik, J.M., Mackor, E.L., Plattecuw, J.C. and van der Wiel, A. (1965) US Patent 3,201,494.
11 Oelderik, J.M. (1968) US Patent 3,394,202.
12 Olah, G.A. (1973) US Patents, 3,708,553 and 3,766,286.
13 Rodewald, P.G. (1976) US Patents, 3,962,133 and 3,984,352.
14 Dooley, K.M. and Gates, B.C. (1985) *Journal of Catalysis*, **96**, 347.
15 Tanabe, K. and Hattori, H. (1976) *Chemistry Letters*, 625.

16 Hattori, H., Takahashi, O., Takagi, M. and Tanabe, K. (1981) *Journal of Catalysis*, **68**, 132.
17 Namba, S., Ihara, K., Sakaguchi, Y. and Yashima, T. (1980) *Journal of the Japan Petroleum Institute*, **23**, 142.
18 Forsman, W.C., Dziemianowicz, T., Leong, K. and Carl, D. (1983) *Synthetic Metals*, **5**, 77.
19 Lalancette, J.M. and Lafontaine, J. (1973) *Journal of the Chemical Society, Chemical Communications*, 815.
20 Laali, K., Muller, M. and Sommer, J. (1980) *Journal of the Chemical Society, Chemical Communications*, 1088.
21 Laali, K. and Sommer, J. (1981) *Nouveau Journal de Chimie*, **5**, 469.
22 Lalancette, J.M., Fournier-Breault, M.J. and Thiffault, R. (1974) *Canadian Journal of Chemistry*, **52**, 589.
23 Olah, G.A., Kaspi, J. and Bukala, J. (1977) *Journal of Organic Chemistry*, **42**, 4187.
24 Olah, G.A. and Kaspi, J. (1977) *Journal of Organic Chemistry*, **42**, 3046.
25 Yoneda, N., Fukuhara, T., Abe, T., Suzuki, A. and Kudo, K. (1981) *Chemistry Letters*, 1485.
26 Ono, Y., Sakuma, S., Tanabe, T. and Kitajima, N. (1978) *Chemistry Letters*, 1061.
27 Ono, Y., Tanabe, T. and Kitajima, N. (1979) *Journal of Catalysis*, **56**, 47.
28 Ono, Y., Yamaguchi, K. and Kitajima, N. (1980) *Journal of Catalysis*, **64**, 13.
29 Heinermann, J.J.L. and Gaaf, J. (1981) *Journal of Molecular Catalysis*, **11**, 215.
30 Arata, K. (1996) *Journal of the Japan Petroleum Institute*, **39**, 185.
31 Tatsumi, T., Matsuhashi, H. and Arata, K. (1996) *Bulletin of the Chemical Society of Japan*, **69**, 1191.
32 Hino, M. and Arata, K. (1980) *Journal of the Chemical Society, Chemical Communications*, 851.
33 Matsuhashi, H., Miyazaki, H. and Arata, K. (2001) *Chemistry Letters*, 452.
34 Matsuhashi, H., Miyazaki, H., Kawamura, Y., Nakamura, H. and Arata, K. (2001) *Chemistry of Materials*, **13**, 3038.
35 Arata, K. and Hino, M. (1984) *Reaction Kinetics and Catalysis Letters*, **25**, 143.
36 Hino, M. and Arata, K. (1979) *Chemistry Letters*, 477.
37 Hino, M. and Arata, K. (1979) *Chemistry Letters*, 1259.
38 Hino, M., Kobayashi, S. and Arata, K. (1979) *Journal of the American Chemical Society*, **101**, 6439.
39 Hino, M. and Arata, K. (1979) *Journal of the Chemical Society, Chemical Communications*, 1148.
40 Matsuhashi, H., Hino, M. and Arata, K. (1988) *Chemical Communications*, 1027.
41 Matsuhashi, H., Hino, M. and Arata, K. (1990) *Applied Catalysis*, **59**, 205.
42 Cutrufello, M.G., Diebold, U. and Gonzalez, R.D. (2005) *Catalysis Letters*, **101**, 5.
43 Arata, K. (1991) *Trends in Physical Chemistry*, **2**, 1, and references cited therein.
44 Matsuhashi, H., Hino, M. and Arata, K. (1991) *Catalysis Letters*, **8**, 269.
45 Arata, K. and Hino, M. (1990) *Applied Catalysis*, **59**, 197.
46 Gawthrope, D.E., Lee, A.F. and Wilson, K. (2004) *Physical Chemistry Chemical Physics*, **6**, 3907.
47 Matsuhashi, H., Sato, D. and Arata, K. (2004) *Reaction Kinetics and Catalysis Letters*, **81**, 183.
48 Arata, K. and Hino, M. (1990) *Materials Chemistry and Physics*, **26**, 213, and references cited therein.
49 Ebitani, K., Konishi, J. and Hattori, H. (1991) *Journal of Catalysis*, **130**, 257.
50 Morterra, C., Cerrato, G., Emanuel, C. and Bolis, V. (1993) *Journal of Catalysis*, **142**, 349.
51 Zhang, C., Miranda, R. and Davis, B.H. (1994) *Catalysis Letters*, **29**, 349.
52 Babou, F., Coudurier, G. and Vedrine, J.C. (1995) *Journal of Catalysis*, **152**, 341.
53 Funamoto, T., Nakagawa, T. and Segawa, K. (2005) *Applied Catalysis A: General*, **286**, 79.
54 Stepanov, A.G., Luzgin, M.V., Arzumanov, S.S., Wang, W., Hunger, M. and Freude, D. (2005) *Catalysis Letters*, **101**, 181.
55 Ma, Z., Zou, Y., Hua, W., He, H. and Gao, Z. (2005) *Topics in Catalysis*, **35**, 141.
56 Li, B. and Gonzalez, R.D. (1998) *Applied Spectroscopy*, **52**, 1488.

57 Li, B. and Gonzalez, R.D. (1998) *Catalysis Today*, **46**, 55.
58 Hino, M., Kobayashi, S. and Arata, K. (1981) *Reaction Kinetics and Catalysis Letters*, **18**, 491.
59 Hino, M. and Arata, K. (1990) *Chemistry Letters*, 1737.
60 Matsuhashi, H. and Katada, N. (2006) *Catalysts and Catalysis*, **48**, 464.
61 (a) Ward, D.A. and Ko, E.I. (1994) *Journal of Catalysis*, **150**, 18.
(b) Ward, D.A. and Ko, E.I. (1995) *Journal of Catalysis*, **157**, 321.
62 Tichit, D., Coq, B., Armendariz, H. and Figueras, F. (1996) *Catalysis Letters*, **38**, 109.
63 Zhuang, Q. and Miller, J.M. (2001) *Canadian Journal of Chemistry*, **79**, 1220.
64 Melada, S., Signoretto, M., Ardizzone, S.A. and Bianchi, C.L. (2001) *Catalysis Letters*, **75**, 199.
65 Ardizzone, S., Bianchi, C.L., Cappelletti, G. and Porta, F. (2004) *Journal of Catalysis*, **227**, 470.
66 Noda, L.K., de Almeida, R.M., Probst, L.F.D. and Goncalves, N.S. (2005) *Journal of Molecular Catalysis A – Chemical*, **225**, 39.
67 Arata, K., Nakamura, H. and Shouji, M. (2000) *Applied Catalysis A: General*, **197**, 213.
68 Arata, K., Matsuhashi, H., Hino, M. and Nakamura, H. (2003) *Catalysis Today*, **81**, 17.
69 Hino, M., Takasaki, S., Furuta, S., Matsuhashi, H. and Arata, K. (2007) *Applied Catalysis A: General*, **321**, 147.
70 (a) Arata, K. and Hino, M. (1988) *Proceedings of the 9th International Congress on Catalysis, Calgary, Canada, 1988* (eds M.J. Phillips and M. Ternan), Vol. 4, 1727.
71 Hino, M. and Arata, K. (1988) *Journal of the Chemical Society, Chemical Communications*, 1259.
72 Hino, M. and Arata, K. (1989) *Chemistry Letters*, 971.
73 Arata, K. and Hino, M., (1993). *Proceedings of the 10th International Congress on Catalysis, Budapest, 1992* (eds L. Guczi, F. Solymosi and P. Tetenyi), Elsevier, Amsterdam, p. 2613.
74 Hino, M. and Arata, K. (1994) *Bulletin of the Chemical Society of Japan*, **67**, 1472.
75 Matsuhashi, H., Kato, K. and Arata, K., (1994) *Studies in Surface Science and Catalysis. Proceedings of the International Symposium on Acid-Base Catalysis II, Kodansha, Tokyo, 1994*, Vol. 90 (eds H. Hattori, M. Misono and Y. Ono), p. 251.
76 Hino, M. and Arata, K. (1992) 70th National Meeting Catal. Soc. Jpn, Niigata, Abstr. no. 4F409.
77 Gregorio, F.D. and Keller, V. (2004) *Journal of Catalysis*, **225**, 45.
78 Lebarbier, V., Clet, G. and Houalla, M. (2006) *Journal of Physical Chemistry B*, **110**, 13905.
79 Benesi, H.A. (1956) *Journal of the American Chemical Society*, **78**, 5490.
80 Cheung, T.K. and Gates, B.C. (1997) *Chemtech*, **27**, 28.
81 Deeba, M. and Hall, W.K. (1979) *Journal of Catalysis*, **60**, 417.
82 Umansky, B., Engelhardt, J. and Hall, W.K. (1991) *Journal of Catalysis*, **127**, 128.
83 Karge, H.G. (1991) *Studies in Surface Science and Catalysis*, **65**, 133.
84 Niwa, M., Iwamoto, M. and Segawa, K. (1983) *Bulletin of the Chemical Society of Japan*, **59**, 3735.
85 Karge, H.G. and Dondur, V. (1990) *Journal of Physical Chemistry*, **94**, 765.
86 Song, X. and Sayari, A. (1994) *Applied Catalysis A: General*, **110**, 121.
87 Katada, N., Igi, H., Kim, J. and Niwa, M. (1997) *Journal of Physical Chemistry B*, **101**, 5969
88 Corma, A., Fornes, V., Juan-Rajadell, M.I. and LopezNieto, J.M. (1994) *Applied Catalysis A: General*, **116**, 151.
89 Hunger, B., Heuchel, M., Clark, L.A. and Snurr, R.Q. (2002) *Journal of Physical Chemistry B*, **106**, 3882.
90 Teunissen, E.H., Jansen, A.P.J. and van Santen, R.A. (1995) *Journal of Physical Chemistry*, **99**, 1873.
91 Parrillo, D.J., Lee, C. and Gorte, R.J. (1994) *Applied Catalysis A: General*, **110**, 67.
92 Fogash, K.B., Yaluris, G., Gonzalez, M.R., Quraipryvan, P., Ward, D.A., Ko, E.I. and Dumesic, J.A. (1995) *Catalysis Letters*, **32**, 241.
93 (a) Katada, N., Endo, J., Notsu, K., Yasunobu, N., Naito, N. and Niwa, M. (2000) *Journal of Physical Chemistry B*, **104**, 10321.

(b) Katada, N., Endo, J., Notsu, K., Yasunobu, N., Naito, N. and Niwa, M. (2000) *Studies in Surface Science and Catalysis*, **130**, 3213.

94 Kim, S.Y., Goodwin, J.G., Jr, Hammache, S., Auroux, A. and Galloway, D. (2001) *Journal of Catalysis*, **201**, 1.

95 Tanabe, K., Misono, M., Ono, Y. and Hattori, H. (1989) *Studies in Surface Science and Catalysis*, **51**, 5.

96 Gao, Z., Chen, J., Hua, W. and Tang, Y. (1994) *Studies in Surface Science and Catalysis*, **90**, 507.

97 Adeeva, V., Lei, G.D. and Sachtler, W.M.H. (1995) *Catalysis Letters*, **33**, 135.

98 Wan, K.T., Khouw, C.B. and Davis, M.E. (1996) *Journal of Catalysis*, **158**, 311.

99 Tabora, J.E. and Davis, R.J. (1996) *Journal of the American Chemical Society*, **118**, 12240.

100 Bardin, B.B. and Davis, R.J. (1998) *Topics in Catalysis*, **6**, 77.

101 Ecormier, M.A., Wilson, K. and Lee, A.F. (2003) *Journal of Catalysis*, **215**, 57.

102 Jin, T., Yamaguchi, T. and Tanabe, K. (1986) *Journal of Physical Chemistry*, **90**, 4797.

103 Kustov, L.M., Kazansky, V.B., Figueras, F. and Tichit, D. (1994) *Journal of Catalysis*, **150**, 143.

104 Adeeva, V., de Haan, J.W., Janchen, J., Lei, G.D., Schunemann, G., van de Ven, L.J.M., Sachtler, W.M.H. and van Santen, R.A. (1995) *Journal of Catalysis*, **151**, 364.

105 Bolis, V., Broyer, M., Barbaglia, A., Busco, C., Foddanu, G.M. and Ugliengo, P. (2003) *Journal of Molecular Catalysis A–Chemical*, **204–5**, 561.

106 Riemer, T., Spielbauer, D., Hunger, M., Mekhemer, G.A.H. and Knozinger, H. (1994) *Journal of the Chemical Society, Chemical Communications*, 1181.

107 Haw, J.F. and Xu, T. (1998) *Advances in Catalysis*, **42**, 115.

108 Farcasiu, D., Ghenciu, A. and Miller, G. (1992) *Journal of Catalysis*, **134**, 118.

109 Biaglow, A.I., Gorte, R.J., Kokotailo, G.T. and White, D.J. (1994) *Journal of Catalysis*, **148**, 779.

110 Semmer, V., Batamack, P., Doremieux-Morin, C., Vincent, R. and Fraissard, J. (1996) *Journal of Catalysis*, **161**, 186.

111 Xu, T., Kob, N., Drago, R.S., Nicholas, J.B. and Haw, J.F. (1997) *Journal of the American Chemical Society*, **119**, 12231.

112 Haw, J.F., Zhang, J., Shimizu, K., Venkatraman, T.N., Luigi, D.P., Song, W., Barich, D.H. and Nicholas, J.B. (2000) *Journal of the American Chemical Society*, **122**, 12561.

113 Coster, D.J., Bendada, A., Chen, F.R. and Fripiat, J.J. (1993) *Journal of Catalysis*, **140**, 497.

114 Gonzalez, M.R., Kobe, J.M., Fogash, K.B. and Dumesic, J.A. (1996) *Journal of Catalysis*, **160**, 290.

115 Ghenciu, A. and Farcasiu, D. (1996) *Journal of Molecular Catalysis A–Chemical*, **109**, 273.

116 Figueras, F., Coq, B., Walter, C. and Carriat, J.-Y. (1997) *Journal of Catalysis*, **169**, 103.

117 Garin, F., Andriamasinoro, D., Abdulsamad, A. and Sommer, J. (1991) *Journal of Catalysis*, **131**, 199.

118 Arata, K. (1996) *Applied Catalysis A: General*, **146**, 3, and references cited therein.

119 Arata, K., Matsuhashi, H., Hino, M. and Nakamura, H. (2003) *Catalysis Today*, **81**, 17, and references cited therein.

120 Sikabwe, E.C., Coelho, M.A., Resasco, D.E. and White, R.L. (1995) *Catalysis Letters*, **34**, 23.

121 Srinivasan, R., Keogh, R.A., Ghenciu, A., Farcasiu, D. and Davis, B.H. (1996) *Journal of Catalysis*, **158**, 502.

122 Stevens, R.W., Jr, Chuang, S.S.C., Davis, B.H. (2003) *Applied Catalysis A: General*, **252**, 57.

123 Matsuhashi, H., Motoi, H. and Arata, K. (1994) *Catalysis Letters*, **26**, 325.

124 Matsuhashi, H. and Arata, K. (2000) *Chemical Communications*, 387.

125 Matsuhashi, H. and Arata, K. (2006) *Catalysis Surveys from Asia*, **10**, 1.

126 Matsuhashi, H., Tanaka, T. and Arata, K. (2001) *Journal of Physical Chemistry B*, **105**, 9669.

127 Matsuhashi, H. and Arata, K. (2004) *Physical Chemistry Chemical Physics*, **6**, 2529.

128 Matsuhashi, H. and Futamura, A. (2006) *Catalysis Today*, **111**, 338.

129 Knozinger, H. (1998) *Topics in Catalysis*, **6**, 107.

130 Adeeva, V., Liu, H.-Y., Xu, B.-Q. and Sachtler, W.M.H. (1998) *Topics in Catalysis*, **6**, 61.
131 Hong, Z., Fogash, K.B. and Dumesic, J.A. (1999) *Catalysis Today*, **51**, 269.
132 Kim, S.Y., Goodwin, J.G., Jr and Farcasiu, D. (2001) *Applied Catalysis A: General*, **207**, 281.
133 Ahmad, R., Melsheimer, J., Jentoft, F.C. and Schlogl, R. (2003) *Journal of Catalysis*, **218**, 365.
134 Boronat, M., Viruela, P. and Corma, A. (1996) *Journal of Physical Chemistry*, **100**, 633.
135 Wakayama, T. and Matsuhashi, H. (2005) *Journal of Molecular Catalysis A–Chemical*, **239**, 32.
136 Matsuhashi, H., Yamagata, K. and Arata, K. (2004) *Chemistry Letters*, **33**, 554.
137 Wakabayashi, F., Kondo, J.N., Domen, K. and Hirose, C. (1995) *Journal of Physical Chemistry*, **99**, 10573.
138 Misono, M., Ono, I., Koyano, G. and Aoshima, A. (2000) *Pure and Applied Chemistry*, **72**, 1305.
139 Wakabayashi, F., Kondo, J., Wada, A., Domen, K. and Hirose, C. (1993) *Journal of Physical Chemistry*, **97**, 10761.
140 Matsuhashi, H., Shibata, H., Nakamura, H. and Arata, K. (1999) *Applied Catalysis A: General*, **187**, 99.
141 (a) Satoh, D., Matsuhashi, H., Nakamura, H. and Arata, K. (2003) *Catalysis Letters*, **89**, 105.
(b) Satoh, D., Matsuhashi, H., Nakamura, H. and Arata, K. (2003) *Physical Chemistry Chemical Physics*, **5**, 4343.
142 Farcasiu, D. and Li, J.Q. (1998) *Applied Catalysis A: General*, **175**, 1.
143 Coman, S., Parvulescu, V., Grange, P. and Parvulescu, V.I. (1999) *Applied Catalysis A: General*, **176**, 45.
144 Sassi, A. and Sommer, J. (1999) *Applied Catalysis A: General*, **188**, 155.
145 Bolis, V., Magnacca, G., Cerrato, G. and Morterra, C. (1997) *Langmuir*, **13**, 888.
146 Tanabe, K., Hattori, H. and Yamaguchi, T. (1990) *Critical Reviews in Surface Chemistry*, **1**, 1.
147 Saur, O., Bensitel, M., Saad, A.B.M., Lavalley, J.C., Tripp, C.P. and Morrow, B.A. (1986) *Journal of Catalysis*, **99**, 104.
148 Bensitel, M., Saur, O., Lavalley, J.C. and Morrow, B.A. (1988) *Materials Chemistry and Physics*, **19**, 147.
149 Morterra, C., Cerrato, G., Pinna, F. and Signoretto, M. (1994) *Journal of Physical Chemistry*, **98**, 12373.
150 Yang, J., Zhang, M., Deng, F., Luo, Q., Yi, D. and Ye, C. (2003) *Chemical Communications*, 884.
151 Lunsford, J.H., Sang, H., Campbell, S.M., Liang, C.-H. and Anthony, R.G. (1994) *Catalysis Letters*, **27**, 305.
152 Clearfield, A., Serrette, G.P.D. and Khazi-Syed, A.H. (1994) *Catalysis Today*, **20**, 295.
153 White, R.L., Sikabwe, E.C., Coelho, M.A. and Resasco, D.E. (1995) *Journal of Catalysis*, **157**, 755.
154 Laizet, J.B., Soiland, A.K., Leglise, J. and Duchet, J.C. (2000) *Topics in Catalysis*, **10**, 89.
155 (a) Hino, M., Kurashige, M., Matsuhashi, H. and Arata, K. (2006) *Thermochimica Acta*, **441**, 35.
(b) Fraenkel, D. (1999) *Chemistry Letters*, 917.
156 Cheung, T.-K. and Gates, B.C. (1997) *Journal of Catalysis*, **168**, 522.
157 Rezgui, S., Jentoft, R.E. and Gates, B.C. (1998) *Catalysis Letters*, **51**, 229.
158 Scheithauer, M., Bossch, E., Schubert, U.A., Knozinger, H., Cheung, T.-K., Jentoft, F.C., Gates, B.C. and Tesche, B. (1998) *Journal of Catalysis*, **177**, 137.
159 Morterra, C., Cerrato, G. and Ciero, S.D. (1997) *Catalysis Letters*, **49**, 25.
160 Song, X., Reddy, K.R. and Sayari, A. (1996) *Journal of Catalysis*, **161**, 206.
161 Sayari, A., Yang, Y. and Song, X. (1997) *Journal of Catalysis*, **167**, 346.
162 Sayari, A. and Yang, Y. (1999) *Journal of Catalysis*, **187**, 186.
163 Grau, J.M. and Parera, J.M. (1997) *Applied Catalysis A: General*, **162**, 17.
164 Larsen, G., Lotero, E., Parra, R.D., Petkovic, L.M., Silva, H.S. and Raghavan, S. (1995) *Applied Catalysis A: General*, **130**, 213.
165 Parvulescu, V., Coman, S., Parvulescu, V.I., Grange, P. and Poncelet, G. (1998) *Journal of Catalysis*, **180**, 66.
166 Yori, J.C. and Parera, J.M. (1995) *Applied Catalysis A: General*, **129**, 83.

167 Perez-Luna, M., Toledo-Antonio, J.A., Montoya, A. and Rosa-Salas, R. (2004) *Catalysis Letters*, **97**, 59.
168 Jentoft, R.E., Hahn, A.H.P., Jentoft, F.C. and Ressler, T. (2005) *Physical Chemistry Chemical Physics*, **7**, 2830.
169 Sohn, J.R., Lim, J.S. and Lee, S.H. (2004) *Chemistry Letters*, **33**, 1490.
170 Lange, F.C., Cheung, T.-K. and Gates, B.C. (1996) *Catalysis Letters*, **41**, 95.
171 Demirci, U.B. and Garin, F. (2001) *Catalysis Letters*, **76**, 45.
172 Moreno, J.A. and Poncelet, G. (2001) *Journal of Catalysis*, **203**, 453.
173 Occelli, M.L., Schiraldi, D.A., Auroux, A., Keogh, R.A. and Davis, B.H. (2001) *Applied Catalysis A: General*, **209**, 165.
174 Hua, W., Goeppert, A. and Sommer, J. (2001) *Applied Catalysis A: General*, **219**, 201.
175 (a) Hino, M. and Arata, K. (2004) *Reaction Kinetics and Catalysis Letters*, **81**, 321.
(b) Wang, J.-H. and Mou, C.-Y. (2005) *Applied Catalysis A: General*, **286**, 128.
(c) Sun, Y., Walspurger, S., Louis, B. and Sommer, J. (2005) *Applied Catalysis A: General*, **292**, 200.
(d) Kim, S.Y., Lohitharn, N., Goodwin, J.G.Jr., , Olindo, R. Pinna, F. and Canton, P. (2006) *Catalysis Communications*, **7**, 209.
176 Perez-Luna, M., Toledo-Antonio, J.A., Hernandez-Beltran, F., Armendariz, H. and Borquez, A.G. (2002) *Catalysis Letters*, **83**, 201.
177 Chen, X.-R., Chen, C.-L., Xu, N.-P., Han, S. and Mou, C.-Y. (2003) *Catalysis Letters*, **85**, 177.
178 Cao, C.-J., Han, S., Chen, C.-L., Mou, N.-P. and Xu, C.-Y. (2003) *Catalysis Communications*, **4**, 511.
179 Cao, C.-J., Yu, X.-Z., Chen, C.-L., Xu, N.-P., Wang, Y.-R. and Mou, C.-Y. (2004) *Reaction Kinetics and Catalysis Letters*, **83**, 85.
180 Signoretto, M., Melada, S., Pinna, F., Polizzi, S., Cerrato, G. and Morterra, C. (2005) *Microporous and Mesoporous Materials*, **81**, 19.
181 Cerrato, G., Morterra, C., Delgado, M.R., Arean, C.O., Signoretto, M., Somma, F. and Pinna, F. (2006) *Microporous and Mesoporous Materials*, **94**, 40.
182 Serra, J.M., Chica, A. and Corma, A. (2003) *Applied Catalysis A: General*, **239**, 35.
183 Vera, C.R., Yori, J.C. and Parera, J.M. (1998) *Applied Catalysis A: General*, **167**, 75.
184 Yori, J.C. and Parera, J.M. (1996) *Applied Catalysis A: General*, **147**, 145.
185 Sparks, D.E., Keogh, R.A. and Davis, B.H. (1996) *Applied Catalysis A: General*, **144**, 205.
186 Bi, M., Pan, W.-P., Lloyd, W.G. and Davis, B.H. (1998) *Catalysis Letters*, **50**, 187.
187 Shishido, T. and Hattori, H. (1996) *Applied Catalysis A: General*, **146**, 157.
188 Shishido, Y., Tanaka, T. and Hattori, H. (1997) *Journal of Catalysis*, **172**, 24.
189 Manoli, J.-M., Potvin, C., Muhler, M., Wild, U., Resofszki, G., Buchholz, T. and Paal, Z. (1998) *Journal of Catalysis*, **178**, 338.
190 Comelli, R.A., Canavese, S.A., Vaudagna, S.R. and Figoli, N.S. (1996) *Applied Catalysis A: General*, **135**, 287.
191 Satoh, N., Hayashi, J. and Hattori, H. (2000) *Applied Catalysis A: General*, **202**, 207.
192 Vijay, S. and Wolf, E.E. (2004) *Applied Catalysis A: General*, **264**, 117.
193 Blekkan, E.A., Johnsen, K.A. and Loften, T. (2005) *Reaction Kinetics and Catalysis Letters*, **86**, 149.
194 Fottinger, K., Zorn, K. and Vinek, H. (2005) *Applied Catalysis A: General*, **284**, 69.
195 Lin, C.-H. and Hsu, C.-Y. (1992) *Journal of the Chemical Society, Chemical Communications*, 1479.
196 Hsu, C.-Y., Heimbuch, C.R., Armes, C.T. and Gates, B.C. (1992) *Journal of the Chemical Society, Chemical Communications*, 1645.
197 Song, X. and Sayari, A. (1996) *Catalysis Reviews–Science and Engineering*, **38**, 329.
198 Hino, M. and Arata, K. (1995) *Catalysis Letters*, **30**, 25.
199 Hino, M. and Arata, K. (1995) *Journal of the Chemical Society, Chemical Communications*, 789.
200 Hino, M. and Arata, K. (1999) *Reaction Kinetics and Catalysis Letters*, **66**, 331.

201 Furuta, S. (2003) *Applied Catalysis A: General*, **251**, 285.

202 Hino, M. and Arata, K. (1998) *Applied Catalysis A: General*, **169**, 151.

203 Hino, M. and Arata, K. (1998) *Applied Catalysis A: General*, **173**, 121.

204 Hino, M. and Arata, K. (1999) *Journal of the Chemical Society, Chemical Communications*, 53.

205 Hwang, G., Bodenbinder, M., Willner, H. and Aubke, F. (1993) *Inorganic Chemistry*, **32**, 4667.

206 Cotton, F.A. and Wilkinson, G. (1988) *Advanced Inorganic Chemistry*, 5th edn, John Wiley & Sons, Inc., New York, p. 922.

207 Olah, G.A. (1963–1964) *Friedel-Crafts and Related Reactions*, Wiley-Interscience, New York, Vol. **1–4**.

208 Olah, G.A. (1973) *Friedel-Crafts Chemistry*, Wiley-Interscience, New York.

209 Nakamura, H. and Arata, K. (2004) *Bulletin of the Chemical Society of Japan*, **77**, 1893.

210 Chiche, B., Finiels, A., Gauthier, C., Geneste, P., Graille, J. and Pioch, D. (1986) *Journal of Organic Chemistry*, **51**, 2128.

211 Nakamura, H., Kashiwara, Y. and Arata, K. (2003) *Bulletin of the Chemical Society of Japan*, **76**, 1071.

212 Quaschning, V., Deutsch, J., Druska, P., Niclas, H.-J. and Kemnitz, E. (1998) *Journal of Catalysis*, **177**, 164.

213 Parida, K., Quaschning, V., Lieske, E. and Kemnitz, E. (2001) *Journal of Materials Chemistry*, **11**, 1903.

214 Kawada, A., Mitamura, S., Matsuo, J., Tsuchiya, T. and Kobayashi, S. (2000) *Bulletin of the Chemical Society of Japan*, **73**, 2325.

215 Patil, P.T., Malshe, K.M., Kumar, P., Dongare, M.K. and Kemnitz, E. (2002) *Catalysis Communications*, **3**, 411.

216 Sakthivel, R., Prescott, H. and Kemnitz, E. (2004) *Journal of Molecular Catalysis A–Chemical*, **223**, 137.

217 Hino, M. and Arata, K. (1985) *Journal of the Chemical Society, Chemical Communications*, 112.

218 Hino, M. and Arata, K. (1996) *Hyoumen*, **34**, 51.

219 Barton, D.G., Soled, S.L. and Iglesia, E. (1998) *Topics in Catalysis*, **6**, 87.

220 (a) Hino, M., Takasaki, S., Furuta, S., Matsuhashi, H. and Arata, K. (2006) *Catalysis Communications*, **7**, 162.
(b) Hino, M., Takasaki, S., Furuta, S., Matsuhashi, H. and Arata, K. (2007) *Applied Catalysis A: General*, **321**, 147.

221 Hino, M., Matsuhashi, H. and Arata, K. (2006) *Catalysis Letters*, **107**, 161, and unpublished data.

222 Muggli, D.S. and Ding, L. (2001) *Applied Catalysis B–Environmental Consulting*, **32**, 181.

223 Ohno, T., Mitsui, T. and Matsumura, M. (2003) *Chemistry Letters*, **32**, 364.

224 Muggli, D.S., Ding, L. and Odland, M.J. (2002) *Catalysis Letters*, **78**, 23.

225 Samantaray, S.K., Mohapatra, P. and Parida, K. (2003) *Journal of Molecular Catalysis A–Chemical*, **198**, 277.

226 Gerlich, M., Kornely, S., Fleischer, M., Meixner, H., Kassing, R. (2003) *Sensors and Actuators B*, **93**, 503.

227 Xie, C., Xu, Z., Yang, Q., Li, N., Zhao, D., Wang, D. and Du, Y. (2004) *Journal of Molecular Catalysis A–Chemical*, **217**, 193.

228 Wang, X., Yu, J.C., Hou, Y. and Fu, X. (2005) *Advanced Materials*, **17**, 99.

229 Nakajima, A., Obata, H., Kameshima, Y. and Okada, K. (2005) *Catalysis Communications*, **6**, 716.

230 Barraud, E., Bosc, F., Keller, N. and Keller, V. (2005) *Chemistry Letters*, **34**, 336.

231 Barraud, E., Bosc, F., Edwards, D., Keller, N. and Keller, V. (2005) *Journal of Catalysis*, **235**, 318.

232 Mohamed, M.M. and Al-Esaimi, M.M. (2006) *Journal of Molecular Catalysis A–Chemical*, **255**, 53.

233 Moos, R., Muller, R., Plog, C., Knezevic, A., Leye, H., Irion, E., Braun, T., Marquardt, K.-J., Binder, K. (2002) *Sensors and Actuators B*, **83**, 181.

234 Shimizu, K. and Satsuma, A. (2006) *Shokubai (Catalysts & Catalysis)*, **48**, 544.

18
Titanium Silicalite-1
Mario G. Clerici

Oxides of metals in Groups IV–VI of the periodic table are generally poor catalysts in oxidations using hydrogen peroxide [1]. Water and other protic solvents necessary to dissolve the oxidant compete for the active site, causing inhibition. Thus, activity and selectivity are low and the decomposition of hydrogen peroxide can be a major side-reaction. An exception is the epoxidation of α,β-unsaturated alcohols and acids, which are able to compete with water for the active site through their oxygenated functionalities. Ti, Mo, V and W compounds, on the other hand, are excellent catalysts in oxidations with organic hydroperoxides, since these allow the use of hydrocarbons and other non-coordinating solvents under water-free conditions.

The absence of co-products needing disposal or further processing makes hydrogen peroxide so attractive as an oxidant that physical and chemical means of avoiding the build-up of water in the oxidation medium were investigated. Early studies envisaged the use of an organic medium, maintained in nearly anhydrous conditions by the addition of dehydrating agents or by the continuous azeotropic distillation of water. This last option was specifically studied for the epoxidation of propene, although never applied industrially probably because of the potential hazards implicit in the method [2]. The alternative was indirect epoxidation carried out in a water-free organic solvent, physically separated from the aqueous hydrogen peroxide reservoir. The true oxidant species was a H_2O_2 carrier, cyclically consumed in the epoxidation medium and regenerated in the aqueous one, as in the epoxidation under phase-transfer conditions or with peracids [3]. In both cases, water was never allowed to come into contact with the epoxidation process.

More promising from an industrial perspective, however, is the separation of the oxidation zone from the aqueous one effected by the catalytic material itself, through the selective adsorption of the reagents. The introduction of Titanium Silicalite-1 (TS-1), in which the hydrophobic properties of the pores protect the active sites from the inhibition of the external aqueous medium, was a demonstration of the concept. The catalyst, the substrate and the aqueous solution of hydrogen peroxide can, in this case, be mixed together, with a great simplification of the process and also a reduction of the hazards. Three commercial processes,

Metal Oxide Catalysis. Edited by S. David Jackson and Justin S. J. Hargreaves
Copyright © 2009 WILEY-VCH Verlag GmbH & Co. KGaA, Weinheim
ISBN: 978-3-527-31815-5

superior to earlier ones in terms of both economic and environmental grounds, prove the viability of the method.

The number of studies on TS-1, and also on other Ti-zeolites, rose exponentially since the first one was published in 1986 and a vast literature now exists [4]. A few review articles deal with the synthesis, the characterization and, to a lesser extent, with the catalysis of such materials [5–9]. It is the aim of this chapter to focus on the last issue, providing a comprehensive picture of the variety of oxidations performed, a critical analysis of the problems and solutions envisaged in the use of hydrogen peroxide and an in-depth discussion of mechanistic aspects.

18.1
Synthesis and Characterization

Taramasso and others first prepared TS-1 at the end of the 1970s, using two different methods for the preparation of the crystallization gel [10]. In the most commonly used "mixed alkoxide method," this was obtained by the controlled hydrolysis of tetraethyltitanate and tetraethylsilicate in the presence of tetrapropylammonium hydroxide. Crystallization under hydrothermal conditions produced crystalline TS-1. In the preparation of the gel, it is crucial to prevent segregation of Ti phases by the correct choice of procedure and control of the conditions. The maximum Ti content, expressed as the atomic ratio Ti/(Ti+Si), was reported to be ca 0.025 [11]. Other information is available in Refs. [6–9] and references therein.

As indicated in the original patent, the absence of alkali metal ions in the synthesis mixture is critical for the incorporation of Ti in the lattice. Their presence as impurities in commercial samples of tetrapropylammonium hydroxide was not recognized in several studies, particularly in early ones, with the result that TS-1 was impure, containing TiO_2 particles and amorphous Ti-silicates. Sometimes, however, the presence of alkali and other impurities is the consequence of a deliberate decision, in an attempt to decrease the cost of production of the catalyst by the use of cheaper sources of Si, Ti and tetrapropylammonium hydroxide [12].

Titanium Silicalite-2 (TS-2), structurally similar to TS-1, could be prepared likewise using tetrabutylammonium hydroxide as the template [13, 14]. Titanium aluminum Beta (Ti,Al-β) was prepared by hydrothermal synthesis from amorphous silica, sodium aluminate, tetraethyltitanate and tetraethylammonium hydroxide [15]. The presence of Al was necessary for the crystallization of the product. Al-free Titanium Beta (Ti-β) could be obtained in the presence of particular templates, such as dibenzyldimethylammonium hydroxide [16]. Titanium Mordenite (Ti-MOR), conversely, was obtained by post-synthesis insertion of Ti to dealuminated Mordenite [17]. Ti-MWW (Ti-MCM-22) was obtained by the synthesis of the lamellar precursor of Ti,B-MCM-22 followed by acid treatment to remove most of the boron and extra-framework Ti and finally calcination to burn out the template and bring about the condensation of lamellae into the three-dimensional MWW structure [18]. Ti is present in a number of different environ-

Table 18.1 Structural properties and composition of some Ti-zeolites.[a]

Name	Structure type	Pore dimensions (Å)	Pore structure	Metal
TS-1	MFI	5.6 × 5.3; 5.5 × 5.1	3 directional	Ti
TS-2	MEL	5.3 × 5.4	3 directional	Ti
Ti-Beta, Ti-β, Ti,Al-β	*BEA[c]	6.4 × 7.6; 5.5 × 5.5	3 directional	Ti, (Al)
Ti-MOR[b]	MOR	6.5 × 7.0; 2.6 × 5.7	1[b] directional	Ti, (Al)
Ti-MWW, Ti-MCM-22	MWW	5.4 × 4.0	2 directional	Ti

a W.M. Meier and D.H. Olson, *Atlas of Zeolite Structure Types*, Butterworth-Heinemann, London, 1992.
b Residual Al can be present in the lattice. The size of the second channel system is too small for catalysis.
c *denotes a well-defined subunit for which pure end members have not been obtained.

ments, that is in the external pockets or cups (0.7 × 0.7 nm), in internal supercages (0.7 × 0.7 × 1.7 nm) and in medium-pore sinusoidal channels. Other Ti-zeolites are not mentioned here, because they are of little catalytic interest.

Ti-MCM-41 and Ti-MCM-48 are mesoporous molecular sieves with ordered porous systems. Owing to their amorphous nature, these cannot be classified as zeolites. Their inclusion, however, is useful in that it extends the range of pore sizes of Ti-molecular sieves to an upper limit of 2–4 nm [19].

Characterization is a crucial step preliminary to any catalytic study, since the selectivity of the catalyst is strictly related to the position of Ti, in atomic dispersion, within the crystal lattice. Extra-framework Ti species, such as TiO_2 particles and amorphous Ti-silicates, indeed, promote H_2O_2 decomposition and radical chain oxidations. Normally, a combination of different techniques is necessary for reliable characterization, for example, UV-Vis, IR and Raman spectroscopies, XRD, EXAFS, XANES, TEM and SEM [8, 20]. Table 18.1 illustrates the main structural features of TS-1 and other Ti-zeolites relevant to this review.

18.2
Hydroxylation of Alkanes

The oxidation of alkanes is a reaction known for the difficulties of its realization under mild conditions. Exceptions are enzymatic and a few organometallic systems, which are, however, unsuitable for industrial applications. The use of molecular oxygen as oxidant usually requires radical initiators or drastic conditions, leading to high selectivities only in a limited number of oxidations. The addition of a reducing agent (Equation 18.1) or the use of hydrogen peroxide and other monooxygen donors (Equation 18.2) allows the reaction to be performed under milder conditions, albeit with yields that are generally too low in relation to the cost of the oxidant. Mixed oxide materials are usually ineffective under the latter conditions. With this background, the efficacy of TS-1, discovered by three groups independently, represented a step-change [21–24].

$$R-H + O_2 + 2H^+ + 2e^- \longrightarrow R-OH + H_2O \tag{18.1}$$

$$R-H + H_2O_2 \longrightarrow R-OH + H_2O \tag{18.2}$$

18.2.1
Titanium Silicalite-1

TS-1 catalyzes the hydroxylation of alkanes with dilute solutions of hydrogen peroxide in water, in a biphasic system of alkane and aqueous H_2O_2, or in aqueous–organic solution. The rate of reaction decreases in the solvent order: t-butanol > t-butanol/water > methanol ≈ acetonitrile ≈ water [24, 25]. The temperature is generally lower than 55 °C in methanol, close to 100 °C in water and of intermediate values in other solvents. Hydroxylation occurs at secondary and tertiary C–H bonds, while primary ones are completely inert (Equations 18.3 and 18.4).

$$CH_3-CH_2-R \longrightarrow \underset{OH}{CH_3-CH-R} + \underset{O}{CH_3-C-R} \tag{18.3}$$

$$\underset{}{\overset{R'}{CH_3-CH-CH_2-R}} \longrightarrow \underset{OH}{\overset{R'}{CH_3-C-CH_2-R}} \tag{18.4}$$

The main side reactions are the decomposition of hydrogen peroxide and, to a lesser extent, the oxidation of alcohol solvent. Their competition with C–H hydroxylation increases with increasing bulkiness of the paraffin, becoming predominant with those having a cross-section close to or larger than the size of the pores.

Linear alkanes yield a mixture of secondary alcohols and ketones (Table 18.2). In methanol, hydroxylation is reduced at more internal positions, while in water and in acetone it occurs randomly. The yield increases with the chain length up

Table 18.2 Hydroxylation of linear alkanes.[a]

Alkane	$t_{1/2}$ (min)[b]	Yield (% based on H_2O_2)[c]	β/γ ratio	γ/δ ratio	Alcohol/ketone ratio
C_3H_8	130	35	–	–	2.0
$n\text{-}C_4H_{10}$	62	69	–	–	1.2
$n\text{-}C_5H_{12}$	80	82	4.5	–	1.0
$n\text{-}C_6H_{14}$	30	86	2.6	–	1.4
$n\text{-}C_7H_{16}$	57	75	1.9	5.2	2.2
$n\text{-}C_8H_{18}$	57	63	2.6	1.9	1.7

a Adapted from Ref. [24]. Copyright 1991, with permission from Elsevier.
  Solvent, methanol 95%; alkane, 0.62 mol l^{-1}; TS-1, 15.7 g l^{-1}; T, 55 °C.
b Time required for 50% H_2O_2 conversion.
c Yields are referred to 98% H_2O_2 conversion.

to a maximum for C_5–C_7 n-alkanes. Values as high as 99% have recently been reported for the hydroxylation of n-hexane, by feeding the oxidant gradually to keep the H_2O_2/alkane ratio low [26]. The rather odd reactivity order, hexane > heptane ≈ octane > butane > pentane > propane, most likely results from the combined effects of electronic factors, diffusivity and adsorption strength. The increase of alkyl chain length actually affects adsorption heats and diffusivity in zeolites in opposite directions. Inductive effects, on the other hand, increase with chain length, gradually levelling off.

The hydroxylation of cyclohexane, of potential interest for the production of cyclohexanone, is exceedingly slow at near room temperature and has low selectivity at 100 °C [27, 28]. Tertiary C–H bonds yield tertiary alcohols, with little or no oxidation observed at the secondary carbons that may be present in the alkyl chain: *t*-C–H ≫ *sec*-C–H (Table 18.3). The steric constraints introduced by alkyl substitution strongly favor the competition of side reactions, at the expense of hydroxylation. On arylalkanes, oxidation occurs on both the aromatic ring and the alkyl chain, with a general preference for the latter. Consistently, the competitive hydroxylation of benzene and n-hexane or cyclohexane mainly occurs on the alkane. However, benzylic methyls, despite the relative weakness of their C–H

Table 18.3 Hydroxylation of branched and substituted alkanes.[a]

Substrate	t (min)	Conversion (% based on H_2O_2)	Yield (% based on H_2O_2)	Products
cyclohexane	150	50	32	cyclohexanol, cyclohexanone
2-methylpropane	180	65	45	*t*-butanol
2-methylbutane	240	65	14	2-methyl-2-butanol
2-methylpentane	300	57	10	2-methyl-pentanol
1-chlorohexane[b]	180	88	37	1-chloro-4-hexanol, 1-chloro-5-hexanol
				1-chloro-4-hexanone, 1-chloro-5-hexanone
methyl heptanoate	120	77	5	5-hydroxyheptanoate, 6-hydroxyheptanoate
				5-oxoheptanoate, 6-oxoheptanoate

a Adapted from Ref. [24]. Copyright 1991, with permission from Elsevier. Solvent, methanol, 95%; alkane, 0.62 mol l^{-1}; TS-1, 15.7 g l^{-1}; *T*, 55 °C.
b Solvent *t*-butanol.

bonds, are inert. The presence of electron-withdrawing groups in 1-chlorohexane and methyl heptanoate strongly decreases hydroxylation while orienting the attack on the remote methylene groups. Consistently, no dihydroxylation is observed under the conditions of Table 18.2, and it is negligible on C_8–C_{12} n-paraffins. Both the absence of consecutive hydroxylation and remote oxyfunctionalization reveal the electrophilic properties of the active species.

Evidence on the latter was obtained by the use of additives, substrate probes and labeled molecules. The addition of small amounts of protonic acids promoted hydroxylation. Alkali metal salts and basic compounds produced the opposite effect, with inhibition or complete deactivation of the catalyst, depending on the amount used. The subsequent addition of hydrochloric acid restored the initial activity, showing that inhibition and deactivation by salts and bases are completely reversible phenomena [24, 29]. On these grounds, the active species could be identified as a fairly acidic Ti hydroperoxide or an oxidant, still unknown, produced by its further transformation [24]. However, the involvement of Ti–OOH as the active species of a heterolytic mechanism is not consistent with the results of the competitive hydroxylation of aromatic and aliphatic C–H bonds, pointing instead to a homolytic pathway.

Two types of substrate probe, cis- and trans-1,3-dimethylcyclopentane and ethyl- and 2-propylcyclopropane, were used to shed light on mechanistic details of the hydroxylation step [30]. In the use of the first two probes, the participation of Ti–OOH species in a concerted mechanism would predict either the retention or the inversion of configuration at the chiral center, while the stereochemistry of a homolytic mechanism would be determined by the competition between the epimerization of the transient tertiary carbon radical and C–O formation (Scheme 18.1). In the hydroxylation of cyclopropyl probes, the cyclopropylcarbinyl radical clock can either rearrange to ring-opened 3-buten-1-yl radical before being trapped or rebound with the hydroxyl carrier to yield the alcohol product directly (Scheme 18.2). With TS-1, nearly equal amounts of trans- and cis-1,3-dimethylcyclopentanol were obtained from the first type of probe, while no rearranged products were obtained with the second ones. These results suggest that the putative radical intermediate is very short-lived, but not short enough to prevent the racemization of the tertiary carbon radical. Incidentally, it was estimated that epimerization and cyclopropyl rearrangement occurred with first order rate constants of 10^9 and ca $10^8 \, s^{-1}$, respectively [30, 31].

The use of radical quenchers and the competitive oxidation of cyclohexane and cyclohexane-d_{12} led to identification of the active species as a Ti-centered radical

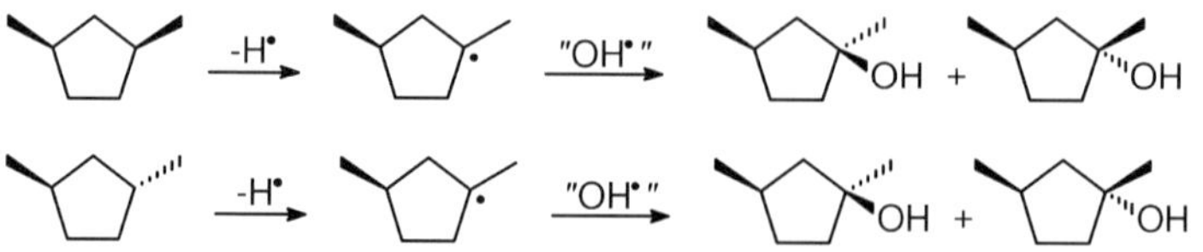

Scheme 18.1 Hydroxylation of sterically pure cis- and trans-1,3-dimethylcyclopentane.

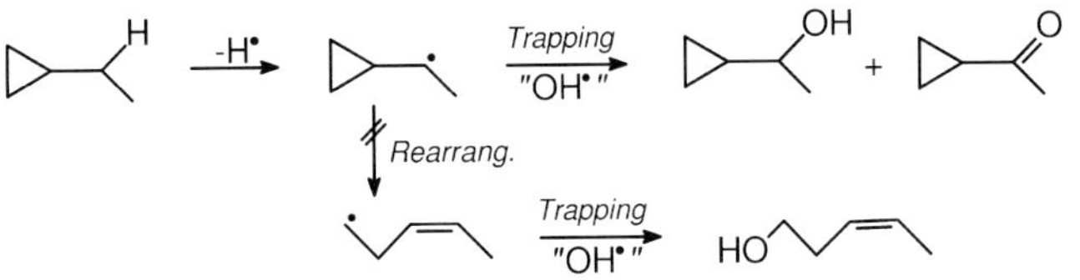

18.2 Hydroxylation of Alkanes

Scheme 18.2 Hydroxylation of ethylcyclopropane.

[24]. Actually, BHT (2,6-di-*tert*-butyl-4-methylphenol), carbon tetrachloride, chloroform and dichloromethane neither affected the hydroxylation rate nor produced chlorinated derivatives, thus excluding a free radical mechanism and the presence of long-lived alkyl radicals, both in the pores of the catalyst and in the external solution. The primary isotopic effect in methanol and *t*-butanol was 4.1 and 4.7, respectively. A k_H/k_D of this magnitude is not compatible with a radical chain oxidation initiated by hydroxyl radicals ($k_H/k_D = 1$–2), while it is fully consistent with substantial C–H bond cleavage in the transition state by a Ti-centered radical.

On the whole, it is conceivable to postulate an active species that is Ti centered, has a radical nature and originates from a Ti–OOH precursor, through some sort of reaction which the latter species may undergo. An early mechanistic proposal, based on a diradical peroxy species, does not match all these conditions (Scheme 18.3a). This requires that the adsorption of hydrogen peroxide on Ti leads to a side-on bonded peroxide and, more importantly, implies the splitting of one strong Ti–O bond instead of the weaker peroxidic O–O one. The unlikelihood of Ti–O cleavage is confirmed by the reaction of a Rh peroxide with carbon dioxide, occurring through the insertion of CO_2 into the O–O bond, to yield the corresponding peroxycarbonate [32].

Scheme 18.3b illustrates a recent mechanistic proposal. This bears similarities to the oxygen-rebound mechanism, proposed for the hydroxylation of alkanes by high-valent transition metal species in biomimetic and enzymatic systems [31, 33, 34]. The generation of the active species and other implications of the mechanism are discussed in detail in Section 18.11.3.

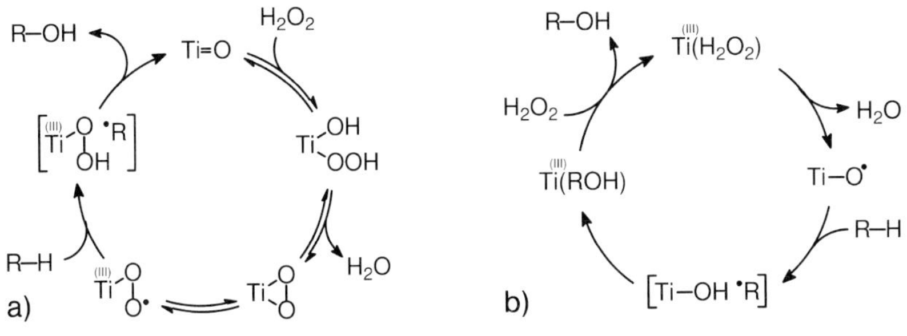

Scheme 18.3 (a) C–H hydroxylation by Ti(η^2–O_2);
(b) C–H hydroxylation by oxygen-rebound-like mechanism.

Table 18.4 Hydroxylation of alkanes on Ti,Al-β and comparison with TS-1 [35].

Alkane	Ti,Al-β		TS-1	
	TON (mol/mol$_{Ti}$)	Selectivity (% based on H$_2$O$_2$)	TON (mol/mol$_{Ti}$)	Selectivity (% based on H$_2$O$_2$)
n-hexane	0.5	32	48.5	100
cyclohexane	2.3	51	a)	–
methylcyclohexane	5.2	88	a)	–

a Products below detection limits.

18.2.2
Other Ti-Zeolites

Only TS-2 and Ti,Al-β, among other Ti-zeolites, catalyze the hydroxylation of alkanes. The close similarity of TS-2 with TS-1 suggests analogous catalytic properties. The somewhat poorer performances reported in the literature probably reflect the poor quality of early catalysts, containing extra-framework impurity Ti species. Ti,Al-β, on the other hand, is definitely inferior to TS-1, except in the hydroxylation of bulky alkanes (Table 18.4) [35]. Interestingly enough, Ti,Al-β catalyzes the hydroxylation of both sec-C—H and t-C—H in the same molecule, though the reactivity of the latter is significantly higher [36]. In 1,2- and 1,3-dimethylcyclohexanes, equatorial C—H bonds are more reactive than those at axial positions, probably because there is less hindrance in the approach to the active species. The direct link between the hydrophobicity of the sample used and catalytic performance is of mechanistic interest [37].

18.3
Hydroxylation of Aromatic Compounds

The synthesis of phenolic compounds, including the industrial production of a major commodity such as phenol, has historically been achieved by the transformation of pre-existing aromatic functional groups. The generally poor selectivity of direct hydroxylation routes is especially evident with molecular oxygen as the oxidant, though sometimes relatively high yields have been claimed. The use of hydrogen peroxide appears more promising, and is characterized by greater selectivity. Actually, in the early 1970s Brichima and Rhône-Poulenc commercialized two industrial processes for the hydroxylation of phenol with H$_2$O$_2$. The development of TS-1 and Fe-ZSM-5 opened new routes to selective hydroxylation processes with hydrogen peroxide (EniChem, hydroxylation of phenol) and nitrous oxide (Solutia, hydroxylation of benzene), respectively.

18.3.1
Hydroxylation of Phenol

18.3.1.1 Titanium Silicalite-1

Normally in a review the hydroxylation of benzene should come before that of phenol. The latter, however, has been the center of interest of most studies, particularly of early ones, whereas benzene has been considered often just for the completeness of the range of substrates. On these grounds, an exception for phenol is justified.

Hydroxylation is a consecutive reaction in which low conversion has to be maintained, to prevent the over-oxidation of catechol and hydroquinone to tars (Equation 18.5). This is normally achieved by operating with an excess of phenol over hydrogen peroxide. The greater the H_2O_2:phenol ratio allowed in the feed, without penalty to the selectivity, the greater the effectiveness of the catalysts (Table 18.5) [5, 38, 39]. A stirred reactor is normally preferred, though the use of the fixed bed type has sometimes been reported [40–42].

$$\text{C}_6\text{H}_5\text{-OH} \xrightarrow{H_2O_2} \text{HO-C}_6\text{H}_4\text{-OH} + \text{o-C}_6\text{H}_4(\text{OH})_2 \xrightarrow{H_2O_2} \text{tar} \quad (18.5)$$

In aqueous acetone and methanol, TS-1 shows superior performance than in t-butanol or just water. The solvent has also a major effect on the catechol:hydroquinone ratio (see below). This varies in the range 0.5–1.3, which is some way from the value of 2 expected for a statistical attack at the *ortho* and *para* positions (Table 18.5, entries 1 and 2). Yet, under practical conditions, the main component of the reaction mixture could be phenol instead of the putative solvent and, under such conditions, the product ratio approaches unity.

The yields based on hydrogen peroxide are comparable to those obtained with acid catalysts, but at six-fold higher conversion (Table 18.4, entries 1 and 3). The most important parameters for the yield are the operating temperature and the purity, concentration and crystal size of the catalyst [5, 43]. Yields increased, at the expense of tar production, up to a maximum of ca 83% at 100 °C, and then

Table 18.5 Hydroxylation of phenol with hydrogen peroxide.

	Catalyst	Ortho/para	Phenol conversion (%)	Selectivity		Ref.
				(% based on H_2O_2)	(% based on C_6H_5OH)	
1	TS-1	0.5–1.3	30	82	92	[5]
2	Co^{+2}, Fe^{+2}	2–2.3	9	66	79	[39]
3	H^+	1.2–1.5	5	85–90	90	[38]

Table 18.6 Henry constants (K_P) for the adsorption of phenol in TS-1.[a]

Solvent	K_P
Water	84.4
Methanol	0.7
Acetone	0.6

a Adapted from Ref. [47]. Copyright 2001, with permission of Elsevier.

decreased rapidly with further temperature rises. Extra-framework Ti phases promote non-productive side reactions, for example decomposition of the oxidant and radical chain oxidations. In general, this applies to any oxidation catalyzed by TS-1, but is crucial for aromatic hydroxylation that requires a relatively high temperature. The hydroxylation of phenol, indeed, was proposed as chemical test to assess the purity of TS-1 [5, 43, 44]. The concentration of TS-1 also affects the yields, owing to greater competition of secondary reactions in the presence of excess H_2O_2 and low catalyst contents. Van der Pool and others originally showed, and other authors confirmed, that the hydroxylation of phenol is diffusion limited for crystal sizes larger than 0.3 μm [45–47].

Experimental data fit well with kinetic expressions that are first order in both phenol and hydrogen peroxide [47]. Observed rate constants were significantly larger in water than in methanol and acetone. This was ascribed to the stronger adsorption of phenol from an aqueous solution in TS-1 (Table 18.6) [47, 48]. Surprisingly enough, in the presence of methanol or acetone the concentration of phenol in the pores was about the same as that in the external solution.

An issue of debate is the relative roles of internal and external sites in the catalytic process. The effects of shape selectivity, clearly present in product distribution, seem to indicate a predominance of intra-porous hydroxylation. However, the different catechol/hydroquinone ratio in methanol (0.5) and acetone (1.3), could indicate a significant contribution of sites located on the outer surface of the crystals, particularly for crystallite sizes <0.3 μm. Tuel and others, studying the time course of the reaction and the solubility of tarry deposits, went further and concluded that catechol and hydroquinone were produced on different sites, external and internal respectively [49]. The effect of acetone and methanol simply reflected their ability to maintain external sites clean from tar deposits, which are soluble in the former and insoluble in the latter. On the other hand, Wilkenhöner and others concluded, with the support of kinetic constants estimated independently for internal and external sites, that catechol was also produced in the pores over the entire reaction profile, albeit at a lower rate [47]. The contribution of the outer surface for crystal sizes close to 0.1 μm ranged from 46% in methanol to 69% in acetone.

Little is known of the active species and the hydroxylation mechanism. It is even unclear whether the mechanism is homolytic or heterolytic. Accordingly, mechanisms of both types have been proposed (Schemes 18.4 and 18.5).

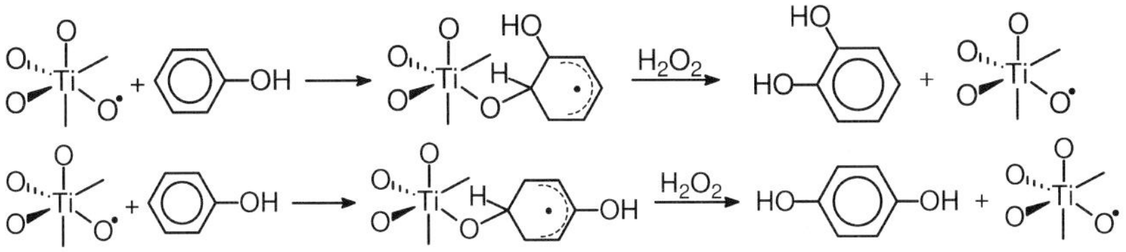

Scheme 18.4 Homolytic hydroxylation of phenol by Ti–O• species.

In the mechanism of Scheme 18.4, hydroxylation is carried out homolytically by the same species proposed for alkane hydroxylation [25]. The oxidation of the cyclohexadienyl intermediate by hydrogen peroxide produces the diphenol and regenerates the active species, closing the catalytic cycle. Scheme 18.5 illustrates the heterolytic routes proposed by Wilkenhöner and others for hydroquinone and catechol production, based on cationic peroxy intermediates [47]. Both types of mechanism, however, are little more than working hypotheses, needing validation by experimental evidence.

18.3.1.2 Other Ti-Zeolites

TS-2 was shown to be almost indistinguishable from TS-1, as predicted by similarity of structures and active sites [46]. Ti-Beta zeolites, with and without Al in the structure, were less effective than TS-1. The yields based on hydrogen peroxide, just above 60%, were typical of rather modest catalysts. Apparently, product selectivity was influenced by the Al content. The relatively hydrophilic Ti,Al-β produced catechol and hydroquinone in nearly equimolar amounts [50]. The Al-free Ti-β showed a higher catechol selectivity, with an *ortho/para* ratio of 2 [47]. In both cases, the greater spaciousness of pores favoured *ortho* hydroxylation. For a useful comparison, the *ortho/para* ratio on medium-pore TS-1 was 0.77 under analogous conditions.

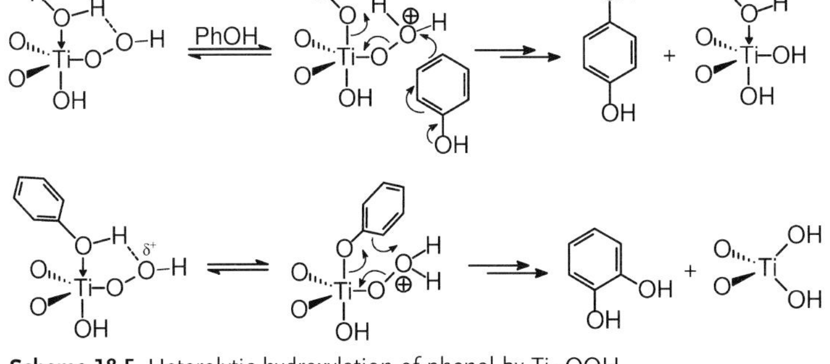

Scheme 18.5 Heterolytic hydroxylation of phenol by Ti–OOH species. (Adapted from Ref. [47]; Copyright 1991, with permission of Elsevier).

18.3.2
Hydroxylation of Benzene

The hydroxylation of benzene on TS-1 produces phenol as the primary product and, by consecutive oxidation, hydroquinone, catechol, benzoquinone and tarry products (Equation 18.6) [5, 30, 51]. The selectivity to phenol generally falls below 50% even at only 3–5% benzene conversion. Hydroxylation with a mixture of hydrogen and oxygen on Pd/TS-1 proved to be even less effective [52].

$$\text{benzene} \xrightarrow{H_2O_2} \text{phenol} \xrightarrow{H_2O_2} \text{hydroquinone} + \text{catechol} \xrightarrow{H_2O_2} \text{tar} \quad (18.6)$$

Recently, the use of sulfolane solvent allowed better kinetic control of the oxidation chain, with an increase of the selectivity to 80% or greater, at ca 8% benzene conversion. The by-products were catechol (7%), hydroquinone (4%), 1,4-benzoquinone (1%) and tar (5%) [53, 54]. According to these authors, a rather stable complex, formed by hydrogen bonding with sulfolane, promoted desorption and hindered the re-adsorption of phenol, protecting it from consecutive oxidation (Equation 18.7). Actually, the rate of oxidation of phenol in the presence of sulfolane was only 1.6 times that of benzene, while it was 10 times higher in the presence of acetone.

$$\text{PhOH} + \text{sulfolane} \rightleftharpoons \text{PhOH} \cdots \text{sulfolane complex} \quad (18.7)$$

The pretreatment of TS-1 with a solution of ammonium fluoride and hydrogen peroxide further increased the conversion (ca 9%) and selectivity. The recovery of catechol and hydroquinone, by their hydrogenation back to phenol, was also considered [55]. It is worth noting that, at a threshold yield value of ca 9%, the hydroxylation of benzene could become competitive with the cumene process, considering that the overall per pass yield in the latter does not exceed 8–9%.

Far greater yields were reported for the hydroxylation of benzene under the so-called triphasic conditions, that is with the solid catalyst, aqueous hydrogen peroxide and an immiscible aromatic phase [56]. Others, however, could not reproduce these results [54].

Benzene hydroxylation on Ti,Al-MOR was studied by three different groups [57–59]. Despite the spaciousness of its pores, the activity of this catalyst was lower than that of TS-1, as expected for a more hydrophilic catalyst. Accordingly, increase in the Al content caused a decrease in the conversion, probably because of reduced adsorption of benzene.

18.3.3
Oxidation of Substituted Benzenes

Electron-withdrawing substituents decrease the rate of oxidation, revealing the electrophilic nature of the oxidant species. Accordingly, negligible or no yields were reported for the hydroxylation of chlorobenzene, nitrobenzene, benzonitrile, benzaldehyde and benzoic acid on TS-1 [5]. Alkyl groups produced contrasting effects: a rate enhancement by electron donation and a rate decrease by steric and transport restrictions. In this regard, the size of the methyl group in toluene virtually compensated for the increase of electron density [5]. For other alkylbenzenes, the nuclear reactivity trend was in the order: toluene > *p*-xylene ≥ ethylbenzene > *p*-methylethylbenzene, showing again the predominance of steric hindrance [5, 30, 59]. Bulkier substituents suppressed hydroxylation, for example in 2-propylbenzene [24].

The oxidation of alkylbenzenes also produced *sec*-alcohols and ketones, with no reaction at tertiary and primary carbons. At variance with the rule that double bonds are preferentially oxidized over other functionalities, side-chain epoxidation had to compete with aromatic hydroxylation in the oxidation of α-methylstyrene, 1-phenyl-2-butene and 4-phenyl-1-butene [60].

Ti-MOR promoted the ring hydroxylation of toluene, ethylbenzene and xylenes with negligible oxidation of the ethyl side chain [59]. In the same study, however, and in contrast to earlier ones, a similar result was also reported for TS-1. No oxidation of benzylic methyls was observed. Cumene yielded mainly the decomposition products of cumyl hydroperoxide. The oxidation of *t*-butylbenzene was negligibly low. The reactivity order, toluene > benzene > ethylbenzene > cumene, reflects the reduced steric constraints in the large pores of mordenite. Accordingly, the rate of hydroxylation of xylene isomers increased in the order *para* < *ortho* < *meta*, in contrast to the sterically controlled one, *ortho* < *meta* ≪ *para*, shown on TS-1. It is worth mentioning that the least hindered *p*-xylene exhibited the same reactivity on either catalyst.

18.4
Oxidation of Olefinic Compounds

18.4.1
Epoxidation of Simple Olefins

The oxidation of olefins with aqueous hydrogen peroxide in methanol can produce several products, by different reaction paths: double bond epoxidation, allylic H-abstraction, epoxide solvolysis, alcohol and glycol oxidation (Scheme 18.6). Normally, oxide catalysts of Group IV–VI metals are poorly selective, because of their acidic properties, the inhibition they are subject to in aqueous media and homolytic side reactions with hydrogen peroxide. The only exception concerns the epoxidation of α,β-unsaturated alcohols and acids, which are able to bind on the

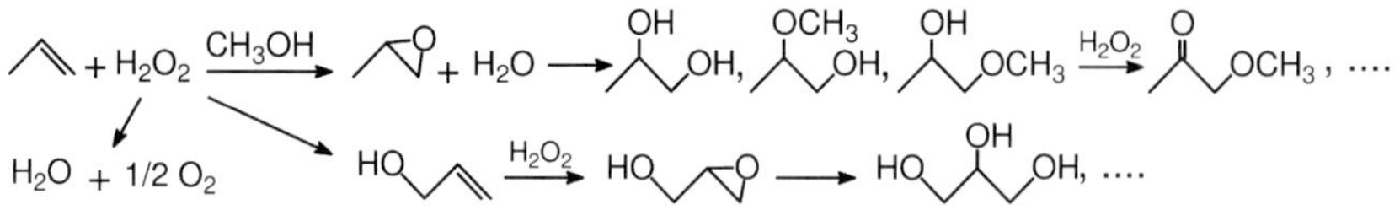

18 Titanium Silicalite-1

Scheme 18.6 Reaction pathways in the oxidation of propene with hydrogen peroxide.

Table 18.7 Epoxidation of linear olefins.[a]

Olefin	T (°C)	t (min)	$t_{1/2}$ (min)[b]	Conversion (%)	Selectivity (% based on H_2O_2)
1-butene	−5	60	–	96	96
1-pentene	25	60	5	94	91
1-hexene	25	70	8	88	90
cyclohexene	25	90	–	9	–
1-octene	45	45	5	81	91
allyl chloride	45	30	7	98	92
allyl alcohol	45[c]	35	16	81	72

a Adapted from Ref. [62]. Copyright 1993, with permission from Elsevier.
  Solvent, methanol; olefin, 0.90 mol l^{-1}; TS-1, 6.2 g l^{-1}.
b $t_{1/2}$ is the time necessary to 50% H_2O_2 conversion.
c TS-1 4.0 g/l

catalyst through the oxygenated functionality. However, TS-1, a mixed oxide of silica and titania similar in composition and Ti dispersion, but not structure, to the Ti/SiO$_2$ Shell catalyst, unexpectedly showed high activity and selectivity. Actually, a most remarkable feature of TS-1 is the smooth and preferential epoxidation of the double bond, over any other functionality that may be present in the molecule.

18.4.1.1 Titanium Silicalite-1

Epoxidation occurs in dilute solutions of aqueous hydrogen peroxide and at near room temperature [61, 62]. The rate is rapid even at H$_2$O$_2$ concentrations of 1% or lower and at temperatures as low as −5 °C. Electron-deficient olefins can also be oxidized at near room temperature (Table 18.7). Preferred solvents are alcohols and, in general, protic or polar media. The rate of epoxidation decreases in the solvent order: methanol > ethanol > i-propanol > acetone > acetonitrile > t-butanol, with a difference greater than one order of magnitude between the two extremes of the series [62, 63]. An exception could be the oxidation of cyclopentene, for which acetone and acetonitrile were reported to be less effective than t-butanol [64]. In the epoxidation of propene, the rates in water were possibly faster than in methanol, though the rapid decay of the catalyst did not allow unequivocal conclusions. As a matter of fact, water contents up to 50 wt% in methanol produced only a moderate reduction of the rate, probably because of the lower solubility of propene in aqueous media (TOF 1–2 s^{-1}, at 40 °C).

The epoxidation of unhindered olefins is nearly quantitative and the incidence of side reactions is negligible [62]. The epoxide selectivity can be higher than 90%. The by-products are merely those produced by the acid solvolysis of the oxirane ring, namely the corresponding glycol and methyl ethers. Basic compounds able to diffuse inside the pores of TS-1, such as sodium acetate, sodium hydroxide and ammonia, decrease the intra-porous acidity generated by Ti–OOH species and, sometimes, by lattice Al impurities. Their addition in parts per million quantities minimizes the cleavage of the epoxide, thus enhancing the selectivity up to 97–98%. Spontaneous hydrolysis, fairly slow but always present in protic media, prevents the achievement of quantitative yields [61]. Further increasing the concentration of the base gradually decreases the catalytic activity up to complete inhibition. Bulky tetrapropylammonium hydroxide, unable to diffuse into the pores of TS-1, does not affect the epoxidation process at any concentration.

The epoxidation rate is related to the electron density of the double bond, increasing with it. Thus, the epoxidation of propene is much faster than that of ethene. The formal substitution of one methyl, chloro or hydroxyl group at the allylic position of propene results into the reactivity order: 1-butene > allyl chloride > allyl alcohol. Methyl substitution on butenes also produces the expected ordering: 2-methyl-2-butene > 2-methyl-1-butene > 3-methyl-1-butene (Table 18.8).

However, since epoxidation occurs within pores of cross-section comparable to that of the olefin, steric restrictions generally prevail over inductive effects, leading to anomalous reactivity orders. They result from restrictions to diffusion in the pores (reactant shape selectivity) and to the approach of the double bond to the active species (transition state shape selectivity). The first is sufficient to explain

Table 18.8 Relative rates in the epoxidation of C_4 and C_5 olefins.[a]

Olefin	TS-1 r_1/r_n [b]	Olefin	TS-1 r_1/r_n [b]	CH_3CO_3H r_1/r_n
1-butene	1.0	1-pentene	1.0	1.0
trans-2-butene	0.59	2-methyl-1-butene	0.13	ca 1.0
isobutene	0.71	3-methyl-1-butene	0.87[c]	20–24
cis-2-butene	0.17[c]	2-methyl-2-butene	0.19	20–24
(cyclic)	2.7[c]	–	–	20–24
–	–	2,3-dimethyl-1-butene	0.79	240

a Data taken from Refs [62, 65].
b Averaged values from competition kinetics.
c Epoxidation occurred with complete retention of configuration.

the rate of epoxidation of cyclohexene, found to be slower than that of 1-hexene by almost two orders of magnitude at near room temperature [62]. The second is apparent in the epoxidation of freely diffusing olefins, for example of the methyl-substituted butenes mentioned above, in which the reactivity order is the expected one, but the relative rates are greatly different from those of a purely electrophilic oxidant (Table 18.8) [65]. Most striking is 2-methyl-2-butene compared to 1-pentene, with a rate somewhat slower on TS-1/H_2O_2 and 240 times faster on peracetic acid. Transition-state shape selectivity in these and other epoxidations is likely to result from steric repulsions exerted by the surface and by the species chemisorbed on Ti on alkyl substituents of the double bond.

The evidence provided so far, that is the quantitative yields (including glycols), the absence of allylic oxidation, the substituent effects and the role of shape selectivity, are consistent with a heterolytic mechanism, involving an electrophilic oxidant species subject to major steric constraints. The retention of configuration in the epoxidation of sterically pure *cis*- and *trans*-butenes is further and convincing support for this proposition. By analogy with many known transition metal peroxides, the active species may be identified among $Ti(\eta^1-OOH)$, $Ti(\eta^2-OOH)$ and $Ti(\eta^2-O_2)$ peroxy species (Figure 18.1a–d). The detection of Ti superoxide radicals (Figure 18.1e), on addition of hydrogen peroxide to TS-1, and the homolytic nature of C—H abstraction do not weaken the idea of a heterolytic mechanism. Based on the author's and others' experience [66] (and in contrast to some other reports [9]), the superoxide species involve only a negligible fraction of Ti-sites.

The competition of one-electron pathways is sometimes detectable in the epoxidations catalyzed by transition metal catalysts [67]. However, in the epoxidation of unhindered olefins on TS-1, the typical radical products are below the detection limits. Their presence could no longer be neglected when the rate of epoxidation is so slow as to become comparable to that of homolytic side reactions, for example with bulky olefins (see also Section 18.11). It is possible that, within these limits only, the epoxide is produced in part through the addition of a radical peroxy intermediate to the double bond [68, 69]. Even so, a homolytic pathway has again been proposed as a generally valid epoxidation mechanism [7].

In early proposals based upon the mechanisms of Mo and V peroxides, $Ti(\eta^2-O_2)$ was thought to be the active species [70, 71]. The epoxidation of the olefin occurred either by a coordinative mechanism, with the olefin adsorbing on Ti prior to the insertion into one Ti-peroxide bond, or by the attack of one peroxy oxygen directly on the double bond. However, the chemical inertness of known $Ti(\eta^2-O_2)$ peroxides, the unlikelihood of olefin adsorption on oxophilic Ti in competition with the protic medium and other evidence led subsequently to the loss

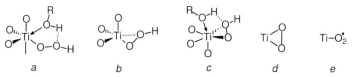

Figure 18.1 Active species proposed for epoxidations on TS-1.

of credibility for this proposal. Currently, a general consensus involving a central role of a Ti–OOH species seems to exist. There is less agreement on the details of its structure, particularly on the presence and the function of a co-adsorbed protic molecule (Figure 18.1a and c) and on the mono- (Figure 18.1a) or di-hapto (Figure 18.1b and c) coordination of the hydroperoxide group. In an early proposal, the latter was monodentate and was stabilized in a five-membered ring by a hydrogen-bonded protic molecule, typically methanol or water (Figure 18.1a) [24, 62]. This proposal was based on adsorption properties of Ti sites, on the reactivity behaviour of organic and inorganic peroxides and on the smooth formation of cyclic structures in hydroperoxides and other compounds. In the alternative species, based on quantum chemical studies, the hydroperoxide group was η^2 coordinated and did not require stabilization (Figure 18.1b) [72]. In another proposal, also based on DFT calculations, η^2 coordination and hydrogen bonding with a protic molecule coexist in the same structure (Figure 18.1c) [73, 74].

Scheme 18.7 illustrates a proposed early mechanism based on the intermediate of Figure 18.1a [62, 75]. The first steps in the catalytic cycle are reversible and envisage the adsorption of one molecule of H_2O_2, the formation of a Ti–OOH species and its hydrogen bonding with one alcohol or water molecule. In the irreversible epoxidation step, the attack on the double bond by the peroxy oxygen vicinal to Ti, leads to the formation of the epoxide, a Ti-alkoxide and a water molecule. Finally, the desorption of the epoxide and the reaction of Ti-alkoxide with water, to form again the initial Ti-site, complete the catalytic cycle. As an alternative to it, H_2O_2 could chemisorb directly on Ti-alkoxide to regenerate the Ti–OOH species. The desorption of the product is an important step for the activity, since the epoxide behaves as an inhibitor.

Scheme 18.8 illustrates the oxygen-transfer steps of species **b** and **c** in Figure 18.1 [72–74]. Both mechanisms are based on DFT studies and are similar to the epoxidations with organic hydroperoxides on Group IV–VI metal catalysts [1]. The

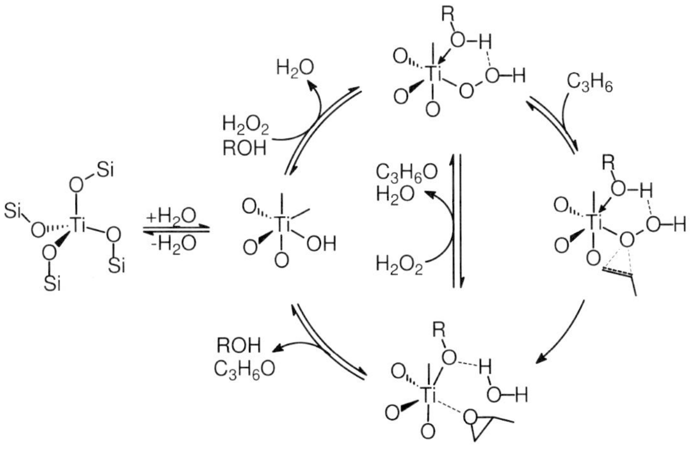

Scheme 18.7 Epoxidation mechanism of species (a) from Figure 18.1.

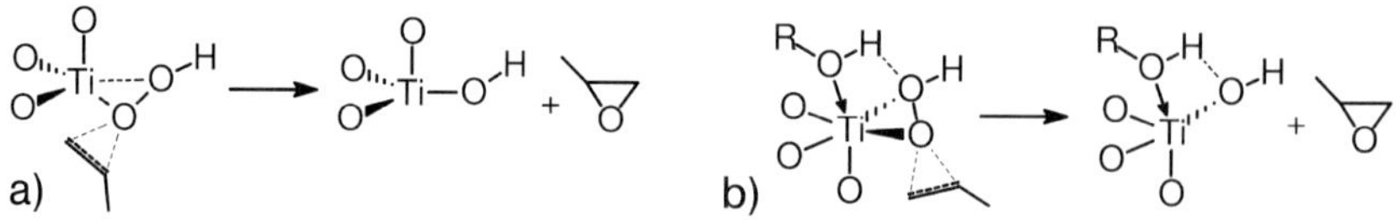

Scheme 18.8 (a) Epoxidation mechanism of species (*b*) from Figure 18.1; (b) Epoxidation mechanism of species (*c*) from Figure 18.1.

coordination of distal peroxy oxygen on Ti facilitates its removal as a titanol. A weakness of Scheme 18.8a is the absence of adsorbates other than hydrogen peroxide on Ti. The hypothesis could be correct for epoxidations with organic hydroperoxides carried out under anhydrous conditions, but not for the use of aqueous hydrogen peroxide. It is fully consistent, however, with epoxidations conducted under reduced pressure, in the conditions studied by Lin and Frei [76]. The same arguments for Scheme 18.7 could apply also to Scheme 18.8b.

18.4.1.2 Other Ti-Zeolites

The number of metal zeolites and their application to the epoxidation of olefins rose in parallel from the late 1980s. TS-2, Ti,Al-β, Ti-β, Ti-MWW and, rarely, Ti-MOR are catalysts that have been studied in some detail [7–9, 35, 77–84]. TS-2 behaves, according to the few studies published, similarly to TS-1. The greater spaciousness of pores in Ti-Beta zeolites and of external cups in Ti-MWW allows the epoxidation, under mild conditions, of olefins unable to diffuse in TS-1 and TS-2, such as methylcyclohexenes, cyclododecene, norbornene, camphene and methyl oleate [80–83]. Steric constraints still prevail over electronic factors, however, as in medium pore Ti-zeolites, even in the epoxidation of linear olefins (Table 18.9). It is generally believed that active sites and epoxidation mechanisms are not significantly different from those of TS-1.

There are major differences that distinguish Ti,Al-β, Ti-β and Ti-MWW. The first concerns the activity and selectivity, which, under optimum conditions for each catalyst, are considerably lower than for TS-1. The activity in methanol decreases in the order, TS-1 > Ti-β > Ti,Al-β ≥ Ti-MWW, in a parallel trend with the decrease of the hydrophobicity. Actually, the density of surface Si–OH species increases in the reverse order: TS-1 < Ti-β < Ti,Al-β ≤ Ti-MWW. The selectivity, in turn, drops owing to a greater incidence of solvolysis and hydrogen peroxide decomposition. Even in the absence of framework Al sites, as in Ti-β, the solvolysis can significantly reduce epoxide yields. Ion exchange with basic compounds or just their addition in the reaction medium is an effective tool to limit the losses of product [84].

A second difference compared to TS-1 concerns the solvent and its effects on kinetics and selectivity. The choice is again restricted to alcohols, ketones and acetonitrile, but the latter is now preferable to methanol for higher rates and lower solvolysis [77, 78]. As a general rule, methanol is the best solvent for oxidations catalyzed by TS-1, whereas acetonitrile is preferable for Ti-β, Ti,Al-β and Ti-MWW

18.4 Oxidation of Olefinic Compounds

Table 18.9 Epoxidation of olefins over Ti-β, Ti,Al-β, TS-1 and CH_3CO_3H. Relative rates.

Olefin	Ti,Al-β[a]	Ti-β[b]	TS-1[a]	CH_3CO_3H[c]
1-hexene	1.0	1.0	1.0	1.0
2-hexene	0.70	0.77	0.63	ca 1
3-hexene	–	1.55	–	20–24
2-methyl-2-pentene	1.1	–	2.2	20–24
4-methyl-2-pentene	0.83	–	–	20–24
cyclohexene	1.6	1.2	–	27
2,4,4-trimethyl-2-pentene	1.6	–	0.63	ca 240
methylenecyclohexane	1.2	0.72	0	(≥240)

a Data from Ref. [83].
b Data from Ref. [81].
c Data from Ref. [65].

[81, 85, 86]. The enhanced activity is likely the result of the effects of different factors, while the greater selectivity has to be related to the reduction of acidity by the mildly basic medium. Interestingly enough, in various instances Ti-β, Ti,Al-β and Ti-MWW turn out to be more active catalysts than TS-1, when acetonitrile is a common solvent. In one study, however, trifluorethanol emerged as the best solvent for Ti,Al-β [84].

A third difference concerns Ti-MWW only. The siting of Ti in different porous environments, that is in external pockets, in internal supercages and in sinusoidal 10-MR channels, leads to active species associated with different diffusional and steric constraints [79]. Thus, the epoxidation of bulky olefins can occur exclusively in external pockets, whereas the linear ones are not subject to site limitations. Ti-MWW is also an unusual catalyst in the epoxidation of stereoisomers. At odds with TS-1 and Ti-Beta zeolites, *trans*-olefins are epoxidized faster than their *cis* analogues [85]. Though the mechanism is still unclear, a better fitting of the trans configuration to the tortuous nature of 10-MR channels could be an explanation.

Ti-Beta zeolites and, even more, mesoporous Ti-silicates can be somewhat unstable to aqueous hydrogen peroxide and to strongly chelating agents. A partial collapse of the lattice and the release of Ti, in the form of TiO_2 particles or soluble Ti peroxides, was sometimes observed under these conditions (see also Section 18.4.2). The structural instability grows in parallel with the hydrophilicity of the surface and the defectiveness of the silica matrix: Ti-β < Ti,Al-β ≪ Ti-MCM-41 [87–89]. For the same reason, the stability of the catalyst is indirectly related to the method of synthesis, as far as this is able to produce materials with a different content of connectivity defects.

Since water is detrimental to catalytic performance, and its presence is unavoidable with hydrogen peroxide as the oxidant, an alternative consists of the use of *t*-butyl hydroperoxide (TBHP), compatible with the pores of Ti-Beta zeolites and with external pockets of Ti-MWW. In the case of the former catalysts, rates are lower than with hydrogen peroxide while with Ti-MWW they are comparable [79, 83]. It should be considered, however, that in this use both large-pore Ti-zeolites and mesoporous Ti-silicates are in competition with the cheaper and easier to prepare Ti/SiO_2.

In summary, the poorer activity, the stronger acidity and sometimes the structural instability are major drawbacks that minimize the chances of application of large-pore Ti-zeolites. The reduced adsorption of alkanes, alkenes and other apolar reagents in favor of that of water, and the defectiveness of the surface, are the principal reasons for this (for a further treatment, the reader is referred to Section 18.11). In relation to these issues, synthetic methods for the production of materials with fewer defective sites and better connectivity, have been developed for both Ti-Beta zeolites and mesoporous Ti-silicates [7–9].

18.4.2
Epoxidation of Unsaturated Alcohols

In principle, the oxidation of allyl alcohols may concern the double bond or the alcohol group. In the absence of steric constraints, TS-1, TS-2, Ti,Al-β, Ti-β and Ti-MWW catalyze the epoxidation with generally high chemoselectivity [81, 88, 90–97]. Carbonyl compounds may form by competitive oxidation, to an extent that depends on steric constraints (Table 18.10). Normally, TS-1 is the most effective

Table 18.10 Oxidation of unsaturated alcohols over TS-1 and Ti-β.

Olefin	TS-1[a]		Ti-β[b]	
	Epoxide (%)	Aldehyde (%)	Epoxide (%)	Aldehyde (%)
⌇⌇⌇OH	100	–	>90	–
⌇⌇⌇OH	100	–	>90	–
⌇⌇⌇OH	90	10	85	11
⌇⌇⌇OH	–	–	89	11
⌇⌇⌇OH	76	24	–	–
⌇⌇⌇OH	82	18	96	–
⌇⌇⌇OH	64	36	–	–

a Data from Ref. [90].
b Data from Ref. [81].

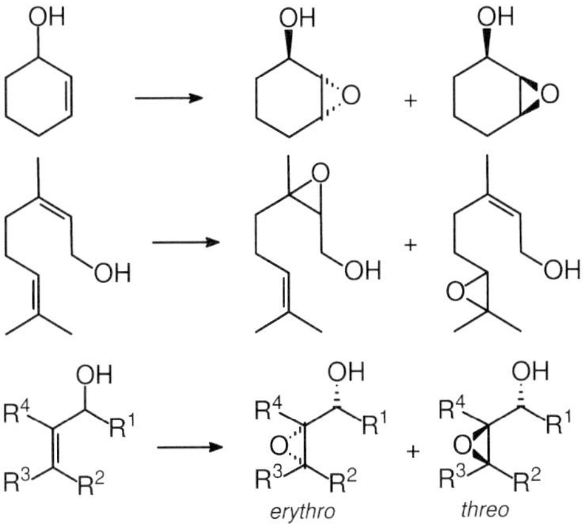

Scheme 18.9 Oxidation of allyl alcohols.

catalyst for unhindered substrates. However, Ti-MWW showed the best activity and selectivity for the synthesis of glycidol in acetonitrile and water: Ti-MWW > TS-1 > Ti-β [96]. Selective poisoning tests showed that the epoxidation occurred in the external pockets of crystallites.

Stereo-, diastereo- and regioselectivity can be very high (Scheme 18.9). This implies OH-assistance in the oxygen-transfer step, through the hydrogen bonding of the alcohol group with the oxidant species, as for vanadium catalysts and peracids. Thus, with TS-1, 3-cyclohexenol and 3-cyclopentenol yielded the *cis* epoxides with 90% selectivity [91]. Chiral allyl alcohols produced, with TS-1 and Ti,Al-β, a mixture of *threo* and *erythro* epoxy alcohols, in a ratio that was apparently independent of constraints produced by the pores [97]. The epoxidation of geraniol and nerol on TS-1 occurred on the vicinal double bond, while the remote and more nucleophilic one was reported to be inert [91]. However, Schofield and others observed epoxidation at either double bond, on TS-1 and Ti,Al-β, in the ratio of ca 2:1 in favor of the distant one [93]. Interestingly enough, the use of methanol completely inhibited epoxidation at the vicinal position, favoring the electron-rich distant one. This suggests that the competition of methanol for adsorption on Ti sites suppresses the coordinative interaction of the allyl alcohol with the active species.

The use of aqueous methanol, through the opening of oxirane rings, can have a detrimental effect on the stability of the catalyst. By-product triols coordinate strongly on Ti, facilitating its removal from the silica matrix, to the point that after an induction period the epoxidation of allyl alcohols on Ti,Al-β and Ti-β became an essentially homogeneous process [88]. In the epoxidation of crotyl alcohol, the trend of Ti leaching was: Ti,Al-β > Ti-β ≫ TS-1, the process being negligible on the latter. The use of acetonitrile and the adoption of anhydrous conditions protected the catalyst from structural degradation, by the reduction of solvolysis. Specific tests revealed that the simultaneous presence of triol and hydrogen peroxide was necessary for leaching to occur, while taken alone they did not produce any significant structural damage.

It is likely that the commercial value of glycidol is the unmentioned background of most studies. However, process parameters were specifically studied in only one case, in which TS-2 was the catalyst. The conversion of the olefin, under optimum conditions, was ca 88% with 100% selectivity [94]. Methallyl alcohol was the subject of a similar study [95].

18.4.3
Epoxidation of Allyl Chloride and other Substituted Olefins

Several patents and two papers deal with the epoxidation of allyl chloride [98, 99]. Actually, a process based on TS-1 would represent an environmentally cleaner alternative to current production of epichlorohydrin. In this regard, one study has addressed the cost of commercial hydrogen peroxide with the *in situ* production of the oxidant, by the use of molecular oxygen and anthrahydroquinone compounds [99]. In a mechanistic study, the kinetic data were interpreted on the basis of Eley–Rideal isotherms, with the rate of reaction being first order on TS-1 and between 0 and 1 on H_2O_2 and C_3H_5Cl (Equation 18.8) [98].

$$r = k_{app} \frac{[H_2O_2] [C_3H_5Cl] [TS\text{-}1]}{1 + K_1 [H_2O_2] + K_2[C_3H_5Cl]} \tag{18.8}$$

Diallyl ether and diallyl carbonate produced monoglycidyl and diglycidyl ethers in a ratio that depended on the conversion. Allyl methacrylate yielded the corresponding glycidyl methacrylate [5]. Ti-MWW, in acetonitrile and acetone solvents, was again reported to be more active and selective than TS-1 [100].

The epoxidation of α,β-unsaturated ketones produced corresponding epoxides and minor amounts of glycols [101]. Methyl substitution on the double bond reduced, or even suppressed, the oxidation. α,β-Unsaturated aldehydes yielded the corresponding carboxylic acids and minor amounts of epoxides. α,β-Unsaturated acids were inert under analogous reaction conditions. In this regard, Ti-zeolites behave differently from homogeneous tungsten catalysts, in which epoxidation does occur through a coordinative mechanism [102]. Consistently, acrylic acid esters were inert to both kinds of catalyst.

18.4.4
Epoxidation with Solvolysis/Rearrangement of Intermediate Epoxide

The presence of acidity leads to a variety of end-products, depending on the catalyst, the olefinic substrate and the reaction conditions (e.g. the sort of solvent, the temperature and the use of acidity moderators). The acid properties of Ti,Al-β in this regard, and the use of basic additives to reduce their effects, have been illustrated [61, 84]. Ti-MOR is also acidic owing to residual Al in the framework [59]. TS-1, with Al, Ga, B or Fe inserted in the framework, behaves similarly [103]. Al-free Ti-β and Ti-MWW are less acidic catalysts than the former ones, but significantly more acidic than TS-1. Most likely, the acid sites in the latter are the Ti–OOH species *a* of Figure 18.1 (see also Section 18.11.3).

In some instances, combined redox/acid catalysis might be desirable, as in the one-pot syntheses of β-phenylacetaldehyde by the oxidation of styrene. The primary product is generally believed to be 1,2-styrene oxide, which, in a second stage, undergoes acid-catalyzed rearrangement to the aldehyde. Acidic zeolites are known to behave as good catalysts in the latter reaction [104]. According to one study, however, 1,2-styrene oxide and β-phenyl acetaldehyde could be formed independently by two competing mechanisms [105].

Derouane and coworkers exploited the dual catalytic function and the spaciousness of pores of Ti,Al-β to prepare *trans*-2-alkoxycycloalkanols, of interest as pharmaceutical intermediates, by the oxidation of cyclohexene and cyclopentene [106]. The one-pot epoxidation/cyclization of unsaturated terpene alcohols, for example linalool and α-terpineol, to mono- and bicyclic derivatives also required the catalysis of large-pore Ti,Al-β (Scheme 18.10) [93, 107].

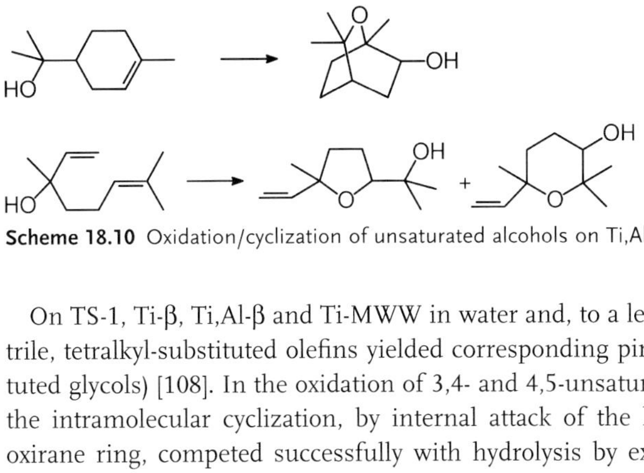

Scheme 18.10 Oxidation/cyclization of unsaturated alcohols on Ti,Al-β.

On TS-1, Ti-β, Ti,Al-β and Ti-MWW in water and, to a lesser extent, in acetonitrile, tetralkyl-substituted olefins yielded corresponding pinacols (tetralkyl-substituted glycols) [108]. In the oxidation of 3,4- and 4,5-unsaturated alcohols on TS-1, the intramolecular cyclization, by internal attack of the hydroxy group on the oxirane ring, competed successfully with hydrolysis by external attack of water (Scheme 18.11) [107]. Cyclization occurred regioselectively on the less remote carbon, with yields often higher than 85%. Oxidation of 5-hexen-1-ol, in which the five-membered ring cannot form, produced the tetrahydropyran derivative. Triol by-products increased with temperature and water content. Conversely, nearly anhydrous conditions and the control of acidity with basic additives turned the oxidation into epoxy alcohol production.

18.5
Oxidation of Alcohol and Other Oxygenated Compounds

The oxidation of primary and secondary alcohols using TS-1 and Ti,Al-β produces corresponding aldehydes and ketones [5, 77, 109–112]. While the latter are sufficiently stable under the reaction conditions, the former can undergo further oxidation to carboxylic acids. Acetals and esters are also produced when operating in

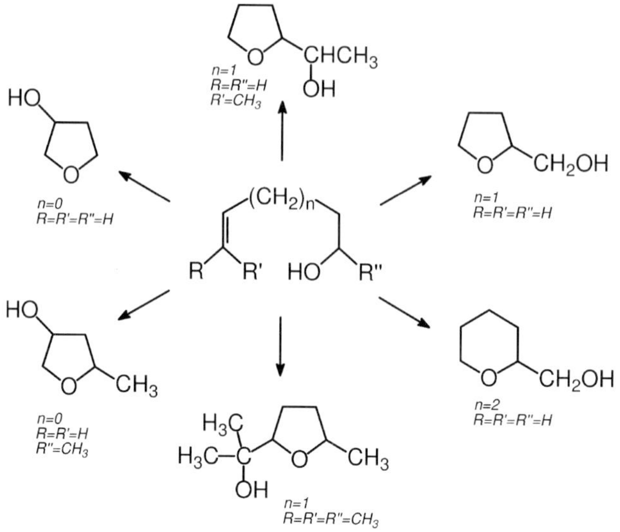

Scheme 18.11 Oxidation of 3,4-, 4,5- and 5,6-unsaturated alcohols on acidic Ti-zeolites.

neat alcohol or in methanol solvent (Equations 18.9 and 18.10). Tertiary alcohols yield corresponding hydroperoxides.

$$RCH_2OH \xrightarrow{H_2O_2} RCHO \xrightarrow{H_2O_2} RCO_2H \quad (18.9)$$

$$RCHO \underset{}{\overset{R'OH}{\rightleftharpoons}} RCH(OR')_2 \xrightarrow{H_2O_2} RCO_2R' \quad (18.10)$$

Primary alcohols react at a much slower rate than secondary ones, suggesting an electrophilic oxidation mechanism (Table 18.11). The relative inertness of methanol justifies its use as an oxidation solvent for TS-1. Within a homologous series, the rate increases with the chain length of the alcohol up to a maximum for octanol, then decreasing again [109]. It is likely that this trend involves the combined effects of electronic factors, deactivation of the catalyst and adsorption phenomena. A true kinetic ordering, only obtainable from initial rates, is substantially lacking for this as for other oxidations. Shape selectivity is apparent in the decrease of the rate of alcohols with the hydroxy group at more internal positions or with vicinal methyl groups, the decrease being larger with the increase of the number of methyl substituents [110, 111].

Two papers deal with the kinetics of the oxidation on TS-1 and Ti,Al-β [110, 111]. Equation 18.11 illustrates the general rate law valid for both catalysts, according to a Langmuir–Hinshelwood model. It is consistent with the competition of alcohol/solvent for Ti sites and the adsorption of the substrate and the oxidant on the same site. The inhibitory effect of water, implicit in the competitive adsorption of the solvent, was proved by oxidation tests performed in acetonitrile containing variable quantities of water [77, 111].

18.5 Oxidation of Alcohol and Other Oxygenated Compounds

Table 18.11 Oxidation of alcohols catalyzed by TS-1.[a]

Alcohol	$t_{1/2}$ (h)[b]	H_2O_2 decomposition (%)[c]
methanol	8.5	13
ethanol	0.7	2.5
1-propanol	1.0	5
1-butanol	1.3	6
1-octanol	3.0	–
2-methyl-1-propanol	3.4[d]	–
2-propanol	0.01	<0.5
2-butanol	0.05	<0.5
2-pentanol	0.06	<0.5
3-pentanol	0.9	5
cyclohexanol	35	50

a Reprinted from Ref. [110]. Copyright 1994, with permission of Elsevier. Solvent, pure alcohol; T, 45 °C; TS-1, 5 wt%; H_2O_2, 0.5 mol l^{-1}.
b Time required for 50% conversion.
c Fraction of H_2O_2 decomposed at total conversion of alcohol.
d The product is t-butyl hydroperoxide.

$$r = \frac{k K_2 K_4 [Ti_0] [ROH] [H_2O_2]}{1 + K_1 [solv] + K_2 [ROH] + [H_2O_2] (K_3 K_1 [solv] + K_4 K_2 [ROH])} \quad (18.11)$$

The oxidation mechanism, based on an early one proposed for a tungsten catalyst, entails a concerted step in which a C—H bond undergoes electrophilic attack by the distal peroxy oxygen of Ti—OOH (Scheme 18.12) [110, 111, 113].

The one-step synthesis of isoamyl butyrate from isoamyl alcohol and n-butyraldehyde, possibly through the formation and subsequent oxidation of an acetal intermediate, could represent a special case in TS-1 catalysis, since the oxidant was molecular oxygen [114]. The authors did not advance any mechanistic hypothesis. n-Butyl hydroperoxide, however, produced in situ by the autoxidation of n-butyraldehyde, could have been the true oxidant, by virtue of its dimensional compatibility with the narrow pores of TS-1.

The oxidation of glycols occurred preferentially at the secondary alcohol group. Thus, propene glycol and 2-phenyl-1,2-ethanediol afforded hydroxyacetone and β-hydroxyacetophenone, respectively, as the major product [115]. Ethene glycol

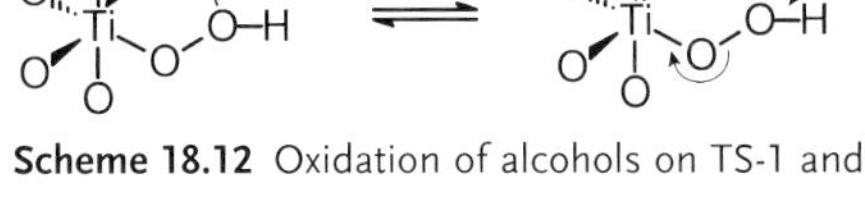

Scheme 18.12 Oxidation of alcohols on TS-1 and Ti,Al-β.

produced glycolic acid. Further oxidation yielded the α-dicarbonyl derivatives and, at prolonged contact time, C–C bond oxidative cleavage.

The oxidation of tetrahydrofuran, tetrahydropyran and dihydropyran produced γ-butyrolactone and δ-valerolactone in fairly good yields. Linear ethers, conversely, produced exclusively the corresponding acids, presumably through the *in situ* hydrolysis of intermediate esters [116].

The Baeyer–Villiger rearrangement of cyclohexanone and acetophenone with TS-1/H_2O_2 proved to be poorly selective [117]. Notably, Ti-β and Sn-β have different chemoselectivities in the oxidation of unsaturated ketones, leading selectively to corresponding epoxides and lactones, respectively [118]. The different oxidation pathways were attributed to the preferential adsorption of hydrogen peroxide on Ti-sites and of the carbonyl group on Sn-sites.

18.6
Ammoximation of Carbonyl Compounds

The word ammoximation defines the reaction of a carbonyl compound with ammonia, under oxidative conditions, to yield the corresponding oxime (Equation 18.12). Early homogeneous and heterogeneous catalysts, for example heteropoly compounds and amorphous Ti/SiO_2, had too poor selectivity to envisage any application. The prospects changed when Roffia and coworkers discovered that TS-1 was an excellent ammoximation catalyst with hydrogen peroxide as oxidant [119]. As Table 18.12 shows, TS-1 and amorphous Ti/SiO_2, albeit with similar Ti contents, lead to completely different results: to quantitative yields for the former and to a mixture of products, with only a minor amount of oxime, for the latter.

$$\text{cyclohexanone} + NH_3 + H_2O_2 \xrightarrow{\text{TS-1}} \text{cyclohexanone oxime} + 2\,H_2O \qquad (18.12)$$

In contrast to the stability shown in other oxidations, TS-1 undergoes a slow but irreversible deactivation. The siliceous matrix of the catalyst, in fact, slowly dissolves in the basic medium, while Ti, released from the framework, aggregates to

Table 18.12 Ammoximation of cyclohexanone on different catalysts.[a]

Cat.	Ti (wt%)	Conv. $C_6H_{10}O$ (%)	Sel. on $C_6H_{10}O$ (%)	Yield on H_2O_2 (%)
TS-1	1.5	99.9	98.2	93.2
Ti/SiO_2	1.5	49.3	9.3	4.4
S-1	–	59.4	0.5	0.3

a Solvent, *t*-butanol/water; *T*, 80 °C; molar ratios, cyclohexanone: NH_3:H_2O_2 1.0:2.0:1.1; t.o.s. 1.5 h.

form TiO$_2$ particles. Accordingly, the activity decreases with time on-stream, approaching that of conventional catalysts at prolonged times [120]. Major side reactions consist of the oxidation of ammonia to N$_2$, N$_2$O, nitrites and nitrates and in the decomposition of H$_2$O$_2$ to molecular O$_2$. Condensation reactions of cyclohexanone in the basic medium and consecutive oxidation of the oxime produce minor amounts of organic by-products. However, the losses of the oxidant are generally moderate, for example <7% under the conditions of Table 18.12, and the decay of the catalyst is not so fast as to preclude the development of an industrially viable process.

The reaction of ammoximation is generally applicable and provides an efficient route to many other oximes in addition to the industrially relevant cyclohexanone oxime. Acetone, butanone, acetophenone, C$_5$–C$_8$ cyclic ketones and methyl-substituted cyclohexanones produced the corresponding oximes with high conversion and selectivity. Even the ammoximation of cyclododecanone and 4-butylcyclohexanone, unable to diffuse in TS-1, occurred with high yields [121–123]. Other ammoximation catalysts are carbon-supported TS-1, TS-2, Ti-MOR and Ti-MWW, with conversions and selectivities close to those of TS-1 [124–128].

Two ammoximation mechanisms were initially proposed, one envisaging the formation of cyclohexanone imine (Equation 18.13) and the other that of hydroxylamine (Equation 18.14) as key intermediates.

$$R_2CO \xrightarrow{NH_3} R_2C=NH \xrightarrow[TS-1]{H_2O_2} R_2C=NOH \qquad (18.13)$$

$$NH_3 \xrightarrow[TS-1]{H_2O_2} NH_2OH \xrightarrow{R_2CO} R_2C=NOH \qquad (18.14)$$

The high yields obtained from ketones unable to diffuse inside the pores of TS-1, namely cyclododecanone and 4-butylcyclohexanone, strongly supports the second mechanism [121, 123]. The imine intermediate, if present, could only play a secondary role in the overall balance of the ammoximation process. In this regard, Mantegazza and others showed that TS-1, under the reaction conditions, catalyzes the oxidation of ammonia to hydroxylamine with yields greater than 60% [129]. The subsequent condensation with the carbonyl group is a fast reaction and does not require catalysis by TS-1. It occurs in the external medium with bulky ketones, while with smaller ones the intra-porous volume also could be involved. In the absence of a carbonyl group acceptor, hydroxylamine undergoes consecutive oxidation to N$_2$, N$_2$O, nitrites and nitrates. On these grounds, the overall ammoximation process could be better defined as an oximation reaction with *in situ* production of hydroxylamine. The relative rates of condensation and consecutive oxidation determine the selectivity of the overall process (Scheme 18.13).

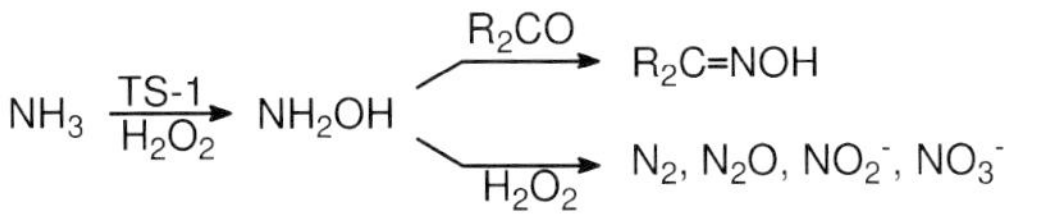

$$NH_3 \xrightarrow[H_2O_2]{TS-1} NH_2OH \begin{array}{c} \xrightarrow{R_2CO} R_2C=NOH \\ \xrightarrow{H_2O_2} N_2, N_2O, NO_2^-, NO_3^- \end{array}$$

Scheme 18.13 Competing reactions of hydroxylamine with cyclohexanone and H$_2$O$_2$.

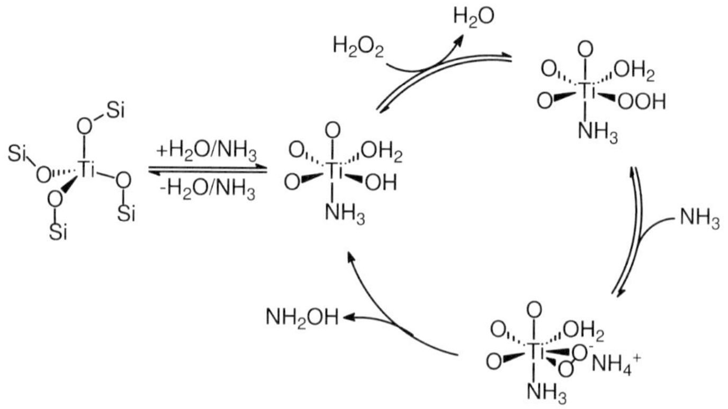

Scheme 18.14 Ammoximation mechanism on TS-1.
(Adapted from Ref. [130], with permission from Elsevier).

The nature of active species and the mode of generation of hydroxylamine are the subject of several studies by various spectroscopic techniques [20, 130]. According to Zecchina and coworkers, the chemisorption of hydrogen peroxide produces the familiar Ti–OOH species, which, however, in the basic medium evolves towards a side-on-bonded anionic peroxide $Ti(\eta^2-O_2)^-$ (Scheme 18.14). The adsorption of water and ammonia completes the coordination sphere of Ti, leading to various species in which either adsorbate can be present with one or two molecules at each time. In Scheme 18.14, the suggestion is made, somewhat arbitrarily, of the mixed adsorption species. A second issue that spectroscopic data do not help solve is whether hydroxylamine is formed by intramolecular reaction between chemisorbed species (in a Languimir–Hinshelwood-type mechanism) or by external attack of physisorbed ammonia on peroxy oxygen (in an Eley–Rideal-type mechanism). A recent kinetic study seems to favor the former pathway [131]. A third issue concerns a possible role of Ti–OOH species in the formation of hydroxylamine. The basic medium, indeed, shifts the acid–base equilibrium towards the formation of the anionic peroxy species, but the presence of Ti–OOH in small amounts cannot be ruled out completely. The greater electrophilicity of Ti–OOH could compensate for its lower concentration.

18.7
Oxidation of N-Compounds

Primary amines, with at least one α-C–H bond, undergo consecutive oxidation to yield unstable alkylnitroso intermediates. These can isomerize to corresponding oximes, dimerize to nitroso dimers and produce nitro derivatives by further oxidation (Scheme 18.15). The selectivity depends on the oxidant system and the reaction conditions. The size restrictions of TS-1 allow a greater selectivity to the

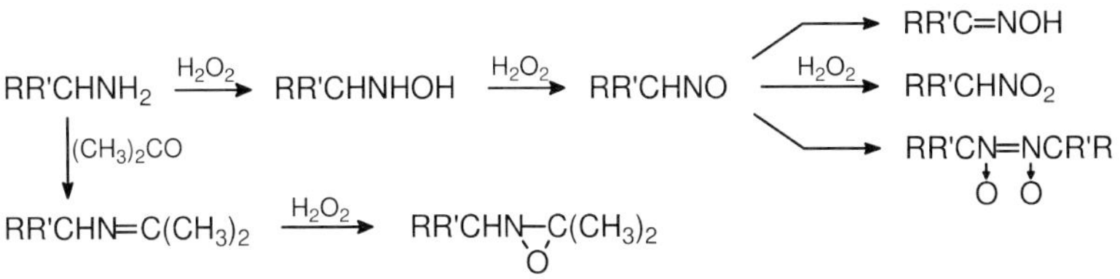

Scheme 18.15

$$RR'CHNH_2 \xrightarrow{H_2O_2} RR'CHNHOH \xrightarrow{H_2O_2} RR'CHNO \xrightarrow{H_2O_2} \begin{cases} RR'C=NOH \\ RR'CHNO_2 \\ RR'CN=NCR'R \\ \quad\;\; \downarrow\;\; \downarrow \\ \quad\;\; O\;\; O \end{cases}$$

$$\downarrow (CH_3)_2CO$$

$$RR'CHN=C(CH_3)_2 \xrightarrow{H_2O_2} RR'CHN\underset{O}{-}C(CH_3)_2$$

Scheme 18.15 Oxidation of primary amines.

desirable oxime than conventional catalysts, at the expense of the bulkier dimer. Methanol and *t*-butanol are suitable solvents, while acetone undergoes side reactions with the amine and the oxidant [132]. The oxidation was broadened to include several aliphatic and benzylic amines, owing to the versatility of alkyl oximes in organic synthesis [133].

The oxidation of secondary amines can produce the corresponding hydroxylamines and nitrones (Equation 18.15) [134, 135]. TS-1 and TS-2 catalyzed the production of the former, at H_2O_2/amine molar ratios below unity. The addition of alkali metal additives improved the selectivity up to 90% [136]. High yields of nitrones could be obtained at greater H_2O_2/amine ratios.

$$RR'CHNHR'' \xrightarrow{H_2O_2} RR'CHNR'' \xrightarrow{H_2O_2} RR'CH=NR'' \qquad (18.15)$$
$$\qquad\qquad\qquad\qquad |\qquad\qquad\qquad\quad\; \downarrow$$
$$\qquad\qquad\qquad\qquad OH\qquad\qquad\qquad\; O$$

The oxidation of aniline and other arylamines yields a wide spectrum of products, among which azoxybenzenes are the most desirable (Scheme 18.16). TS-1, Ti-ZSM-5, TS-2, Ti,Al-β and Ti-MCM-41, were studied as catalysts, with hydrogen peroxide and *t*-butyl hydroperoxide as the oxidants [137–140]. The best yields, up to 95%, were reported for Ti,Al-β and mesoporous Ti-silicates, with high aniline/H_2O_2 molar ratios to minimize over-oxidation to nitrobenzene. Anilines, with either electron-donating or -withdrawing groups, and 1-naphthylamine behaved similarly. Owing to diffusional restrictions, TS-1 was less effective than large-pore and mesoporous catalysts.

The oxidation of aromatic and aliphatic oximes and of tosylhydrazones on TS-1 with excess H_2O_2 was reported to regenerate corresponding aldehydes [141, 142]. These were obtained also by the one-pot oxidation of primary amines under analogous conditions [143].

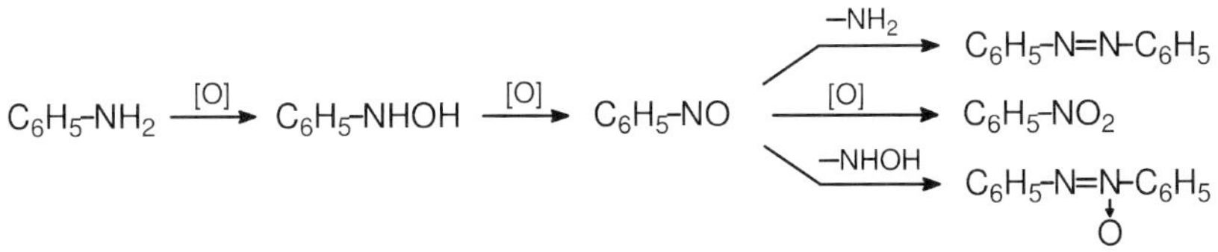

$$C_6H_5-NH_2 \xrightarrow{[O]} C_6H_5-NHOH \xrightarrow{[O]} C_6H_5-NO \begin{cases} \xrightarrow{-NH_2} C_6H_5-N=N-C_6H_5 \\ \xrightarrow{[O]} C_6H_5-NO_2 \\ \xrightarrow{-NHOH} C_6H_5-N=N-C_6H_5 \\ \qquad\qquad\qquad\quad \downarrow \\ \qquad\qquad\qquad\quad O \end{cases}$$

Scheme 18.16 Oxidation of aniline.

18.8
Oxidation of S-Compounds

The oxidation of thioethers on TS-1, Ti-β and Ti,Al-β produced corresponding sulfoxides and, more slowly, sulfones by consecutive oxidation [144–149]. Allyl methyl thioether was oxidized selectively on sulfur, with no reaction of the double bond [148]. The kinetic law was obtained for the oxidation of dibutylsulfoxide on Ti,Al-β [147].

The oxidation of thiophene sulfides combined with a selective separation process of oxidized products, for example by adsorption on inert materials, might offer an alternative to deep hydrotreating for desulfurization of oil fractions. On these grounds, the oxidation of disulfides has gained considerable interest [149]. Owing to the nature of sulfur impurities, large-pore and even more mesoporous Ti-silicates would be appropriate catalysts.

18.9
Industrial Processes Catalyzed by TS-1

Three oxidation processes are already commercial. Other processes could be developed in the future, for example for the production of epichlorohydrin and phenol, especially in the event that new production strategies reduced the cost of hydrogen peroxide (see below). It is unlikely, however, that Ti-zeolites other than TS-1 could find application in the foreseeable future.

18.9.1
Hydroxylation of Phenol to Catechol and Hydroquinone

In 1986 the EniChem process replaced the Brichima process, with significant materials and energy savings (Table 18.5, entries 1 and 2). It operates with excess phenol, fed with H_2O_2 into a slurry of TS-1 in aqueous acetone, at ca 100 °C [5]. While the conversion of the oxidant is driven to completeness, excess phenol is recovered by distillation and recycled. The catalyst is structurally stable under operating conditions, being deactivated only by pore plugging and active site fouling. The regeneration is conveniently carried out thermally under flowing air.

With an overall capacity of $10\,000\,t\,a^{-1}$, it is a modest process when compared to recent applications of TS-1 in ammoximation and propene epoxidation. The introduction of digital photography, which no longer needs hydroquinone for the development of silver emulsions, risks the continuity of the process, unless the selectivity to catechol is greatly improved or new uses are developed for hydroquinone.

18.9.2
Salt-Free Production of Cyclohexanone Oxime

A demonstration plant, built by EniChem at the industrial site of Porto Marghera, near Venice, began operation in 1994 ($12\,000\,t\,a^{-1}$). The process was operated continuously in the liquid phase in a stirred reactor. Cyclohexanone, ammonia and hydrogen peroxide were fed, at 80–90 °C and slight over-pressure, into a slurry of TS-1 in aqueous *t*-butanol. The effluent was separated from the catalyst through a filter, and sent to a distillation column. Excess ammonia and the solvent, recovered overhead, were recycled in the reactor, while the aqueous solution of oxime, taken from the bottom, underwent further purification. Periodic purging and make-up operations of the catalyst were necessary to compensate for irreversible decay.

The first commercial application, made possible by an agreement between EniChem and Sumitomo, went on-stream in 2003 in Japan, within the context of an integrated process for the production of ε-caprolactam by a new salt-free technology (ca 60 000 t a^{-1}). Actually, besides the ammoximation step, no major by-product is produced even in the gas-phase rearrangement carried out on silicalite-1 as the catalyst. On the whole, the ammonium sulfate is no longer a burden and the gaseous emissions too are drastically reduced.

The truly innovative nature of the EniChem process over earlier ones is apparent. The preparation of hydroxylamine in the latter case necessitates multi-step operation, often ending with major co-production of inorganic salts, as in the Raschig process. The ammoximation of cyclohexanone, with its *in situ* generation of the intermediate, reduces significantly the investment and operation costs while improving the environmental compatibility. The new process represents a good example of how to combine profitability and environmental concern.

18.9.3
Propene Oxide Synthesis (HPPO)

The industrial process for propene oxide manufacture is commonly referred to as the HPPO (hydrogen peroxide propene oxide) process. EniChem set up a prototype plant in 2002 [150]. BASF/Dow Chemicals and Degussa, in turn, have the construction of commercial plants already in progress or at the planning stage [151].

In the EniChem process (6 t day^{-1}), a dilute solution of hydrogen peroxide and a gaseous stream of propene were fed into a stirred reactor containing TS-1 and aqueous methanol [150]. The process was operated continuously, at moderate pressure and temperature. The selectivity could be as high as 98%. Propene, the epoxide and methanol were recovered by distillation of the effluent, while glycol by-products accumulated in the aqueous solution at the bottom of the column. The latter, depending on the selectivity of epoxidation, was either processed to recover the glycols or sent directly to the biological plant, for final treatment before disposal. The stability of the catalyst to deactivation, achieved by the appropriate control of operating conditions, was very high and did not require frequent regeneration. According to Romano, the new process is characterized by lower environmental impact, simpler process design and relatively lower investment and operating costs than conventional ones [150].

Less is known of the processes developed by other companies. BASF/Dow Chemical, in a joint venture, started the construction of a 300 000 t a^{-1} plant at the BASF site in Antwerp (Belgium) at the end of 2006 [151]. Hydrogen peroxide will be supplied by a 230 000 t a^{-1} plant based on Solvay's anthraquinone technology and located at the same site in Antwerp. The operative start-up is scheduled for early 2008. Both plants will be the world's largest single train ones in operation. HPPO technology is also under evaluation for other facilities in the USA and Asia.

Degussa, in turn, has recently announced the commercialization of a HPPO process, jointly developed with the engineering company Uhde. Parallel to this, Degussa with Headwaters is also working on the direct synthesis of H_2O_2, for which a demonstration plant was completed in 2006 [151]. According to the news release, hydrogen peroxide will be obtained in the new process as a dilute methanol solution to be used directly in the epoxidation of propene.

18.10
Problems in the Use of H_2O_2 and Possible Solutions

Hydrogen peroxide is currently produced by the anthraquinone autoxidation (AO) process. It is a relatively expensive chemical, used for bleaching in the pulp/paper (ca 57%) and textile industries (6%), for waste water and effluent treatment (5%) and for laundry products (ca 12%). Only a minor amount (ca 10%) is currently consumed for the production of fine chemicals (the figures refer to the early 2000s). The cost and the production technology, while compatible with its present uses, fit less well with petrochemical applications. For instance, the cost of commercial hydrogen peroxide (0.69 $/lb, 100% H_2O_2) is comparable, on a weight basis, to that of propene oxide (0.70 $/lb) [152]. The annual capacity of the plants, mainly comprising the ranges of 10 000–20 000 and 40 000–70 000 tons, with a very few slightly over 100 000 t a^{-1}, is undersized for average petrochemical production. The shipment of large amounts of hydrogen peroxide could represent a problem from the point of view of logistics and safety. In addition, the anthraquinone route involves complex technology, not easy to access (Figure 18.2). Thus, the applications of TS-1 in the petrochemical industry were, from the beginning, strong incentives for the investigation of potential alternatives for commercial hydrogen peroxide production. Direct synthesis from the elements,

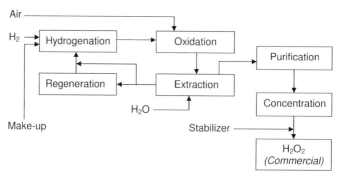

Figure 18.2 Simplified block diagram of anthrahydroquinone autoxidation (AO).

in situ production and process integration have been the most studied solutions. Unless otherwise specified, HPPO is the process of reference in the following discussion.

18.10.1
Direct Synthesis of Hydrogen Peroxide

The catalyzed reaction of the elements produces hydrogen peroxide directly, in competition with water (Equation 18.16). Most selective catalysts are based on supported Pd, often modified by other metals. Methanol is increasingly studied as a reaction medium, from the perspective of the direct use of the solution obtained in the HPPO process. For high selectivities, the reaction normally necessitates relatively low temperatures, often room temperature or lower, and the presence of mineral acids and other additives [153].

$$H_2 + O_2 \xrightarrow{Pd} H_2O_2 \qquad (18.16)$$

The direct synthesis is in principle a one-step process, which is simpler and easier to scale-up for large capacity plants than the AO route. However, it also suffers from major drawbacks, for example continuing low yields and the hazards of mixing flammable gases. Various technical solutions are envisaged in the patent literature to minimize the risks of explosion and fire, such as the dilution of hydrogen and oxygen with an inert fluid.

18.10.2
In Situ Production of Hydrogen Peroxide

In situ generation removes dependence on external supplies. The availability of hydrogen at the chemical site is the only requisite for its feasibility. The approach generally envisaged in the literature is of two types [154–158]. The first requires that TS-1 is modified by a metal species, generally Pd, able to catalyze the direct synthesis. H_2, O_2 and propene are fed to the reactor to produce propene oxide and water directly (Equation 18.17). On a microscopic scale, hydrogen peroxide is formed on metal particles, then diffuses in TS-1 to react with propene on Ti sites. The high activity of TS-1 in very dilute solutions of hydrogen peroxide prevents the accumulation of the oxidant in contact with metal particles that would otherwise catalyze its decomposition. A major drawback of the method relates to the excellent hydrogenation properties of Pd, leading to significant losses of propene as propane.

$$H_2 + O_2 + C_3H_6 \xrightarrow{Pd/TS\text{-}1} C_3H_6O + H_2O \qquad (18.17)$$

In early studies, carried out batch-wise on Pd-Pt/TS-1, the maximum yield was 11.7%, with 46% selectivity based on propene [156]. By operating in a continuous fixed-bed reactor, Jenzer and others showed that the selectivity could initially be as high as 99% at 3.5% conversion. The catalyst, however, underwent rapid deactivation with a parallel decrease of the selectivity. Methyl formate from the oxidation of methanol, and acrolein, acetone and other compounds from that of propene became the main products after few hours on-stream [157]. The use of supercritical carbon dioxide, without the addition of methanol/water, allowed the reduction of side reactions [158]. Other studies envisage *in situ* generation by an electrochemical cell [159] and gas-phase epoxidation on Ag/TS-1 at 150 °C [160].

A second approach consists of the autoxidation of an organic carrier, in the presence of propene and TS-1 (Equations 18.18 and 18.19) [154, 155, 161, 162]. The solvent is a crucial aspect in the use of alkylanthrahydroquinone compounds [154]. It should be able to solubilize both the oxidized and the hydrogenated forms of the carrier and possess a molecular size larger than the pores of TS-1. On this basis, the various components of the working solution cannot interfere in the epoxidation mechanism or undergo oxidative degradation at the Ti sites. Normally, the conditions are met by the use of a mixture of two or more exotic solvents, to which methanol is also added to increase the rate of epoxidation. Specific tests proved the feasibility of the method, albeit with somewhat lower yields than when carried out with *ex situ* hydrogen peroxide [154]. A subsequent study, while confirming early results, provided useful information on the reactor and operative conditions [161]. The use of a gas-lift loop reactor, allowed the process to be carried out satisfactorily even in the presence of two liquid phases, one richer in methanol/water and the other in the working solution. The best yields were 82%, with 85% selectivity based on alkylanthrahydroquinone.

$$C_3H_6 + O_2 + QH_2 \xrightarrow{TS\text{-}1} C_3H_6O + H_2O + Q \tag{18.18}$$

$$Q + H_2 \xrightarrow{Pd} QH_2 \tag{18.19}$$

Major shortcomings also affect this *in situ* route to propene oxide. The most obvious one is the even greater complexity than for the AO process, owing to the modification required to accommodate the epoxidation of propene. This explains why patents have been filed on anthraquinone compounds and other organic carriers, soluble in aqueous methanol. The aim was a radical simplification of the working solution, trying to eventually use the simple methanol/water medium for the whole series of reactions. Examples of soluble carriers are anthraquinone compounds carrying hydrophilic alkyl ammonium or sulfonic substituent groups and secondary alcohols.

18.10.3
Process Integration

The integration of propene oxide and hydrogen peroxide processes appears, from a short-term perspective, a more realistic approach than *in situ* generation. It is also favored by the location of both facilities at the same chemical site. Basically, process integration envisages a simplification of H_2O_2 recovery that can even

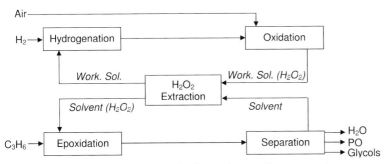

Figure 18.3 Integrated process scheme for the production of propene oxide.

result, at best, in the complete elimination of the purification, concentration and stabilization steps [154]. In this hypothesis, the aqueous H_2O_2 extract obtained is fed, without further processing, into the epoxidation reactor. Impurities from the working solution, which may have been extracted with the H_2O_2, are hindered from diffusing to Ti sites by their relatively large molecular size. Drawbacks could ensue, for example for the activity of the catalyst, from inorganic additives that may be required by the AO process [161]. Beside anthraquinone compounds, old and new carriers such as isopropanol, once used industrially for the production of hydrogen peroxide and 1-phenylethanol, have also been studied from the perspective of process integration [29, 163].

A variant envisages the use of aqueous methanol for both the extraction of hydrogen peroxide and the epoxidation of propene as shown, in a simplified form, by Figure 18.3. Thus, no extra water is added to the epoxidation reactor, which decreases separation and waste-treatment costs. On the laboratory scale, the quantity of methanol dissolving into the working solution is small and apparently does not interfere with the AO cycle of reactions [154, 161].

In principle, process integration applies even more to direct synthesis of hydrogen peroxide, further improving the advantages over the anthraquinone route. Methanol can be used to replace water as the solvent and the dilute methanol solution obtained fed into the epoxidation reactor. Minimal purification may be required, for example for the removal of hydrogen bromide and other additives that may have been needed to increase the selectivity.

18.10.4
Miscellanea

The relatively small capacity of the phenol hydroxylation process does not justify the development of alternatives to the use of commercial hydrogen peroxide. The direct synthesis of phenol, for which process integration might be crucial, would be a different matter.

The ammoximation process is not very compatible with *in situ* generation. From one side, the possible reaction of ammonia with the working solution and the complex recovery of cyclohexanone oxime rule out the use of organic carriers, and, from the other, no catalyst of direct synthesis is known that is sufficiently active and selective in a basic medium. Process integration, however, is conceivable and is present in the patent literature. In an early patent, a mixture of cyclohexanone and hydrogen peroxide was produced by the autoxidation of cyclohexanol and fed directly into the ammoximation reactor [164]. Hydrogen peroxide from the autoxidation of anthrahydroquinones was extracted with aqueous methanol or *t*-butanol and used as obtained [165, 166].

18.11
Adsorption, Active Species and Oxidation Mechanisms

In previous sections, mechanistic proposals have been illustrated with a sometimes limited discussion of active species and of the relations existing between them. A deeper insight is also needed into adsorption phenomena and their role in the network of reactions catalyzed by TS-1. In this section, these issues will be examined more closely, with the aim of identifying the chemical and physical factors determining the direction of the oxidation pathway from the perspective of a general mechanistic scheme.

18.11.1
Adsorption and Catalytic Performances

It is clear from the first part of this chapter that the external medium has a major effect on the catalytic properties of TS-1 and other Ti-zeolites. Early rationalization suggested that the adsorption of a protic molecule, ROH, had a stabilizing effect for Ti–OOH species. In this view, water and alcohol solvents influenced the catalytic performance through the electronic and steric effects exerted in the oxidation step by the ROH molecule present in the active species (Section 18.4.1.1) [62].

Subsequently, Langhendries and others shed light on the key role the external medium has in the adsorption of olefins, alkanes and alcohols [167]. They showed that partition coefficients, defined as the ratio of intra-porous to extra-porous concentrations, decreased in the order of decreasing solvent polarity, $CH_3OH > C_2H_5OH > 1\text{-}C_3H_7OH$, $CH_3CN > CH_3COCH_3$, and for a given solvent with decreasing hydrophobicity of the molecular sieve, TS-1 > Ti,Al-β > Ti-MCM-41. Partition coefficients increased also with the chain length of the adsorbate. The reactivity orders in the epoxidation of a series of olefins were fully consistent with the trends. On these grounds, the authors concluded that the adsorption phenomena were sufficient to explain the effect of the solvent on kinetics, without the need to invoke other roles.

Actually, the study provides an experimental demonstration of the primary role the surface has in the catalytic performance of Ti-zeolites. TS-1 is superior to Ti,Al-β because its surface is more hydrophobic and, therefore, the intra-porous concentration of the olefins and other apolar reagents is greater, rather than because the activity of Ti-sites is substantially different. In this regard, the micro-

Table 18.13 Solvent effects in the oxidation of n-hexane on TS-1.[a]

Solvent	K_s [b]	$V_0 \cdot 10^{-4} \, (m_{H_2O_2} \cdot s^{-1})$ [c]
water	–	1.1 (two liquid phases)
methanol	16.5	1.2
acetonitrile	2.5	0.9
i-propanol	0.8	–
t-butanol/H$_2$O (27%)	–	2.4
t-butanol	–	4.3

a Reprinted from Ref. [169]. Copyright 2006, with permission of Thomas Vieth Publisher.
b Partition coefficients of n-hexane.
c Rates at 15–30% conversion.

porous nature of the catalyst has a major role in that it enhances both attractive and repulsive interactions between surface and adsorbate and, through them, the selective adsorption of some compounds over others. It is useful, in this regard, to recall the concept of zeolites as solid solvents [168]. Nonetheless, the partition coefficients alone are unable to account for the catalytic behavior of Ti-zeolites in all circumstances. So, the role of a protic adsorbate on Ti–OOH species is again of assistance [25]. Three pieces of evidence in the literature are interesting in this regard:

- If a correlation was apparent between adsorption and epoxidation rates within a homogeneous series of solvents, this no longer held when the solvents were considered all together and no distinction was made between protic and aprotic ones [25, 167].
- In a study by Corma and coworkers, the rate of epoxidation of 1-hexene on Ti,Al-β matched, for a homogeneous series of solvents, the trend of adsorption. However, it was twice as fast in acetonitrile than in methanol, in contrast to partition coefficients which are ordered in the reverse direction [77, 167]. The relationship for cyclohexanol oxidation was more complex, the rate increasing with the polarity of aprotic solvents and decreasing with polarity increase in protic ones [77].
- The hydroxylation of n-hexane on TS-1, in contrast to the epoxidation of propene, reached its maximum rate in the least polar solvent, t-butanol (Table 18.13). Acetonitrile behaved quite similarly to methanol and water [24, 25, 169]. On the assumption that t-butanol was comparable to i-propanol for the effects on adsorption, a clear relationship between rates and partition coefficients was lacking. Considering that hydroxylation and epoxidation involve different active species and mechanisms, a diverse role of the solvent in the two active species could contribute to the differences, whereas the partition coefficients alone could not.

In principle, the solvent can affect catalysis in a multiplicity of modes, for example through the adsorption of the reagents, the stabilization of Ti–OOH species or interaction with oxidation intermediates [25]. It is worth noting that the first two phenomena produce effects in the same direction, and their individual contributions are not easily distinguished. For instance, the rate of epoxidation in t-butanol may be slower than in methanol owing to smaller partition coefficients of the olefin, to more sterically hindered active species and to less electrophilic peroxidic oxygen.

18.11.2
The Structure of Ti–OOH Species

The structural and chemical properties of active species in TS-1 are strictly related to the tetrahedral configuration that Ti is forced to assume in the silicalite framework, to its preference for the octahedral one, which can be achieved by the adsorption of polar molecules, and to the reversible cleavage of one Ti–OSi bond by a protic molecule [8, 170, 171]. On this basis, Scheme 18.17 illustrates the formation of Ti–OH and Ti–OOH species, by the chemisorption of water and hydrogen peroxide, respectively. The use of an alcoholic solution leads to analogous results. Other Ti-species generally predicted by theoretical studies, for example, bearing an undissociated H_2O_2 molecule weakly adsorbed on Ti or consisting of Ti–O–O–Si peroxides, are not very compatible with the chemistry of TS-1 and are not considered further in this context.

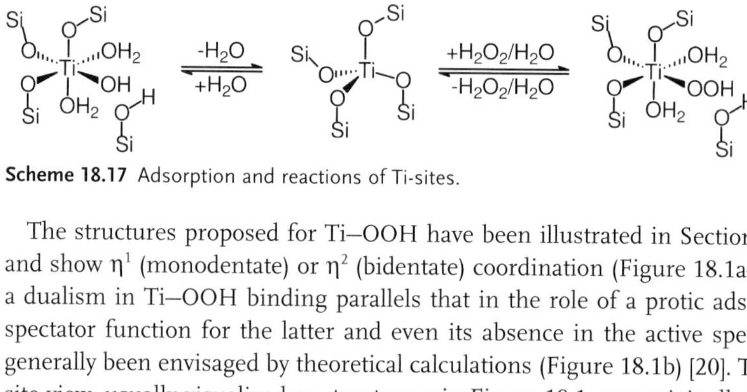

Scheme 18.17 Adsorption and reactions of Ti-sites.

The structures proposed for Ti–OOH have been illustrated in Section 18.4.1.1 and show η^1 (monodentate) or η^2 (bidentate) coordination (Figure 18.1a–c). Such a dualism in Ti–OOH binding parallels that in the role of a protic adsorbate. A spectator function for the latter and even its absence in the active species have generally been envisaged by theoretical calculations (Figure 18.1b) [20]. The opposite view, usually visualized as structure a in Figure 18.1, was originally based on the interpretation of early experimental results [24, 61, 62, 171]. Currently, the likelihood of structures a and c in Figure 18.1, entailing an active role for the protic adsorbate, can rely upon additional evidence in the literature.

- Freely diffusing olefins reveal large steric constraints in the approach to peroxidic oxygen, in the epoxidation on TS-1 and Ti-Beta zeolites (Tables 18.8 and 18.9) [62, 81, 83]. For the active species b in Figure 18.1, the repulsive effect exerted by the surface on the alkyl substituents is the only possible explanation. For species a and c, the R group would produce additional, and probably greater, steric constraints. Whether the surface alone is sufficient to account for the magnitude of the effect is a matter of debate. However, the epoxidation of 2-butene compared to that of terminal olefins shows that the steric effects are large even for *cis*-isomers, in which both methyl groups can point away from the surface. Clearly, steric effects should be larger for *trans*-2-butene, as experimentally found, for the opposite reason.

- The addition of modest quantities of water to acetonitrile solvent increases the rate of oxidation of 1-hexene but decreases that of cyclohexanol [77]. The contrasting effects were attributed to the adsorption of water on Ti–OOH to form the cyclic structure of *a* (Figure 18.1). In the epoxidation mechanism (R=H), this was reputed to be the active species and, therefore, the addition of water promoted the rate through its formation. In the oxidation of the alcohol, a species with the same structure (R = cyclohexyl) was the oxidation intermediate and water, in competition with cyclohexanol for adsorption on Ti (R=H), could only decrease the rate. It should be added that the study was performed with Ti,Al-β as the catalyst, but, as mentioned in Section 18.4.1.2, the same active species is believed to apply generally [77–79].

- The diffusivity and adsorption capacity of aromatic molecules, in TS-1 and in Ti-MOR, were both reduced by the simultaneous presence of H_2O_2 and H_2O, in a linear relationship with the Ti content [172]. No such effects were produced by the use of H_2O alone under analogous conditions or by that of H_2O_2/H_2O mixtures with Ti-free zeolites. The underlying rationale, in the authors' view, is the formation of bulky peroxy species, with the structure *a* of Figure 18.1, protruding into the pores. Their presence hinders the movement of adsorbing reagents and reduces the free volume available for them in the pores. No such bulky species can be formed when water alone adsorbs on Ti-sites.

The evidence above and a number of other studies, of various strength, based upon IR and UV-Vis spectroscopy, together with XANES and EXAFS studies and the kinetic behavior of the oxidation of alcohols and allyl chloride (Equations 18.8 and 18.11), demonstrate the co-adsorption of H_2O_2 and, at least one further molecule of water or alcohol on Ti. In some cases, they also confirm the formation of cyclic structures like *a* and *c* in Figure 18.1, meaning that structure *b* is only possible under vacuum conditions. Their formation under the reaction conditions also favor, albeit not proving, their participation in the oxygen-transfer step to the double bond. Actually, the steric constraints in the epoxidation of linear olefins and the promoter or inhibitor effect of water, according to the type of substrate, point in this direction. The opposite view, based on theoretical models, is weakened by the variety of significantly different active species they lead to.

18.11.3
Reactive Intermediates and Oxidation Mechanisms

The heterolytic mechanism and the Ti–OOH nature of the active species in the epoxidation of olefins have the support of several chemical, spectroscopic and theoretical studies. For the hydroxylation of alkanes, some indirect evidence favors homolytic H-abstraction by a Ti-centered radical, the nature of which, however, remains unknown. Scarce information exists on the oxidation mechanisms of arenes and alcohols, for which mechanistic proposals have been advanced by analogy with other oxidations. Even less is available for N- and S-compounds and for the decomposition of hydrogen peroxide. On the whole, the understanding of TS-1 catalysis is still limited, with epoxidation the only exception. Nonetheless, two deductions are possible from available information. Ti–OOH can directly perform the oxidation of certain substrates, for example, olefins and, maybe, O-, S- and N-compounds, or undergo a redox process, converting into a radical oxidant species. As a result, competitive heterolytic and homolytic pathways can co-exist in TS-1. Precedents in this regard can be found in the literature [67].

The epoxidation species received adequate attention in previous sections, with Figure 18.1 and Schemes 18.7, 18.8 and 18.17 illustrating the mode of generation, the possible structures and the mechanisms of oxygen transfer proposed in the literature. Conversely, the species responsible for the cleavage of saturated C–H bonds necessitates at this stage a deeper insight, particularly in the light of Scheme 18.3b, based on the atypical Ti–O$^{\bullet}$ radical. Actually, no precedent exists in the chemistry of Ti for such a species. Nonetheless, other conceivable mechanisms based on the participation of the species a–d of Figure 18.1, of Ti–O$_2^{\bullet}$ superoxide and of a Ti^{III}–OO$^{\bullet}$ diradical are even less credible. Actually, it is hard to reconcile C–H hydroxylation with Ti(η^2–O$_2$) and Ti–OOH species, considering that the reaction is homolytic whereas the oxidant is not. The implausibility of diradical Ti^{III}–OO$^{\bullet}$ has been discussed (Section 18.2.1). The fact that no major role is assigned to Ti–O$_2^{\bullet}$, is not only because this represents a negligible fraction of overall Ti content [66]. Actually, the reactivity of metal superoxides with alkanes was shown to be generally low or zero [1]. The generation of Ti–O$_2^{\bullet}$ requires the intervention of an OH$^{\bullet}$ radical, whose origin remains unknown. No evidence of its reactivity was detected in the oxidation of olefins, in which TiO$_2^{\bullet}$ is expected to add across the double bond (e.g. the stereo scrambling of cis and trans olefins by addition–elimination).

It is useful to note, at this stage, that TS-1 shows similarities in C–H hydroxylation with biomimetic catalysts and simple Fe^{II} ions dissolved in acetonitrile, for example the homolytic nature of the reaction, the electrophilic character of the active species, a significant primary H/D isotope effect, the competition between hydroxylation and H$_2$O$_2$ decomposition and, often, the stereoscrambling ability. Similarities exist also with enzymatic systems, notably with cytochrome P-450 and the methane monooxygenase (MMO) type of enzymes [31, 173]. On these grounds, it is tempting to extend further the analogy, proposing Ti–O$^{\bullet}$ as the active species related to the Fe^{IV}–O$^{\bullet}$ of biomimetic and enzymatic systems and Fe^{III}–O$^{\bullet}$ of Fe^{II}/H$_2$O$_2$/CH$_3$CN (Fe^{IV}–O$^{\bullet}$ and Fe^{III}–O$^{\bullet}$ are radical notations of Fe^V–O and Fe^{IV}–O, respectively). Indirect support for Ti–O$^{\bullet}$ comes from a study by Bonchio and others on the hydroxylation of benzene by the peroxide VO(O$_2$)(H$_2$O)$_2$L (L = picolinic acid), in which evidence exists for the participation of a V^V–O$^{\bullet}$ species, instead of the diradical V^{IV}–OO$^{\bullet}$ previously suggested by others [174]. However, while the generation of high-valent iron oxo species from d^5 Fe^{III}–OOH

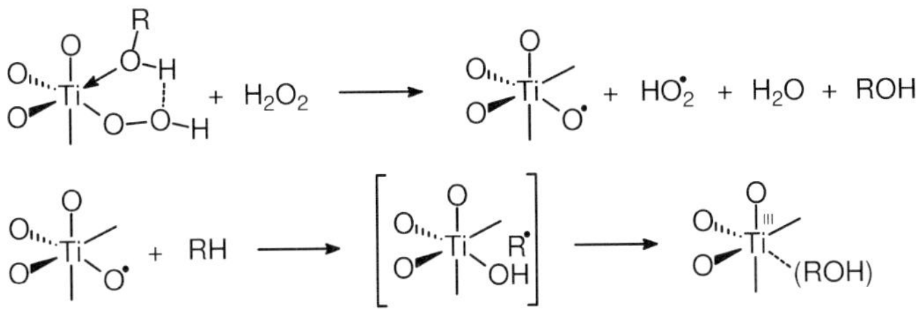

Scheme 18.18 Generation of Ti–O• active species and C–H hydroxylation.

and d^6 Fe^{II}–OOH hydroperoxides is straightforward, an external reducing agent would be required for that of Ti–O• from d^0 Ti^{IV}–OOH (Scheme 18.18). A redox process had been proposed also for V^V–O• formation from precursor $VO(O_2)(H_2O)_2L$.

The regeneration of the active species from Ti^{III} represents a second problem to solve. Actually, the one-electron oxidation of low-valent metal ions with H_2O_2 normally produces free hydroxyl radicals, which, however, are not detected with TS-1. A possible solution is suggested by the redox behavior of Fe^{II} perchlorate in acetonitrile solvent [33, 34]. At odds with aqueous media promoting Fenton-like reactions, acetonitrile favored the formation of an iron-bound oxidant, able to homolytically hydroxylate paraffinic C–H bonds. A parallel is apparent between the aprotic medium of Fe^{II} and the intra-porous volume of TS-1, in which surface Si–OH and protic adsorbates are negligible under the hydroxylation conditions. Thus, hydrogen peroxide could regenerate Ti–O• from Ti^{III}, as it does with Fe^{IV}=O from Fe^{II}, without being decomposed into free OH• radicals (Scheme 18.19). Whether this process passes through Ti^{III}–OOH, is of little interest at this stage. Scheme 18.19 only means that H_2O_2 regenerates the active species, by the oxidation of transient Ti^{III} intermediate, without decomposition into free hydroxyl radicals. The organophilic nature of the intra-porous volume of TS-1 promotes the process. It is remarkable, in this regard, that biomimetic catalysts are generally used in aprotic solvents and that the active species of MMO is thought to be located in a hydrophobic pocket of the enzyme.

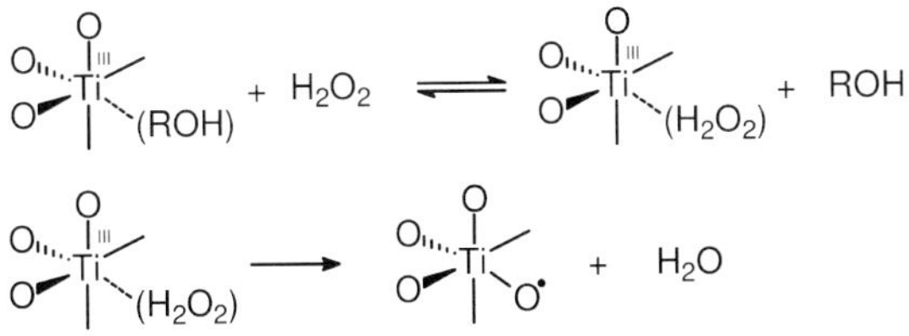

Scheme 18.19 Regeneration of Ti–O• species from Ti^{III} intermediate by hydrogen peroxide.

Ti–O• is also an excellent candidate as an initiator of H_2O_2 decomposition (Scheme 18.20). The subsequent decomposition of HO_2• species into molecular oxygen could follow, at least in part, known pathways [1]. The details of oxygen evolution, however, is an issue that goes beyond the scope of this chapter.

Scheme 18.20 Decomposition of hydrogen peroxide initiated by Ti–O• species.

The acidic character of Ti–OOH species is the final property to be addressed. An acid–base equilibrium is established under reaction conditions, with mildly basic compounds, including water and alcohols, adsorbed in the pores (Scheme 18.21) [24, 171, 175]. The addition of basic compounds in small amounts improves the selectivity of epoxidation through its effect on the equilibrium. On increasing the local basicity, this shifts completely to the right, with the formation of anionic η^2-peroxides [176]. Since $Ti(\eta^2-O_2)^-$ species are characterized by low electrophilicity, their formation has a profound effect on the reactivity of TS-1, with inhibition and eventually deactivation in most reactions (ammoximation is one exception). Protonic acids regenerate initial Ti–OOH species. Owing to the equilibrium nature of the reaction, any attempt to prepare anionic TS-1 peroxides in a pure state failed. Hydrogen peroxide and the base gradually leached into water washings, to be eventually completely removed with them [176].

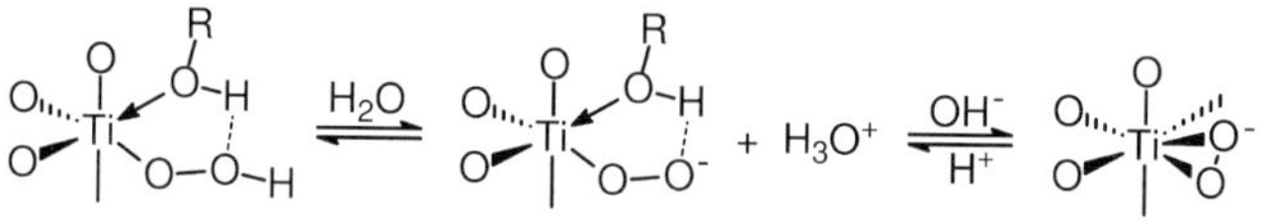

Scheme 18.21 Acid–base properties of Ti–OOH hydroperoxide.

18.11.4
Proposal for a General Mechanistic Scheme

In conclusion, different active species and different mechanisms are operative in TS-1, in a network of competing reactions. Scheme 18.22 is a tentative rationalization, in a simplified form, of information presented in previous sections on most studied reactions. Different factors determine the direction of the oxidation pathway, for example the temperature, the solvent, the nature of the surface and the substrate itself [169].

Scheme 18.22 pivots upon species **a** of Figure 18.1 (although the participation of **c** is equally conceivable). Each oxidation path is in competition with the others, becoming predominant under specific conditions. For instance, unhindered olefins in a methanol solution of hydrogen peroxide undergo fast epoxidation even at below room temperature (mechanism A). The oxidation of allylic C–H (mechanism B), the oxidation of CH_3OH (mechanism C) and the decomposition of H_2O_2 (reactions D), although equally possible, are too slow to be detectable to a significant extent. At the same time, the acid–base equilibrium can be controlled by the addition of parts per million quantities of basic compounds, to minimize the solvolysis of the product. On raising the temperature and contact time to assist the

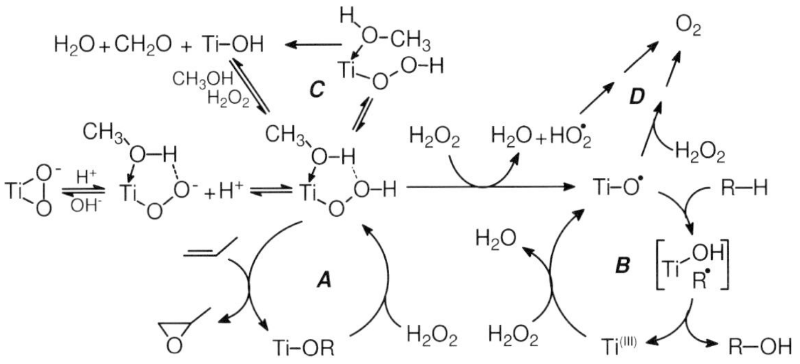

Scheme 18.22 Reactive intermediates and oxidation pathways in TS-1/H_2O_2. (Adapted from Ref. [169]; Copyright 2006, with permission of Thomas Vieth Publisher).

epoxidation of sterically hindered olefins, the reaction of Ti–OOH with H_2O_2, and therefore the homolytic pathways B and D, may compete significantly, to the point that with the least reactive olefins these can prevail over epoxidation A. The literature offers several examples in which epoxidation, allylic oxidation and H_2O_2 decomposition are all present, in different proportions, even in the oxidations with TS-1 and particularly in those with Ti-Beta catalysts.

The hydroxylation of alkanes is considerably slower than epoxidation and necessitates significantly higher temperatures to occur at a comparable rate. As a matter of fact, H_2O_2 decomposition D is never negligible in the hydroxylation of alkanes, its competition being significant even in the fastest one observed on TS-1, the hydroxylation of n-hexane (Table 18.2). It is also a side reaction in the generation of Ti–O$^\bullet$ species, but for the most part it should originate from the direct reaction of H_2O_2 (first reaction of Schemes 18.18 and 18.20). With the least activated hydrocarbons, for example with bulky alkanes, the decomposition of H_2O_2 becomes the main reaction. Analogous considerations hold also for the competition of the oxidation of methanol solvent, a side reaction that occurs to a significant extent with the least reactive substrates, together with decomposition of the oxidant. Table 18.11 shows, in this regard, why other alcohols are inappropriate for use as solvents.

The relative weight of different pathways in Scheme 18.22 is determined also by the combination of surface/solvent, through the control of the substrate-to-oxidant ratio in the pores. At a fixed composition of external medium, the H_2O_2-to-hydrocarbon and the H_2O-to-hydrocarbon ratios increase in the order TS-1 < Ti-β < Ti,Al-β ≪ Ti-MCM-41, owing to increasingly hydrophilic pores. Thus, the competition of H_2O_2 with the hydrocarbon for reaction with Ti–OOH increases in the same order and with it the generation of Ti–O$^\bullet$ radicals in the pores. Allylic oxidation (path B) and H_2O_2 decomposition (path D) increase accordingly, at the expense of epoxidation (path A). As a matter of fact, the epoxidation of olefins on large-pore catalysts is generally slower and less selective than on TS-1, becoming negligible on mesoporous Ti-silicates, owing to a lower adsorption of the substrate and a greater incidence of such side reactions.

Analogous considerations on the effect of the surface/solvent couple also apply to the oxidation of alkanes and aromatics. The lower reactivity of C–H bonds combines with decreasing alkane/H_2O_2 ratios in the order TS-1 > Ti-β > Ti,Al-β ≫ Ti-MCM-41, favoring the competition of hydrogen peroxide for Ti–O· and, consequently, its decomposition over alkane hydroxylation in a similar trend. Actually, only a few alkanes are oxidized with relatively good yields on TS-1, while their hydroxylation on large-pore zeolites is normally an exception. With hydrophilic Ti-MCM-41, alcohols are likely produced by a free radical chain oxidation of the alkane occurring in the external medium. The hydroxylation of aromatics competes in a similar way with solvent oxidation and hydrogen peroxide decomposition. It is interesting to notice, in this regard, that oxygen evolution is significantly smaller in the hydroxylation of phenol than in that of alkanes, a possible indication of the quenching effect of phenolic species on the generation and the reactions of $HO_2^·$ radicals.

18.12
Conclusions

Progress in the area of Ti-zeolites, since early papers published at the end of the 1980s, has been impressive. Synthesis, catalysis, characterization and modeling have been the subject of numerous studies. Most synthetic efforts concentrate on methods to obtain TS-1 by the use of cheaper template and silica sources. The lower grade TS-1 thus obtained is generally intended for the epoxidation of propene, in which the catalyst inventory is relatively high and a lower cost for it could compensate for moderate losses of selectivity. The search for large-pore Ti-zeolites is also the subject of many studies. Though their number in the patent literature is now of the order of several tens, only a few have been studied effectively and, moreover, these show little applicative potential. Currently, no new Ti-zeolite, sufficiently effective to at least supplement TS-1 in the oxidation of bulky substrates, is on the horizon. Combining the spaciousness of pores with a highly hydrophobic surface appears a task of maximum difficulty.

Characterization has confirmed, with plenty of data, the isomorphous substitution and tetrahedral structure of Ti, the reversible splitting of one Ti–OSi bond by water and methanol and the formation of Ti–OOH active species. Mechanistic proposals have been advanced on these and on chemical grounds, albeit limited for the most part to the epoxidation reaction. Modeling has in several cases provided valuable support in this regard. For further mechanistic insights, *in situ* studies might be helped by the recent availability of gaseous H_2O_2 in a pure state [175].

Catalytic studies have extended the applications of TS-1 from early aromatic hydroxylation to the oxidation of olefins, alkanes, alcohols, amines and thioethers, and to other minor reactions. A combination of three factors is the basis of its effectiveness: the high activity of Ti peroxy species, the organophilic properties of the surface and the size of the pores in the range of molecular dimensions. For the role of propene oxide in the chemical industry, most of the research interest has focused on epoxidation, leading eventually to the development of a new process. On the whole, three oxidation processes are now commercial – HPPO, ammoximation and phenol hydroxylation – with prospects for other applications, particularly in the event that cheaper hydrogen peroxide becomes available in the future.

References

1 Sheldon, R.A. and Kochi, J.K. (1981) *Metal-Catalyzed Oxidations of Organic Compounds*, Academic Press, New York.
2 Schirmann, J.P. and Delavarenne, S.Y. (1979) *Hydrogen Peroxide in Organic Chemistry*, Edition et Documentation Industrielle, Paris.
3 Venturello, C., Alneri, E. and Ricci, M. (1983) *Journal of Organic Chemistry*, **48**, 3831.
4 Perego, G., Bellussi, G., Corno, C., Taramasso, M., Buonomo, F. and Esposito, A. (1986) *New Developments in Zeolite Science Technology* (eds Y. Murakami, A. Iijima and J.W. Ward), Kodansha, Tokyo, p. 129.
5 Romano, U., Esposito, A., Maspero, F., Neri, C. and Clerici, M.G. (1990) *La Chimica e l'Industria*, **72**, 610.
6 Bellussi, G. and Rigutto, M.S. (1994) *Advanced Zeolite Science and Applications* (eds J.C. Jansen, M. Stöcker, H.G. Karge and J. Weitkamp), Studies in Surface Science and Catalysis, Vol. **85**, Elsevier, Amsterdam, p. 177.
7 Notari, B. (1996) *Advances in Catalysis* (eds D.D. Eley, W.O. Haag and B. Gates), Academic Press, New York, Vol. **41**, p. 253.
8 Perego, G., Millini, R. and Bellussi, G. (1998) *Synthesis and Characterization of Molecular Sieves Containing Transition Metals in the Framework* (eds H.G. Karge and J. Weitkamp), Molecular Sieves Science and Technology, Vol. **1**, Springer Verlag, Heidelberg, p. 187.
9 Ratnasamy, P., Srinivas, D. and Knözinger, H. (2004) *Advances in Catalysis* (eds B. Gates and H. Knözinger), Academic Press, New York, Vol. **48**, p. 1.
10 Taramasso, M., Perego, G. and Notari, B. (1983) US 4,410,501, to Snamprogetti S.p.A.
11 Millini, R., Previde Massara, E., Perego, G. and Millini, R. (1992) *Journal of Catalysis*, **137**, 497.
12 Wang, X., Guo, X. and Li, G. (2002) *Catalysis Today*, **74**, 65.
13 Bellussi, G., Carati, A., Clerici, M.G., Esposito, A., Millini, R. and Buonomo, F. (1989) Belgian Patent 1,001,038.
14 Reddy, J.S., Kumar, R. and Ratnasamy, P. (1990) *Applied Catalysis*, **58**, L1.
15 Blasco, T., Camblor, M.A., Corma, A. and Pérez-Pariente, J. (1993) *Journal of the American Chemical Society*, **115**, 11806.
16 van der Waal, J.C., Rigutto, M.S. and van Bekkum, H. (1994) *Journal of the Chemical Society D – Chemical Communications*, 1241.
17 Wu, P., Komatsu, T. and Yashima, T. (1966) *Journal of Physical Chemistry*, **100**, 10316.
18 Wu, P., Tatsumi, T., Komatsu, T. and Yashima, T. (2001) *Journal of Physical Chemistry B*, **105**, 2897.
19 Ciesla, U. and Schüth, F. (1999) *Microporous and Mesoporous Materials*, **27**, 131.
20 Bordiga, S., Damin, A., Bonino, F. and Lamberti, C. (2005) *Surface and Interfacial Organometallic Chemistry and Catalysis* (eds V. Copéret and B. Chaudret), Topics in Organometallic Chemistry, Vol. 16, Springer-Verlag, Heidelberg, p. 37.
21 Clerici, M.G. and Bellussi, G. (1989) European Patents 315,247 and 315,248.
22 Tatsumi, T., Nakamura, M., Negishi, S. and Tominaga, H. (1990) *Chemical Communications*, 476.

23 Huybrechts, D.R.C., De Bruycker, L. and Jacobs, P.A. (1990) *Nature*, **345**, 240.
24 Clerici, M.G. (1991) *Applied Catalysis*, **68**, 249.
25 Clerici, M.G. (2001) *Topics in Catalysis*, **15**, 257.
26 Halasz, I., Agarwal, M., Senderov, E. and Marcus, B. (2003) *Applied Catalysis A: General*, **241**, 167.
27 Spinacé, E.V., Pastore, H.O. and Schuchardt, U. (1995) *Journal of Catalysis*, **157**, 631.
28 Tao, J., Tang, D., Li, Q., Yu, Z. and Min, E. (2001) *Journal of Natural Gas Chemistry*, **10**, 295.
29 Saxton, R.J. (1999) *Topics in Catalysis*, **9**, 43.
30 Khouw, C.B., Dartt, C.B., Labinger, J.A. and Davis, M.E. (1994) *Journal of Catalysis*, **149**, 195.
31 Costas, M., Chen, K. and Que, L. (2000) *Coordination Chemistry Reviews*, **200-2**, 517.
32 Aresta, M., Tommasi, I., Quaranta, E., Fragale, C., Mascetti, J., Tranquille, M., Galan, F. and Fouassier, M. (1996) *Inorganic Chemistry*, **35**, 4254.
33 Groves, J.T. and van der Puy, M. (1974) *Journal of the American Chemical Society*, **96**, 5274.
34 Groves, J.T. and McClusky, G.A. (1976) *Journal of the American Chemical Society*, **98**, 859.
35 Corma, A., Camblor, M.A., Esteve, P., Martínez, A. and Pérez-Pariente, J.P. (1994) *Journal of Catalysis*, **145**, 151.
36 Jappar, N., Xia, Q. and Tatsumi, T. (1998) *Journal of Catalysis*, **180**, 13.
37 Tatsumi, T. (2000) *Research on Chemical Intermediates*, **26**, 7.
38 Varagnat, J. (1976) *Industrial & Engineering Chemistry Product Research and Development*, **15**, 212.
39 Maggioni, P. and Minisci, F. (1977) *La Chimica e l'Industria*, **59**, 239.
40 Tendulkar, S.B., Tambe, S.S., Chandra, I., Rao, P.V., Naik, R.V. and Kulkarni, B.D. (1998) *Industrial and Engineering Chemistry Research*, **37**, 2081.
41 Liu, H., Lua, G., Guo, Y.L., Guo, Y. and Wang, J. (2004) *Catalysis Today*, **93-5**, 353.
42 Liu, H., Lua, G., Guo, Y.L., Guo, Y. and Wang, J. (2005) *Chemical Engineering Journal*, **108**, 187.
43 Martens, J.A., Buskens, P., Jacobs, P.A., van der Pol, A., van Hooff, J.H.C., Ferrini, C., Kouwenhoven, H.W., Kooyman, P.J. and van Bekkum, H. (1993) *Applied Catalysis A: General*, **99**, 71.
44 Kraushaar-Czarnetzki, B. and van Hooff, J.H.C. (1989) *Catalysis Letters*, **2**, 43.
45 van der Pol, A.J.H.P., Verduyn, A.J. and van Hooff, J.H.C. (1992) *Applied Catalysis A: General*, **92**, 113.
46 Tuel, A. and Ben Taarit, Y. (1993) *Applied Catalysis A: General*, **102**, 69.
47 Wilkenhöner, U., Langhendries, G., van Laar, F., Baron, G.V., Gammon, D.W., Jacobs, P.A. and van Steen, E. (2001) *Journal of Catalysis*, **203**, 201.
48 Wilkenhöner, U., Duncan, W.L., Möller, K.P. and van Steen, E. (2004) *Microporous and Mesoporous Materials*, **69**, 181.
49 Tuel, A., Moussa-Khouzami, S., Ben Taarit, Y. and Naccache, C. (1991) *Applied Catalysis A: General*, **68**, 45.
50 Camblor, M.A., Constantini, M., Corma, A., Esteve, P., Gilbert, L., Martinez, A. and Valencia, S. (1995) *Applied Catalysis A: General*, **133**, L185.
51 Thangaray, A., Kumar, R. and Ratnasamy, P. (1990) *Applied Catalysis*, **57**, L1.
52 Tatsumi, T., Yuasa, K. and Tominaga, H. (1992) *Journal of the Chemical Society D – Chemical Communications*, 1446.
53 Bianchi, D., Balducci, L., Bortolo, R., D'Aloisio, R., Ricci, M., Tassinari, R. and Ungarelli, R. (2003) *Angewandte Chemie–International Edition in English*, **42**, 4937.
54 Bianchi, D., Balducci, L., Bortolo, R., D'Aloisio, R., Ricci, M., Spanò, G., Tassinari, R., Tonini, C. and Ungarelli, R. (2007) *Advanced Synthesis Catalysis*, **349**, 979.
55 Bianchi, D., Bortolo, R., Buzzoni, R., Cesana, A., Dalloro, L. and D'Aloisio, R. (2002) European Patent 1,424,320.
56 Bhaumik, A., Mukherjee, P. and Kumar, R. (1998) *Journal of Catalysis*, **178**, 101.
57 Kim, G.J., Cho, B.R. and Kim, J.H. (1993) *Catalysis Letters*, **22**, 259.

58 Bhelhekar, A.A., Das, T.K., Chaudhari, K., Hegde, S.G. and Chandwadkar, A.J. (1998) *Recent Advances in Basic and Applied Catalysis* (eds T.S.R. Prasada Rao and G. Murali Dhar), Studies in Surface Science and Catalysis, Vol. **113**, Elsevier, Amsterdam, p. 195.

59 Wu, P., Komatsu, T. and Yashima, T. (1997) *Progress in Zeolites and Microporous Materials* (eds H. Chon, S.-Ihm and Y.S. Uh), Studies in Surface Science and Catalysis, vol. **105**, Elsevier, Amsterdam, p. 663.

60 Clerici, M.G. (2001) *Fine Chemicals through Heterogeneous Catalysis* (eds R.A. Sheldon and H. van Bekkum), Wiley-VCH, Weinheim, p. 538.

61 Clerici, M.G., Bellussi, G. and Romano, U. (1991) *Journal of Catalysis*, **129**, 159.

62 Clerici, M.G. and Ingallina, P. (1993) *Journal of Catalysis*, **140**, 71.

63 Liu, X., Wang, X., Guo, X. and Li, G. (2004) *Catalysis Today*, **93**, 505.

64 Hulea, V., Dumitriu, E., Patcas, F., Ropot, R., Graffin, P. and Moreau, P. (1998) *Applied Catalysis A: General*, **170**, 169.

65 Swern, D. (1947) *Journal of the American Chemical Society*, **69**, 1692.

66 Bonoldi, L., Busetto, C., Cangin, A., Marra, G., Ranghino, G., Salvalaggio, M., Spanò, G. and Giamello, E. (2002) *Spectrochimica Acta, Part A*, **58**, 1143.

67 Sheldon, R.A. and van Doorn, J.A. (1973) *Journal of Catalysis*, **31**, 427.

68 Koelewijn, P. (1972) *Recueil des Travaux Chimiques des Pays-Bas*, **91**, 759.

69 Sheldon, R.A. and van Doorn, J.A. (1973) *Recueil des Travaux Chimiques des Pays-Bas*, **92**, 253.

70 Huybrechts, D.R.C., Buskens, P.L. and Jacobs, P.A. (1992) *Journal of Molecular Catalysis*, **71**, 129.

71 Notari, B. (1993) *Catalysis Today*, **16**, 163.

72 Karlsen, E. and Schöffel, K. (1996) *Catalysis Today*, **32**, 107.

73 Sinclair, P.E. and Catlow, C.R.A. (1999) *The Journal of Physical Chemistry B*, **103**, 1084.

74 Barker, C.M., Kaltsoyannis, N. and Catlow, C.R.A. (2001) *Zeolites and Mesoporous Materials at the Dawn of the 21st Century* (eds A. Galarneau, F. Di Renzo, F. Fajula and J. Vedrine), Studies in Surface Science and Catalysis, Vol. **135**, Elsevier, Amsterdam, p. 260.

75 Neurock, M. and Manzer, L.E. (1996) *Chemical Communications*, 1133.

76 Lin, W. and Frei, H. (2002) *Journal of the American Chemical Society*, **124**, 9292.

77 Corma, A., Esteve, P. and Martínez, A. (1996) *Journal of Catalysis*, **161**, 11.

78 van der Waal, J.C. and van Bekkum, H. (1997) *Journal of Molecular Catalysis A–Chemical*, **124**, 137.

79 Wu, P., Tatsumi, T., Komatsu, T. and Yashima, T. (2001) *Journal of Catalysis*, **202**, 245.

80 Wu, P., Nuntasri, D., Ruan, J., Liu, Y., He, M., Fan, W., Terasaki, O. and Tatsumi, T. (2004) *Journal of Physical Chemistry B*, **108**, 19126.

81 van der Waal, J.C., Rigutto, M.S. and van Bekkum, H. (1998) *Applied Catalysis A: General*, **167**, 331.

82 Camblor, M.A., Corma, A., Esteve, P., Martínez, A. and Valencia, S. (1997) *Chemical Communications*, 795.

83 Corma, A., Esteve, P., Martínez, A. and Valencia, S. (1995) *Journal of Catalysis*, **152**, 18.

84 Sato, T., Dakka, J. and Sheldon, R.A. (1994) *Zeolites and Related Microporous Materials: State of the Art 1994* (eds J. Weitkamp, H.G. Karge, H. Pfeifer and W. Hölderich), Studies in Surface Science and Catalysis, Vol. **84**, Elsevier, Amsterdam, p. 1853.

85 Wu, P. and Tatsumi, T. (2002) *Journal of Physical Chemistry B*, **106**, 748.

86 Wu, P., Nuntasri, D., Liu, Y., Wu, H., Jiang, Y., Fan, W., He, M. and Tatsumi, T. (2006) *Catalysis Today*, **117**, 199.

87 Carati, A., Flego, C., Previde Massara, E., Millini, R., Carluccio, L., Parker, W.O. Jr and Bellussi, G. (1999) *Microporous and Mesoporous Materials*, **30**, 137.

88 Davies, L.J., McMorn, P., Bethell, D., Bulman Page, P.C., King, F., Hancock, F.E. and Hutchings, G.J. (2001) *Journal of Catalysis*, **198**, 319.

89 Chen, L.Y., Chuah, C.K. and Jaenicke, S. (1998) *Catalysis Letters*, **50**, 107.

90 Tatsumi, T., Yako, M., Nakamura, M. and Yuhara, Y. (1993) *Journal of Molecular Catalysis*, **78**, L41.

91 Kumar, R., Pais, G.C.G., Pandey, B. and Kumar, P. (1995) *Journal of the Chemical Society D – Chemical Communications*, 1315.
92 Hutchings, G.J. and Lee, D.F. (1994) *Journal of the Chemical Society D – Chemical Communications*, 1095.
93 Schofield, L.J., Kerton, O.J., McMora, P., Bethell, D., Ellwood, S. and Hutchings, G.J. (2002) *Journal of the Chemical Society – Perkin Transactions 2*, 1475.
94 Wróblewska, A. (2006) *Applied Catalysis A: General*, **309**, 192.
95 Wróblewska, A., Rzepkowska, M. and Milchert, E. (2005) *Applied Catalysis A: General*, **294**, 244.
96 Wu, P. and Tatsumi, T. (2003) *Journal of Catalysis*, **214**, 317.
97 Adam, W., Corma, A., Reddy, T.I. and Renz, M. (1997) *Journal of Organic Chemistry.*, **62**, 3631.
98 Gao, H., Lu, G., Suo, J. and Li, S. (1996) *Applied Catalysis A: General*, **138**, 27.
99 Wang, Q., Mi, Z., Wang, Y. and Wang, L. (2005) *Journal of Molecular Catalysis A*, **229**, 71.
100 Wu, P., Liu, Y., He, M. and Tatsumi, T. (2004) *Journal of Catalysis*, **228**, 183.
101 Sasidharan, M., Wu, P. and Tatsumi, T. (2002) *Journal of Catalysis*, **205**, 332.
102 Payne, G.B. and Williams, P.H. (1959) *Journal of Organic Chemistry*, **24**, 54.
103 Bellussi, G., Carati, A., Clerici, M.G. and Esposito, E. (1991) *Preparation of Catalysts V* (eds G. Poncelet, P.A. Jacobs, P. Grange and B. Delmon), Studies in Surface Science and Catalysis, Vol. **63**, Elsevier, Amsterdam, p. 421.
104 Smith, K. and Al-Shamali, M. (1999) *Proceedings of the 12th International Zeolite Conference, July 5–10, 1998, Baltimore* (eds E. Bei, M.M.J. Treacy, B. K. Marcus, M.E. Bisher and J.B. Higgins), Materials Research Society, Warrendale, PA, p. 1129.
105 Zhuang, J., Yang, G., Ma, D., Lan, X., Liu, X., Han, X., Bao, X. and Mueller, U. (2004) *Angewandte Chemie – International Edition in English*, **43**, 6377.
106 Derouane, E.G., Hutchings, G.J., Mbafor, W.F. and Roberts, S.M. (1998) *New Journal of Chemistry*, 797.
107 Bhaumik, A. and Tatsumi, T. (1999) *Journal of Catalysis*, **182**, 349.
108 Sasidharan, M., Wu, P. and Tatsumi, T. (2002) *Journal of Catalysis*, **209**, 260.
109 van der Pol, A.J.H. and van Hoof, J.H.C. (1993) *Applied Catalysis A: General*, **106**, 97.
110 Maspero, F. and Romano, U. (1994) *Journal of Catalysis*, **146**, 476.
111 Corma, A., Esteve, P. and Martínez, A. (1996) *Applied Catalysis A: General*, **143**, 87.
112 Hayashi, H., Kikawa, K., Murai, Y., Shigemoto, N., Sugiyama, S. and Kawashioro, K. (1996) *Catalysis Letters*, **36**, 99.
113 Jacobson, S.E., Muccigrosso, D.A. and Mares, F. (1979) *Journal of Organic Chemistry*, **44**, 921.
114 Zhao, R., Ding, Y., Peng, Z., Wang, X. and Suo, J. (1993) *Catalysis Letters*, **87**, 81.
115 Sheldon, R.A. and Dakka, J. (1993) *Erdöl Erdgas Kohle*, **109**, 520.
116 Sasidharan, M., Suresh, S. and Sudalai, A. (1995) *Tetrahedron Letters*, **36**, 9071.
117 Bhaumik, A., Kumar, P. and Kumar, R. (1996) *Catalysis Letters*, **40**, 47.
118 Corma, A., Nemeth, L.T., Renz, M. and Valencia, S. (2001) *Nature*, **412**, 423.
119 Roffia, P., Leofanti, G., Cesana, A., Mantegazza, M., Padovan, M., Petrini, G., Tonti, S. and Gervasutti, P. (1990) *New Developments in Selective Oxidations* (eds G. Centi and F. Trifirò), Studies in Surface Science and Catalysis, Vol. **55**, Elesevier, Amsterdam, p. 43.
120 Petrini, G., Cesana, A., De Alberti, G., Genoni, F., Leofanti, G., Padovan, M., Paparatto, G. and Roffia, P. (1991) *Catalysts Deactivation 1991* (eds C.H. Bartholomew and J.B. Butt), Studies in Surface Science and Catalysis, Vol. **68**, Elsevier, Amsterdam, p. 761.
121 Zecchina, A., Spoto, G., Bordiga, S., Geobaldo, F., Petrini, G., Leofanti, G., Padovan, M., Mantegazza, M. and Roffia, P. (1993) *New Frontiers in Catalysis* (eds L. Guczi, F. Solymosi and P. Tétényi), Studies in Surface Science and Catalysis, Vol. **75**, Elsevier, Amsterdam, p. 719.
122 Tatsumi, T. and Jappar, N. (1996) *Journal of Catalysis*, **161**, 560.

123 Zhang, Y., Wang, Y., Bu, Y., Wang, L., Mi, Z., Wu, W., Min, E., Fu, S. and Zhu, Z. (2006) *Reaction Kinetics and Catalysis Letters*, **87**, 25.

124 Sudhakar Reddy, J., Sivasanker, S. and Ratnasamy, P. (1991) *Journal of Molecular Catalysis*, **69**, 383.

125 Birke, P., Kraak, P., Pester, R., Schödel, R. and Vogt, F. (1994) *Zeolites and Microporous Crystals* (eds T. Hattori and T. Yashima), Studies in Surface Science and Catalysis, Vol. **83**, Elsevier, Amsterdam, p. 425.

126 Le Bars, J., Dakka, J. and Sheldon, R.A. (1996) *Applied Catalysis A: General*, **136**, 69.

127 Wu, P., Komatsu, T. and Yashima, T. (1997) *Journal of Catalysis*, **168**, 400.

128 Song, F., Liu, Y., Wu, H., He, M., Wu, P. and Tatsumi, T. (2006) *Journal of Catalysis*, **237**, 359.

129 Mantegazza, M., Leofanti, G., Petrini, G., Padovan, M., Zecchina, A. and Bordiga, S. (1994) *New Development in Selective Oxidation II* (eds V. Cortés-Corberán and S. Vic Bellon), Studies in Surface Science and Catalysis, Vol. **82**, Elsevier, Amsterdam, p. 541.

130 Zecchina, A., Bordiga, S., Lamberti, C., Ricchiardi, G., Scarano, D., Petrini, G., Leofanti, G. and Mantegazza, M. (1996) *Catalysis Today*, **32**, 97.

131 Kul'kova, N., Kotova, V.G., Kvyathovskaya, M.Yu. and Murzin, D.Yu. (1997) *Chemical Engineering and Technology*, **20**, 43.

132 Sudhakar Reddy, J. and Jacobs, P.A. (1993) *Journal of the Chemical Society – Perkin Transactions 1*, 2665.

133 Reni, J., Ravindranathan, T. and Sudalai, A. (1995) *Tetrahedron Letters*, **36**, 1903.

134 Reni, J., Sudalai, A. and Ravindranathan, T. (1995) *Synlett*, 1177.

135 Sudhakar Reddy, J. and Jacobs, P.A. (1996) *Catalysis Letters*, **37**, 213.

136 Jorda, E., Tuel, A., Teissier, R. and Kervenal, J. (1999) *Proceedings of the 12th International Zeolite Conference, July 5–10, 1998, Baltimore* (eds E. Bei, M.M.J. Treacy, B.K. Marcus, M.E. Bisher and J.B. Higgins), Materials Research Society, Warrendale, PA, p. 1269.

137 Sonawane, H.R., Pol, A.V., Moghe, P.P., Biswas, S.S. and Sudalai, A. (1994) *Journal of the Chemical Society D – Chemical Communications*, 1215.

138 Gontier, S. and Tuel, A. (1994) *Applied Catalysis A: General*, **118**, 173.

139 (a) Gontier, S. and Tuel, A. (1995) *Journal of Catalysis*, **157**, 124; (b) Gontier, S. and Tuel, A. (1995) *Catalysis Letters*, **31**, 103.

140 Selvam, T. and Ramaswamy, A.V. (1995) *Catalysis Letters*, **31**, 103.

141 Reni, J., Sudalai, A. and Ravindranathan, T. (1993) *Journal of the Chemical Society D – Chemical Communications*, 1553.

142 Reni, J., Sudalai, A. and Ravindranathan, T. (1994) *Tetrahedron Letters*, **35**, 5493.

143 Suresh, S., Joseph, R., Jayachandran, B., Pol, A.V., Vinod, M.P. and Sudalai, A. (1995) *Tetrahedron*, **41**, 11305.

144 Reddy, R.S., Sudhakar Reddy, J., Kumar, R. and Kumar, P. (1993) *Journal of the Chemical Society D – Chemical Communications*, 84.

145 Hulea, V., Moreau, P. and Di Renzo, F. (1996) *Journal of Molecular Catalysis A*, **111**, 325.

146 Hulea, V. and Moreau, P. (1996) *Journal of Molecular Catalysis A*, **113**, 499.

147 Moreau, P., Hulea, V., Gomez, S., Brunel, D. and Di Renzo, F. (1997) *Applied Catalysis A: General*, **155**, 253.

148 Robinson, D.J., Davies, L., McGuire, N., Lee, D.F., McMorn, P., Willock, D.J., Watson, G.W., Bulman Page, P.C., Bethell, D. and Hutchings, G.J. (2000) *Physical Chemistry Chemical Physics*, **2**, 1523.

149 Hulea, V., Fajula, F. and Bousquet, J. (2001) *Journal of Catalysis*, **198**, 179.

150 Romano, U. (2001) *La Chimica e l'Industria*, **83**, 30.

151 Tullo, A.H. and Short, P.L. (2006) *Chemical and Engineering News*, **84** (*41*), 22.

152 Chemical Market Reporter (2004) 20–27 December.

153 Campos Martin, J.M., Blanco Brieva, G. and Fierro, J.L.G. (2006) *Angewandte Chemie – International Edition in English*, **45**, 6962.

154 Clerici, M.G. and Ingallina, P. (1996) *Green Chemistry, Designing Chemistry for the Environment* (eds P.T. Anastas and T.C. Williamson), ACS Symposium

Series, Vol. **626**, American Chemical Society, Washington, DC, p. 58.
155 Clerici, M.G. and Ingallina, P. (1998) *Catalysis Today*, **41**, 351.
156 Meiers, R., Dingerdissen, U. and Hölderich, W.F. (1998) *Journal of Catalysis*, **176**, 376.
157 Jenzer, G., Mallat, T., Maciejewski, M., Eigenmann, F. and Baiker, A. (2001) *Applied Catalysis A: General*, **208**, 125.
158 Danciu, T., Beckman, E.J., Hancu, D., Cochran, R.N., Grey, R., Hajnik, D.M. and Jewson, J. (2003) *Angewandte Chemie – International Edition in English*, **42**, 1140.
159 Zimmer, A., Mönter, D. and Reschetilowski, W. (2003) *Journal of Applied Electrochemistry*, **33**, 933.
160 Wang, R., Guo, X., Wang, X., Hao, J., Li, G. and Xiu, J. (2004) *Applied Catalysis A: General*, **261**, 7.
161 Wang, C., Wang, B., Meng, X. and Mi, Z. (2002) *Catalysis Today*, **74**, 15.
162 Wang, Q., Mi, Z., Wang, Y. and Wang, L. (2005) *Journal of Molecular Catalysis A*, **229**, 71.
163 Liang, X., Mi, Z., Wu, Y., Wang, L. and Xing, E. (2003) *Reaction Kinetics and Catalysis Letters*, **80**, 207.
164 Roffia, P., Paparatto, G., Cesana, A. and Tauszik, G. (1990) US Patent 4,894,478.
165 Clerici, M.G. and Ingallina, P. (1993) US Patent 5,252,758.
166 Liu, T., Meng, X., Wang, Y., Liang, X., Mi, Z., Qi, X., Li, S., Wu, W., Min, E. and Fu, S. (2004) *Industrial and Engineering Chemistry Research*, **43**, 166.
167 Langhendries, G., De Vos, D.E., Baron, G.V. and Jacobs, P.A. (1999) *Journal of Catalysis*, **187**, 453.
168 Derouane, E.G. (1998) *Journal of Molecular Catalysis A*, **134**, 29.
169 Clerici, M.G. (2006) *Oil Gas European Magazine*, **32**, 77.
170 Boccuti, M.R., Rao, K.M., Zecchina, A., Leofanti, G. and Petrini, G. (1989) *Structures and Reactivity of Surfaces* (eds C. Morterra, A. Zecchina and G. Costa), Studies in Surface Science and Catalysis, Vol. 48, Elsevier, Amsterdam, p. 133.
171 Bellussi, G., Carati, A., Clerici, M.G., Maddinelli, G. and Millini, R. (1992) *Journal of Catalysis*, **133**, 220.
172 Wu, P., Komatsu, T. and Yashima, T. (1998) *Journal of Physical Chemistry B*, **102**, 9297.
173 Merkx, M., Kopp, D.A., Sazinsky, M.H., Blazyk, J.L., Müller, J. and Lippard, S.J. (2001) *Angewandte Chemie – International Edition in English*, **40**, 2782.
174 Bonchio, M., Conte, V., Di Furia, F. and Modena, G. (1989) *Journal of Organic Chemistry*, **54**, 4368.
175 Prestipino, C., Bonino, F., Usseglio, S., Damin, A., Tasso, A., Clerici, M.G., Bordiga, S., D'Acapito, F., Zecchina, A. and Lamberti, C. (2004) *ChemPhysChem*, **5**, 1799.
176 Clerici, M.G., Ingallina, P. and Millini, R. (1993) *Proceedings of the 9th Zeolite Conference, Montreal 1992* (eds R. von Ballmoos, J.B. Higgins and M.M.J. Treacy), Butterworth-Heinemann, Boston, p. 363.

19
Oxide Materials in Photocatalytic Processes
Richard P.K. Wells

19.1
Introduction

In 1972, Fujishima and Honda reported the photocatalytic splitting of water using TiO_2 electrodes [1]. This event marked the beginning of a new era in the study of photocatalysis. Since then, considerable research effort has been invested in understanding the physicochemical processes occurring, and in the continued development of active heterogeneous photocatalysts based on metal oxide semiconductors such as TiO_2 [2]. The breadth of the subject, in terms of both the materials used and their applications, is immense, and it is not the aim of this chapter to deal with all aspects of the science; rather it is to present the basic principles of the process and to review progress in the modification of metal oxides in order to enhance their photoreactivity in the visible part of the spectrum. There are many excellent reviews in the literature dealing with the basic concepts of the photocatalytic process and the reader is referred in particular to those by Hoffmann and coworkers [3], Mills and coworkers [4], and Kamat [5]. By far the majority of research has focused on TiO_2 as the material of choice and much effort has gone into enhancing the ability of visible light to activate the material rather than merely the ultraviolet component of the spectrum [6, 7]. In terms of application, environmental remediation has been the target of most research, owing to the ability of TiO_2 to function as a slurry in aqueous media, the annual rate of publication in this area being of the order of four hundred reports [8, 9]. More recent uses include the abatement of contamination within gaseous environments, with particular emphasis on the removal of chlorinated and sulfurous compounds in air [10], and the incorporation of photocatalytically mediated steps in synthetic organic chemistry [11, 12].

19.2
Basic Principles of Heterogeneous Photocatalysis

Heterogeneous photocatalysis was defined by Palmisano and Sclafani [13] as "a catalytic process during which one or more reaction steps occur by means of electron–hole pairs photogenerated on the surface of semiconducting materials illuminated by light of a certain energy." The term photocatalysis in general, refers to any chemical process in which the external energy source is derived from radiation in the ultraviolet-visible range [14]. The basic mechanisms of heterogeneous photocatalysis have been investigated by many researchers [3, 9, 15–17] and can be represented schematically by the band gap model (Figure 19.1).

The band gap is characteristic of the electronic structure of the semiconductor and is defined as the energy difference (ΔE_g) between the valence and conduction bands; that is the highest energy band with all electronic levels occupied and the lowest energy band without electrons [18]. Table 19.1 shows some typical metal oxide and chalcogenide photocatalysts, together with their respective band gaps [9].

When energy greater than ΔE_g is applied to the semiconductor surface, valence band electrons are promoted to the conduction band, creating electron–hole pairs (**1**). Migration of the pairs to the semiconductor surface (**2**) allows the occurrence of redox reactions with adsorbates with suitable redox potentials (**3**). Thermodynamically, oxidation will occur if the redox potential of the valence band is more positive than that of the adsorbates. Similarly, conduction band electrons can reduce adsorbed species if their redox potential is more negative than that of the adsorbates. Clearly, the most likely outcome of the formation of electron–hole pairs is their simple recombination, with subsequent release of thermal energy and/or light (**4**). Recombination proceeding within the same timescale as the redox

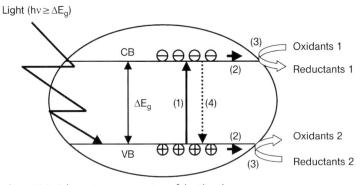

Figure 19.1 Schematic representation of the "band gap model". (1) Photoinduced electron–hole creation; (2) charge migration to the surface; (3) redox reactions; (4) recombination. VB and CB represent valence and conduction band respectively.

Table 19.1 Band gap energy of various photocatalysts.

Photocatalyst	Band gap energy (eV)	Photocatalyst	Band gap energy (eV)
Si	1.1	ZnO	3.2
TiO_2 (rutile)	3.0	TiO_2 (anatase)	3.2
WO_3	2.7	CdS	2.4
ZnS	3.7	$SrTiO_3$	3.4
SnO_2	3.5	WSe_2	1.2
Fe_2O_3	2.2	α-Fe_2O_3	3.1

processes being promoted reduces photocatalytic activity considerably. Therefore, for electron–hole pairs, whose lifetime can normally be considered to be of the order of femtoseconds, separation is a vital parameter when determining activity [19]. Before any reaction or recombination step can take place the charge carriers must be trapped, and defects play an important role in this process. Defects can be either intrinsic, for example oxygen vacancies in nanostructured reducible oxides, or extrinsic, such as dopants or impurities. Minimizing the likelihood of recombination has led to the preparation of nanostructured oxide materials [14] together with the development of materials capable of enhanced photocatalytic activity through spatial structuring [20] and encapsulation within host materials such as zeolites [21].

19.3
Traditional Photocatalysts

TiO_2 is by far the most widely investigated metal oxide, mainly because of its chemical stability, non-toxicity, and well-positioned valence and conduction bands. Two crystalline phases have been investigated, rutile and anatase. As shown in Figure 19.2, the valence band redox potential of both forms are more positive than that of the •OH/⁻OH redox couple, resulting in the oxidation of adsorbed water and hydroxyl groups to highly reactive hydroxyl radicals on both irradiated forms of the oxide. This reaction is the basis behind the immense research effort into the oxidation of organic contaminants in aqueous waste streams [3, 9, 22–24]. Of the two phases of TiO_2, anatase is widely regarded as being the most photocatalytically active [25, 26]; this is due to a variety of factors, the main one being the more negative redox potential of the anatase conduction band, making it a more efficient reducing agent than rutile. For example, molecular oxygen can be reduced to superoxide radicals by illuminated anatase, but not by illuminated rutile. This also results in a greater amount of electron–hole recombination and hence lowers the efficiency in the latter material.

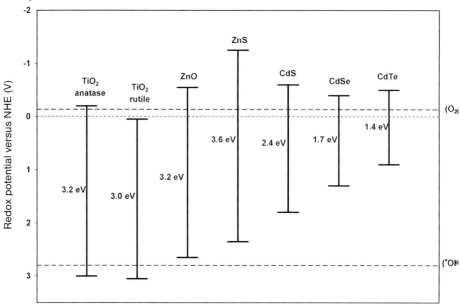

Figure 19.2 Band gaps, together with valence and conduction band edges of common semiconductors, placed alongside the standard redox potentials (versus normal hydrogen electrode (NHE)) of the O_2/O_2^- and •OH/$^-$OH redox couple.

The origin of the differing band gaps between the various polymorphs of TiO_2 is in their structure. Anatase is the least thermodynamically stable, although from energy calculations this is the predominant phase when the particle size is small (<10 nm) [27]. The crystalline structure is made up of TiO_6 octahedral chains differing by the distortion of each octahedron and the assembly pattern of octahedral chains [14]. The Ti–Ti bond lengths in anatase are greater than those in rutile, whereas Ti–O distances are shorter [28]. These structural differences lead to differing mass densities and hence differing electronic structures of the bands. Anatase is 9% less dense than rutile, resulting in a greater degree of delocalization of the Ti 3d states, and therefore a narrower 3d band. Also the O 2p–Ti 3d hybridization of the two structures differs, with rutile exhibiting a greater degree of covalent mixing and hence anatase exhibiting a valence and conduction band with more pronounced O 2p–Ti 3d character [29]. These structural and electronic differences are responsible for the different reactivities of the materials.

Although anatase presents the highest photoactivities reported, a commercial material, Degussa P25, is commonly used as a reference catalyst. This product is a mixture of anatase and rutile in a ratio of 80:20 and displays an unexpectedly high photoactivity, the source of which is still a matter of some debate. Bickley and coworkers reported a model based upon there being a small amount of amorphous TiO_2 present in the mix [30]. The coexistence of the phases leads to a synergistic effect that improves the photoreactivity of the catalyst. A similar junction effect has been reported for an anatase–brookite mixture, which presents a high activity for methanol photo-oxidation [31].

The principal use of TiO_2 as a photocatalyst has been in the degradation of organic material in either aqueous or gaseous waste streams. Two oxidation reactions can take place at the TiO_2 valence band. Firstly, adsorbed organic material (RH) can be directly oxidized by TiO_2 (h⁺) producing cationic radicals RH–⁺ (1). Secondly, the oxidation of water and/or hydroxyl groups can give rise to the formation of highly reactive hydroxyl radicals, •OH (2-3) that can subsequently initiate further compound oxidation.

$$TiO_2(h^+) + RH_{ads} \rightarrow TiO_2 + RH^{\cdot+}_{ads} \rightleftharpoons TiO_2 + R^{\cdot}_{ads} + H^+ \quad (19.1)$$

$$TiO_2(h^+) + H_2O_{ads} \rightarrow TiO_2 + H_2O^{\cdot+}_{ads} \rightleftharpoons TiO_2 + {}^{\bullet}OH_{ads} + H^+ \quad (19.2)$$

$$TiO_2(h^+) + {}^-OH_{ads} \rightarrow TiO_2 + {}^{\bullet}OH_{ads} \quad (19.3)$$

The main reaction occurring at the TiO_2 conduction band is the reduction of adsorbed molecular oxygen, resulting in the formation of superoxide radicals ($O_2^{\cdot-}$). Such superoxide radicals, can give rise to the formation of additional •OH radicals through a series of reactions (Equations 19.5 to 19.7).

$$TiO_2(e^-) + O_{2,ads} \rightarrow TiO_2 + O_2^{\cdot-}{}_{ads} \quad (19.4)$$

$$O_2^{\cdot-}{}_{ads} + H^+ \rightleftharpoons HO_2^{\cdot}{}_{ads} \quad (19.5)$$

$$2\,HO_2^{\cdot}{}_{ads} \rightleftharpoons H_2O_{2,ads} + O_{2,ads} \quad (19.6)$$

$$TiO_2(e^-) + H_2O_{2,ads} \rightarrow TiO_2 + OH_{ads} + {}^-OH_{ads} \quad (19.7)$$

However, photocatalytic reduction of organic compounds is generally less important than their oxidation because the reduction potential of a $TiO_2(e^-)$ is lower than the oxidation potential of a $TiO_2(h^+)$. Also, most of the reducible substrates cannot compete kinetically with oxygen as an electron scavenger [15].

Of the other potential photocatalysts listed in Table 19.1, ZnO has limited use as it is relatively unstable when illuminated in aqueous solution, particularly at low pH, tending to form Zn^{2+} (aq) after oxidation by the valence band positive holes. However, at higher pH it has been shown to be as active as TiO_2 for the degradation of phenol [32]. Also, for applications in the gas phase, such as the degradation of ethanal, ZnO thin films have been shown to be highly active [33]. WO_3, although active in a portion of the visible spectrum is considered in general to be less active than TiO_2 [34]. CeO_2 presents many of the characteristics of a suitable photocatalyst, having a band gap of 3.1 eV [35] and being relatively inexpensive and non-toxic; it is also highly absorbing in the UV spectrum and has been shown to have significant activity for the photocatalytic oxidation of water [36].

19.4
Improving Photocatalytic Activity

Because the band gap of TiO_2 is 3.2 eV it is defined as a "wide band gap semiconductor", with onset of the optical absorption band at about 350 nm [20]. Hence, only 5% of the solar energy can be absorbed by TiO_2. For this reason, considerable research effort has been invested into increasing the photocatalytic activity of TiO_2, particularly by enhancing its ability to be activated by visible light. In general, two approaches have been taken. One is to modify the band gap by addition of a secondary component, such as the adsorption of an organic dye or small amounts of another, usually metallic element. Alternatively, the basic concepts of physics can be used, and by changing the preparation methods producing photocatalytic particles of such dimensions that their band gaps are altered by quantum effects. Both methods have been shown to have a considerable effect on the reactivity, as outlined in the following section.

19.4.1
Visible Light Sensitization by Adsorption of Organic and Inorganic Dyes

Considerable research effort has been invested in the combination of photocatalysts with organic and inorganic dyes such as Erythrosin B [37] and *cis*-di(thiocyanato)-N,N-bis(2,2′-bipyridyl dicarboxylate)Ru(II) [38]. A huge number of potential dye-sensitizers have been reported in the literature, but the ones with optimum performance appear to be transition metal complexes derived from polypyridines [Ru(II)], porphyrins, or phthalocyanines [Zn(II), Mg(II) or Al(III)] as ligands [39, 40]. The principle behind these systems is that the dyes absorb the visible radiation and form excited states. From these states, electrons are injected into the conduction band of the semiconductor oxide, thereby inducing a visible response in the oxide. For effective function, the adsorbed dye needs to have an excited state located above the bottom of the semiconductor band [41] and to have a strong interaction with the oxide such that a fast and efficient electron junction is produced. Assuming these considerations are met the results of such sensitization are (i) an increased efficiency of the excitation process, and (ii) an expansion of the absorption spectrum of the semiconductor via excitation through the sensitizer. However, such dyes are complex molecules and are not a low-cost option. This, together with the fact that their thermal and/or photochemical stability is poor, can lead to poor lifetimes for photocatalysts prepared in this manner [42]. However, if oxygen is excluded from the system and the oxidation state of the dye is quenched with an appropriate electrolyte, such a strategy has been demonstrated to be successful [43].

19.4.2
Visible Light Sensitization by Anion Doping

More recently, visible light sensitization has been shown to be possible by doping with small quantities of C [44–47], N [48–50], F [51, 52], S [53, 54], and to a lesser extent with B [55, 56], Cl and Br [57], P [58], and I [59]. The work by Asahi and coworkers [48] was the first to illustrate the effect that nitrogen doping has on the photocatalytic properties of TiO_2. Nitrogen-doped TiO_2 has since been shown to catalyze the visible-light-initiated oxidation of organic substrates, and extensive investigations have been conducted in order to characterize the photoelectrochemical properties of $TiO_{2-x}N_x$ with the goal of developing materials suitable for the promotion of the water splitting reaction. Photoelectrochemical characterization, together with theoretical calculations of the $TiO_{2-x}N_x$ materials indicate that the addition of anions introduces states above the valence band of TiO_2, owing to the mixing of the O 2p and N 2p states [60]. However, how much of the effect of anion doping is down to the narrowing of the band gap as predicted by Batzill and coworkers [60] is a matter of some debate. Such doping of TiO_2 not only changes the electronic properties of the oxide but also can induce the formation of oxygen defect sites [61]. These sites can then act as charge trapping centers, reducing the likelihood of electron–hole recombination. Irie and coworkers [62] suggested that the N 2p levels are separated from the valence band (formed from O 2p states), rather than mixed, thereby forming an isolated narrow band responsible for the enhanced light sensitivity of the nitrided oxides. Also, some researchers have suggested that the nitrogen exists as NO and that no direct Ti–N bonds are formed, in contrast to the assumption that the nitrogen substitutes for the oxygen in the lattice. Substantiation of this proposal has recently been published by Reyes-Garcia and coworkers, who have shown a range of nitrogen-containing species within TiO_2 by a combination of solid-state NMR and ESR spectroscopies, but observed no evidence for direct substitution of lattice oxygen [62]. However, the substitution of C for lattice O within TiO [2] has been suggested by density functional theory methods [63]. The resulting structure, being structurally the same as the parent oxides, has enhanced photoactivity in the visible spectrum owing to the resulting unoccupied impurity state occurring in the band gap. The degree of band gap narrowing is variable and depends upon the synthesis technique but values of 2.32 eV [44] and 2.0 eV [64] have been reported in materials prepared by simple thermal treatment of the carbon-modified sample in air. Some reports have suggested that optimum performance of modified TiO_2 materials can be obtained by treatment with a combination of dopants. Significant visible response has been reported using a combination of N, together with F [51] and C [65].

19.4.3
Visible Light Sensitization by Metal Ion Implantation Techniques

Another method for improving the efficiency of semiconductor oxides is the addition of transition metal cations within the photocatalyst structure [6]. Such modification of the structure has several effects, although the primary aim of such preparations is the expansion of the absorption into the visible spectrum. In certain cases, the metal dopants can act as electron/hole trappers, leading to an increase in the charge carrier lifetimes and by doing so reducing the likelihood of recombination processes [7, 66]. This occurs through the following processes:

$$M^{n+} + e_{cb}^- \to M^{(n-1)+} \quad \text{(electron trapping)} \tag{19.8}$$

$$M^{n+} + h_{vb}^+ \to M^{(n+1)+} \quad \text{(hole trapping)} \tag{19.9}$$

If the $M^{n+}/M^{(n-1)+}$ pair is located below the conduction band edge and the energy level for $M^{n+}/M^{(n+1)+}$ above the valence edge then the trapping of electrons and holes would take place, affecting the lifetime of the charge carriers. The effect of the addition of metallic dopants is therefore a complex one and differentiating the individual effects is not simple. The efficiency of a metal ion dopant depends on whether it serves as a mediator of interfacial charge transfer or as a recombination center. Its efficiency as an electron/hole trap is related to several factors including the dopant concentration, its energy level within the semiconductor lattice, its electronic configuration, its distribution within the semiconductor, and the electron donor concentration. When metal ions are incorporated into the TiO_2 by conventional methods such as impregnation, a small absorption band is visible between 400 and 550 nm as a shoulder due to the formation of impurity levels in the band gap of the parent oxide [21] (Figure 19.3). However, in order to directly modify the band gap of the semiconductor, rather than simply insert an additional impurity energy level, a modified preparation technique can be used that applies transition metal ions at high acceleration energy (50–200 keV) [6, 42, 67, 68]. When the semiconductor is bombarded by such high-energy transition metal ions, the ions have been shown by a variety of spectroscopic techniques, including SIMS, EXAFS, and ESR, to be implanted into the lattice without destroying the underlying surface structure of the oxide. Rather than producing a small shoulder in the absorption band a much more significant and smooth shift is observed to higher wavelength, indicating that the resulting band gaps of the materials prepared in

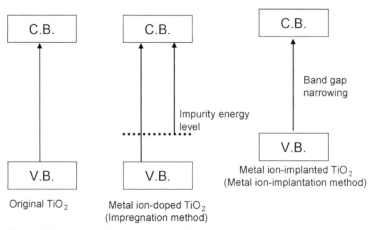

Figure 19.3 Schematic band structures of TiO_2, impregnated TiO_2, and ion implanted TiO_2.

this manner are significantly smaller than those of either the parent semiconductor or the parent oxide modified by conventional impregnation techniques.

Materials prepared by simple impregnation methods have been shown to exhibit a significantly lower photocatalytic activity than their parent oxides when illuminated solely with UV light. This has been accounted for by the dopants outside the structural framework acting as recombination centers for electron–hole pairs. Materials prepared by the ion implantation method showed no such decrease [21].

The techniques discussed above have all involved the preparation of powdered photocatalysts with enhanced visible light activity. A method recently developed for the preparation of thin-film photocatalysts with similar enhanced visible light activity is the deposition by radio-frequency magnetron sputtering, in which a TiO_2 plate is used as the source material and pure Ar as the sputtering gas [68–71]. It has been shown that the degree of absorption in the visible region is dependent upon both the temperature of preparation and the Ar pressure [72]. Similarly, nitrogen-doped materials have been prepared by the use of a mixed N_2/Ar sputtering gas mixture [73, 74]. With low concentrations of substituted nitrogen (<2%), a shoulder was observed in the UV-Vis absorption spectrum, but once concentrations above 2% were substituted a smooth shift in the absorption spectrum towards the visible region was observed. The band gaps of such materials were estimated to be between 2.25 and 2.58 eV. Such nitrided TiO_2 thin films showed good activity for the degradation of propan-2-ol diluted in water under visible irradiation [73].

19.4.4
Physical Methods to Enhance Photocatalytic Activity

Quantum physics dictates that by reducing particle size to the nanometer scale, a point is reached beyond which further decrease in the particle size will lead to variation in the position of the valence and conduction bands, and in the gap between the two [75, 76]. A variety of strategies have been suggested for controlling particle size and structure in order to produce particles in such a size regime and Table 19.2 includes some of them, together with the size range of the particles formed if such strategies are followed.

Table 19.2 Examples of preparation strategy together with the resulting particle scales.

Length scale	Examples
<1 nm	TiO_2 clusters encapsulated in zeolites
1–10 nm	TiO_2 nanoparticles forming periodic mesoporous powders
10 nm–10 μm	TiO_2 nanotubes
>200 μm	TiO_2 membranes/photonic crystals/photonic sponges

TiO$_2$ can be prepared on the nanometer scale [20], but the particles thus formed tend to agglomerate. One strategy to avoid aggregation of the particles has been to incorporate the TiO$_2$ clusters into the pores of a zeolite. Faujasite and zeolite β have been used extensively for this purpose [77]. Zeolites do not absorb in the UV-Vis region of the spectrum and hence it is possible to irradiate guest species anchored within the micropores of the host material. Two strategies have been explored for the incorporation of the TiO$_2$ into the zeolite, although both use molecular titanium sources with a single titanium atom that subsequently oligomerizes to form Ti–O–Ti structures. In the first approach, TiCl$_3$ or TiCl$_4$ vapors are absorbed into a partially or totally dehydrated zeolite structure, after which oligomerization is promoted by thermal treatment in the presence of moisture [78, 79]. The level of dehydration of the zeolite prior to the addition of the titanium source can give a degree of control as to where the TiO$_2$ structures locate. The greater the degree of dehydration the deeper into the microporous structure the TiCl$_3$ or TiCl$_4$ vapors can enter. The main problem with this technique is the explosive and pyrophoric nature of TiCl$_3$ and the highly corrosive HCl vapors formed upon hydrolysis which can cause subsequent damage to the zeolite framework [80]. The second approach is based upon the ion-exchange of the Na$^+$ counter-ions present in synthetic zeolites by (Ti=O)$^{2+}$ titanyl salts. This approach has the significant advantage that the titanium salts are non-corrosive and the ion-exchange process can be carried out in aqueous solution [81]. Hydrolysis and oligomerization can then be carried out by treatment of the titanyl salts at approximately 150 °C.

$$TiCl_4 + 2\, H_2O \rightarrow TiO_2 + 4\, HCl \quad (19.10)$$

$$TiO^{2+} + H_2O \rightarrow TiO_2 + 2\, H^+ \quad (19.11)$$

When low loadings of TiCl$_4$ were used, it was shown by UV-Vis spectroscopy that isolated titanium atoms are formed, bonded to the zeolite surface either mono, bi- or tripodally [82]. Increasing the loading led to the formation of aggregated moieties with Ti–O–Ti linkages present. The photoreactivity of materials produced was shown to be significantly higher than Degussa P25 TiO$_2$ for the partial oxidation of thianthrene [83].

Incorporation of titanium oxide species within the framework of mesoporous silicas has been shown to produce highly efficient photocatalytic materials. Extremely careful preparation conditions [84] leads to highly structured materials comprising anatase nanoparticles of dimension between 5 and 10 nm. The channeled structure, together with the hydrophobic/hydrophilic character, are also key features controlling their enhanced photoreactivity. The photocatalytic activity of such mesoporous catalysts has been studied for the degradation of phenol in aqueous solutions [85]. It was observed that for structured mesoporous materials with low Ti content, the turnover frequency was four times greater than that for standard P25.

A development in the use of structured TiO$_2$ has been the production of new one-dimensional materials such as nanotubes, nanofibers, and nanowires [86–

88]. TiO$_2$ nanotubes have a relatively high surface area compared with bulk titania powder but, most interestingly, time-resolved diffuse reflectance spectroscopy has shown that charge recombination is disfavored by the tubular morphology. Such nanotubes can be relatively easily prepared starting from conventional nanoparticles such as P25. Such particles are digested in strong base in an autoclave at 150 °C for several hours [89]. Annealing of the resulting material at 400 °C for 3 h results in nanotubes composed exclusively of anatase. Laser flash photolysis has shown that photogenerated holes on nanotubes have an average half-life approximately six times that of conventional TiO$_2$ of an equivalent BET area [89].

Only a small percentage of the solar flux can be absorbed by powdered metal oxide semiconductors. Hence considerable research effort has been invested in trying to modify the composition of such materials by chemical means, in order to enhance absorption in the visible region, and several examples of methods developed with this aim have been discussed earlier. In addition to such strategies, one physical approach is to increase the optimum light path, leading to entrapment of the light in the material. This methodology enhances light absorption by increasing the light path through the material in what are known as photonic crystals [90, 91]. Since the light is trapped within the crystal there is a significant enhancement in the likelihood of electron–hole pair formation, particularly at the onset of the absorption band where the specific absorption of the material is very low. The fundamental concept of a photonic crystal is the structuring of the space creating a periodic dielectric constant field in the scale of the visible light wavelength [92]. This spatial structuring produces coherent Bragg diffraction patterns that forbid light of the corresponding wavelength propagating through the material. At the Bragg diffraction frequencies, photons propagate with greatly reduced velocity and are known as "slow" photons. Crucially, if the slow photons exhibit energy which overlaps the absorption band of the material then the enhancement of the light absorption occurs as a consequence of the increase of the effective optical path length through the material. Ozin and coworkers prepared an inverse opal constituted of anatase nanocrystals ordered around monodisperse empty spheres of dimension between 280 and 500 nm. The system was investigated for the degradation of methylene blue. Using white light illumination, it was observed that the optimum void dimension was 300 nm, for which a photocatalytic activity enhancement of 2.3 was observed with respect to TiO$_2$ nanoparticles [93].

A further development of the opal structure was developed by Meseguer and coworkers, who expanded the concept to form a photonic sponge [94]. In this configuration, rather than having a monodispersion of spherical voids of a single dimension, the sponge has an appropriate distribution of differing size spheres, resulting in the trapping of an entire region of energy within a particular spectral region. Hence, with appropriate preparation, a sponge can be prepared that traps most of the visible spectrum. Such materials, when tested for the degradation of succinonitrile, have been reported to have their photocatalytic activity enhanced by an order of magnitude in comparison to TiO$_2$ nanoparticles [20].

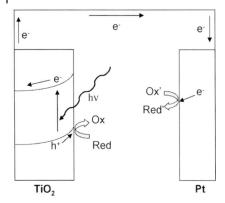

Figure 19.4 Schematic representation of a photoelectrocatalytic process.

19.4.5
Potential-Assisted Photocatalysis

Potential-assisted photocatalysis combines the advantages of photocatalysis with those of electrocatalysis. It is a combined technology that improves the efficiency of photocatalysis by the application of an external potential. This potential difference can either be simply a result of the resting potentials of the differing electrodes, or the application of an external electrically applied bias. This concept was first introduced by Honda and Fujishima, as they demonstrated the photoelectrocatalysis of water under the influence of an anodic bias [1].

Figure 19.4 shows a schematic representation of a photoelectrochemical cell. Photo-generated electrons are driven through the external circuit as a result of the potential difference across the cell. Hence, oxidation and reduction reactions occur at different electrodes of the device. Photoelectrochemically active semiconductor devices have been prepared from colloidal ZnO, TiO_2, SnO_2, and WO_3 [95]. In a photoelectrochemical cell the photocurrent observed provides a direct measurement of the reaction rate.

19.5
Conclusions

In a heterogeneous photocatalytic system, the efficiency for any chemical transformation depends upon the photoactivity of the adsorbate molecule and the catalyst substrate. In the case of a reactive semiconductor metal oxide, the catalyst can either provide energy levels to mediate electron transfer between adsorbate molecules (by temporarily accommodating an electron) or alternatively by acting as both an electron donor (derived from the photogenerated electron in the conduction band) and an electron acceptor (derived from the photogenerated hole in the valence band). Hence the band structure of the substrate plays a crucial role in the reactivity of the oxide. A small change in the electronic structure of the material can lead to a significant change in the band structure. Such electronic modification can be achieved in a variety of ways, including the addition of dye sensitizers or cationic and anionic dopants, or alternatively the physical manipulation of the particles at a nanometer scale resulting in the alteration of the band structure as a result of quantum effects. Such methods, or a combination of methods, can be used to alter the absorption characteristics of the oxides resulting in their activity in the visible region of the spectrum, a vital characteristic for commercial viability. The ability to use fundamental design principles based upon modern experimental and theoretical techniques coupled with the ever-increasing cost of, and demand for energy, means semiconductor photocatalysis is a rich area for both scientific and technological development in the twenty-first century.

References

1 Fujishima, A. and Honda, K. (1972) *Nature*, **37**, 238.
2 Linsebigler, A.L., Lu, G. and Yates, J.T. (1995) *Chemical Reviews*, **95**, 735.
3 Hoffmann, M.R., Martin, S.T., Choi, W. and Bahnemann, D.W. (1995) *Chemical Reviews*, **95**, 69.
4 Mills, A., Davies, R.H. and Worsley, D. (1993) *Chemical Society Reviews*, **22**, 417.
5 Kamat, P.V. (1993) *Chemical Reviews*, **93**, 267.
6 Anpo, M. (1997) *Catalysis Survey from Japan*, **1**, 169.
7 Carlson, T. and Griffin, G.L. (1986) *Journal of Physical Chemistry*, **90**, 5896.
8 Blake, D.M. (2001) Bibliography of work on the heterogeneous photocatalytic removal of hazardous compounds from water and air (Update Number 4), Technical Report, National Renewable Energy Laboratory, Golden, CO.
9 Bhatkhande, D.S., Pangarkar, V.G. and Breenackers, A.A.C.M. (2001) *Journal of Chemical Technology and Biotechnology (Oxford, Oxfordshire: 1986)*, **77**, 102.
10 Demeestere, K., Dewulf, J. and Van Langenhove, H. (2007) *Critical Reviews in Environmental Science and Technology*, **37**, 489.
11 Fagnoni, M., Dondi, D., Ravelli, D. and Albini, A. (2007) *Chemical Reviews*, **107**, 2725.
12 Palmisano, G., Augugliari, V., Pagliaro, M. and Palmisano, L. (2007) *Chemical Communications*, 425.
13 Palmisano, L. and Sclafani, A. (1997) *Heterogeneous Photocatalysis*, John Wiley & Sons, Ltd, Chichester, p. 109.
14 Colón-Ibánez, G., Belver-Coldeira, C. and Fernández-García, M. (2007) *Synthesis, Properties and Applications of Oxide Nanomaterials*, John Wiley & Sons, Inc., Hoboken, NJ, p. 491.
15 Fox, M.A. and Dulay, M.T. (1993) *Chemical Reviews*, **93**, 341.
16 Zhao, J. and Yang, X. (2003) *Building and Environment*, **38**, 645.
17 Peral, J. and Ollis, D.F. (1997) *Journal of Molecular Catalysis A – Chemical*, **115**, 347.
18 Xu, Y. and Schoonen, M.A.A. (2000) *American Mineralogist*, **85**, 543.
19 Cao, L., Huang, A., Spiess, F. and Suib, S.L. (1999) *Journal of Catalysis*, **188**, 48.
20 Aprile, C., Corma, A. and Garcia, H. (2008) *Physical Chemistry Chemical Physics*, **10**, 769.
21 Kitano, M., Masaya, M., Ueshima, M. and Anpo, M. (2007) *Applied Catalysis A: General*, **325**, 1.
22 Mills, G. and Hoffmann, M.R. (1993) *Environmental Science and Technology*, **27**, 1681.
23 D'Oliveira, J.C. and Al-Sayyed, G. and Pcihat, P. (1990) *Environmental Science and Technology*, **24**, 990.

24 Janus, M. and Morawski, A.W. (2007) *Applied Catalysis B – Environmental*, **75**, 118.
25 Sclafini, A. and Herrmann, J.M. (1996) *Journal of Physical Chemistry*, **100**, 13655.
26 Carp, O., Huisman, C.L. and Rellr, A. (2004) *Progress in Solid State Chemistry*, **32**, 33.
27 Dietbold, U., Ruzycki, N., Herman, G.S. and Selloni, A. (2003) *Catalysis Today*, **85**, 93.
28 Burdett, J.K., Hughbands, T., Gordon, J.M., Richardson, J.W. and Smith, J. (1987) *Journal of the American Chemical Society*, **109**, 3639.
29 Asahi, R., Taga, Y., Mannstadt, W. and Freeman, A.J. (2002) *Physical Review B*, **65**, 224112.
30 Bickley, R.I., González-Correño, T., Lee, J.S., Palmisano, L. and Tilley, R.J.D. (1991) *Journal of Solid State Chemistry*, **92**, 178.
31 Ozawa, T., Iwasaki, M., Tada, H., Akita, T., Tanaka, K. and Ito, S. (2005) *Journal of Colloid and Interface Science*, **281**, 510.
32 Dindar, B. and Icli, S. (2001) *Journal of Photochemistry and Photobiology A: Chemistry*, **140**, 263.
33 Yamaguchi, Y., Yamazaki, M., Yoshihara, S. and Shirakashi, T. (1998) *Journal of Electroanalytical Chemistry*, **442**, 1.
34 Khalil, L.B., Mourad, W.E. and Rophael, M.W. (1998) *Applied Catalysis B – Environmental*, **17**, 267.
35 Bensalem, A., Shafeev, G. and Bozon-Verduraz, F. (1993) *Catalysis Letters*, **18**, 165.
36 Bamwenda, G.R., Uesigi, T., Abe, Y., Sayama, K. and Arakawa, H. (2001) *Applied Catalysis A: General*, **205**, 117.
37 Kamat, P.V. and Fox, M.A. (1983) *Chemical Physics Letters*, **102**, 379.
38 O'Regan, B. and Grätzel, M. (1991) *Nature*, **353**, 737.
39 Choi, Y.M., Choi, W.U., Lee, C.H., Hyeon, T. and Lee, H.I. (2001) *Environmental Science and Technology*, **35**, 966.
40 Cheung, S., Fung, A. and Lam, M. (1998) *Chemosphere*, **36**, 2461.
41 Zhang, J.Z. (2000) *Journal of Physical Chemistry B*, **104**, 7239.
42 Anpo, M. (2004) *Bulletin of the Chemical Society of Japans*, **77**, 1427.
43 Jin, Z., Zhang, X., Lu, G. and Li, S. (2006) *Journal of Molecular Catalysis A – Chemical*, **259**, 275.
44 Khan, S.U.M., Al-Shahry, M. and Ingler, W.B. (2002) *Science*, **297**, 2243.
45 Sakthivel, S. and Kisch, H. (2003) *Angewandte Chemie – International Edition*, **42**, 4908.
46 Tachikawa, T., Tojo, S., Kawai, K., Endo, M., Fujitsuka, M., Ohno, T., Nishijima, K., Miyamoto, Z. and Majima, T. (2004) *Journal of Physical Chemistry B*, **108**, 19299.
47 Li, Y., Hwang, D., Lee, N.H. and Kim, S.-J. (2005) *Chemical Physics Letters*, **404**, 25.
48 Ashai, R., Morikawa, T., Ohwaki, T., Aoki, K. and Taga, Y. (2001) *Science*, **293**, 269.
49 Sakthivel, S., Janczarek, M. and Kisch, H. (2004) *Journal of Physical Chemistry B*, **108**, 19384.
50 Di Valentin, C., Pacchioni, G., Selloni, A., Livaghi, S. and Giamello, E. (2005) *Journal of Physical Chemistry B*, **109**, 11414.
51 Li, D., Haneda, H., Hishita, S. and Ohashi, N. (2005) *Chemistry of Materials*, **17**, 2596.
52 Li, D., Haneda, H., Labhsetwar, N.K., Hishita, S. and Ohashi, N. (2005) *Chemical Physics Letters*, **401**, 579.
53 Umebayashi, T., Yamaki, T., Itoh, H. and Asai, K. (2002) *Applied Physics Letters*, **81**, 454.
54 Demeestere, K., Dewulf, J., Ohno, T., Salgado, P.H. and Langenhove, H.V. (2005) *Applied Catalysis B – Environmental*, **61**, 140.
55 Chu, S.Z., Inoue, S., Wada, K., Li, D. and Suzuki, J. (2005) *Langmuir*, **21**, 8035.
56 Zhao, W., Ma, W., Chen, C., Zhao, J. and Shuai, Z. (2004) *Journal of the American Chemical Society*, **126**, 4782.
57 Luo, H., Takata, T., Lee, Y., Zhao, J., Domen, K. and Yan, Y. (2004) *Chemistry of Materials*, **16**, 846.
58 Lin, L., Lin, W., Zhu, Y., Zhao, B. and Xie, Y. (2005) *Chemistry Letters*, **34**, 284.
59 Hong, X., Wang, Z., Cai, W., Lu, F., Zhang, J., Yang, Y., Ma, N. and Liu, Y. (2005) *Chemistry of Materials*, **17**, 1548.
60 Batzill, M., Morales, E.H. and Diebold, U. (2006) *Physical Review Letters*, **96**, 26103.
61 Ihara, T., Miyoshi, M., Iriyama, Y., Matsumoto, O. and Sugihara, S. (2003) *Applied Catalysis B – Environmental*, **42**, 403.

62 Reyes-Garcia, E.A., Sun, Y., Reyes-Gil, K. and Raftery, D. (2007) *Journal of Physical Chemistry*, **111**, 2738.

63 Irie, H., Watanabe, Y. and Hashimoto, K. (2003) *Journal of Physical Chemistry B*, **107**, 5483.

64 Barborini, E., Conti, A.M., Kholmanov, L., Piseri, P., Podestà, A., Milani, P., Cepek, C., Sakho, O., Macovez, R. and Sancrotti, M. (2005) *Advanced Materials*, **17**, 1842.

65 Noguchi, D., Kawamata, Y. and Nagatomo, T. (2005) *Journal of the Electrochemical Society*, **152**, D124.

66 Sclafani, A., Mozzanega, M.N. and Pichat, P.J. (1991) *Journal of Photochemistry and Photobiology A: Chemistry*, **59**, 181.

67 Anpo, M., Dohshi, S., Kitano, M., Hu, Y., Takeuchi, M. and Matsuoka, M. (2005) *Annual Review of Materials Research*, **35**, 1.

68 Anpo, M. and Takeuchi, M. (2003) *Journal of Catalysis*, **216**, 505.

69 Kitano, M., Takeuchi, M., Matsuoka, M., Thomas, J.M. and Anpo, M. (2005) *Chemistry Letters*, **34**, 616.

70 Matsuoka, M., Kitano, M., Takeuchi, M. and Thomas, J.M. (2005) *Topics in Catalysis*, **35**, 305.

71 Kikuchi, H., Kitano, M., Takeuchi, M., Matsuoaka, M., Anpo, M. and Kamat, P.V. (2006) *Journal of Physical Chemistry B*, **110**, 5537.

72 Kitano, M., Takeuchi, M., Matsuoka, M., Thomas, J.M. and Anpo, M. (2007) *Catalysis Today*, **122**, 51.

73 Kitano, M., Funatsu, K., Matsuoka, M., Ueshima, M. and Anpo, M. (2007) *Journal of Physical Chemistry B*, **110**, 25266.

74 Kitano, M., Kudo, T., Matsuoka, M., Ueshima, M. and Anpo, M. (2007) *Materials Science Forum*, **544–5**, 107.

75 Yamashita, H., Takeuchi, M. and Anpo, M. (2004) *Encyclopedia of Nanoscience and Nanotechnology*, Vol. 10 (ed. H.S. Nalwa), American Science Publishers, Los Angeles, CA, p. 639.

76 Anpo, M., Yamashita, H., Ichihashi, Y., Fujii, Y. and Honda, M. (1997) *Journal of Physical Chemistry B*, **101**, 2632.

77 Ikeue, K., Yamashita, H., Anpo, M. and Takewaki, T. (2001) *Journal of Physical Chemistry B*, **105**, 8350.

78 Grubert, G., Stockenhuber, M., Tkachenko, O.P. and Wark, M. (2002) *Chemistry of Materials*, **14**, 2458.

79 Bossmann, S.H., Turro, C., Schnabel, C., Pokhrel, M.R., Payawan, L.M., Baumeister, B. and Woerner, M. (2001) *Journal of Physical Chemistry B*, **105**, 5374.

80 Kaipas, T. and Griffiths, R.F. (2005) *Journal of Hazardous Materials*, **119**, 41.

81 Liu, X., Iu, K. and Thomas, J.K. (1993) *Journal of the Chemical Society–Faraday Transactions*, **89**, 1861.

82 Klaas, J., Schultz-Ekloff, G. and Jaeger, N.I. (1997) *Journal of Physical Chemistry B*, **101**, 1305.

83 Cosa, G., Galletero, M.S., Fernandez, L., Marquez, F., Garcia, H. and Scaiano, J.C. (2002) *New Journal of Chemistry*, **26**, 1448.

84 Bartl, M.H., Boettcher, S.W., Frindell, K.L. and Stucky, G.D. (2005) *Accounts of Chemical Research*, **38**, 263.

85 Alvaro, M., Aprile, C., Benitez, M., Carbonell, E. and Garcia, H. (2006) *Journal of Physical Chemistry B*, **110**, 6661.

86 McCann, J.T., Marquez, M. and Xia, Y. (2006) *Nano Letters*, **6**, 2868.

87 Zhong, Z., Ang, T.P., Luo, J., Gan, H.C. and Gedanken, A. (2005) *Chemistry of Materials*, **17**, 6814.

88 Xiong, C. and Balkus, K.J. (2005) *Chemistry of Materials*, **17**, 5136.

89 Tachikawa, T., Fujitsuka, M. and Majima, T. (2007) *Journal of Physical Chemistry C*, **111**, 5259.

90 Hall, N. (2003) *Chemical Communications*, 2639.

91 Baba, T. (2007) *National Photonics*, **1**, 11.

92 Ozin, G. and Yang, S.M. (2001) *Advanced Functional Materials*, **11**, 95.

93 Chen, J.I.L., von Freymann, G., Kitaev, V. and Ozin, G.A. (2006) *Advanced Materials*, **18**, 1915.

94 Ramiro-Manzano, F., Atienzar, P., Rodriguez, I., Meseguer, F., Garcia, H. and Corma, A. (2007) *Chemical Communications*, 242.

95 Vinodgopal, K. and Kamat, P.K. (1995) *Solar Energy Materials and Solar Cells*, **38**, 401.

20
Catalytic Ammoxidation of Hydrocarbons on Mixed Oxides
Fabrizio Cavani, Gabriele Centi, and Philippe Marion

20.1
Introduction

Ammoxidation, sometimes also termed oxidative ammonolysis, describes the process of catalytic oxidation of hydrocarbons (particularly alkenes, alkanes, alkyl-aromatics and alkyl-pyridines) to organic nitriles in the presence of ammonia, typically using mixed oxide catalysts:

$$R-CH_3 + 1.5\,O_2 + NH_3 \rightarrow R-CN + 3\,H_2O \tag{20.1}$$

The most important example of this process is the synthesis of acrylonitrile from propene. Acrylonitrile is a large-volume commodity chemical (within the top twenty). The world's production capacity for acrylonitrile was 6.14 million tons per annum in 2005 and the output reached 5.24 million tons, an increase of 0.5% over 2004. The operating rate of production units was more than 85%, lower than the operating rate of 90% in 2004. The drop in utilization of capacity has been higher than the growth rate of the new capacity in recent years, and is essentially related to the market situation and relatively old plant technology. There is a supply deficit of acrylonitrile in the world today, but profitability in the acrylonitrile sector is still low. The mean spread, that is the difference between the acrylonitrile price and the raw material cost, between 2002 and 2006 was about $150/metric ton. Figure 20.1 reports the world acrylonitrile demand (by use) over the 1995–2006 period and projection up to 2010 [1, 2]. Figure 20.1 also shows that acrylonitrile is a chemical intermediate used mainly in acrylic fibres, ABS (acrylonitrile–butadiene–styrene), SAN (styrene–acrylonitrile) and NBR (nitrile–butadiene–rubber).

There is a mean annual increase in world demand of about 3%, driven mainly by ABS/SAN resin and other applications. Acrylonitrile is also used to produce adiponitrile (for manufacture of hexamethylenediamine used in Nylon-6,6 fibers and resins) and acrylamide for water-treatment polymers. Approximately 52% of the total EU production of acrylonitrile is used in production of fibres, 15% in production of ABS and SAN resins, 15% in the production of acrylamide and adiponitrile, and 18% for other uses [2].

Metal Oxide Catalysis. Edited by S. David Jackson and Justin S. J. Hargreaves
Copyright © 2009 WILEY-VCH Verlag GmbH & Co. KGaA, Weinheim
ISBN: 978-3-527-31815-5

772 | 20 Catalytic Ammoxidation of Hydrocarbons on Mixed Oxides

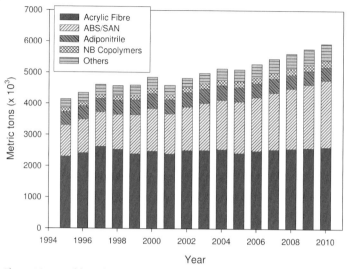

Figure 20.1 World acrylonitrile demand by derivative (data estimated from 2006 to year 2010). Data from PCI Acrylonitrile Ltd [1].

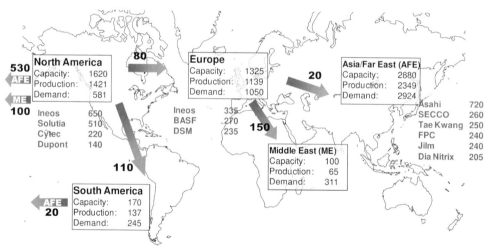

Figure 20.2 World acrylonitrile trade flow by year 2005 and major manufacturers' capacity. Data from PCI Acrylonitrile Ltd [1].

The global market for acrylonitrile is currently quite balanced. After a period of oversupply in 2001, brought about by the start-up of two large production units, many closures in Western countries offset capacity growth and brought the market into better balance. In general, a shift of capacity from Western to Asian countries has occurred between 2004 and 2008. Figure 20.2 shows the world acrylonitrile

Table 20.1 Propene demand by application (year 2005).

Application	Use (thousand metric tons)	%
Polypropene	39 289	60
Acrylonitrile	5 684	9
Propene oxide	4 780	7
Cumene	3 629	6
2-Ethylhexanol	2 424	4
Butanols	2 261	4
Isopropanol	1 350	2
Oligomers	1 327	2
Others	3 566	6

trade flow in 2005 [1] with an indication also of the major manufacturers' capacity. North America is still the largest acrylonitrile exporter and Asia is the major import area. However, recent and future expansion of production in Asia will result in a shrinking export market for North America. Costs of acrylonitrile production are highly dependent upon propene and ammonia prices. Since 2006, prices for acrylonitrile have increased substantially (by more than one-third), mainly because of rising feedstock costs. For this reason, and also because of weak demand, acrylonitrile margins remain poor.

The cost of propene is mainly driven from competing areas of application (Table 20.1). Acrylonitrile accounts for only about 9% of propene demand and, because of the expansion of the market for other applications, a further increase of propene cost is expected. Global propene demand grew from 16.4 million tons in 1980 to around 30 million tons in 1990 and about 52 million tons in 2000, with thus an average annual growth of about 6% per year, which is much higher than the average annual growth of demand for acrylonitrile. Expected demand for propene by 2010 will be 81 million tons with demand growing faster than supply. Propene supply/demand conditions and pricing are strongly dependent on refinery production and the supply/demand balance, operating rates and feedstock in the ethane industry. Globally, more than 25% of the new crackers currently planned are based on ethane, and therefore will produce little propene. Propene production by dehydrogenation of propane, although increasing, is still an expensive route. Therefore, higher prices for propene are forecast in the next decade.

Owing to the increasing cost of propene, interest in using alternative cheaper raw materials, particularly propane, is expanding rapidly. Asahi Kasei Chemicals Corporation (AKC) has begun commercial operation of the world's first propane process for acrylonitrile production, with the start-up and validation of a propane-process line at the Ulsan Plant of its Korean subsidiary Tongsuh Petrochemical Corporation. An existing 70 000 ton/year acrylonitrile line was modified to use the propane process. Propene, however, will still remain the main raw material to manufacture acrylonitrile for several years. The most used process is Innovene's (Ineos) acrylonitrile technology, known in the industry as the SOHIO acrylonitrile process, which is used in the manufacture of over 90% of the world's acrylonitrile. In 1996, this process, which involves the oxidation of propene and ammonia to acrylonitrile in a fluid-bed reactor, was designated the eleventh National Historic Chemical Landmark by the American Chemical Society.

The SOHIO process involves the catalytic oxidative reaction of propene with ammonia in the vapor phase [3]. Approximately stoichiometric amounts of propene and ammonia, combined with air, are passed through the reactor in a single-pass operation with a residence time of just a few seconds.

$$H_2C=CHCH_3 + 3/2\, O_2 + NH_3 \to H_2C=CHCN + 3\, H_2O$$
$$\Delta H° = -515\, kJ\, mol^{-1} \qquad (20.2)$$

The reaction is highly exothermic and the heat of reaction is generally used to make high-pressure steam, utilized downstream in separation and purification operations. The main useful by-products from the process are HCN (about 0.1 kg per kg of acrylonitrile), which is used primarily in the manufacture of methyl methacrylate, and acetonitrile (about 0.03 kg per kg of acrylonitrile), a common industrial solvent. Smaller quantities of carbon oxides and nitrogen (from ammonia combustion) are also obtained. Unreacted ammonia in the reactor effluent is neutralized with sulfuric acid. The resulting ammonium sulfate can be recovered for use as a fertilizer.

Since the reactions generating by-products (carbon oxide formation and ammonia combustion) are themselves highly exothermic, the total exothermicity of the reaction is around 530–660 kJ mol^{-1}, making control of the reaction temperature critical.

Prior to 1960, acrylonitrile was produced commercially by processes based on either ethylene oxide and hydrogen cyanide or acetylene and hydrogen cyanide. In the late 1950s, Standard Oil (later Sohio and then BP) [4] and Distillers [5] developed a heterogeneous vapor-phase catalytic process for acrylonitrile by selective oxidation of propene and ammonia using a catalyst based on bismuth, tin and antimony salts of molybdic and phosphomolybdic acids and bismuth phosphotungstate. The $Bi_9PMo_{12}O_{52}/SiO_2$ system was first reported in 1955 for propene oxidation. The system works either in a cyclic oxidant process mode, or as a genuine redox catalyst. The first attempt to use it in a cyclic oxidant mode was abandoned after demonstration pilot operations, because about 200 kg of the solid catalyst had to be circulated to produce 1 kg of acrolein [6]. However, the same system operates successfully as a redox catalyst and was first commercialized for the production of acrolein from propene in 1957 (Degussa licensee), and subsequently commercialized internally in 1959 for the production of acrylonitrile from propene (licensed worldwide).

Around the same time, but slightly later, Edison (later Montedison) was also developing a similar process, but based on different catalysts (tellurium-cerium oxides) [7]. Although at that time the performance of these catalysts were equal or slightly superior to bismuth molybdenum oxides, the process was never commercialized.

As a consequence of the introduction of the new catalytic ammoxidation process, a steep rise in the production of acrylonitrile occurred starting from the year 1960 [6]. Further step changes were observed as a consequence of the introduction of each new generation of more effective catalysts (the acrylonitrile yield has been increased over the past 40 years from 50 to over 80%). The substitution of the inefficient and expensive process of IG Farben (HCN + acetylene) to produce acrylonitrile by the highly efficient and environmentally friendly SOHIO processs (propene + ammonia + air) can be considered as one of the first examples of 'green' sustainable chemistry. At the time of writing (2008), over 90% of the worldwide production of acrylonitrile is made using the Sohio ammoxidation process [8].

20.2
Propene Ammoxidation to Acrylonitrile

Commercial catalysts are made of multicomponent Bi-Fe-Ni-Co molybdates, also containing several additives: Cr, Mg, Rb, K, Cs, P, B, Ce, Sb and Mn are those more frequently mentioned in the several patents issued. The mixed oxide is dispersed in silica (50 wt%) for fluid-bed reactor applications. Owing to the high exothermicity of the reaction and by-product formation, most of the plants use a fluidized-bed technology. A typical empirical composition is $(K,Cs)_{0.1}(Ni,Mg,Mn)_{7.5}(Fe,Cr)_{2.3}Bi_{0.5}Mo_{12}O_x/SiO_2$ [9, 10]. The role of the various elements in the multi-component catalyst, in relation to the multi-step reaction mechanism, are discussed in detail by Grasselli [10]. Production improvements since 1980 have stemmed largely from the development of several generations of increasingly more efficient catalysts. In addition to mixed metal oxide catalysts based on multimetal molybdates (MMM), other types of commercially used catalysts were based on iron antimony oxide, uranium antimony oxide (in the past), and tellurium molybdenum oxide. MMM-based catalysts are the most commonly used nowadays.

Even though the literature on this topic has been mainly focussed on the structural and chemical-physical properties of Bi molybdates, and on the reactivity of its various polymorphs (the α, β and γ structures), the industrial catalyst consists of several divalent and trivalent metal molybdates. Indeed, Bi is present in minor amounts in catalyst formulations. The two classes of molybdate contribute differently to catalytic performance: (1) trivalent Bi/Fe/Cr molybdates, having the Scheelite-type structure, contain the catalytically active elements while (2) divalent Ni/Co/Fe/Mg molybdates, having the Wolframite-type structure, mainly enhance the catalyst re-oxidation rate.

The catalyst also contains excess Mo with respect to the stoichiometric requirement for the formation of the molybdates. The excess Mo is considered important for catalyst performance because:

- it provides a molecular bridge between the co-operating molybdates if their crystalline match is not perfect;

- it provides Mo to those phases partly depleted of it because of the redox cycle, especially under more reducing conditions (Mo is lost in the form of volatile $MoO(OH)_2$). For instance, owing to the reduction of Fe^{3+} to Fe^{2+}, part of the Mo is excluded from the molybdate, and is finally lost in volatile form.

Excess MoO_3, when present in the catalyst composition from the beginning, or when added during reaction, migrates along hydrated silica surfaces towards Mo-lean catalytically active phases. Often, Mo-enriched make-up MMM catalyst is preferentially added in place of MoO_3 during reactor operation. Fundamental understanding of these complex catalysts and the surface-reaction mechanism of propene ammoxidation has contributed substantially to the development of new catalyst generations (currently at the fourth generation). Detailed mechanisms for selective ammoxidation of propene over bismuth molybdate and antimonate catalysts have been proposed [11]. The rate-determining step is abstraction of an α-hydrogen of propene by an oxygen in the catalyst to form a π-allyl complex on the surface [11, 12]. Lattice oxygens from the catalyst participate in further hydrogen abstraction, followed by oxygen insertion to produce acrolein in the absence of ammonia, or nitrogen insertion to form acrylonitrile when ammonia is present [13]. The oxygen removed from the catalyst in these steps is replenished by gas-phase oxygen, which is incorporated into the catalyst structure at a surface site separate from the site of propene reaction. In the ammoxidation reaction, ammonia is activated by an exchange with oxygen ions to form isoelectronic NH^{2-} moieties, which are inserted into the allyl intermediate to produce acrylonitrile.

The active site on the surface of a selective propene ammoxidation catalyst contains three critical functionalities associated with the specific metal components of the catalyst [14]: an α-H abstraction component such as Bi^{3+}, Sb^{3+} or Te^{4+}; an olefin chemisorption and oxygen or nitrogen insertion component such as Mo^{6+} or Sb^{5+}; and a redox couple, such as Fe^{2+}/Fe^{3+} or Ce^{3+}/Ce^{4+}, to enhance transfer of lattice oxygen between the bulk and surface of the catalyst. Moreover, in general, it may be considered that the large improvement in the selectivity of these catalysts derives from the application of seven principles ('seven pillars') [6]: lattice oxygen, metal–oxygen bond strength, host structure, redox activity, multi-functionality of active sites, site isolation and phase co-operation.

The process schematic of propene ammoxidation is shown in Figure 20.3 [15, 16]. A single-pass configuration is possible, because over 95 wt% conversion can occur with selectivities to acrylonitrile which nowadays are well above 80%. Air, ammonia and propene are sent to a fluidized-bed reactor, which may contain up to 70–80 tons of catalyst in the form of fine spherical particles (<40 μm in diameter) highly resistant to mechanical attrition. The purity of reactants is very high (>90% for propene and >99.5% for ammonia). The ammonia to propene molar ratio is in the range 1.05 to 1.2 and the O_2/propene ratio in the range 1.9–2.1; typically, oxygen-enriched air is used in industrial operation. Reaction temperature is in the 420–450 °C interval, residence time is between 3 and 8 s, with a linear gas velocity from 0.2 and 0.5 m s^{-1} and pressure between 1.5 and 3 atm. Since the rate of acry-

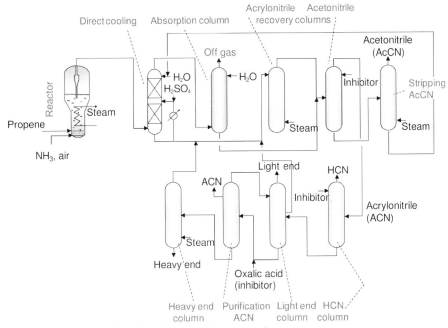

Figure 20.3 Schematic flow sheet of the SOHIO process of propene ammoxidation to acrylonitrile. Adapted from [11].

lonitrile synthesis is first order with respect to propene, with higher orders being shown towards the by-products, an increase in pressure has a negative effect on selectivity. However, positive pressure is necessary to maintain the correct fluidization characteristics in the reactor.

The feed enters the reactor through separate inlets, in order to minimize homogeneous reactions and prevent local flammable compositions developing, although owing to the effect of the solid in inhibiting radical propagation within the fluidized-bed reactor, it is possible to work inside the flammability region in the catalytic bed. The fluidized-bed reactor contains many steam coils to keep the reaction temperature constant, avoid bubble coalescence and reduce gas backmixing. By proper design of these coils a virtually plug-flow regime may be possible. An expansion area to homogenize the gas velocity profiles, and thus minimize the entrapment of solid particles, and cyclones, to recover the finest particles, are integrated into the fluidized-bed process.

The hot reactor effluent is sent to a water absorber where it is quenched countercurrently, while unreacted ammonia is neutralized with sulfuric acid. The resulting ammonium sulfate can be recovered and used as a fertilizer. The off-gases containing N_2, carbon oxides and unreacted hydrocarbon are sent to incineration. The solution of acetonitrile/acrylonitrile is a heteroazeotrope. After settling, an aqueous and an organic phase are obtained. The first is refluxed, while the latter, rich in acrylonitrile and HCN, is sent to the purification step. The aqueous acetonitrile recovered at the bottom is further concentrated by azeotropic distillation. The crude acrylonitrile is first purified in two in-series columns to separate HCN and impurities (acetone, acetaldehyde, propionaldehyde, acrolein) and finally further purified under vacuum. Final polymer grade must have a purity higher than 99.4%. Disposal of the process impurities includes deep-well disposal (not sustainable), wet air oxidation, ammonium sulfate separation, biological treatment and incineration [17].

20.3
Propane Ammoxidation to Acrylonitrile

In the current process technology for the manufacture of acrylonitrile, the propene feedstock cost represents about 67% of the full (or fixed) cost of production [18]. The price differential between propene and propane depends on many factors, but can be estimated to be on average $360 per ton during 2007 [19], compared with around $320 per ton in 2004 [18]. This price differential makes propane ammoxidation competitive using the currently available catalysts. In fact some plants have already started to be revamped. For example, at Asahi Kasei Corporation an existing 70 000 ton/year acrylonitrile line was modified to use the propane process. Production began in January 2007 [20a]).

The reaction conditions claimed by the various companies are substantially different. As shown in Figure 20.4, sometimes propane-rich conditions have been claimed, as in earlier patents from Standard Oil, while in other cases propane-lean conditions are preferred. In the former case, the conversion of propane is low, and therefore recycling of the unconverted paraffin becomes necessary. Mitsubishi was the first to claim the use of hydrocarbon-lean conditions, that is conditions in which very high propane conversions can be reached. In more recent patents, BP has claimed the use of analogous conditions, using oxygen-rich feed, and propane as the limiting reactant. However, the lower activity of antimonates makes it necessary to use temperatures which are approximately 50 °C higher than those employed with the Mitsubishi catalyst.

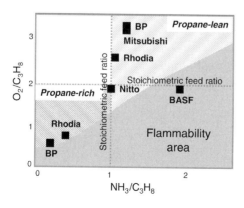

Figure 20.4 Feed composition in propane ammoxidation claimed by different companies.

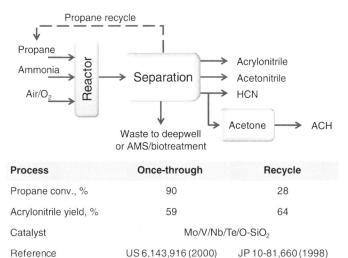

Figure 20.5 Once-through and recycle processes for propane to acrylonitrile (ACH: acetone cyanohydrins). Adapted from [18a].

A once-through process is the preferred option [18a] (Figure 20.5). The catalyst and process conditions are tailored to achieve as near total feedstock conversion as possible. Products are separated and fresh feedstock is introduced into the reactor on a continuous basis. In addition to providing the lowest capital cost for a new-build plant, this also provides the best possibility for retrofitting existing plants by simply replacing catalyst and feedstock. This option is preferable to the combination of an alkane dehydrogenation unit with an existing propene-to-acrylonitrile process, owing to the significant capital cost involved. The significant improvement in selectivity at high conversion using V/Mo/Nb/Te oxide catalysts, compared to the earlier generation of catalysts based on V-Sb oxide, also makes the once-through option preferable to the recycle option (Figure 20.5) [18a]. Although selectivity is higher when operating at lower propane conversion, additional equipment is required for recovery and recycling of the unreacted feedstock. Nevertheless, if selectivities to acrylonitrile could exceed 80% at a propane conversion lower than about 30% using new generation catalysts, the recycle process could become an interesting option. The recycling of unconverted propane is also an option when high alkane conversion is reached, because this allows not only the complete recovery of propane but also propene recycling (one side-product of the reaction, which is a precursor of acrylonitrile), along with carbon dioxide, which may act as a ballast for the reaction. The Mitsubishi process makes use of the BOC-PSA technology for the removal of N_2 (both present in the feed and generated in the reactor by ammonia combustion), while the purge stream is incinerated [20b]. One Mitsubishi patent claims the staged feeding of ammonia along the catalytic bed [20c]; this is a necessary option if the catalyst is very active in ammonia

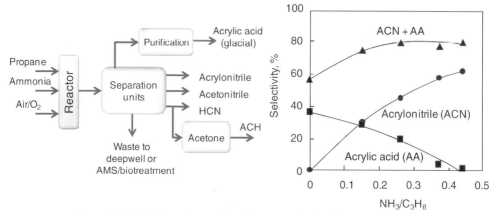

Figure 20.6 Multiple product option with propane feedstock (US Patent 6,166,241 (2000)). Adapted from [18a].

combustion to N_2, in order to avoid a lack of ammonia in the final part of the reactor, which would favor combustion of the hydrocarbons and the formation of propene.

More complex process concepts have also been proposed which attempt to optimize the processes. One option is the use of multiple reaction steps operating at different reaction conditions [18a]. Three sequential process steps are used, each operated with increasing conversion of propane. Product is separated after each step and the unreacted propane is sent to the next reactor with fresh ammonia and air/oxygen. An overall propane conversion of 95% with overall acrylonitrile yield of 63% is achieved in three steps. Another approach is to increase the process flexibility in synthesizing multiple products. By changing the ratio of propane/ammonia in the feed to the reactor, acrylonitrile and acrylic acid could be simultaneously produced with variable yields, depending on NH_3/C_3H_8 ratio (Figure 20.6). Even if the complexity of product purification and recovery sections increases the process costs, this option provides the opportunity to optimize the operation of the plant based on changes in feedstock and product price and demand.

A number of reviews have discussed the catalyst and catalytic reaction chemistry in propane ammoxidation [21–25]. So far, two main catalytic systems have been proposed in the literature. They are based on V-antimonates with rutile structure or multi-component molybdates (Mo/V/Nb/Te/O). Among the antimonates are compositions of the Al/Sb/V/W/O system which give the highest acrylonitrile yields reported (about 39%) [26]. The most promising catalyst discovered thus far is the Mo/V/Nb/Te/O system, which has been shown to be both active and selective, giving acrylonitrile yields up to 62% [6b, 27]. Mitsubishi Kasei [27] originally developed this catalyst composition which gives the highest yield of acrylonitrile to date, although the long-term stability is still unclear. The optimal catalyst composition is $MoV_{0.3}Te_{0.23}Nb_{0.12}O_x$, which gives a yield of acrylonitrile of about 50%; this can be further increased to nearly 60% by the addition of Sb, B or Ce. Key characteristics of the catalyst are not only its composition, but also the procedure of synthesis and activation and the modality of doping. These catalysts are prepared by hydrothermal synthesis, which results in nucleation and growth of M1 and M2 phases with well defined crystal morphologies.

Several articles on synthesis and characterization of Mo/V/M/O (M = Te, Sb, Al) catalysts have been published recently [28]. The effects of metal oxide precursor sources, synthesis conditions and post-synthesis treatments are subjects of current studies. Asahi [29] has modified the composition of the Mitsubishi catalyst, by incorporation of Sb in place of Te in the M1 phase; this catalyst is more stable than the original one, and hence provides longer catalyst lifetime.

Standard Oil (later BP America, now Ineos) developed V–Sb-based catalysts which show a high selectivity to acrylonitrile especially when using high propane concentration [30, 31]. Single and dual catalyst compositions were patented, with the second catalyst having the function of converting the propene intermediate to acrylonitrile [30]. Claimed catalyst compositions are mixed oxides such as (i) $Cr_aMo_bTe_cM_d$, where M is Mg, Ti, Sb, Fe, V, W, Cu, La, P, Ce or Nb; (ii) VSb_aM_b, where M is one or more of Sn, Ti, Fe, Mn or Ga; (iii) $Bi_aFe_bMo_cA_dB_e$, where A is one or more alkali or alkaline metals, boron, W, Sn or La and B is one or more of the elements Cr, Sb, Pb, P, Cu, Ni, Co, Mn or Mg; and (iv) $Bi_aFeMo_{12}V_bD_cE_dF_eG_f$, where D is one or more of the alkali metals; E is one or more of Mn, Cr, Cu, Zn, Cd or La; F is one or more of P, As, Sb, F, Te, W, B, Sn, Pb or Se; and G is one or more of Co, Ni or an alkaline earth metal. The gas feed composition was usually propane/ammonia/oxygen/water in a 5/1/2/11 molar ratio, with excess alkane and water used as a diluent.

A third catalytic system, based on vanadium aluminum oxynitrides (VALON), has also been proposed [32]. Maximum acrylonitrile yield was about 30%, but with an acrylonitrile productivity four times higher than V-Sb-W-Al-O catalysts and one order of magnitude higher than Mo/V/Nb/Te/O catalysts [33].

Other companies have studied and developed proprietary formulations, but in general catalytic systems belong either to the antimonate family (Standard Oil, Rhodia, BASF, Nitto, Monsanto) [34–38] or to the molybdate family (Mitsubishi, Asahi).

All catalysts claimed are 'multi-functional' systems. Indeed, the formation of acrylonitrile from propane occurs mainly via the intermediate formation of propene, which is then transformed to acrylonitrile via the allylic intermediate. It follows that the catalyst possesses different kinds of active site: one site that is able to activate the paraffin and oxidehydrogenates it to the olefin, and one site that (amm)oxidizes the adsorbed olefinic intermediate. This second step must be very rapid to limit, as much as possible, the desorption of the olefin. In order to develop an effective cooperation between the two sites, it is necessary to have systems in which they are in close proximity. The multi-functionality is achieved either through the combination of two different compounds (phase-cooperation), or through the presence of different elements inside a single crystalline structure. In antimonate-based systems, the cooperation between the metal antimonate (having the rutile crystalline structure), responsible for propane oxidative dehydrogenation to propene and propene activation, and antimony oxide, active in allylic ammoxidation, is made more efficient through the dispersion of the latter compound over

20 Catalytic Ammoxidation of Hydrocarbons on Mixed Oxides

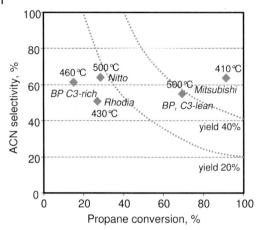

Figure 20.7 A summary of best performances in propane ammoxidation claimed by different companies.

the former one; in this case, an optimal ratio between the two compounds exists, which makes the cooperation more efficient. In metal molybdates (the Mitsubishi catalyst), one single crystalline structure contains the elements required for the activation and oxidative dehydrogenation of the hydrocarbon (vanadium), and those active for the transformation of the olefin and the allylic insertion of the NH^{2-} species (molybdenum and either tellurium or antimony). Niobium has the role of improving the structural stability of the compound.

The selectivity to acrylonitrile versus propane conversion is plotted in Figure 20.7 for the best performances reported in patents issued by different companies, and the temperature of reaction is also reported. In general, best per-pass yields obtained with propane-rich conditions are between 20 and 25%. The yield to acrylonitrile reported is close to 60% at 87–89% propane conversion, with a selectivity of 60–64% at 410°C, under propane-lean conditions, reported in Mitsubishi patents with a catalyst made of mixed molybdate.

20.3.1
Mo/V/Te/Sb/(Nb)/O Catalysts

The elements constituting this catalyst are the basis for several crystalline structures, as illustrated in Figure 20.8, which summarizes the main mixed oxides that can be obtained by combination of V, Mo, Nb and Te(Sb). The figure shows the various bi-component systems in a triangular diagram of composition, as well as the areas which include tri- and multi-component systems, exhibiting superior performance in ethane oxidation, and in propane oxidation and ammoxidation. Many of the reported structures are related; for example orthorhombic $Mo_{5-x}(V,Nb)_xO_{14}$ solid solutions are isostructural with $\theta\text{-}Mo_5O_{14}$ and with several bi-component Sb/Mo, Nb/Mo and Te/Mo mixed oxides. The M1 phase is

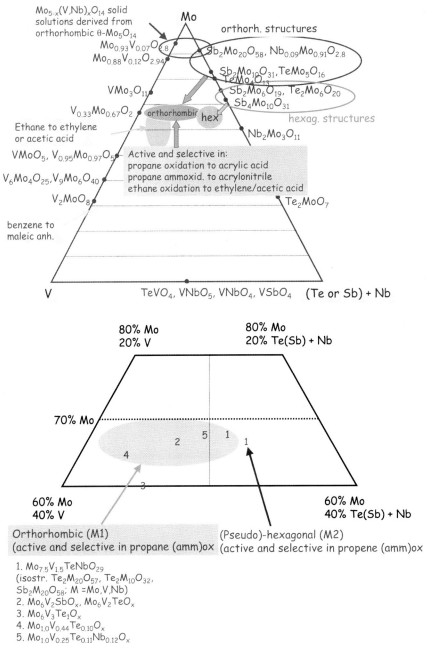

Figure 20.8 Top: Triangular diagram of composition for the Mo/V/Nb(+Te or Sb)/O system, indicating the stoichiometry of compounds which form, and the area of existence for systems active and selective in alkanes oxidation and ammoxidation. Bottom: detail of the top Figure.

structurally derived from the latter oxides, and the M2 phase is related to other Mo-leaner Te/Mo and Sb/Mo mixed oxides.

Two main preparation methods have been used to synthesize Mo/V/Te/(Nb)/O catalysts active in propane ammoxidation: (a) the dry-up and (b) the hydrothermal synthesis. The dry-up method involves mixing aqueous slurries of metal oxide precursors followed by a gradual evaporation of the combined aqueous slurry. Solvent evaporation leads to nucleation and growth of precursor metal oxide phases, which require further heat treatment to obtain active catalysts.

For the synthesis of Mo/V/Te/Nb/O catalysts, ammonium paramolybdate, vanadyl sulfate, ammonium niobium oxalate and tellurium oxide or telluric acid are mixed as aqueous solutions or slurries and dried at about 150 °C. These mixed slurries are then calcined at 500–650 °C in a N_2 stream [39]. In addition to the dominant M1 and M2 phases, impurity phases, for example MoO_3, $Mo_6V_9O_{40}$ and $Mo_3Nb_2O_{11}$, are also observed [28c]. It should be noted that Mo-V-Te-O catalysts cannot be obtained by the dry-up method.

The preferable method for catalyst preparation is the application of hydrothermal conditions [28a, 40], because they allow the syntheses of single-phase complex metal oxides, such as binary Mo/V/O, ternary Mo/V/Te/O and quaternary Mo/V/Te/Nb/O, which have various crystal structures [40g, 40h] and performance [41]. $Mo_{1.0}V_{0.44}Te_{0.1}O_x$ has an orthorhombic structure, $Mo_{1.0}V_{0.81}Te_{0.64}O_x$ a hexagonal structure and $Mo_{1.0}V_{0.25}O_x$ a tetragonal structure. All these basically have the same layered structure, in which networks of corner-shared MO_6 (M = Mo, V) octahedra form slabs and the octahedra between the slabs also share corner oxygen forming linear infinite chains of octahedra along the c-direction [41]. The structures differ in the arrangement of octahedra within the slabs. In the orthorhombic structure, the MO_6 (M = Mo, V) network in the slab is constructed with pentagonal and hexagonal rings of octahedra. The pentagonal bipyramidal sites may be occupied by Mo and V. Te, as the third constituent element, is exclusively located in hexagonal channels, whereas the heptagonal channels remain empty. The other structures do not possess the heptagonal rings and showed much worse performance, indicating that catalysts with the particular arrangement of MO_6 (M = Mo, V) octahedra forming slabs with pentagonal, hexagonal and heptagonal rings in the (001) plane of an orthorhombic structure is the active and selective phase in propane and ethane conversion [41]. Mo and V are indispensable elements for the structure formation. Te, located in the central position of the hexagonal ring, promoted the conversion of intermediate propene. Introduced Nb occupied the same structural position as V and the resulting catalyst clearly showed improved selectively.

Mo/V/Te/(Nb)/O catalysts owe their activity and selectivity towards propane conversion to the essential presence of so-called M1 (orthorhombic) and M2 (pseudo-hexagonal) phases [42]. The M1 phase alone is capable of selective transformation of propane [6b], and the presence of the M2 phase is claimed to improve the selectivity under more demanding conditions, such as high conversion [43]. Significantly improved acrylonitrile yields from propane above those obtained with pure M1 phase were observed using as catalyst a physical mixture with a composition of 50 wt% M1/50 wt% M2 and a surface area ratio of about 4/1. The two phases must be thoroughly ground and brought into intimate contact with each other on a micro-/nano-scale for synergy to occur.

The orthorhombic and hexagonal structures of the M1 and M2 phases was first recognized by workers at CNRS and Elf Atochem [44] using electron microscopy. They were even able to propose schematic structures for these materials, but these models do not indicate which cation is found at a particular site and thus provide no knowledge of cation co-ordination or valence state.

The M1 and M2 compounds are not simple phases. They have large unit cells and do not form single crystals. Using combinatorial synthesis, it was possible to find ways to prepare the M1 and M2 phases separately [45] and subsequently perform structural analysis combining neutron and X-ray powder diffraction data. The result for the M1 phase shows that the material has a layer structure with two sets of channels, one approximately hexagonal, the other roughly heptagonal. The channels allow the accommodation of large metal cations. This is where the large Te^{4+} cation sits. The large channel provides access to the cation's lone pair electrons, which is likely quite important in the ammoxidation reaction (Figure 20.9).

The MoVNbTeO$_x$ system for propane ammoxidation comprises three crystalline phases: orthorhombic $Mo_{7.8}V_{1.2}NbTe_{0.94}O_{28.9}$ (**M1**) (*Pba2*: $a = 21.1337$ Å; $b = 26.6440$ Å; $c = 4.01415$ Å; $z = 4$), pseudo-hexagonal $Mo_{4.67}V_{1.33}Te_{1.82}O_{19.82}$ (**M2**) (*Pmm2*: $a = 12.6294$ Å; $b = 7.29156$ Å; $c = 4.02010$ Å; $z = 4$) and a trace of monoclinic $TeMo_5O_{16}$ (*P21/C*: $a = 10.0349$ Å; $b = 14.430$ Å; $c = 8.1599$ Å; $\beta = 90.781°$; $z = 1$) [45c]. The catalytically active and selective centers reside on the surface of the basal plane of the M1 phase and are composed of an assembly of five metal oxide octahedra ($2V^{5+}_{0.32}/Mo^{6+}_{0.68}$, $1V^{4+}_{0.62}/Mo^{5+}_{0.38}$, $1Mo^{6+}_{0.5}/Mo^{5+}_{0.5}$) and two tellurium–oxygen sites ($2Te^{4+}_{0.94}$), which are stabilized and structurally isolated from each other (site isolation) by four Nb^{5+} centers, each surrounded by five molybdenum–oxygen octahedra (Figure 20.8b). The V^{5+} surface sites, distinguished through their (V^{5+} = O↔V^{4+}–O) resonance structure, are the paraffin activating sites capable of methylene-H abstraction; the Te^{4+} sites (lone pair of electrons) are required for the α-H abstraction of the chemisorbed propene molecule, once formed; and the adjacent Mo^{6+} sites for NH insertion into the π-allylic surface intermediate. Within this structure are contained all key catalytic elements needed to transform propane directly to acrylonitrile, strategically arranged and within bonding distance of each other, generating the active center of the M1 phase [45c].

Under mild operating conditions, the M1 phase alone is enough to effectively convert propane directly to acrylonitrile. Under demanding conditions symbiosis between the M1 and M2 phases occurs, with the latter serving as a co-catalyst or mop-up phase to the former, transforming unconverted, desorbed propene to acrylonitrile. The M2 phase is incapable of propane activation, lacking V^{5+} sites, but is a good propene ammoxidation catalyst. A maximum acrylonitrile yield from propane of 61.8% (86% conversion, 72% selectivity at 420 °C) was achieved with a nominal catalyst composition of $Mo_{0.6}V_{0.187}Te_{0.14}Nb_{0.085}O_x$, identified by a combinatorial methodology, which is composed of 60% M1, 40% M2 and a trace of $TeMo_5O_{16}$ [45c]. The Te environment is consistent with that observed by Millet and coworkers using EXAFS in mixed-phase samples [46].

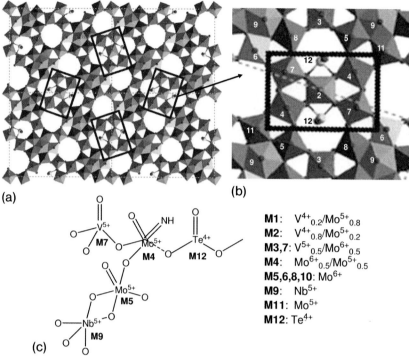

Figure 20.9 (a) Unit cell of the M1(orthorhombic) phase along the c axis with the regions considered to be the active centers in propane ammoxidation shown in rectangular boxes. (b) Enlarged view of the proposed active center in M1 phase ($Mo_{7.5}V_{1.5}NbTeO_{39}$) in [001] projection with indication (numbers) of the different atoms summarized below in the scheme of the active site. (c) Elaborated from [6a, 45].

20.3.2
Rutile-Type Antimonate Catalysts

Several papers have been published dealing with the study of metal antimonates having the rutile structure. In particular, V/Sb/O and Fe/Sb/O systems have been the subject of many investigations, aimed at understanding the nature of these mixed oxides, and at the identification of the active species.

Indeed, the preparation of a truly stoichiometric metal antimonate, $MSbO_4$ (M = metal), is a difficult task. The method of preparation employed affects the nature of the catalysts prepared, but in general non-stoichiometry is a particular feature of these systems [47–50]. The most striking case is the V/Sb/O system, for which the composition $V_{0.92}Sb_{0.92}O_4$ (*quasi*-$VSbO_4$) has been reported for the catalyst with the V/Sb atomic ratio equal to 1/1 [49]. This cation-deficient structure, which has 0.04 cationic positions unoccupied per O^{2-} anion, contains Sb^{5+}, while vanadium is present both as V^{4+} and as V^{3+}. Electroneutrality is preserved for the composition $V^{3+}_{0.28}V^{4+}_{0.64}Sb^{5+}_{0.92}O_4$ [51].

However, the ratio between V^{3+} and V^{4+} can vary depending on the method of preparation; of particular importance are the atmosphere and temperature of thermal treatment. The preparation procedure will affect the number of cationic vacancies (for V_{tot}/Sb ratio equal to 1.0) within the $V_{1-y}Sb_{1-y}O_4$ series, which is theoretically comprised between the two limiting compounds $V^{3+}_1V^{4+}_0Sb^{5+}_1O_4$ (the stoichiometric compound, with no cationic vacancies and only V^{3+}) and $V^{3+}_0V^{4+}_{0.89}Sb_{0.89}O_4$ (with only V^{4+}, and the maximum of 0.22 vacant cation positions), with y varying between 0 and 0.11 [47, 48, 52]. This series of catalysts is preferentially formed by treatment in air. Therefore, the presence of cation vacancies is a consequence of V having an oxidation state higher than 3+ which explains why V/Sb/O systems give such large deviations from stoichiometry, at least under conditions that are not the most stable from the thermodynamic point of view.

Another possibility is to have a ratio between Sb^{5+} and V different from 1.0. The $V_{0.9+y}Sb_{0.9}O_4$ non-stoichiometric series has been described, with y theoretically ranging from 0 to 0.2, and correspondingly with a degree of vacant cation positions ranging from 0.2 (in $V_{0.9}Sb_{0.9}O_4$, with $V^{3+}_{0.1}V^{4+}_{0.8}$, where the ratio V_{tot}/Sb is 1.0) to 0 ($V_{1.1}Sb_{0.9}O_4$, with full occupancy of the cation sites, and with vanadium as $V^{3+}_{0.9}V^{4+}_{0.2}$) [53]. In practice, vacant positions in $V_{0.9}Sb_{0.9}O_4$ (0.2 per unit formula) are progressively occupied by V ions, with a corresponding increase of the V/Sb ratio and of the V^{3+}/V^{4+} ratio. The latter is mainly affected by the atmosphere of thermal treatment, and therefore this parameter finally affects the extent of cation site occupancy in the rutile structure.

Berry and coworkers [47] described a compound of composition $VSb_{1-y}O_{4-2y}$, where $y \approx 0.1$, which does not formally contain cation vacancies, and which is obtained when the thermal treatment of the catalyst is carried out with oxygen-impure nitrogen. In practice, it is assumed that the ratio between V^{3+}/Sb^{5+} is equal to 1.0, and the solid solution is enriched with V^{4+} ions (and obviously with additional O^{2-} ions); which represents a solid solution between $VSbO_4$ and VO_2 (both rutile compounds). Only in the presence of oxygen-free nitrogen is it possible to obtain monophasic compounds belonging to the series $VSb_{1-y}O_{4-1.5y}$. For $y = 0.1$ the compound $V_{1.039}Sb_{0.935}O_4$ is formed, with a small degree of cation site unoccupancy (0.026 vacancies per unit formula), a V/Sb ratio higher than 1.0 and again the presence of V^{4+}.

In most cases described in the literature, the rutile V/Sb/O system has been reported to possess either equimolar amounts of V and Sb, or an excess of V. Moreover, Sb is also present exclusively as Sb^{5+}. Excess Sb, if preparations are carried out with an Sb/V atomic ratio higher than 2 and calcination is performed in air, forms either α-Sb_2O_4 or β-Sb_2O_4 (the latter at calcination temperatures higher than 800°C). Incorporation of small amounts of V in these Sb oxides is also likely, lowering the temperature of the $\alpha \rightarrow \beta$ transformation. Antimony oxide is also present in the form of amorphous oxide dispersed over the rutile, as suggested by IR spectroscopy, for Sb/V ratios between 1 and 2 [54].

The case of FeSbO$_4$ is different. It is known that the structure can host an excess of Sb. Berry and coworkers have demonstrated that on increasing the Sb/Fe ratio, an increase in cell volume of the tetragonal structure occurs [55]. Moreover, instead of a random distribution of iron and antimony in the cation sites, cation ordering has been found, with development of a tri-rutile-like structure. The change from the mono- to the tri-rutile-like ordering, from the Fe^{3+}SbO$_4$ stoichiometric compound to the Fe^{2+}Sb$_2$O$_6$ (with an Sb/Fe atomic ratio equal to 2) limiting compound occurs with the reduction of Fe^{3+} to Fe^{2+}. Therefore, it is possible that the formation of Sb-enriched solid solutions preferentially occurs with metal ions which can exist in stable 3+/2+ oxidation states. Indeed, the formation of very small amounts of crystalline (VO)$^{2+}$Sb$_2$O$_6$ has sometimes been observed [54a]. Fe/Sb/O systems also have been studied as catalysts for propane and propene ammoxidation [34b, 36, 56, 57].

The degree of structural defects can be checked by means of IR spectroscopy. Specifically, IR bands at 880 and 1015 cm^{-1} are attributed to the Sb–O–Sb and Sb–O–V^{4+} stretching vibrations involving co-ordinatively unsaturated O^{2-} ions adjacent to cationic vacancies [58]. Cationic vacancies play an important role in the catalytic performance of rutile-type mixed oxides [59–61]. For instance, it was found that an increased concentration of cationic vacancies and isolated V^{4+} species in V/Sb/(Fe)/O systems, due to the introduction of increasing amounts of Fe in the lattice, led to a proportionally higher activity [60c]. The formation of V^{4+} in Cr/V/Sb mixed oxides had the similar effect of increasing propane conversion considerably [62].

The IR band at around 820–840 cm^{-1}, also observed in all antimonates, is not typical of M–O stretching vibrations in rutile single oxides, and is also not observed in any antimony oxide [63]. This band has been attributed to the stretching vibration of Sb^{5+}–O–Sb^{3+} [50, 64], present on the surface of the rutile, and possibly constituting an active site for the reaction. Indeed, this vibration could be due to the presence of SbO$_x$-rich domains in the outer zones of non-stoichiometric rutile crystallites. Non-stoichiometric FeSbO$_4$ prepared by thermal treatment at 800 °C with excess antimony in the structure, is characterized by a tetragonal cell larger than that of stoichiometric FeSbO$_4$ and by the presence of the IR band at 820 cm^{-1}. Treatment of this compound at 1000 °C leads to (i) a decrease in the cell volume, which becomes similar to that of stoichiometric rutile and (ii) a strong decrease in the intensity of the mentioned IR absorption band [65]. Therefore, it is possible that the development to a well-crystallized, ordered rutile structure occurs as a consequence of a cation redistribution in the lattice at high temperature, with the disappearance of features related to Sb-enrichment.

The best performances among the variety of compositions which have been claimed for V/Sb/O-based systems are summarized in Table 20.2 [66]. It can be seen that Standard Oil (then BP and now Ineos) has claimed the same catalyst type for both propane-lean conditions, with high conversion of the paraffin, and propane-rich conditions, with low propane conversion.

In this system, the main catalyst component is the *quasi*-VSbO$_4$ rutile, which activates the paraffin and transforms it into the olefin-like adsorbed intermediate.

Table 20.2 Performance of some V/Sb/O-based catalysts described in the literature.

Catalyst formula	T (°C)	C_3/NH_3/O_2/H_2O/inert (mol%)	C_3H_8 conversion (%)	Selectivity to AN (%)	Ref.
$VSb_5W_{0.5}Te_{0.5}Sn_{0.5}O_x$–$SiO_2$	500	6.5/13/12.9/19.4/48.4	68.8	56.7	[31a]
$VSb_{1.4}Sn_{0.2}Ti_{0.2}O_x$	460	51/10.2/28.6/10.2/0	14.5	61.9	[29b]
$VSb_{1.4}Sn_{0.2}Ti_{0.1}O_x$	480	6.4/7.7/18.6/0/67.3	40.3	47.5	[31h]
$VSb_5Bi_{0.5}Fe_5O_x$–Al_2O_3	440	7.5/15/15/20/42.5	39	77	[34d]
$VSb_5Sn_5O_x$	450	8/8/20/0/64	30	49	[37b]

The intermediate may either desorb to yield propene, or be transformed to acrylonitrile over the SbO_x 'overlayers', the amount of which is a function of the excess Sb with respect to the rutile formula [22b].

In systems developed by Rhodia [34, 37] the main component is SnO_2 (cassiterite), which is inactive in propane ammoxidation, while it acts as the carrier for the V/Sb/O and SbO_x active components. Tin oxide is also able to disperse these components, through the dissolution of V and Sb ions, thus yielding a multi-functional catalyst where the two active compounds can effectively co-operate in the reaction.

Other antimonates have been studied as catalysts for this reaction [23, 67–70]. In the case of the Ga/Sb/O system [23, 67], a decrease in the Ga/Sb ratio (from $Ga_1Sb_1O_4$ to $Ga_1Sb_{49}O_{124}$) leads to a progressive decrease in activity and increase in selectivity to acrylonitrile. The best yield has been obtained at 550 °C with a catalyst having composition of $Ga_1Sb_{49}O_{124}$, with 28.3% propane conversion and 35.3% selectivity to acrylonitrile. The performance was improved by adding Ni, P and W as dopants. Recently, V^{5+}-doped $Al_{0.5}Ga_{0.5}PO_4$ was found to be active in propane ammoxidation, although its performance was not outstanding [71].

Rutile-type Cr antimonate, $CrSbO_4$, is fairly active and selective in propane ammoxidation [72] and adding V considerably improves the catalytic activity. The (Cr + V)/Sb atomic ratio is the main compositional parameter affecting the nature of these crystalline compounds [59a, 62, 73]. When almost equimolar ratios were used (e.g. in Cr/V/Sb 1/1/1), the corresponding XRD pattern showed the presence of a single, well-crystallized, rutile-type three-component $CrVSbO_6$ compound, which in practice corresponds to an equimolar solid solution between $CrSbO_4$ and VO_2 (both characterized by the rutile-type structure). In Cr/V/Sb/O samples with atomic ratios of Cr/V/Sb 1/x/1, the general formula $Cr_1V_xSb_1O_{4+2x}$ was extrapolated, with compositions ranging from $CrSbO_4$ to $CrVSbO_6$. In these catalysts, V was present mainly as V^{4+}, and samples were extremely active in propane ammoxidation but poorly selective to acrylonitrile, leading to the predominant formation of propene and carbon oxides. When samples were instead prepared with an atomic ratio (Cr + V)/Sb < 1, V was present mainly in the rutile mixed oxide as a V^{3+} species. The main peculiarity of these Cr-based antimonates is their ability to host excess Sb in the structure, thereby developing non-stoichiometric compounds, especially in samples with lower V content. The formation of cationic vacancies was evidenced by an intense Raman band at 880–920 cm^{-1}, especially in samples having very low (Cr + V)/Sb ratio (i.e. a large excess of Sb) and higher relative amounts of V (i.e. a Cr/V ratio close to 1).

Also with these systems, an excess of Sb with respect to the stoichiometric requirement for the rutile formation is necessary, in order to reach good selectivity to acrylonitrile and low selectivity to carbon oxides.

An important dopant for rutile-based mixed oxides is Nb oxide [59a]. The Nb/Sb/O system is active in the ammoxidation of ethane to acetonitrile [74], and the combination of Bi/Mo/O and alumina-supported Nb_2O_5 gives good performance in the ammoxidation of isobutane to methacrylonitrile [75]. Nb is one component of the V/Nb/Sb/O catalyst for propane ammoxidation developed by Nitto Chemical Industries [35]. When used as the support for V/Sb mixed oxide, Nb_2O_5 formed new phases by reaction with V and Sb under catalytic reaction conditions [76]. These phases, of unclear nature, affected catalytic performance in propane ammoxidation. When Nb was instead added as a promoter for the alumina-supported V/Sb/O system [77], the interaction between the active components V, Sb and Nb led to an improvement of catalytic performance with respect to undoped V/Sb/O. However, the Nb–Sb interaction may also lead to the development of $SbNbO_4$, which is not active for propane ammoxidation, and, in addition, removes Sb and Nb sites that would have been co-ordinated with V to form efficient V–Sb or V–Nb species. Nb also develops mixed (Cr)V/Nb/Sb oxides with the rutile structure, and hence modifies the properties of Sb cations in allylic ammoxidation [59a].

The incorporation of other elements, either trivalent or tetravalent, in the rutile lattice may substantially improve the performance of V antimonate [21]. This is the case for Al^{3+}, Ti^{4+} and W^{4+}. In the Al-Sb-V–O system, the active phase was identified as having the composition $Al_{1-x}SbV_xO_4$ with $0 < x < 0.5$ [78]. (Al,V)SbO_4 can be described as a solid solution between rutile-type $AlSbO_4$ and quasi-$VSbO_4$. A yield to acrylonitrile of almost 40% has been reported for an Al-Sb-V-W oxide catalyst, using feed ratios of propane, oxygen and ammonia that are stoichiometric with respect to the formation of acrylonitrile [26b]. A comprehensive study of the Al-Sb-V-W-O system has shown that the active phase in this system is $Sb_{0.9}V_{0.9-x}W_xO_4$, which is a solid solution between $V_{0.92}Sb_{0.92}O_4$ and rutile WO_2. Titanium substitution in rutile V/Sb/O produced an improved selectivity to acrylonitrile [79]; Ti^{4+} replaced for both V^{4+} and V^{3+}/Sb^{5+} pairs, forming solid solutions of continuous composition.

It was demonstrated that the better catalytic properties of (Al,V)SbO_4 and of quasi-Sb(V,W)O_4 compared with the pure rutile V/Sb/O phase could be rationalized in terms of the 'site-isolation' theory. According to this theory, which was originally formulated by Callahan and Grasselli [6c, 80], a catalyst that is selective for partial oxidation can be developed by creating structural isolation of the active site. Thus, the improvement of the selectivity to acrylonitrile by substitution of some of the V atoms with Al or W can be explained by the fact that substitution creates isolation of the remaining vanadium atoms to a suitable degree, preventing combustion of the reactants and of the desired product.

Recently, an *operando* Raman technique was used to study the relationship between surface changes during reaction and catalytic performance using VSbO$_4$-based catalysts [81]. It was observed that the active rutile VSbO$_4$ phase forms during catalytic operation by reaction-induced interaction between surface vanadium and antimony oxide species.

20.4
Alkylaromatic Ammoxidation

Various aromatic nitriles such as benzonitrile, phthalonitrile, isophthalonitrile, terephthalonitrile and nicotinonitrile have applications as chemical intermediates. Nicotinonitrile can be hydrolyzed to nicotinoamide or nicotinic acid, which are used in the synthesis of vitamin B. Isophthalonitrile is used in the production of herbicides/fungicides. Phthalonitrile is an intermediate for phthalocyanine pigments. In addition, substituted aromatic nitriles have applications in the synthesis of several intermediates for applications as fine chemicals and in the life sciences.

20.4.1
Alkylbenzenes and Substituted Alkylbenzenes

Various catalysts have been reported to be active in this class of reaction (Scheme 20.1): (i) vanadium supported on TiO$_2$ (preferably in the anatase form) or ZrO$_2$ and modified by various dopants such as Cs, P, W; (ii) multi-component molybdate; (iii) doped vanadium antimonate; (iv) bulk V-P-O; (v) metal-containing (in particular Cu and Mo) zeolites; (vi) supported heteropolyacids (PV$_3$Mo$_{12}$O$_x$ on silica) and (vii) metal phosphates [82–91]. Other proposed catalysts are AgAlBO$_x$ and perovskite materials such as YBaCu$_3$H$_6$ and metal fluorides such as CeF$_3$, AlF$_3$ and MgF$_2$ [92]. However, the best catalysts belong to groups (i), (iii) and (iv), even though it should be noted that, depending on the specific substituents on the aromatic, a different ranking of these catalysts can be observed.

Various reviews have discussed in detail the characteristics and performance of these catalysts and the reaction mechanism in alkylaromatic ammoxidation [93–95]. Over V$_2$O$_5$/Al$_2$O$_3$ catalysts, the reaction mechanism proceeds via the

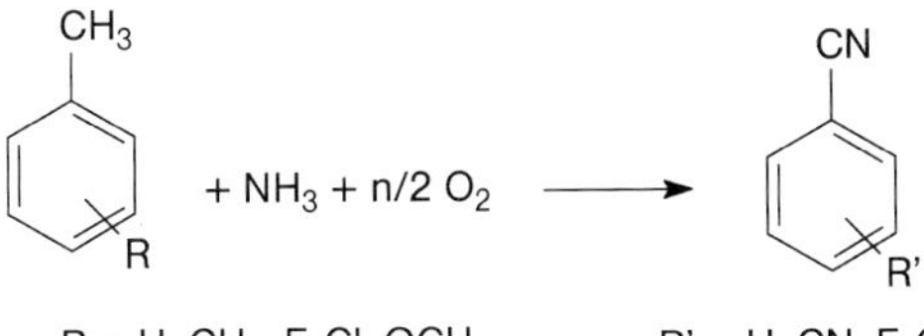

CH$_3$-C$_6$H$_4$-R + NH$_3$ + n/2 O$_2$ → CN-C$_6$H$_4$-R'

R = H, CH$_3$, F, Cl, OCH$_3$ R' = H, CN, F, Cl, OCH$_3$

Scheme 20.1 Synthesis of aromatic nitriles by catalytic ammoxidation.

reaction of surface ammonium ions with benzoate ions, which are detected as surface species by IR spectroscopy [96]. Adsorbed benzylamine and benzaldehyde species were identified as reaction intermediates [97], while a kinetic study by Otamiri and Andersson [98] suggests that vanadium imido (V=NH) or hydroxylamino (V—NHOH) species may play a role as nitrogen insertion sites on the catalyst surface. On vanadium phosphate catalysts, in particular $(VO)_2P_2O_7$, the preferred role of a benzaldehyde-like species was also pointed out, whereas the reaction pathway via benzamide or benzylamine intermediates was suggested to be unlikely [99].

Adsorbed ammonia can exist in three different kinds of nitrogen-containing surface species such as a protonated cation (NH_4^+), and co-ordinately adsorbed NH_3 and $-NH_2$ groups. In addition, vanadium imido or hydroxylamino could be also present, especially upon interaction of ammonia with the catalyst at high temperature. All these species may potentially act as N-insertion species, and could be present and react on the catalyst surface during catalytic reaction. Therefore, unique mechanistic conclusions are difficult to obtain because multiple reaction pathways are clearly present, the relative reaction rates of which depend on the reaction conditions. However, in general terms it may be proposed that owing to the stronger chemisorption of the aromatic ring, the mechanism of reaction in alkylaromatic ammoxidation is different from that observed in propene and propane ammoxidation.

Besides selectivity to aromatic nitriles, the minimization of the direct ammonia oxidation to N_2 side-reaction is a critical factor, because otherwise runaway conditions may be possible. There is always competition between NH_3 oxidation and ammoxidation on the catalyst surface. Furthermore, contact between ammonia and the catalyst surface, particularly at high temperatures, causes a partial reduction of the oxide surface because of NH_3 oxidation to N_2. Therefore, control of the rate of unselective oxidation of ammonia to N_2 is an important factor in determining the selectivity of the nitrile product, because this side reaction limits the availability of the surface ammonia species that are necessary for nitrile synthesis.

Typical performance is a selectivity higher than 90% at a conversion in the 50–90% range, mainly depending on the type of substrate. Substituted pyridines yield better selectivities at high conversion than the equivalent alkylaromatics. The nature and position of the substituents in substituted alkylaromatics also play an important role in determining selectivity and activity. The commercial application of this technology is mainly hindered by the relatively small plant necessary for these products as compared to full-scale processes. The further implementation of the process of alkylaromatic catalytic ammoxidation would thus require the development of multi-purpose small-size continuous plants using small fluidized bed-reactors (to better control temperature and allow easier substitution of the catalyst).

Aromatic nitriles are relatively stable under oxidation conditions even at higher temperatures compared to the equivalent oxy products. Therefore these compounds are often used as intermediates for subsequent hydrogenation or reduction steps (Scheme 20.2).

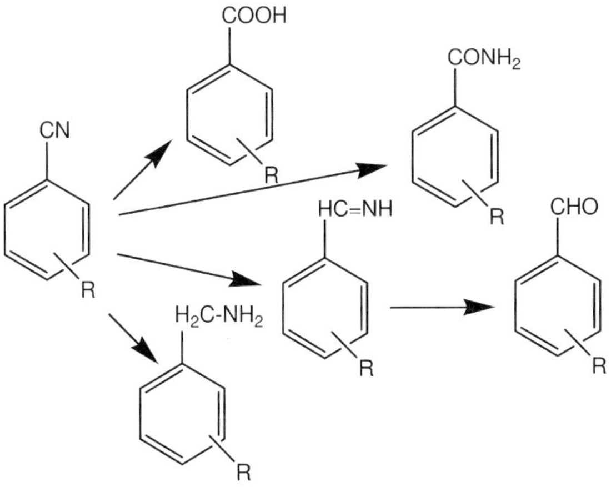

20.4 Alkylaromatic Ammoxidation

Scheme 20.2 Substituted aromatic nitriles as intermediates for aromatic acids, amides, imines, aldehydes and amines. Elaborated from [82a].

The reactivity of an alkylbenzene in the ammoxidation reaction increases with increasing size of the alkyl groups in the side chain. The rate of ammoxidation of alkylbenzenes mainly depends on the chemical nature, size and structure of substituent [92a]. The ammoxidation of substituted toluenes such as 1,2-,1,3- or 1,4-methyl-substituted toluenes can result in mono- or dinitriles [100]. There are a few reports on shape-selective ammoxidation, one such example being the ammoxidation of m- and p-xylenes over Cu-containing ZSM-5 zeolites to their corresponding mono- and dinitriles [87, 88]. Besides dialkyl-substitutions, higher methylated benzenes (e.g. mesitylene) can also be ammoxidized. However, with increase in the number of methyl groups the number of products will also increase. Furthermore, ammoxidations of substituted methylaromatics such as α-methylstyrene to atroponitrile [101a], and β-methylstyrene or allylbenzene to cinnamonitrile [101b] have also been reported in the literature. In these cases also, processes have never reached the commercial stage.

A commercially interesting area of application of alkylaromatic conversion is that of ammoxidation of halogen-substituted toluenes. The activity and the nitrile selectivity of the ammoxidations of substituted toluenes depend strongly on the position of the substituent, because it influences the accessibility of the methyl group and especially the strength of the chemisorption and the stabilization of reaction intermediates. Electron-withdrawing substituents enhance the nitrile yield, whereas electron-donating substituents cause an increased total oxidation. Therefore, halogen-substituted toluenes are easily converted into nitriles [102]. The conversion rate of p-halotoluenes was found to be nearly independent of the nature of the halogen substituent, but the selectivity decreases in the sequence p-Cl > p-Br >> p-I over V-P-O catalysts. The ammoxidation of the isomeric chlorotoluenes shows a dependence of the position of substitution in conversion ($p \gg o > m$) and selectivity ($p > o > m$.)

The reaction of dichloro-substituted toluenes is much more influenced by the geometric position of the substituents. Closer proximity of the substituents to the methyl group results in lower conversions and nitrile selectivities (2,6-di-Cl < 2,5-di-Cl < 2,3-di-Cl < 2,4-di-Cl < 3,4-di-Cl) [102b, 103]. 2,6-Dichlorobenzonitrile is of particular industrial importance for the production of herbicides and pesticides, and for the preparation of special kinds of engineering plastics with high thermal resistance. Halogen-substituted xylenes can also be ammoxidized to their corresponding nitriles, for example 3,4,5,6-tetrachlorophthalodinitrile is formed from the corresponding chloro-substituted xylene in 45% yield [104]. Recently, especially in China, there has been renewed interest in these reactions.

The ammoxidation of toluenes substituted with electron-donating groups, for example hydroxy- and alkoxy-substituted toluenes is rather less selective. However, under carefully chosen conditions (choice of the catalyst, feed composition, reaction conditions) adequate yields of nitriles can be achieved. Stability of the catalyst performances is typically an issue.

Aromatic imides are another type of product which can be synthesized by catalytic ammoxidation. o-Xylene is converted over vanadium-titanium oxide catalysts to tolunitrile and then, depending on catalyst composition and reaction conditions, phthalimide or phthalonitrile can be selectively synthesized (Scheme 20.3) [94].

The product distribution between nitrile and imide depends upon the reaction conditions and the nature of the catalyst used [105]. The influence of various reaction parameters such as (i) reaction temperature; (ii) water vapor addition in the feed gas; (iii) NH_3/o-xylene mole ratio and (iv) space velocity, were studied [105]. The removal of water vapor from the feed gas has a highly pronounced promotional effect on the selectivity of phthalonitrile. The nitrile selectivity increased from 2.1 to 34% at the expense of phthalimide (which decreased from 53 to 9%) with the complete removal of water vapor in the reactant feed mixture. This observation gives an indication that phthalonitrile being formed in the reaction is further hydrolyzed to phthalimide via the amide intermediate in presence of water

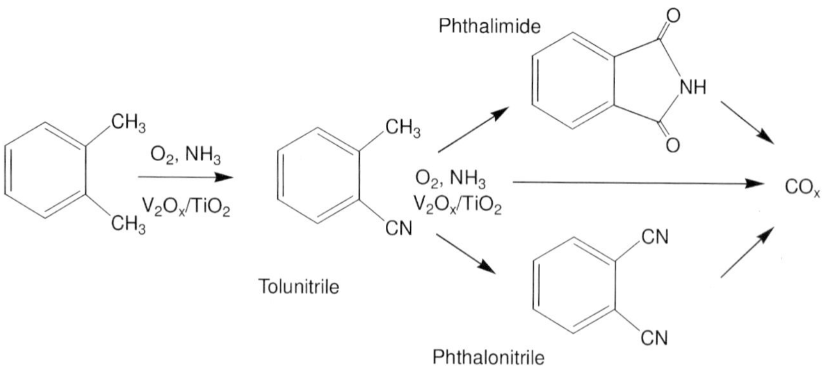

Scheme 20.3 Reaction network in o-xylene conversion to phthalimide and phthalonitrile. Adapted from [84].

vapor. Another interesting aspect is that titania-supported catalysts always gave significantly higher selectivities of phthalimide compared to phthalonitrile. Phthalimide yields of up to 80% were reached on a VPO/TiO$_2$ catalyst [105].

20.4.2
Alkylaromatics Containing Hetero-Groups

Another industrially relevant area of application of catalytic ammoxidation is the synthesis of cyano-pyridine. A relevant example is the new green process for producing nicotinic acid (niacin) developed by Lonza Ltd. Co-enzyme I (nicotinamide-adenine dinucleotide NAD) and co-enzyme II (nicotinamide-adenine dinucleotide phosphate NADP) are required by all living cells. They enable both the conversion of carbohydrates into energy and the metabolism of proteins and fats. Both nicotinamide and nicotinic acid are building blocks for these co-enzymes. Since the human body produces neither nicotinic acid nor the amide, it is dependent on intake via foodstuffs. Old methods to produce nicotinic acid cause large production of waste and use toxic reactants. For example, the classic method of preparing nicotinic acid was by oxidizing nicotine with potassium dichromate, which produces over 9 tons of waste per ton of nicotinic acid, and Cr^{VI} is carcinogenic and environmentally threatening.

A new route to prepare nicotinic acid starts from 2-methylglutaronitrile, a major side-product in the adiponitrile process and, as such, a readily available starting-material. It is easily hydrogenated to 2-methylpentanediamine, which is then condensed to methyl piperidine and dehydrogenated to 3-picoline. The gas-phase ammoxidation of the latter to cyanopyridine is followed by hydrolysis to either nicotinamide or nicotinic acid (Scheme 20.4). The cyanopyridine route for the production of nicotinic acid has the advantage of a significantly better selectivity with respect to the direct oxidation route from 3-picoline owing to the easy decar-

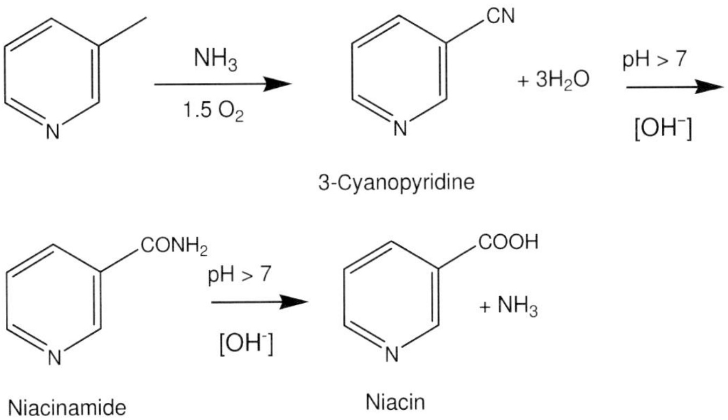

Scheme 20.4 Ammoxidation of 3-picoline and hydrolysis of cyanopyridine to niacinamide (nicotinamide) and niacin (nicotinic acid). Adapted from [106].

borboxylation and further oxidation of nicotinic acid during gas-phase oxidation. On the other hand, the cyanopyridine route involves more reaction steps and consumption of ammonia, even if it can be in part recycled.

In addition to ammoxidation of β-picoline (3-picoline), the conversion of γ-picoline (4-picoline) and α-picoline is also industrially relevant for the manufacture of various interesting intermediates starting from the corresponding cyanopyridines. Koei Chemical Company, for example, offers a range of cyanopyridine and cyanopyrazines produced by vapor-phase catalytic ammoxidation. The reaction and related catalysts have been known for a long time. In general, V antimonate-based catalysts active in propane ammoxidation also show quite good performance in methylpyridine and m-xylene ammoxidation to cyanopyridine and isophthalonitrile, respectively [107]. Vanadium on a variety of supports, such as TiO_2 [108], ceria [109] and CeF_3 [110] is active in the reaction, with yields up to about 80%. More recent proposed catalysts include (i) a vanadium-incorporated ammonium salt of 12-molybdophosphoric acid supported on titania or zirconia [111]; (ii) V_2O_5 supported on zirconium phosphate [112]; and (iii) metal phosphates (α- and β-$VOPO_4$) [113]. Despite the large research effort, no significant improvement in catalytic performance with respect to earlier investigated catalysts has been achieved.

The reaction mechanism on a vanadia-titania catalyst has been re-investigated using FTIR spectroscopy [114], although the conclusions were not markedly different from earlier proposals. The interaction of methylpyrazine with the catalyst surface involves a consecutive transformation of co-ordinatively bound methylpyrazine into oxygenated surface compounds, namely an aldehyde-like complex and an asymmetric carboxylate. The main reaction product, amidopyrazine, is formed through the interaction of the surface oxy-intermediates with adsorbed ammonia species.

In conclusion, research and industrial interest in the ammoxidation of methylpyridine and methylpyrazine is still active. This process could offer a greener alternative to current processes in the production of intermediates for a variety of specialty chemicals, especially in rapidly developing locations such as China where a growing interest in the reaction has been noted [115, 116].

20.4.3
Ammonolysis vs Ammoxidation

The ammoxidation of methylaromatic compounds has been shown to be a convenient route to produce many nitriles required for further syntheses of side-chain functionalized products. This method is versatile and can be carried out very easily because of the stability, and undamaged desorption, of the nitriles formed under severe gas-phase conditions. A series of interesting new patents on catalysts for the ammoxidation of alkylaromatics have been issued, indicating that there is still commercial interest in this reaction [117].

Another approach developed by Lummus Co is used commercially for isophthalonitrile production from m-xylene [118]. Oxidation is carried out in the absence of gaseous oxygen (ammonolysis) using lattice oxygen from the catalyst, which is regenerated in a separate vessel. The approach is analogous to the Dupont process

[119] for maleic anhydride production from *n*-butane. A fluidized catalyst is circulated continuously from a reaction zone to a regeneration zone where it is brought into contact with the aromatic + ammonia feed and air, respectively.

The reactor effluent contains the nitrile product, intermediates, unconverted feedstock for recycling, ammonia, water vapor, nitrogen, carbon oxides and traces of HCN. Separation is carried out by condensation to recover most of the organic components, which are separated (depending on the specific substrate converted) by distillation, extraction or crystallization. Off-gases are scrubbed with basic and then acidic solutions to eliminate CO_2 and NH_3, respectively, and then recycled.

The advantages of ammonolysis over ammoxidation are safer operation and higher selectivities, but fixed capital and running costs are higher and, furthermore, the process may be adapted only with difficulty to multi-purpose plants that require easy catalyst loading/unloading operations.

20.5
Ammoxidation of Unconventional Molecules

The literature on ammoxidation is very wide. The majority of papers and patents published in this field deal with propene and propane ammoxidation to acrylonitrile, of isobutane and isobutene ammoxidation to methacrylonitrile and methylaromatics and methylpyridines (picoline) ammoxidation to the corresponding cyano-containing compounds, as discussed in the previous sections. A small amount of literature deals with the ammoxidation of the following molecules:

- linear C_4 aliphatic hydrocarbons (*n*-butane, butenes, butadiene) to maleonitrile and fumaronitrile;
- cyclohexanol and cyclohexanone to adiponitrile;
- *n*-hexane and cyclohexane to adiponitrile;
- benzene to C_4 and C_6 unsaturated dinitriles.
- ethane ammoxidation to acetonitrile

Since these results are far less well known than those reported for C_3 and alkylaromatic ammoxidation, they are discussed in a more detail below.

20.5.1
The Ammoxidation of C_4 Hydrocarbons

Furuoya and coworkers [120] investigated the ammoxidation of butenes, *n*-butane and butadiene using different types of catalyst. The best results were obtained with a TiO_2-supported mixed active phase catalyst containing V, W, Cr and P, and using butadiene as the substrate. The approximate composition of the catalyst giving the best performance is (atomic ratios): $V_1W_{1.1}Cr_1P_{22}Si_{0.1}O_{x-87}TiO_2$. There is an equimolar amount of V, W and Cr, with a large excess of P. The formation of metal phosphates during the calcination treatment is highly likely.

The highest yield reported to a mixture of fumaronitrile (*trans*-1,4-dicyano-2-butene) and maleonitrile (*cis*-1,4-dicyano-2-butene) is 66.8% at 560°C (conversion

Table 20.3 Summary of results for butadiene ammoxidation with the catalyst V/W/Cr/P/O-TiO$_2$[120]

Feed composition C$_4$/NH$_3$/air	T (°C)	τ (s)	Butadiene conversion (%)	Yield to MN + FN (%)
1/5/94	552	1.8	96.8	49.3
1/5/94	517	3.6	nr	49.1
0.5/2.5/97	560	1.2	nr (100%)	66.8
0.5/2.8/96.7	498	1.2	99.7	60.7

FN: fumaronitrile; MN: maleonitrile. The two compounds are reported to form with a ratio close to 1. τ = contact time, nr = not reproted.

is not reported, but extrapolation from the other results suggests it is likely total), with a contact time of 1.2 s, a dilute butadiene stream (0.5 mol%) and a butadiene/NH$_3$/O$_2$ feed ratio of 1/5/20. By-products were acetic acid, HCN, CO, CO$_2$ and traces of crotononitrile (these compounds were not detailed). The effects of important reaction parameters are summarized in Table 20.3.

The catalytic system appears to give interesting results. It is worth noting that in order to obtain a good performance, it is necessary to combine the properties of several elements. In fact, some of the corresponding bi-component systems (V/P, W/P, W/Cr) although active in butadiene conversion, were less selective to nitriles with the best yield being lower than 20%. However, from the data reported, it is very difficult to identify the role of each component in the reaction. In general, one might believe that V, W and Cr should play the same role of activation of the hydrocarbon, while P should play the role of an enhancer of surface acidity. The latter property is indeed very important, owing to the basic characteristics of both the reactant and the products. However, from the data reported it appears that the important step is not the activation of butadiene, but rather its transformation to the nitriles. This means that the adsorption properties of the catalyst and the insertion of N into the molecule are the key steps for the performance.

The considerations reported above suggest that the mechanism of reaction might not be the same as the well known allylic insertion of a nucleophilic NH^{2-} species onto the activated hydrocarbon, to generate the precursor of the cyano group. Indeed, none of the elements included in catalyst formulation is able to produce the M=NH species (which is generated by Mo and Sb in propene ammoxidation catalysts).

In view of this, the mechanism might consist of the following steps:

- Co-ordination of butadiene to the metal (V or W), with electron transfer from the diolefin to the metal ion, and intramolecular rearrangement of the residual double bond.
- Sequential attack of ammonia from the gas phase onto the activated hydrocarbon at the C1 and C4 atoms, with formation of amino groups and then nitriles (possibly via radical reactions).
- Desorption of the compound and re-oxidation of the reduced metal ion by O$_2$.

Therefore, the co-ordinating properties (intrinsic Lewis acidity) and the redox properties (electron transfer) of the metal ion are important for the first step, and for the transformation of the adsorbed hydrocarbon into the nitrile rather than to carbon oxides and HCN. It is also worth noting that when a conventional alkene ammoxidation catalyst is used, such as Bi/Mo/O, the performance is much worse (yield 4.5% at 450 °C).

An alternative mechanism might include the following steps: (i) transformation of butadiene to maleic anhydride (catalyzed by the V/P/O elements included in the catalyst); (ii) hydrolysis of the anhydride to the diacid (owing to the presence of water in the reaction environment, and catalyzed by the presence of P); (iii) transformation of the diacid to the diamide by reaction with ammonia and (iv) (oxy)dehydrogenation of the diamide to the dinitrile.

The best performance is obtained with very dilute hydrocarbon streams. This might be due to the fact that an excessive concentration of butadiene might lead to a high catalyst surface coverage and over-reduction, with a consequent modification of the adsorption–activation properties. The high selectivity achieved at high temperature indicates that the two cyan groups are very stable.

The catalyst based on V/W/Cr/P/O-TiO$_2$ also gives good performance with n-butene and n-butane reactants (Table 20.4).

The catalyst showed comparable selectivities from the various C$_4$ hydrocarbons. In the case of n-butane the yield was lower owing to the lower hydrocarbon conversion, but the selectivity remained close to 50% (46%). In this case, vanadium played the additional role of oxydehydrogenation of butane to butenes. The reactivity of butene was lower than that of butadiene (both were higher than that of n-butane) which indicates that the mechanism requires the oxy-dehydrogenation of butene to yield butadiene, which is the reactive intermediate that undergoes ammoxidation.

A recent patent by DSM [121] claims catalysts very similar to those reported by Furuoya [120]. Indeed, the catalyst preparation is the same as the one described by Furuoya [120b]. The only difference concerns the addition of silica in relevant amounts, in the form of silica sol. This is claimed to give better reproducibility in performance for catalysts prepared in the powder form (by spray-drying). The best performance reported is 95% conversion and 58% yield to nitriles (fumaronitrile + maleonitrile + succinonitrile, the saturated dinitrile). The authors report the use of a low O$_2$ concentration, with a feed composition of 0.50% butadiene, 2.50% ammonia and 4.4% oxygen (the balance being nitrogen).

Table 20.4 Ammoxidation of C$_4$ hydrocarbons with the V/W/Cr/P/O-TiO$_2$ system [120].

Feed composition C$_4$/NH$_3$/air	T (°C)	τ (s)	C$_4$ conversion (%)	Yield to MN + FN (%)
n-butene: 0.7/2.6/96.7	531	1.2	96.5	56.8
n-butane: 0.6/3.7/95.7	598	1.2	57.0	26.2

Table 20.5 Summary of performance of catalysts reported in [122] for butadiene ammoxidation.

Catalyst	T (°C), τ (s)	Butadiene conversion (%)	Selectivity to FN + MN (%)	Selectivity to crotonitrile (%)	Selectivity to CO, (%)
V/Mo/O	450, 0.04	20	18	13	18
Sb/Fe/O	450, 0.04	16	9	11	ng
Bi/Mo/P/O	450, 0.04	9	51	11	18

Feed: butadiene/oxygen/ammonia/nitrogen/steam (molar ratios): 2/19/4/74/1.
FN: fumaronitrile; MN: maleonitrile.

Colleuille and coworkers [122] investigated catalysts for butadiene ammoxidation which are similar to those also studied in the ammoxidation of benzene (see below). Table 20.5 summarizes the results reported. The main products were fumaronitrile and maleonitrile, crotonitrile (the unsaturated mononitrile, 1-cyanopropene, with the two trans and cis isomers) and CO_x, with traces of acrylonitrile and furan. The residence time used was very low; in this case the best performance was obtained with a typical propene ammoxidation catalyst, made of Bi/Mo/P/O under conditions of low butadiene conversion.

Very interesting new types of catalyst, based on Re-Sb-O compounds ($SbRe_2O_6$, $SbOReO_4 \cdot 2H_2O$ and $Sb_4Re_2O_{13}$), were shown to be selective in the ammoxidation of isobutylene to methacrylonitrile [123]. The Re-based catalysts were active for methacrylonitrile synthesis with selectivities of 47.9–83.6% at 400°C, depending on composition. Bulk Sb oxides (Sb_2O_3 and Sb_2O_4) showed no activity. Among these catalysts, $SbRe_2O_6$ was most active and selective (83.6%) for methacrylonitrile formation at 400°C. No structural change in the bulk and surface of $SbRe_2O_6$ after i-C_4H_8 ammoxidation was evidenced by X-ray diffraction, X-ray photoelectron spectroscopy, scanning electron microscopy or confocal laser Raman microspectroscopy. The good performance of $SbRe_2O_6$ may be ascribed to its specific crystal structure composed of alternate octahedral $(Re_2O_6)^{3-}$ and $(SbO)^+$ layers. Pulse reaction studies suggested that adsorbed NH_3 species on the $SbRe_2O_6$ catalyst facilitated the adsorption and subsequent activation of isobutylene.

The same selective behavior in isobutane conversion to methacrylonitrile was reported, although at very low conversion (selectivity to methacrylonitrile at an isobutane conversion of 3 to 4% was about 45 to 50%), [123b].

20.5.2
The Ammoxidation of Cyclohexanol and Cyclohexanone

In one of the most controversial papers published in this field [124], a catalyst made of an alumina-supported V/P/Sb/O is reported to give very good performance in the ammoxidation of cyclohexanol and cyclohexanone to adiponitrile.

The elements claimed are the same as those reported in References [120, 121]. The preparation procedure employed is known to lead to the formation of $VOPO_4$, rather than $(VO)_2P_2O_7$. The presence of Sb, however, may lead to a modification of the structural features. Indeed, the authors claim the presence of vanadyl pyrophosphate as the major phase present in catalysts, with a minor amount of vanadium phosphate. The atomic ratio between the components of the γ-alumina-supported active phase was V/Sb/P 1/1.9/1.18. The reaction conditions were 425 °C (at which the best yields were reported), and a feed ratio of reactant/air/ammonia of 0.6–1.0/4.2/1.5. The following results were claimed under these conditions:

- from cyclohexanol, conversion 78%, selectivity to adiponitrile 75%, selectivity to hexanenitrile 18%. The by-products were CO, CO_2 and cracking products,
- from cyclohexanone, conversion 60%, selectivity to adiponitrile 48%, selectivity to hexanenitrile 30%. The by-products were CO, CO_2 and cracking products.

This performance is outstanding, but it has never been successfully reproduced by other teams active in this research field.

The reaction of cyclohexanol ammoxidation has been investigated by Chen and Lee, with a V_2O_5 catalyst [125]. The feed mixture comprised cyclohexanol/oxygen/ammonia in the ratio 1.2/9/15. A large excess of ammonia was thus used. Cyclohexanone, adiponitrile, adipic acid, benzene and CO_2 formed. The best yield to adiponitrile was 4%, with a conversion of approximately 52% (as inferred from the sum of the yields), at $W/F = 0.45\,g\,s\,cm^{-3}$ and 365 °C. At a higher W/F value, the maximum yield to cyclohexanone was reached (19%). All these products underwent consecutive reaction to benzene and CO_2. In a pulse reactor, the yield to adiponitrile was 6.3%, at 96% reactant conversion. Cyclohexanone was also used in the pulse reactor, but the predominant product was CO_2. The authors proposed that the reaction pathway includes the dehydrogenation of cyclohexanol to cyclohexanone, its oxidative cleavage to adipic acid and subsequent transformation to the dinitrile. The formation of benzene occurs via dehydration to cyclohexene, followed by dehydrogenation. The ammoxidation of cyclohexene exclusively yielded benzene.

The ammoxidation of cyclohexanol or cyclohexanone is also reported in one patent [126], using a cyclohexanone/air/ammonia/H_2O ratio of 1/10/2.5/10 (the presence of added steam is noteworthy), 450 °C, and a contact time of 1 s. Several catalysts were used, as reported in Table 20.6. The products obtained were aniline, phenol and adiponitrile.

The nature of the products formed is rather unusual. The formation of aniline implies a reaction between the ketone and ammonia to yield the cyclohexanoneimine. The latter then rearranges with aromatization, to yield aniline. Aromatization occurs with cyclohexanone, leading to the formation of phenol. Therefore, the formation of the aromatic ring is quicker than the opening of the aliphatic cycle. This is the key point of the reaction involving cyclic reactants: the competition between the parallel reactions of ring opening and aromatization controls the selectivity.

Table 20.6 A summary of catalyst performance reported in [126] for cyclohexanone ammoxidation.

Catalyst	Comment	Yield aniline (%)	Yield phenol (%)	Yield ADN (%)
K/Ni/Co/Fe/Bi/P/Mo/O-silica	Typical catalyst for propene ammoxidation	3.1	11.6	tr
V/W/Mo/O-silica	Typical catalyst for electrophilic oxidation	10.2	19.2	0.6
P/Mo/O-silica	Heteropolycompound	8.9	10.0	–
W/O-silica		5.4	7.8	–
Fe/V/Sb/O-silica	Catalyst for propane and propene ammoxidation	0.1	18.9	0.3
Sb/Mo/O	Catalyst for allylic (amm)oxidation	–	3.4	–

ADN: adiponitrile

20.5.3
The Ammoxidation of Cyclohexane and *n*-Hexane

Some papers and patents report on the ammoxidation of cyclohexane in the gas phase [124, 127–129] or in the liquid phase [130]. In Reference [124], which was discussed in the previous section, the ammoxidation of cyclohexane is reported to give 50% adiponitrile selectivity and 35% hexanenitrile selectivity for a conversion close to 70%, at 425 °C.

Osipova and coworkers [127] used a catalyst based on Ti/Sb mixed oxide, calcined at 750 °C. Under these conditions, titania is in the rutile form, and can host up to 7 mol% Sb_2O_5 in solid solution. Higher Sb content led to the development of also an equimolar Ti/Sb compound ($TiSbO_4$), while in systems with more than 50 mol% Sb_2O_5, the additional formation of α-Sb_2O_4 was found. A catalyst with the composition 70% Sb_2O_5–30%TiO_2 gave the best performance.

Catalytic tests were run in a pulse reactor, at 400 °C, with a cyclohexane/oxygen/ammonia feed composition in mol% of 3/6/4 (the balance being He). The main products obtained were adiponitrile (ADN) and benzene, with an overall selectivity of more than 90% (the cyclohexane conversion is not reported). The rates of benzene and ADN formation are plotted in Figure 20.10 as functions of the Sb_2O_5 content of catalysts. It is shown that the overall formation of benzene considerably decreased on increasing the amount of Sb in catalysts. The formation of ADN decreased, but the decrease was less pronounced than that of

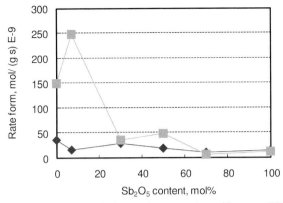

Figure 20.10 Rates of adiponitrile (ADN) (◆) and benzene (■) formation as functions of the content of Sb_2O_5 in catalysts for cyclohexane ammoxidation. Elaborated from [127a].

benzene. Moreover, the formation of benzene was much greater than that of ADN for low Sb contents, while the two rates became comparable for intermediate and large values of Sb content; therefore, the selectivity to ADN increased on increasing the Sb content in catalyst.

Osipova and coworkers [127] also found that under non-steady conditions (those utilized in the pulse reactor), a considerable fraction of ADN remained adsorbed on the catalyst. A rough estimation for the catalyst with 30% Sb_2O_5 led to the conclusion that the overall amount of ADN produced in the single pulse approximately corresponded to 1/3 of the reacted cyclohexane. In particular, ADN bound strongly to the catalyst surface and made heavy (polymeric compounds), at least at 300 °C and in the presence of ammonia. Adsorbed ammonia contributed to the polymerization of ADN at the catalyst surface.

Simon and Germain [128] investigated several catalytic systems and found that cyclohexane conversion is high (70%) even in the absence of catalyst at 460 °C (feed cyclohexane/oxygen/ammonia 1/3.6/1.5) yielding mainly cyclohexene, benzene and CO_x, with minor amounts of other compounds.

The presence of a catalyst led to the formation of C_4 dinitriles (maleonitrile, fumaronitrile, succinonitrile), C_5 dinitriles (glutaronitrile) and C_6 dinitriles (muconitrile, adiponitrile), but the yield of these compounds was very low. In the best case, with a V/Mo/O catalyst (atomic ratio V/Mo 4/1; phase: V_2O_5), the yield to maleonitrile was 1.9% and 0.8% to fumaronitrile, 17% to benzene, 23% to CO_x, with traces of mucononitrile, at a conversion of 57% at 460 °C. With the same catalyst, the initial selectivity (extrapolated at zero conversion) to C_4 nitriles was approx 5% (negligible to other nitriles), while the predominant primary products were benzene and carbon oxides. For temperatures lower than 420 °C the predominant product was cyclohexene, while at higher temperatures benzene and CO_x prevailed (Figure 20.11).

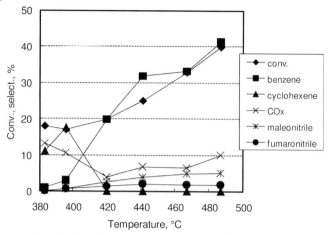

Figure 20.11 Effect of temperature on catalytic performance of V/Mo/O catalyst in cyclohexane ammoxidation [128].

The performance was very similar with a Ti/Mo/O catalyst (atomic ratio Ti/Mo 1.14/1; phases TiO_2 and MoO_3) and with a Bi/Mo/O catalyst (atomic ratio Bi/Mo 0.9/1.) Other catalysts (Sn/Mo/O, Sb/Mo/O, Sn/Sb/Fe/O) were more selective than V/Mo/O to either benzene or CO_x.

The mechanism of reaction included the rapid oxy-dehydrogenation of cyclohexane to cyclohexene and benzene. The latter was then transformed to successive products (see the section on benzene ammoxidation). The opening of the ring may occur either for cyclohexene or for benzene itself.

Table 20.7 summarizes the performance of catalysts described in a patent [129a] which reports the formation of saturated dinitriles (C_4, C_5 and C_6). All catalysts described contain Sb oxide as the main component. In References [129b, 129c], the same catalysts are described as those reported in Reference [129a], but with the addition of a halogenated compound in the feed stream.

The addition of halogenated compounds increased the conversion of cyclohexane, while the selectivities remained substantially unchanged. This indicates the presence of a radical-based mechanism for the activation of cyclohexane. In fact, it is known that the radical chain reaction is enhanced in the presence of halogens, owing to their tendency to give rise to X• species.

The only example reported in the literature concerning n-hexane ammoxidation is, once again, Reference [124]. In this paper, for a conversion of approx 12%, a selectivity to adiponitrile close to 40% and to hexanenitrile of approx 30% are reported, at 425 °C and with a molar feed ratio hexane/air/ammonia of 0.6–1.0/4.2/1.5. To our knowledge, these interesting results have never been confirmed.

Table 20.7 A summary of catalytic performance reported in [129a] for cyclohexane ammoxidation.

Catalyst	T (°C), τ (s), feed	Cyclohexane conversion (%)	Selectivity to ADN (%)	Selectivity to GN (%)	Selectivity to SN (%)
Sb/Sn/O 2/1	450, 1.95,[a]	19.3	23.7	14.2	3.5
"	550, 1.95,[a]	27.2	14.1	10.6	2.2
"	450, 3.9,[a]	21.4	22.9	13.7	3.9
Sb/Sn/O 4/1	445, 2.0,[b]	12.4	18.7	18.3	5.1
Sb/Sn/O 1/2	470, 2.0,[b]	14.3	14.1	15.7	1.3
Sb/Sn/O 3/1	430, 1.5,[c]	19.5	21.7	12.4	5.1
Sb/U/O 2/1	435, 2.0,[b]	21.3	24.3	10.3	4.0
Sb/Ti/O 2/1	440, 2.0,[d]	24.2	19.7	13.5	5.9

a feed composition (mol%) cyclohexane/oxygen/ammonia 5/10/6.6 (balance N_2).
b feed composition (mol%) cyclohexane/oxygen/ammonia 5/10/10 (balance N_2).
c feed composition (mol%) cyclohexane/oxygen/ammonia 8.4/10/14 (balance N_2).
d feed composition (mol%) cyclohexane/oxygen/ammonia 7/14.7/14 (balance N_2).
GN: glutaronitrile; SN: succinonitrile; ADN: adiponitrile.

20.5.4
The Ammoxidation of Benzene

The ammoxidation of benzene is described in some papers and patents [124, 128, 131, 132]. Whilst benzene has been reported to remain unconverted in the presence of ammonia and oxygen with a V_2O_5 catalyst [124], it was efficiently transformed to nitriles with catalysts based on mixed molybdates [128].

Figure 20.12 reports the performance of the best catalyst, based on V/Mo/O (V/Mo atomic ratio 4/1) at 465 °C, with a feed composition of benzene/oxygen/ammonia/nitrogen 1/4.4/1.6/17.6 (molar ratios). Among the various systems investigated, those giving the best performance were based on V/Mo/O and Ti/Mo/O which gave quite similar performance. Other systems were active but not selective to nitriles (i.e. Sn/Mo/O), or were neither active nor selective (i.e. Bi/Mo/O, Sb/Mo/O, Sn/Sb/Fe/O). The C_6 unsaturated dinitriles (mucononitrile) included the three isomers: *cis-cis* dicyanobutadiene, *cis-trans* dicyanobutadiene and *trans-trans* dicyanobutadiene.

The data apparently indicate that the C_6 dinitrile was a primary product, and the overall selectivity at low benzene conversion (less than 7%) was around 20%; the selectivity to mucononitriles (the three isomers) rapidly declined when the benzene conversion was increased above 7–8%, and finally became nil. The C_4 unsaturated dinitriles, maleonitrile and fumaronitrile, reached a maximum selectivity of around 20% and 10% respectively, at conversions lower than 10%. It is not clear whether they were primary or secondary products, because of the errors made in yield and selectivity calculation for low benzene conversions (and also errors made in data extrapolation from figures). It is worth noting that the C balance was much lower than 100%. Therefore, it is likely that some products were not detected.

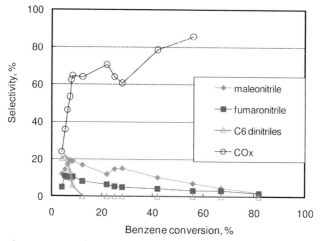

Figure 20.12 Catalytic performance of V/Mo/O catalyst in benzene ammoxidation [128].

The C_4 dinitriles were more stable than the C_6 dinitriles. Although their selectivity decreased with increasing benzene conversion, they were still present at high benzene conversion. CO_x formed mainly by decomposition of the nitriles, but a primary contribution (benzene combustion) cannot be excluded. The authors also mention a slight increase of the selectivity to C_6 dinitriles on increasing the reaction temperature (giving 5% more when raising from 440 to 500 °C).

The authors made a hypothesis about the mechanism. They assumed that benzene undergoes H• abstraction to generate a mono-radical which then reacts with an $NH_2^•$ fragment to generate aniline (however, the latter is not found among the reaction products) and di-radicals. The *ortho* di-radical is transformed to *o*-quinone diimine, the *para* to *p*-quinone diimine. The opening of the ring in the former case generated the C_6 dinitriles, and in the latter case maleonitrile + $2CO_2$. The *meta* di-radical is supposed to give CO_x. In the case of benzene oxidation, the same authors reported that only *p*-quinone and maleic anhydride were found as the products (oxidation of the *para* di-radical), while *o*-quinone and muconic acid (oxidation of the *ortho* di-radical) were not found.

Unger [131] investigated several catalysts for the ammoxidation of benzene (Table 20.8). The report shows that the performance of catalysts was not much different from that reported in Reference [128]. However, in the only example where the conversion was reported, the selectivity to mucononitrile was 17% (for a yield of 7%), at 40% conversion. In this case, therefore, the compound can be saved from consecutive degradation. It is likely that one key factor in saving mucononitrile from consecutive degradation is the reactor configuration and catalytic bed arrangement. This is also clearly demonstrated in Reference [128].

Reference [132] also describes catalysts and methods for the ammoxidation of benzene to mucononitrile. The performance seems to be greatly affected by the type of reactor used and the reactor conditions.

Table 20.8 Summary of results reported in [131] for benzene ammoxidation.

Catalyst	Benzene conversion (%)	T (°C), τ (s), feed comp	Selectivity to mucononitrile (%)	Selectivity to maleic anhydride (%)	Selectivity to maleonitrile (%)
Bi/Mo/P/O-BPO$_4$	15–20	450, 1.4,[a]	80[g]	18	0
Co/Mo/O-SiC	ng	450, 1.4,[a]	95[g]	5	0
Co/Mo/O-Al$_2$O$_3$	ng	450, 1.5,[b]	75[g]	20	0
V/P/O-Al$_2$O$_3$	ng	450, 1.5,[c]	70[g] (mainly cis-cis)	30	0
V/P/O-Al$_2$O$_3$	ng	450, –,[d]	78[g]	12	10
Bi/Mo/P/O-BPO$_4$	ng	450, –,[e]	99[g]	0	0
V/Mo/O- BPO$_4$	40	450, –,[f]	17[h]	ng	ng

a Feed: benzene/oxygen/ammonia/nitrogen 30/120/70/200.
b Feed: benzene/oxygen/ammonia/nitrogen 30/60/60/250.
c Feed: benzene/oxygen/ammonia/nitrogen 30/70/70/200.
d Feed: benzene/oxygen/ammonia/nitrogen 15/90/110/425.
e Feed: benzene/oxygen/ammonia/nitrogen 17/60/80/450.
f Feed: benzene/oxygen/ammonia/nitrogen 15/50/50/200.
g the selectivity is expressed only in reference to the condensable (high-boiling) products.
h this is a true selectivity.
ng = not given.

Catalysts based on Fe/Sb/O, Sn/Sb/O and U/Sb/O were fairly active and selective to C$_4$ unsaturated dinitriles, but non-selective to mucononitrile. Catalysts made of Bi/Mo/P/O, alumina-supported Mo/O and alumina-supported W/O had low activities.

20.5.5
Ammoxidation of C$_2$ Hydrocarbons

Although industrial interest in the synthesis of acetonitrile directly from C$_2$ hydrocarbons is currently limited, with acetonitrile being mainly produced as a by-product in acrylonitrile production, there are a number of indications regarding the future need of direct production of acetonitrile by C$_2$ hydrocarbon (ethane, in particular) ammoxidation. In fact, acetonitrile is used as a solvent and also as an intermediate in the production of many chemicals, ranging from pesticides to perfumes. Production trends for acetonitrile generally follow those of acrylonitrile, but the growth rate for acetonitrile use is higher than that of acrylonitrile. The four main producers of acetonitrile in the United States are: Ineos, DuPont, J.T. Baker Chemical and Sterling Chemicals.

Acetonitrile can be produced by catalytic ammoxidation of ethane and propane over Nb-Sb mixed oxides supported on alumina, with selectivities to acetonitrile of about 50–55% at alkane conversions of around 30% [133]. In both cases, CO forms in approximately a 1:1 molar ratio with acetonitrile, owing to a parallel reaction from a common intermediate. When feeding n-butane, the selectivity to acetonitrile halves. Bondareva and coworkers [134] also studied ethane ammoxidation over similar types of catalyst (V/Mo/Nb/O).

A different type of catalyst is constituted by Co ion-exchanged zeolites and mesoporous materials (MCM-49). Co-ZSM-5 was found to be selective in ethane ammoxidation [135]. A good correlation between the acidity and the catalytic activity was observed. The strength of ammonia bonding to the catalyst appears to have a crucial effect on the activity of Co-ZSM-5. Li and Armor [136] reported that dealuminated zeolite was active for the ammoxidation of ethane to acetonitrile. Pan and coworkers [137] instead studied ion-exchanged Co-Na-MCM-49 and Co-H-MCM-49 materials for the same reaction, reporting that the presence of ammonia in the feed considerably improved the selectivity and total yield of ethylene and acetonitrile.

It should be mentioned finally that acetonitrile could be also prepared by the catalytic ammoxidation of bioethanol over vanadium-alumino-phosphate (VAPO) catalysts [138], which is an alternative starting from biomass-derived raw materials.

20.5.6
Conclusions on the Ammoxidation of Unconventional Molecules

Analysis of the literature published in the field of catalytic, gas-phase ammoxidation of 'unconventional' molecules allows the following conclusions to be drawn:

1. The mechanism of catalytic ammoxidation generally accepted in literature for activated methyl groups (allylic position in olefins, or side position in alkylaromatics) includes the generation of a $-CH_2^{\bullet}$ radical as the first, rate-determining step. The data analyzed in the present report confirm that the generation of a radical species is also the key step for the ammoxidation of other molecules, such as butenes, cyclohexane, cyclohexanol and n-hexane. The first step is less evident in the case of butadiene or benzene (and also in the case of ethylene ammoxidation to acetonitrile). From benzene, the formation of the Ph· radical is hypothesized in literature. By analogy, the formation of a $-C=CH\cdot$ species can be postulated from butadiene. Catalysts able to perform this type of attack are a function of the type of molecule used, but in general V oxide is one key component of the catalysts.

2. Once the radical species has been generated, this is followed by π-allyl complex formation in the case of substrates in which the double bond is in the α-position with respect to the radical. Insertion of N into the substrate generally occurs

by insertion of the NH^{2-} species from an M=NH species (isoelectronic with the M=O species); M ions known to generate this moiety are Mo^{6+} and Sb^{5+}. However, it seems that the mechanism may indeed include the insertion of an $NH_2^{\cdot}$ species. The formation of the latter does not require the presence of metal ions able to generate the M=NH species. This may explain why with some molecules investigated, the performance of catalysts that do not include either Mo or Sb is at least comparable (and often better) than that of catalysts including these elements. So, the key-point is the generation of radical species that can either react to generate the nitrile precursors or react to yield the by-products. It is worth mentioning that the same hypothesis of a radical mechanism and reaction with $NH_2^{\cdot}$ has also been made for ammoxidation of alkylaromatics [97c].

3. An alternative mechanism may include first an O-insertion step, followed by the transformation of the oxidized compound into the cyano-containing compound. An example might be the oxidation of butadiene to maleic anhydride followed by hydrolysis to the acid, the formation of the diamide and the oxydehydrogenation to the dinitrile.

4. There are several contradictions in literature concerning the nature of the products obtained. Table 20.9 tries to summarize the literature information, with an indication of the best catalyst reported and of the best yield to nitriles. The main discrepancy concerns the nature of nitriles obtained. In general, it is possible to infer that starting from more reactive C_6 compounds (e.g. cyclohexane), which may also undergo a relevant number of transformations,

Table 20.9 A summary of literature data on 'unconventional' ammoxidation reactions.

Reactant	Major products	Best catalyst	Best yield (%)
Butadiene	maleonitrile + fumaronitrile (crotonitrile)	V/W/Cr/P/O-TiO$_2$	67
n-Butane	maleonitrile + fumaronitrile (crotonitrile)	V/W/Cr/P/O-TiO$_2$	26
Cyclohexanol	adiponitrile + hexanenitrile (aniline, phenol)	V/Sb/P/O-Al$_2$O$_3$	72[a]
Cyclohexanone	adiponitrile + hexanenitrile (aniline, phenol)	V/Sb/P/O-Al$_2$O$_3$	47[a]
Cyclohexane	adiponitrile + glutaronitrile + succinonitrile + maleonitrile + fumaronitrile (benzene, cyclohexene)	Ti/Sb/O	10
n-Hexane	adiponitrile + hexanenitrile	V/Sb/P/O-Al$_2$O$_3$	8[a]
Benzene	mucononitrile + maleonitrile + fumaronitrile	V/Mo/O	9

a Reference [124] is very controversial.

the formation of both saturated and unsaturated nitriles (mono and di) with the number of C atoms ranging from 4 to 6 is possible. This is a further indication of the radical nature of the mechanism of the reaction. No author has reported the formation of C_1, C_2 or C_3 nitriles (cyanhydric acid, acetonitrile and acrylonitrile), but we believe that this is highly likely. When the reaction is instead carried out over benzene, the number of nitriles that can form is lower.

20.6
Use of Other Oxidants for Ammoxidation Reactions

There is very limited literature data on the possible use of other oxidants in ammoxidation reactions, because air (O_2) is the preferable source from an economic point of view. However, it may be interesting to cite direct propane ammoxidation with N_2O and O_2 over steam-activated Fe-silicalite zeolite [139]. Yields of acrylonitrile and acetonitrile below 5% were obtained using N_2O or O_2 as the oxidant. Cofeeding N_2O and O_2 boosts the performance of Fe-silicalite compared to the individual oxidants, leading to acetonitrile yields of 14% and acrylonitrile yields of 11% (propane conversions of 40% and product selectivities of 25–30%). The beneficial effect of O_2 on the propane ammoxidation with N_2O contrasts with other N_2O-mediated selective oxidations over iron-containing zeolites (e.g. hydroxylation of benzene and oxidative dehydrogenation of propane), where a small amount of O_2 in the feed dramatically reduces the selectivity to the desired product. It is shown that the productivity of acrylonitrile, and especially acetonitrile, expressed as mol(product) h^{-1} kg(cat)$^{-1}$, is significantly higher over Fe-silicalite than over active propane ammoxidation catalysts reported in the literature.

20.7
Conclusions

Catalytic vapor-phase ammoxidation on mixed oxides is an important class of industrial processes. Propene ammoxidation to acrylonitrile is a well established process for the synthesis of this widely used monomer and intermediate. Over the 40 years since its commercial introduction, the yield to acrylonitrile has nearly doubled to over 80% with the fourth generation of catalysts. This is due to the intensive research effort and understanding of the several factors underpinning catalytic activity. Commercial catalysts contain over 20 elements, the presence of all of which is necessary to optimize the catalytic behavior.

A new current challenge is the shift from propene to propane as feedstock. Since the early 1990s it has been necessary to develop new generations of multi-component catalysts to improve the performance. However, both technical and economic conditions (the price differential between propane and propene) now exist for the commercial introduction of direct propane to acrylonitrile processes. Initial examples already exist and rapid growth is expected in the near future, both in new plants and in retrofitted current ones.

The ammoxidation of (substituted) alkylaromatics, with or without heteroatoms, is also a well established reaction of interest for the production of fine and specialty chemicals. Although commercial, the number of processes is still limited. However, there is a growing interest in new market areas such as China, and new processes may be expected in the near future.

The ammoxidation process could be also used in the conversion of less-conventional molecules. Various examples are discussed in detail, and it can be remarked that owing to the relatively limited research effort, there is still a relatively large degree of possible improvement. The need for more reproducible results in some cases is also stressed. Nevertheless, this field could offer new interesting commercial opportunities.

Although still limited, there is also some interest in using biomass-derived raw materials (bio-ethanol, glycerine) in ammoxidation processes. These processes could be of value only in the context of valorization of side streams in bio-refinery plants. However, owing to the growing interest in the latter, it may be expected that some opportunities will arise for the ammoxidation of biomass-derived side-products in the near future.

This overview of the catalysts and catalytic processes of vapor-phase ammoxidation has briefly summarized the state of the art and some perspectives in the field, although it was not possible to present a comprehensive review owing to the very large volume of literature and patents that has appeared in recent years. In some cases, as for C_3 and akylaromatic ammoxidation, various reviews are available and these have been referenced.

In conclusion, this overview demonstrates that the ammoxidation field is still a quite attractive area from both fundamental and commercial points of view. It also stresses that the same concepts and catalysts used for more known areas (such as acrylonitrile synthesis) could also be successfully applied in the conversion of less-conventional molecules.

References

1 Garmston, S. (2005) Acrylonitrile and derivatives – World Supply/Demand Report 2005, PCI Acrylonitrile Ltd (http://www.pci-acrylo.com) (accessed Sept. 2007).

2 Association of petrochemical producers in Europe (2007) (http://www.petrochemistry.net) (accessed Sept. 2007).

3 Brazdil, J.F. (2001) Acrylonitrile, in *Kirk-Othmer Encyclopedia of Chemical Technology*, Vol. 1, John Wiley & Sons, Ltd, p. 397.

4 Idol, J.D. and Heights, S. (1959) US Patent 2,904,580, assigned to The Standard Oil Co. US.

5 (a) Barclay, J.L., Bream, J.B., Hadley, D.J. and Stewart, D.G. (1959) Brit. Patent 876,446.
(b) Barclay, J.L., Bream, J.B., Hadley, D.J. and Stewart, D.G. (1964) US Patent 3,152,170, assigned to Distillers Company Ltd, UK.

6 (a) Grasselli, R.K. (2002) *Topics in Catalysis*, **21**, 79.

(b) Grasselli, R.K., Burrington, J.D., Buttrey, D.J., DeSanto, P., Jr, Lugmair, C.G., Volpe, A.F. Jr, and Weingand, T. (2003) *Topics in Catalysis*, **23**, 5.
(c) Grasselli, R.K. (2001) *Topics in Catalysis*, **15**, 93.

7 Giordano, N., Caporali, G. and Ferlazzo, N. (1965, priority 1961) US Patent 3,226,421, assigned to Edison, Italy.

8 Callahan, J.L., Grasselli, R.K., Milberger, E.C. and Strecker, H.A. (1970) *Industrial Engineering Chemical Product: Research and Development*, **9**, 134,

9 Grasselli, R.K. and Burrington, J.D. (1981) *Advanced Synthesis Catalysis*, **30**, 133–63.

10 Grasselli, R.K. (1999) *Handbook of Heterogeneous Catalysis*, Vol. 5 (G. Ertl, H. Knözinger and J. Weitkamp), Wiley-VCH Verlag GmbH, pp. 2302–26.

11 Burrington, J.D., Kartisek, C.T. and Grasselli, R.K. (1984) *Journal of Catalysis*, **87**, 363.

12 (a) Adams, C.R. and Jennings, T.J. (1963) *Journal of Catalysis*, **2**, 63.
(b) Adams, C.R. and Jennings, T.J. (1964) *Journal of Catalysis*, **3**, 549.

13 (a) Keulks, G.W. (1970) *Journal of Catalysis*, **19**, 232.
(b) Brazdil, J.F., Suresh, D.D. and Grasselli, R.K. (1980) *Journal of Catalysis*, **66**, 347.

14 Grasselli, R.K. (1985) *Applied Catalysis*, **15**, 127.

15 Chauvel, A. and Lefebvre, G. (1989) *Petrochemical Processes*, Vol. 2, Editions Technip, Paris, pp. 219–32.

16 (a) Centi, G., Cavani, F. and Trifirò, F. (2001) *Selective Oxidation by Heterogeneous Catalysis. Recent Developments*, Series: Fundamental and Applied Catalysis (eds M.V. Twigg and M.S. Spencer), Plenum Publishing Corporation, New York & London.
(b) Arpentinier, P., Cavani, F. and Trifirò, F. (2001) *The Technology of Catalytic Oxidations*, Editions Technip, Paris.

17 *Chemical and Engineering News*, **67** (2), 23 (1989).

18 (a) Brazdil, J.F. (2006) *Topics in Catalysis*, **38**, 289.
(b) Morgan, M. (2002) *Hydrocarbon Engineering*, 14, October.

19 Cmaiglobal Monomers Market report (http://www.cmaiglobal.com/MarketReports/samples/mmr.pdf) (accessed Sept. 2007)

20 (a) Japan Corporate News (http://www.japancorp.net/Article.Asp?Art_ID=14097) (accessed Sept. 2007).
(b) Ramachandran, R. and Dao, L. (1994) Eur. Patent 646,558, assigned to the BOC Group.
(c) Ushikubo, T., Oshima, K., Ihara, T. and Amatsu, H. (1996) US Patent 5,534,650, assigned to Mitsubishi Chemical Co.

21 Andersson, A., Hansen, S. and Wickman, A. (2001) *Topics in Catalysis*, **15**, 103.

22 (a) Centi, G. and Perathoner, S. (1998) *CHEMTECH*, **28**, 13.
(b) Centi, G., Perathoner, S. and Trifiro, F. (1997) *Applied Catalysis A: General*, **157**, 143.
(c) Centi, G., Trifirò, F. and Grasselli, R.K. (1990) *La Chimica and L'Industria (Milan)*, **72**, 617.
(d) Catani, R., Centi, G., Trifirò, F. and Grasselli, R.K. (1992) *Industrial and Engineering Chemistry Research*, **31**, 107.
(e) Centi, G., Grasselli, R.K. and Trifirò, F. (1992) *Catalysis Today*, **13**, 661.
(f) Centi, G. and Mazzoli, P. (1996) *Catalysis Today*, **28**, 351.
(g) Centi, G., Mazzoli, P. and Perathoner, S. (1997) *Applied Catalysis*, A, 165–273.
(h) Cavani, F. and Trifirò, F. (2003) *Basic Principles in Applied Catalysis* (ed. M. Baerns), Series in Chemical Physics, 75, Springer, Berlin, p. 21.
(i) Ballarini, N., Cavani, F. and Trifirò, F. (2005), The Valorization of Alkanes by Oxidation: Still a Bridge between Scientific Challenges and Industrial Needs, Proceedings of the DGMK-Conference Oxidation and Functionalization: Classical and alternative routes and sources, Milan, 12–14 October 2005, ISBN 3-936418-39-X, 19–33.

23 Sokolovskii, V.D., Davydov, A.A. and Ovsitser, O.Yu. (1995) *Catalysis Reviews – Science and Engineering*, **37**, 425.

24 Moro-Oka, Y. and Ueda, W. (1994) *Catalysis*, **11**, 223.

25 Prada, S.R. and Grange, P. (2003) *Belg. Oil, Gas*, **29** (3), 145, Hamburg, Germany.

26 (a) Guttmann, A.T., Grasselli, R.K. and Brazdil, J.F. (1988) US Patent 4,746,641, assigned to Standard Oil Company, OH, USA.
(b) Nilsson, J., Landa-Cánovas, A.R., Hansen, S. and Andersson, A. (1999) *Journal of Catalysis*, **186**, 442.

27 (a) Ushikubo, T., Oshima, K., Kayo, A., Umezawa, T., Kiyono, K. and Sawaki, I. (1992) European Patent 529,853, assigned to Mitsubishi Chemical Corporation, Tokyo, Japan.
(b) Komada, S., Hinago, H., Kaneta, M. and Watanabe, M. (1998) European Patent 895,809, assigned to Asahi Kasei Kogyo Kabushiki Kaisha, Osaka, Japan.

28 (a) Ueda, W. and Oshihara, K. (2000) *Applied Catalysis A: General*, **200**, 135.
(b) Oshihara, K., Hisano, T. and Ueda, W. (2001) *Topics in Catalysis*, **15**, 153.
(c) Botella, P., Lopez Nieto, J.M. and Solsona, B. (2002) *Catalysis Letters*, **78**, 383.

29 Hamada, K. and Komada, S. (1999) US Patent 5,907,052, assigned to Asahi Kasei Kogyo Kabushiki Kaisha.

30 (a) Guttman, A.T., Grasselli, R.K. and Brazdil, J.F. (1988) US Patent 4,746,641, assigned to the Standard Oil Company, OH, USA.
(b) Lynch, C.S., Glaeser, L.C., Brazdil, J.F. and Toft, M.A. (1992) US Patent 5,094,989, assigned to the Standard Oil Company, OH, USA.

31 (a) Guttmann, A.T., Grasselli, R.K. and Brazdil, J.F. (1988) US Patent 4,788,317.
(b) Bartek, J.P. and Guttmann, A.T. (1989) US Patent 4,797,381.
(c) Glaeser, L.C., Brazdil, J.F. and Toft, M.A. (1989) US Patent 4,837,191.
(d) Seely, M.J., Friedrich, M.S. and Suresh, D.D. (1990) US Patent 4,978,764.
(e) Suresh, D.D., Seeley, M.J., Nappier, J.R. and Friedrich, M.S. (1992) US Patent 5,171,876.
(f) Brazdil, J.F., Glaeser, L.C. and Toft, M.A. (1992) US Patent 5,079,207.
(g) Bartek, J.P., Ebner, A.M. and Brazdil, J.R. (1993) US Patent 5,198,580.
(h) Brazdil, J.F. and Cavalcanti, F.A.P. (1996) US Patent 5,576,469.
(i) Brazdil, J.F. and Cavalcanti, F.A.P. (1996) US Patent 5,498,588. All patents assigned to Standard Oil Company (Cleveland, OH, US).

32 (a) Florea, M., Prada Silvy, R. and Grange, P. (2003) *Catalysis Letters*, **87**, 63.
(b) Prada Silvy, R., Florea, M., Blangenois, N. and Grange, P. (2003) *American Institute of Chemical Engineers*, **49**, 2228.
(c) Florea, M., Prada Silvy, R. and Grange, P. (2005) *Applied Catalysis A: General*, **286**, 1.
(d) Olea, M., Florea, M., Sack, I., Prada Silvy, R., Gaigneaux, E.M., Marin, G.B. and Grange, P. (2005) *Journal of Catalysis*, **232**, 152.

33 (a) Hatano, M. and Kayo, A. (1988) European Patent 318,295.
(b) Ushikubo, T., Oshima, K., Umezawa, T. and Kiyono, K. (1992) European Patent 512,846.
(c) Ushikubo, T., Oshima, K., Kayo, A., Umezawa, T., Kiyono, K. and Sawaki, I. (1992) European Patent 529,853. All patents assigned to Mitsubishi Chemical Co.

34 (a) Albonetti, S., Blanchard, G., Burattin, P., Cavani, F. and Trifirò, F. (1996) European Patent 723,934.
(b) Albonetti, S., Blanchard, G., Burattin, P., Cavani, F. and Trifirò, F. (1997) European Patent 932,662.
(c) Blanchard, G., Burattin, P., Cavani, F., Masetti, S. and Trifirò, F. (1997) WO Patent 97/23,287 A1.
(d) Blanchard, G. and Ferre, G. (1994) US Patent 5,336,804. All patents assigned to Rhodia.

35 Mimura, Y., Ohyachi, K. and Matsuura, I. (1999) *Science and Technology in Catalysis 1998*, Kodansha, Tokyo, p. 69.

36 Bowker, M., Kerwin, P. and Eichhorn, H.-D. (1997) UK Patent 2,302,291, assigned to BASF.

37 (a) Albonetti, S., Blanchard, G., Burattin, P., Cassidy, T.J., Masetti, S. and Trifirò, F. (1997) *Catalysis Letters*, **45**, 119.
(b) Albonetti, S., Blanchard, G., Burattin, P., Cavani, F., Masetti, S. and Trifirò, F. (1998) *Catalysis Today*, **42**, 283.

38 Shaikh, S., Bethke, K. and Mamedov, E. (2006) *Topics in Catalysis*, **38**, 241.

39 (a) Ushikubo, T., Nakamura, H., Koyasu, Y. and Wajki, S. (1995) US Patent 5,380,933 to Mitsubishi Chemical Corp.

(b) Lin, M. and Linsen, M.W. (2001) US Patent 6,180,825, to Rohm and Haas Company (USA).
(c) Ushikubo, T., Koyasu, Y., Nakamura, H. and Wajiki, S. (1998) JP 10,045,664, to Mitsubishi Chemical Corp.

40 (a) Oshihara, K., Hisano, T. and Ueda, W. (2001) *Topics in Catalysis*, **15**, 153.
(b) Ueda, W., Chen, N.F. and Oshihara, K. (1999) *Chemical Communications*, 517.
(c) Oshihara, K., Hisano, T., Kayashima, Y. and Ueda, W. (2001) *Studies in Surface Science and Catalysis*, **136**, 93.
(d) Oshihara, K., Nakamura, Y., Sakuma, M. and Ueda, W. (2001) *Catalysis Today*, **71**, 153.
(e) Ueda, W., Oshihara, K., Vitry, D., Hisano, T. and Kayashima, Y. (2002) *Catalysis Surveys from Japan*, **6**, 33.
(f) Vitry, D., Morikawa, Y., Dubois, J.-L. and Ueda, W. (2003) *Topics in Catalysis*, **23**, 47.
(g) Vitry, D., Morikawa, Y., Dubois, J.-L. and Ueda, W. (2003) *Applied Catalysis A: General*, **251**, 411.
(h) Katou, T., Vitry, D. and Ueda, W. (2003) *Chemistry Letters*, **32**, 1028.

41 (a) Vitry, D., Dubois, J.-L. and Ueda, W. (2004) *Journal of Molecular Catalysis A–Chemical*, **220**, 67.
(b) Watanabe, N. and Ueda, W. (2006) *Industrial & Engineering Chemistry Research*, **45**, 607.
(c) Merzouki, M., Taouk, B., Monceaux, L., Bordes, E. and Courtine, P. (1992) *Studies in Surface Science and Catalysis*, **72**, 165.
(d) Roussel, M., Bouchard, M., Bordes-Richard, E., Karim, K. and Al-Sayari, S. (2005) *Catalysis Today*, **99**, 77.

42 Ushikubo, T., Oshima, K., Kayou, A. and Hatano, M. (1997) *Studies in Surface Science and Catalysis*, **112**, 473.

43 (a) Holmberg, J., Grasselli, R.K. and Andersson, A. (2004) *Applied Catalysis A: General*, **270**, 121.
(b) Holmberg, J., Haeggblad, R. and Andersson, A. (2006) *Journal of Catalysis*, **243**, 350.

44 Aouine, M., Dubois, J.L. and Millet, J.M.M. (2001) *Chemical Communications*, **13**, 1180.

45 (a) DeSanto, P., Jr, Buttrey, D.J., Grasselli, R.K., Lugmair, C.G. and Volpe, A.F., Jr, Toby, B.H. and Vogt, T. (2003) *Topics in Catalysis*, **23**, 23.
(b) DeSanto, P., Jr, Buttrey, D.J., Grasselli, R.K., Lugmair, C.G. and Volpe, A.F., Jr, Toby, B.H. and Vogt, T. (2004) *Zeitschrift fuer Kristallographie*, **219**, 152.
(c) Grasselli, R.K., Buttrey, D.J., DeSanto, P., Jr, Burrington, J.D., Lugmair, C.G., Volpe, A.F., Jr, and Weingan, T. (2004) *Catalysis Today*, **91–2**, 251.

46 Millet, J.M.M., Roussel, H., Pigamo, A., Dubois, J.L. and Jumas, J.C. (2002) *Applied Catalysis A: General*, **1–2**, 23277.

47 Berry, F.J., Brett, M.E. and Patterson, W.R. (1983) *Journal of the Chemical Society–Dalton Transactions*, 9–13.

48 Teller, R.G., Antonio, M.R., Brazdil, J.F., Grasselli, R.K. (1986) *Journal of Solid State Chemistry*, **64**, 249.

49 Birchall, T. and Sleight, A.E. (1976) *Inorganic Chemistry*, **15**, 868.

50 Centi, G. and Trifirò, F. (1986) *Catalysis Reviews–Science and Engineering*, **28**, 165.

51 Hansen, S., Ståhl, K., Nilsson, R. and Andersson, A. (1993) *Journal of Solid State Chemistry*, **102**, 340.

52 Berry, F.J., Brett, M.E. and Patterson, W.R. (1982) *Journal of the Chemical Society D–Chemical Communications*, 695.

53 Landa-Canovas, A., Nilsson, J., Hansen, S., Ståhl, K. and Andersson, A. (1995) *Journal of Solid State Chemistry*, **116**, 369.

54 (a) Centi, G. and Perathoner, S. (1995) *Studies in Surface Science and Catalysis*, **91**, 59.
(b) Centi, G. and Mazzoli, P. (1996) *Catalysis Today*, **28**, 351.

55 (a) Berry, F.J., Brett, M.E., Marbrow, R.A. and Patterson, W.R. (1984) *Journal of the Chemical Society–Dalton Transactions Trans*, 985.
(b) Berry, F.J., Holden, J.G. and Loretto, M.H. (1987) *Journal of the Chemical Society–Faraday Transactions*, **1**, 83–615.
(c) Berry, F.J., Holden, J.G. and Loretto, M.H. (1986) *Solid State Communications*, **59**, 397.

56 (a) Allen, M.D. and Bowker, M. (1995) *Catalysis Letters*, **33**, 269.
(b) Bowker, M., Bricknell, C.R. and Kerwin, P. (1996) *Applied Catalysis A: General*, **136**, 205.

(c) Poulston, S., Price, N.J., Weeks, C., Allen, M.A., Parlett, P., Steinberg, M. and Bowker, M. (1998) *Journal of Catalysis*, **178**, 658.
57 Magagula, Z. and van Steen, E. (1999) *Catalysis Today*, **49**, 155.
58 Nilsson, R., Lindblad, T. and Andersson, A. (1994) *Journal of Catalysis*, **148**, 501.
59 (a) Ballarini, N., Cavani, F., Cimini, M., Trifirò, F., Millet, J.M.M., Cornaro, U. and Catani, R. (2006) *Journal of Catalysis*, **241**, 255.
(b) Cimini, M., Millet, J.M.M. and Cavani, F. (2004) *Journal of Solid State Chemistry*, **177**, 1045.
(c) Cimini, M., Millet, J.M.M., Ballarini, N., Cavani, F., Ciardelli, C. and Ferrari, C. (2004) *Catalysis Today*, **91**, 259.
60 (a) Roussel, H., Mehlomakulu, B., Belhadj, F., van Steen, E. and Millet, J.M.M. (2002) *Journal of Catalysis*, **205**, 97.
(b) Nguyen, D.L., Ben Taarit, Y. and Millet, J.M.M. (2003) *Catalysis Letters*, **90**, 65.
(c) Millet, J.M.M., Marcu, J.C. and Herrmann, J.M. (2005) *Journal of Molecular Catalysis A–Chemical*, **226**, 111.
61 Xiong, G., Sullivan, V.S., Stair, P.C., Zajac, G.W., Trail, S.S., Kaduk, J.A., Golab, J.T. and Brazdil, J.F. (2005) *Journal of Catalysis*, **230**, 317.
62 Ballarini, N., Cavani, F., Cimini, M., Trifirò, F., Catani, R., Cornaro, U. and Ghisletti, D. (2003) *Applied Catalysis A: General*, **251**, 49.
63 Cody, C.A., DiCarlo, L. and Darlington, R.K. (1979) *Inorganic Chemistry*, **18** (6), 1572.
64 Sala, F. and Trifirò, F. (1974) *Journal of Catalysis*, **34**, 68.
65 Carbucicchio, M., Centi, G. and Trifirò, F. (1985) *Journal of Catalysis*, **91**, 85.
66 Grasselli, R.K. (1999) *Catalysis Today*, **49**, 141.
67 Osipova, Z.G. and Sokolovskii, V.D. (1979) *Kinetics Catalysis*, **20**, 910.
68 Sasaki, Y., Sutsumi, H. and Miyaki, K. (1992) US Patent 5,139,988, assigned to Nitto Chem. Ind. Co.
69 Kahney, R.H. and McMinn, T.D. (1975) US Patent 4,000,178, assigned to Monsanto Co.
70 Albonetti, S., Blanchard, G., Burattin, P., Masetti, S. and Trifirò, F. (1997) *Studies in Surface Science and Catalysis*, **110**, 403.
71 Soria, M.A., Delsarte, S., Gaigneaux, E.M. and Ruiz, P. (2007) *Applied Catalysis A: General*, **325**, 296.
72 Ballarini, N., Cavani, F., Giunchi, C., Masetti, S., Trifirò, F., Ghisletti, D., Cornaro, U. and Catani, R. (2001) *Topics in Catalysis*, **15**, 111.
73 (a) Ballarini, N., Catani, R., Cavani, F., Cornaro, U., Ghisletti, D., Millini, R., Stocchi, B. and Trifirò, F. (2001) *Studies in Surface Science and Catalysis*, **136**, 135.
(b) Ballarini, N., Cavani, F., Ghisletti, D., Catani, R. and Cornaro, U. (2003) *Catalysis Today*, **78**, 237.
74 Catani, R. and Centi, G. (1991) *Journal of the Chemical Society D–Chemical Communications*, 1081.
75 Matsuura, I., Oda, H. and Oshida, K. (1993) *Catalysis Today*, **16**, 547.
76 (a) Guerrero-Perez, M.O., Fierro, J.L.G. and Bañares, M.A. (2006) *Topics in Catalysis*, **41**, 43.
(b) Guerrero-Perez, M.O., Fierro, J.L.G. and Bañares, M.A. (2003) *Catalysis Today*, **78**, 387.
(c) Guerrero-Perez, M.O., Fierro, J.L.G. and Bañares, M.A. (2003) *Physical Chemistry Chemical Physics*, **5**, 4032.
77 (a) Guerrero-Perez, M.O., Fierro, J.L.G. and Banares, M.A. (2006) *Catalysis Today*, **118**, 366.
(b) Guerrero-Perez, M.O., Martinez-Huerta, M.V., Fierro, J.L.G. and Banares, M.A. (2006) *Applied Catalysis A: General*, **298**, 1.
78 Nilsson, J., Landa-Canovas, A.R., Hansen, S. and Andersson, A. (1996) *Journal of Catalysis*, **160**, 244.
79 Wickman, A., Wallenberg, L.R. and Andersson, A. (2000) *Journal of Catalysis*, **194**, 153.
80 Callahan, J.L. and Grasselli, R.K. (1963) *American Institute of Chemical Engineers*, **9**, 755.
81 Guerrero-Perez, M.O. and Banares, M.A. (2007) *Journal of Physical Chemistry C*, **111**, 1315.
82 (a) Cavalli, P., Cavani, F., Manenti, I. and Trifiro, F. (1987) *Catalysis Today*, **1**, 245.

(b) Sanati, M. and Andersson, A. (1990) *Journal of Molecular Catalysis*, **59**, 233.
(c) Andersson, A., Andersson, S.L.T., Centi, G., Grasselli, R.K., Sanati, M. and Trifiro, F. (1994) *Applied Catalysis A: General*, 43.

83 (a) Narayana, K.V., Venugopal, A., Rama Rao, K.S., Venkat Rao, V., Masthan, S. Khaja and Kanta Rao, P. (1997) *Applied Catalysis A: General*, **150**, 269.
(b) Narayana, K.V., Venugopal, A., Rama Rao, K.S., Khaja Masthan, S., Venkat Rao, V. and Kanta Rao, P. (1997) *Applied Catalysis A: General*, **167**, 11.
(c) Kanta Rao, P., Rama Rao, K.S., Masthan, S.K., Narayana, K.V., Rajiah, T. and Rao, V.V. (1997) *Applied Catalysis, A*, 163–23.

84 Makedonski, L., Nikolov, V., Nikolov, N. and Blaskov, U. (1999) *Reaction Kinetics and Catalysis Letters*, **66**, 237.

85 Martin, A., Wolf, G.-U., Steinicke, U. and Luecke, B. (1998) *Journal of the Chemical Soceity – Faraday Transactions*, **94**, 2227.

86 Otamiri, J.C., Andersson, S.L.T. and Andersson, A. (1990) *Applied Catalysis, A*, 65–159.

87 Cavani, F., Trifiro, F., Jiru, P., Habersberger, K. and Tvaruzkova, Z. (1988) *Zeolites*, **8**, 12.

88 (a) Beschmann, K., Fuchs, S. and Hahn, T. (1998) *Chemie Ingenieur Technik*, **70**, 1436.
(b) Beschmann, K. and Riekert, L. (1993) *Chemie Ingenieur Technik*, **70**, 1251.

89 Kim, S.H. and Chon, H. (1992) *Applied Catalysis A: General*, **85**, 47.

90 Fu, J., Ferino, I., Monaci, R., Rombi, E., Salinas, V. and Forni, L. (1997) *Applied Catalysis A: General*, **154**, 241.

91 Srilakshmi, C., Lingaiah, N., Nagaraju, P., Sai Prasad, P.S., Narayana, K.V., Martin, A. and Luecke, B. (2006) *Applied Catalysis A: General*, **309**, 247.

92 (a) Luecke, B., Narayana, K.V., Martin, A. and Jaehnisch, K. (2004) *Advanced Synthesis & Catalysis*, **346**, 1407.
(b) Lucke, B. and Martin, A. (2001) *Fine Chemicals through Heterogeneous Catalysis* (eds R.A. Sheldon and H. Van Bekkum), Wiley-VCH Verlag GmbH, Weinheim, Germany, p. 527.
(c) Martin, A. and Lucke, B. (2000) *Catalysis Today*, **57**, 61.
(d) Lucke, B. and Martin, A. (1999) Recent Advances in the Oxidation and Ammoxidation of Aromatics, *Proceedings of the DGMK-Conference: The future role of aromatics in refining and petrochemistry*, Erlangen, 13–15 October 1999, p. 139.
(e) Martin, A. and Luecke, B. (1996) *Catalysis Today*, **32**, 279.

93 (a) Narayana, K.V., Khaja Masthan, S., Venkat Rao, V., Raju, David, B. and Kanta Rao, P. (2002) *Catalysis Communications*, **3**, 173.
(b) Narayana, K.V., Khaja Masthan, S., Venkat Rao, V., Raju, David, B. and Kanta Rao, P. (2002) *Catalysis Letters*, **48**, 27.

94 Centi, G. (1996) *Applied Catalysis A: General*, **147**, 267.

95 Rizaev, R.G., Mamedov, E.A., Vislovskii, V.P. and Sheinin, V.E. (1992) *Applied Catalysis A: General*, **83**, 103.

96 (a) Murakami, J.Y., Niwa, M., Hattori, T., Osawa, S., Igushi, S. and Ando, H. (1977) *Journal of Catalysis*, **49**, 83.
(b) Niwa, M., Ando, H. and Murakami, H. (1977) *Journal of Catalysis*, **49**, 92.

97 (a) Cavalli, P., Cavani, F., Manenti, F. and Trifiro, F. (1987) *Industrial and Engineering Chemistry Research*, **26**, 639.
(b) Centi, G., Marchi, F. and Perathoner, S. (1997) *Applied Catalysis A: General*, **149**, 225.
(c) Cavalli, P., Cavani, F., Manenti, I., Trifiro, F. and El-Sawi, M. (1987) *Industrial and Engineering Chemistry Research*, **26**, 804.
(d) Busca, G., Cavani, F. and Trifiro, F. (1987) *Journal of Catalysis*, **106**, 471.

98 Otamiri, J. and Andersson, A. (1988) *Catalysis Today*, **3**, 211–23.

99 Zhang, Y., Martin, A., Berndt, H., Luecke, B. and Meisel, M. (1997) *Journal of Molecular Catalysis A – Chemical*, **118**, 205.

100 (a) Rizayev, R.G., Mamedov, E.A., Vislovskii, V.P. and Sheinin, V.E. (1992) *Applied Catalysis A: General*, **83**, 103.
(b) Sze, M.C. and Gelbein, A.P. (1979) *Hydrocarbon Processing*, **55**, 103.

101 (a) Grasselli, R.K., Burrington, J.D., Suresh, D.D., Friedrich, M.S. and Hazle, M.A. (1981) *Journal of Catalysis*, **68**, 109.
(b) Grasselli, R.K., Burrington, J.D., Suresh, D.D., Friedrich, M.S. and Hazle, M.A. (1981) *Journal of Catalysis*, **41**, 317.

102 (a) Martin, A. and Luecke, B. (1996) *Catalysis Today*, **32**, 279.
(b) Martin, A., Luecke, B., Wolf, G.-U. and Meisel, M. (1994) *Chemie Ingenieur Technik*, **66** 948.

103 (a) Martin, A., Luecke, B., Wolf, G.-U. and Meisel, M. (1995) *Catalysis Letters* **33**, 349.
(b) Dropka, N., Kalevaru, V.N., Martin, A., Linke, D. and Luecke, B. (2006) *Journal of Catalysis*, 240–8.

104 Shapovalov, A.A. and Kh.Sembaev, D. (1996) *Izvestiya Ministerstva Nauki-Akademii Nauk Respubliki Kazakhstan, Seriya Khimicheskaya*, **2**, 89.

105 Narayana, K.V., Martin, A., Bentrup, U., Luecke, B. and Sans, J. (2004) *Applied Catalysis A: General*, **270**, 57.

106 Chuck, R. (2005) *Applied Catalysis A: General*, **280**, 75.

107 Brazdil, J.F. and Bartek, J.P., Jr (1998) US Patent 5,854,172, assigned to the Standard Oil Company, US.

108 (a) Narayana, K.V., Venugopal, A., Rama Rao, K.S., Khaja Masthan, S., Venkat Rao, V. and Kanta Rao, R. (1998) *Applied Catalysis A: General*, **167**, 11.
(b) Narayana, K.V., Venugopal, A., Rama Rao, K.S., Khaja Masthan, S., Venkat Rao, V. and Kanta Rao, R. (1997) *Applied Catalysis A: General*, **150**, 269.

109 Kanta Rao, P., Rama Rao, K.S., Khaja Masthan, S., Narayana, K.V., Rajiah, T. and Venkat Rao, V. (1997) *Applied Catalysis A: General*, **163**, 123.

110 Narayana, K.V., David Raju, B., Khaja Masthan, S., Venkat Rao, V., Kanta Rao, P. and Martin, A. (2004) *Journal of Molecular Catalysis A–Chemical*, **223**, 321.

111 (a) Reddy, K.M., Lingaiah, N., Rao, K.N., Rahman, N., Prasad, P.S.S. and Suryanarayana, I. (2005) *Applied Catalysis A: General*, **296**, 108.
(b) Srilakshmi, C., Lingaiah, N., Suryanarayana, I., Prasad, P.S.S., Ramesh, K., Anderson, B.G. and Niemantsverdriet, J.W. (2005) *Applied Catalysis A: General*, **296**, 54.
(c) Rao, K.N., Reddy, K.M., Lingaiah, N., Suryanarayana, I. and Prasad, P.S.S. (2006) *Applied Catalysis A: General*, **300**, 139.

112 Srilakshmi, C., Ramesh, K., Nagaraju, P., Lingaiah, N. and Prasad, P.S.S. (2006) *Catalysis Letters*, **106**, 115.

113 Srilakshmi, C., Lingaiah, N., Nagaraju, P., Prasad, P.S.S., Narayana, K.V., Martin, A. and Luecke, B. (2006) *Applied Catalysis A: General*, **309**, 247.

114 Bondareva, V.M., Andrushkevich, T.V., Paukshtis, E.A., Paukshtis, N.A., Budneva, A.A. and Parmon, V.N. (2007) *Journal of Molecular Catalysis A–Chemical*, **269**, 240.

115 Hong, C. and Li, Y. (2006) *Chinese Journal of Chemistry Engineering*, **14**, 670.

116 Ni, K., Chen, F., Fang, W. and Zhong, Y. (2007) *Ziran Kexueban*, **30**, 75.

117 (a) Huber, S., Petzoldt, J., Rosowski, F. and Hibst, H. (2007) Ger. Offen., DE 2005-102005033826.
(b) Huber, S., Hugo, R., Dahmen, K., Preiss, T. and Hibst, H. (2007) Hartmut PCT Int. Appl., WO 2007009921. Both patents assigned to Basf Aktiengesellschaft, Germany.

118 Sze, M.C. and Gelbein, A.P. (1976) *Hydrocarbon Processing*, **55**, 103.

119 Contractor, R.M. and Sleight, A.W. (1987) *Catalysis Today*, **1**, 587.

120 (a) Furuoya, I. (1999) *Studies in Surface Science and Catalysis*, **121**, 343.
(b) Furuoya, I. and Kitazawa, Y. (1984) US Patent 4,436,671, assigned to Takeda Chem Ind.

121 Peters, A. and Schevelier, P.A. (2006) WO Patent 2006/053786 A1, assigned to DSM IP Assets.

122 (a) Colleuille, Y. and Perron, R. (1972) UK Patent 1,394,207, assigned to Rhone-Poulenc.
(b) Colleuille, Y. and Perron, R. (1971) French Patent 2,151,704.

123 (a) Liu, H., Imoto, H., Shido, T., Iwasawa, Y. (2001) *Journal of Catalysis*, **200**, 69.

(b) Liu, H., Shido, T. and Iwasawa, Y. (2000) *Chemical Communications*, 1881.
124 Reddy, B.M. and Manohar, B. (1993) *Journal of the Chemical Society D – Chemical Communications*, 330.
125 Chen, W.-S. and Lee, M.-D. (1992) *Reaction Kinetics and Catalysis Letters*, **47**, 187.
126 Grasselli, R.K. and Suresh, D.D. (1981) US Patent 4,271,091, assigned to Standard Oil Co.
127 (a) Ovsitser, O.Yu., Osipova, Z.G. and Sokolovskii, V.D. (1989) *Reaction Kinetics and Catalysis Letters*, **38**, 91.
(b) Ovsitser, O.Yu., Davydov, A.A., Osipova, Z.G. and Sokolovskii, V.D. (1989) *Reaction Kinetics and Catalysis Letters*, **38**, 125.
(c) Ovsitser, O.Yu., Davydov, A.A., Osipova, Z.G. and Sokolovskii, V.D. (1989) *Reaction Kinetics and Catalysis Letters*, **40**, 307.
128 Simon, G. and Germain, J.-E. (1980) *Bulletin de la Societe Chimique de France*, 3–4, 149.
129 (a) Burnett, C., Dewing, J. and Jubb, A.H. (1971) US Patent 3,627,817.
(b) Burnett, C. and Dewing, J. (1974) US Patent 3,818,066.
(c) Burnett, C. and Dewing, J. (1968) UK Patent 1,195,037, All patents assigned to ICI.
130 Mee, A. (1973) UK Patent 1,455,830, assigned to ICI.
131 Unger, M.O. (1971) US Patent 3,579,559, assigned to du Pont de Nemours.
132 Colleuille, Y. and Perron, R. (1971) UK Patent 1,338,952, assigned to Rhone Poulenc.
133 (a) Centi, G. and Perathoner, S. (1998) *Studies in Surface Science and Catalysis*, **119** (Natural Gas Conversion V), 569.
(b) Catani, R. and Centi, G. (1991) *Journal of the Chemical Society D–Chemical Communications*, 1081.
134 Bondareva, V.M., Andrushkevich, T.V., Aleshina, G.I., Maksimovskaya, R.I., Plyasova, L.M., Dovlitova, L.S. and Burgina, E.B. (2006) *Reaction Kinetics and Catalysis Letters*, **88**, 183.
135 Boubaker, H. Ben, Fessi, S., Ghorbel, A., Marceau, E. and Che, M. (2004) *Studies in Surface Science and Catalysis*, **154C** (Recent Advances in the Science and Technology of Zeolites and Related Materials), 2655.
136 Li, Y. and Armor, J.N. (1999) *Applied Catalysis A: General*, **188**, 211.
137 Pan, W., Jia, M., Lian, H., Shang, Y., Wu, T. and Zhang, W. (2005) *Reaction Kinetics and Catalysis Letters*, **86**, 67.
138 Kulkarni, S.J., Rao, R.R., Subrahmanyam, M., Navis, S.F., Rao, P.K. and Rao, A.V.R. (2003) Indian Patent IN 191083 and IN 191169.
139 Perez-Ramirez, J., Blangenois, N. and Ruiz, P. (2005) *Catalysis Letters*, **104**, 163.

21
Base Catalysis with Metal Oxides

Khalaf AlGhamdi, Justin S. J. Hargreaves, and S. David Jackson

21.1
Introduction

With increasing recognition of the complimentarity of their behavior to that of acid catalysts, the study of base catalysts is becoming more widespread. Whilst, for example, both acid and base catalysts of appropriate strength can be applied to analogous processes, there are pronounced differences in the product distribution, which is a consequence of the fundamentally different pathways involved. An illustrative example of this behavior is the alkylaromatic side-chain alkylation reaction, where it is observed that in the presence of a basic catalyst the side chain of the alkyl group is alkylated rather than the aromatic ring, as would be expected with acid catalysis [1]. In a survey of industrial processes published in 1999, Tanabe and Holderich [2] identified 10 major base-catalyzed processes as compared to 103 acid-catalyzed processes and 14 solid acid–base-catalyzed processes. In this chapter we concentrate our attention on heterogeneous base catalysts, and the examples and discussion given are intended to be illustrative of the general field rather than providing an extensive bibliography. Additional reviews detailing this topic can be found elsewhere, for example References [3–8]. Within this chapter, we exclusively direct our attention to oxide, or oxide-based, catalytic systems.

Base catalysts are extremely sensitive to the atmosphere in which they have been stored and generally require some sort of activation procedure, usually heating, prior to application. For materials stored in air, the formation of surface carbonates via reaction with atmospheric carbon dioxide is problematic. Figure 21.1 illustrates the effect of pre-treatment upon base-catalyzed *n*-butene isomerization [9]. It can be seen that activity increases up to a maximum of circa 650 °C.

It is, of course, important to recognize that in general base catalysts exhibit bifunctional behavior and are therefore, strictly, amphoteric. The same general point can also be made concerning acid catalysts, and therefore the distinction between acidic and basic solid catalysts is made on the basis of dominant behavior, although in some cases, for example ZrO_2, there may be no single dominant behavior [10]. An illustrative example of this is the alkaline earth metal oxides, classical base catalysts, where oxide ions behave as bases and the metal cations

Metal Oxide Catalysis. Edited by S. David Jackson and Justin S. J. Hargreaves
Copyright © 2009 WILEY-VCH Verlag GmbH & Co. KGaA, Weinheim
ISBN: 978-3-527-31815-5

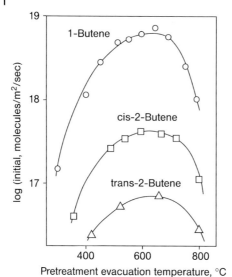

Figure 21.1 The development of base-catalyzed activity at 0 °C over La_2O_3 as a function of pre-treatment temperature. (Reproduced from [9] with permission).

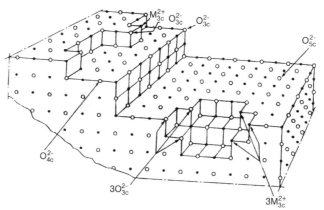

Figure 21.2 A schematic illustrating a variety of different co-ordination sites on the MgO (100) surface. (Reproduced from [12] with permission).

can function as Lewis acids. Given the distribution of different co-ordination sites present, as discussed elsewhere in this volume and shown schematically in Figure 21.2 [11, 12], it can be anticipated that a range of base strengths will be exhibited, and the stronger sites can be successively accessed by increasing severity of pre-treatment. Furthermore, such considerations are indicative that there would be a significant dependence of activity for a base-catalyzed reaction upon catalyst morphology and/or particle size. This aspect has seldom been emphasized, although a study of structure sensitivity in a base-catalyzed reaction is given by Chizallet and coworkers [13]. In this work, a detailed study of the effect of surface morphology upon the conversion of 2-methylbut-3-yn-2-ol is presented and structure sensitivity is reported. $Mg_{3C}O_{4C}$ and $Mg_{4C}O_{3C}$ ion pairs, combining a strongly basic O^{2-} site with a Lewis acidic site able to stabilize the resultant anion, were shown to be active for the deprotonation of various probe molecules. High surface area forms of MgO have been applied to Michael addition of malonates to enones and a range of Michael addition and Knoevenagel condensation reactions by Kantam and coworkers [14] and Xu and coworkers [15] respectively.

In considering their application, it is, of course, crucial to be able to characterise a base catalyst both in terms of the strength of sites and their number. Unlike acids, the distinction between Brønsted and Lewis sites does not have to be made for bases. Historically, as for acid catalysts, base site strengths have been measured by application of indicator methods. This procedure relies upon the observation of color changes of indicators of known strength. Considering the general Brønsted acidic indicator A–H, and the base catalyst B, when the strength of the base is high enough the following process can occur:

$$A-H + B \rightarrow A^- + B-H^+$$

A Hammett base scale, H_-, can then be defined:

$$H_- = pK_a + \log([A^-]/[AH])$$

where pK_a refers to the acid strength of the indicator. Consequently, by applying a range of indicators of varying strength, the base site strength characteristics of a solid can be determined. A selection of indicators of different acid strength is given in Table 21.1. Indicators are generally applied as solutions in non-polar solvents such as benzene and isooctane. As for the case of acidity determination, the determination of base strengths by indicator methods is subject to severe limitations. Using the naked eye, color changes only become perceptible when circa 10%

Table 21.1 A selection of indicators used to probe base strength. (Adapted from [3]).

Indicators	Acid form color	Base form color	pK_a
Bromothymol blue	yellow	green	7.2
2,4,6-Trinitroaniline	yellow	reddish-orange	12.2
2,4-Dinitroaniline	yellow	violet	15.0
4-Chloro-2-nitroaniline	yellow	orange	17.2
4-Nitroaniline	yellow	orange	18.4
4-Chloroaniline	colorless	pink	26.5[a]

a Estimated value.

Table 21.2 Base strength required to remove a proton from a R_1–CH_2–R_2 reactant molecule. (Adapted from [16]).

R_1	R_2	pK_a	Base required
CH_3	CH_3	42	Superbase
CH_3	$CH=CH_2$	35.5	Superbase
C_6H_5	H	35	Superbase
C_6H_5	C_6H_5	33	Superbase
CH_3	CN	25	Strong base
CH_3	COOR	24.5	Strong base
CH_3	$COCH_3$	20	Strong base
CH_3	COH	19.7	Strong base
COOR	COOR	11.5	Medium base
CN	CN	11.2	Medium base
CH_3	NO_2	10.6	Medium base
COR	COR	9	Medium base
COH	COH	5	Mild base
NO_2	NO_2	3.6	Mild base

of the adsorbed layer of indicator is adsorbed, giving a limiting accuracy, and hence UV-Visible spectroscopic methods would be preferred. Questions can also be raised concerning equilibration time effects, solvent effects, and color changes induced simply by adsorption. Consequently, although the application of indicator methods is still common in the literature, it is generally backed up by the use of additional methods, many of which also give site strength distributions, as described below.

Despite the fact that the additional methods do not generate H_0 values directly as such, base strength is most generally described and classified in terms of H_0 value. For example, "superbases" are defined as systems which contain sites with $H_0 \geq 26$. Recently, Kelly and King [16] have defined the minimum base strength requirements for different classes of reaction. The scale is reproduced in Table 21.2. It can be seen that the minimum site strength requirement for the base-catalyzed heterolytic activation of propane would have a pK_a of 42. To date, few solid bases have been identified as possessing such high strength, although Na/NaOH/Al_2O_3 is reported to be the strongest known superbase applied commercially, with a value of 37 [16]. Such materials would, of course, be extremely susceptible to poisoning.

If indicator methods are used, site distributions can be determined by UV-Visible spectroscopic procedures. Historically, a common way of determining distributions was via titration methods, although these are now less widespread.

Examination of the reaction pathways of probe molecules has also been applied to the characterization of basicity. In this method, a suitable probe molecule is selected with the intention that base sites will convert it uniquely to a given product. The conversion of isopropanol has been widely applied in this context, for example Reference [17] and references therein, and it is generally considered that base sites will yield acetone via catalyzed dehydrogenation, whereas acid sites would yield propene via catalyzed dehydration. Whilst this type of test is normally used in a qualitative way – that is, the production of acetone demonstrating the presence of base sites – in principle careful determination of the kinetics *may* yield parameters that relate to base site strength (activation energies) and site density (pre-exponential factors.) Within the literature, a whole range of different probe molecules have been applied. Among the most popular has been methanol [18]. A potential pitfall with this method is the possible existence of alternative non-base-catalyzed pathways to yield the signature product, and consideration must also be given to the accessibility of surface sites to the probe molecule. In the former context, attention is drawn to the case of solid superacids, where suggestions of bimolecular and also one-electron oxidation pathways, as opposed to acid-catalyzed unimolecular isomerization pathways, have been raised, throwing into doubt the possibility that some solids are superacidic [19].

Providing the mode of adsorption is known, namely that it is truly an acid–base interaction, thermal methods are useful in quantifying the number and base strength of sites. Perhaps the simplest method is the thermal desorption of a probe molecule. Most commonly CO_2 has been used in this respect and a sample temperature-programmed desorption (TPD) profile is shown in Figure 21.3 [20]. Again, it may be possible to derive information concerning the number and strength of sites from this method, although appropriate caution must be applied. For example, in one study [21], it was shown that the CO_2 desorption temperature

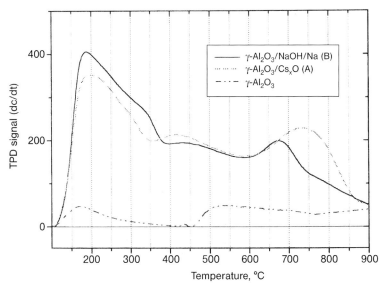

Figure 21.3 CO_2 TPD profiles on selected base catalysts. (Reproduced from [20] with permission).

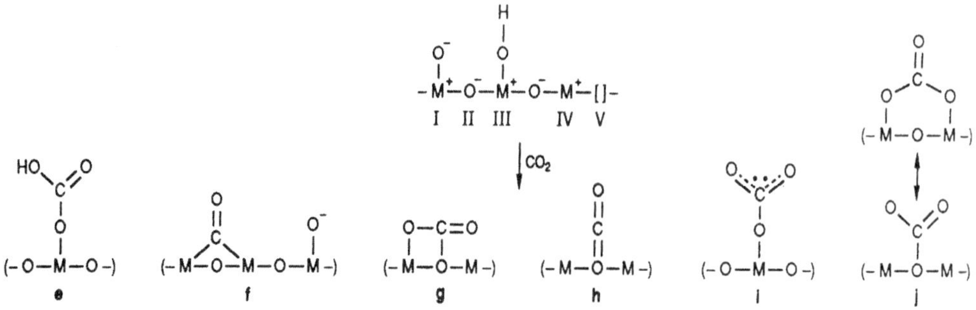

Figure 21.4 Different interactions of CO_2 with a metal oxide surface. (Reproduced from [22] with permission).

matched the decomposition temperature of the bulk metal hydrogencarbonate. The activation energy of desorption can be determined by investigation of the influence of temperature ramp rate on the temperature of the desorption maxima of profiles. In this method it is important that the adsorbate molecules do not modify the surface, and it is prudent to check that they desorb intact (i.e. that the true surface acid-based interaction is being probed and not a temperature-programmed reaction.) Again, in this respect, parallels can be drawn with the literature relating to the determination of acid site strength. Even with a molecule as simple as CO_2, a variety of different interactions could be envisaged as shown in Figure 21.4 [22].

For a selected range of heterogeneous bases, Martin and Duprez [23] have demonstrated that there is generally a good level of agreement between the determination of base strength via the cyclohexanol probe reaction and CO_2 chemisorption.

Although more complex experimentally, the application of microcalorimetry allows the determination of both base site strength and site distribution. In this method, the heat of adsorption of an acidic probe molecule, often CO_2, is measured directly as a function of surface coverage. An example of this type of measurement is given in Figure 21.5, which has been taken from the work of Auroux and Gervasini [22]. Again, it is prudent to verify the nature of the interaction via spectroscopic observation and, as for TPD, attention must be paid to aspects such as sample pre-treatment which can, for example, dramatically alter the degree of surface hydration and therefore the nature of interaction with the probe molecule. In general, the combination of thermal methods and spectroscopy, in particular Fourier-transform infrared FTIR spectroscopy, represents a very powerful tool for the characterization of surface acid-base characteristics.

Spectroscopic methods have been used in isolation for the determination of base characteristics. A number of studies, for example Reference [24], have correlated the O 1s binding energy measured by X-ray photoelectron spectroscopy (XPS) to the determination of base site strength. It has been proposed that decreasing binding energy relates to an increased ability for electron pair donation and hence

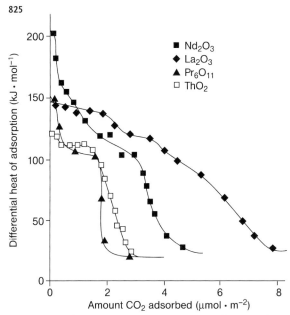

Figure 21.5 Differential heats of adsorption as a function of CO_2 coverage on various rare earth metal oxides. (Reproduced from [22] with permission).

stronger basicity. However, whilst such relationships have been established between materials of similar type, application to materials of different type must be treated with caution. In addition, it is generally the case that catalysts exhibit a range of differing site strengths and therefore spectral deconvolution is required. An example of the application of this type of method to solid "superbase" catalysts prepared from γ-Al_2O_3 is given in work by Tanaka and coworkers [25]. Table 21.3 presents the corresponding binding energies for the materials studied. N 1s XPS spectra of adsorbed pyrrole, a probe molecule, have also been applied to the measurement of basicity with a degree of success [2].

Pyrrole has also been applied as a probe molecule in FTIR specroscopic studies. Upon interaction with a base site, the N—H stretching vibration is found to shift to lower wavenumber and in alkali metal-exchanged zeolites this behavior has been found to correlate with both N 1s XPS data and the negative charge calculated from Sanderson electronegativities [4, 26].

21.2
Catalysts and Catalytic Processes

In this section, we briefly discuss some of the salient points concerning various different types of solid base catalyst. This section is not exhaustive and is designed

Table 21.3 Observed O 1s binding energies of a selection base catalysts. (Adapted from [25]).

Sample	O 1s (eV)
γ-Al_2O_3	531.1
NaOH/γ-Al_2O_3	530.3
Na/NaOH/γ-Al_2O_3	529.5
Na/NaOH/γ-Al_2O_3 exposed to air	531.5
α-Al_2O_3	530.1
NaOH/α-Al_2O_3	530.0
Na/NaOH/α-Al_2O_3	529.9
KOH/γ-Al_2O_3	530.4
K/KOH/γ-Al_2O_3	529.1
RbOH/γ-Al_2O_3	530.6
Rb/RbOH/γ-Al_2O_3	529.7
Rb/RbOH/α-Al_2O_3	530.2

to draw attention to some aspects of various classes of base, rather than providing detailed discussion of individual materials.

21.2.1
Alkali Metal Oxides

Alkali metal oxides have principally been studied in their supported form, most notably as catalyst modifiers. In these cases, the basic properties are often the key to the modifier action yet the potential for a direct base catalysis role is rarely considered. The literature concerning the use of alkali metal oxides as base catalysts is less extensive. The base strength increases down the group with $Li_2O < Na_2O < K_2O < Cs_2O$, hence many of the processes use cesium as the preferred base to optimize the basic strength of the catalyst. Production of supported alkali metal oxides usually involves impregnation with a simple alkali metal salt, typically the nitrate, followed by calcination at high temperatures, for example Reference [27]. It is rare for any characterization to be performed that confirms that it is indeed the oxide that is formed. In the presence of a hydroxylated support surface it is likely that more than a single species is formed, namely a combination of oxide and hydroxide. Indeed, from the study by Canning and coworkers [21], in which the alkali metal hydrogencarbonate was formed from carbon dioxide adsorption, one can postulate that a surface hydroxide was present. However, for this review we will consider that, unless there is specific evidence to the contrary, the material produced after high temperature calcination will be principally oxide and hence will fall within the scope of this discussion. This need for high temperature activation to produce the oxide, and the sensitivity of the oxide to poisons such as water and carbon dioxide, could potentially limit the industrial applicability of these materials. Nevertheless, supported alkali metal oxide catalysts are being actively

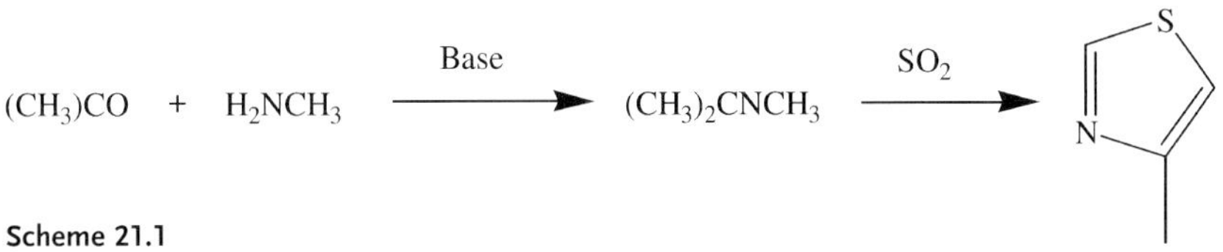

$(CH_3)CO$ + H_2NCH_3 →[Base] $(CH_3)_2CNCH_3$ →[SO_2] [4-methylthiazole structure]

Scheme 21.1

researched by industry, as witnessed by reports in the patent literature. The synthesis of aziridine from N-methylethanolamine is reported in BASF [28] and Nippon Shokubai patents [29] using cesium and potassium oxides in conjunction with phosphoric acid supported on silica or glass fabric. Merck have patented a zeolite-supported cesium catalyst [30] for the production of 4-methylthiazole (Scheme 21.1).

4-Methylthiazole is a fungicide and a pharmaceutical intermediate. The catalyst gives a selectivity of >60% at a conversion of >85% but deactivates with time on stream such that half the activity is lost within two weeks.

The Lucite ALPHA process [31] for the production of methyl methacrylate via an aldol reaction between methyl propionate and formaldehyde, also uses a supported cesium oxide catalyst, which requires various additives to minimize catalyst deactivation. An issue that is rarely discussed in the academic literature, possibly because of the nature of catalyst testing in academia, is that of deactivation due to volatilization. Alkali metal oxides may exhibit appreciable volatility under reaction temperatures >400 °C and so there can be a loss of activity over time through volatilization. To overcome this deactivation mechanism, an AMOCO patent [32] describes adding an alkali metal compound into the process gas stream so that the alkali metal compound is deposited on the catalyst during operation to compensate for any loss. It was also noted in the Lucite patent [31] that alkali metal oxides under some reaction conditions (those where water is in the feedstream) might enhance the loss in surface area of the silica support. To counteract such effects the incorporation of modifiers, such as boron, aluminum, magnesium, zirconium or hafnium may be added to the catalysts to retard the rate of surface area decrease. In other studies [33], cesium oxide was thought to sinter under a water-containing feed but could re-distribute when the water was removed.

The use of alkali metal oxide catalysts for aldol condensation reactions has been examined for the production of 2-ethylhexenal from butanal [34]. When coupled to a hydrogenation catalyst the system can produce the plasticizer alcohol 2-ethylhexanol directly. When isobutyraldehyde was used as the feed to a silica-supported sodium oxide catalyst, no products were formed but a significant amount of carbon was deposited on the catalyst and in the reactor (Scheme 21.2).

These results indicated that an aldehyde with a methyl branch α- to the carbonyl group passed over a Na_2O/SiO_2 catalyst was unable to undergo an aldol reaction but did lay down carbon. This has implications for a combined $Pd/Na_2O/SiO_2$ system, as the 2-ethylhexenal that is formed as an intermediate has an ethyl branch α- to the carbonyl group and may lead to product poisoning of the base sites of the catalyst, in a manner similar to that found with isobutyraldehyde over a Na_2O/SiO_2 catalyst.

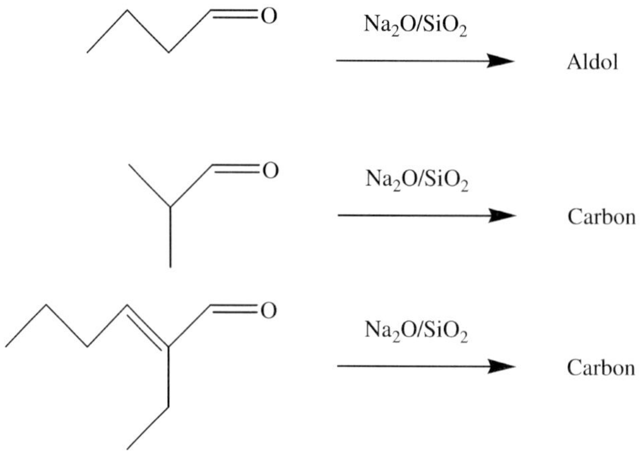

Scheme 21.2

The conversion of acetone to methyl isobutyl ketone (MIBK) also uses a combination of base catalysis with a hydrogenation catalyst [35]. The base component converts the acetone to diacetone alcohol (DAA) via an aldol reaction, which is then dehydrated by the silica to give mesityl oxide (MO). The final step is the hydrogenation of the MO to MIBK over the metal component. The action of the base catalyst in the absence of the hydrogenating metal has been studied [36]. As well as the aldol condensation reactions shown below, the cesium oxide also hydrogenated MO to MIBK, albeit at a low level (Scheme 21.3).

Deuterium studies showed that the hydrogenation was not affected by gas-phase deuterium but used protium left on the basic site from the exchange reaction of acetone and deuterium. Further studies [37] revealed that activation energies calculated for MO and MIBK showed a trend following the notional base strength, with the Na_2O/silica catalyst having the highest activation energy and the Cs_2O/silica catalyst having the lowest activation energy, as shown in Table 21.4.

The variation in the MO values, although following the same trend as MIBK, is, however, within the error limits of the measurement and so this variation is not statistically significant. The variation for the MIBK activation energies is significant. Also, these values are higher than those for MO formation and so can be related to the addition of hydrogen rather than the aldol condensation.

In Figure 21.6 it can be seen that there is a direct relationship between the activation energy and the ionization potential of the alkali metal. An outline mechanism for the hydrogenation of MO is shown in Scheme 21.4.

Table 21.4 Activation energies for MO and MIBK (kJ mol^{-1}).

	NaOH/SiO$_2$	KOH/SiO$_2$	CsOH/SiO$_2$
MO	33	25	23
MIBK	47	36	28

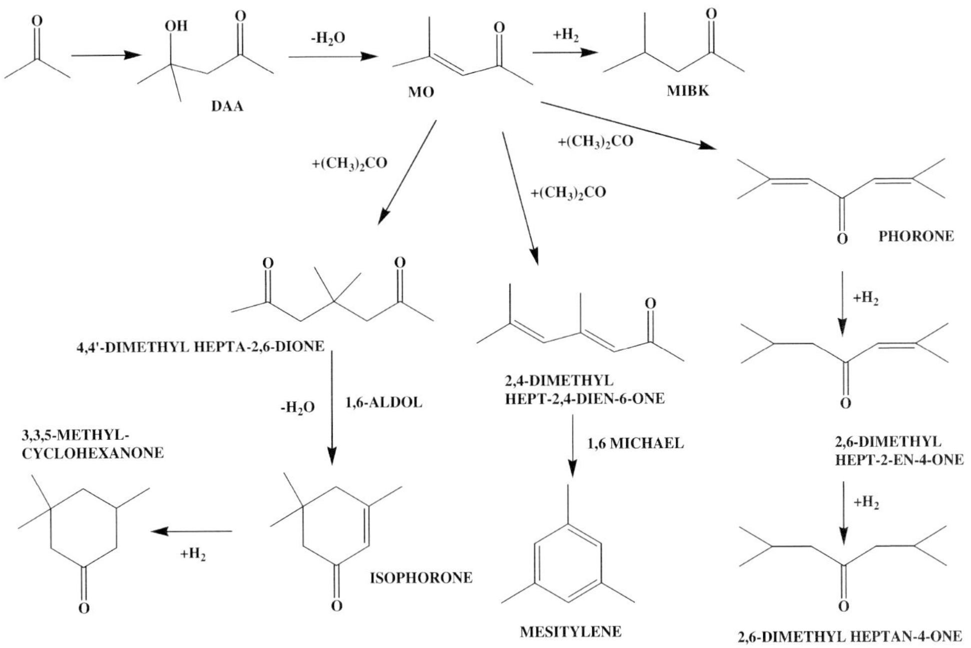

Scheme 21.3

Gorzawski and Holderich [38] examined the generation of a $Cs_xO/\gamma\text{-}Al_2O_3$ superbase from the decomposition of $CsOAc/\gamma\text{-}Al_2O_3$. After heating in vacuum to 700–750 °C the resulting material had an H_0 value ≥37. The catalyst was used for the transesterification of methyl benzoate and dimethyl terephthalate with ethylene glycol. A selectivity of 90% at a conversion of 87% was obtained for the methyl

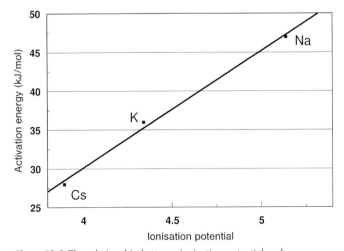

Figure 21.6 The relationship between ionization potential and hydrogenation activation energy.

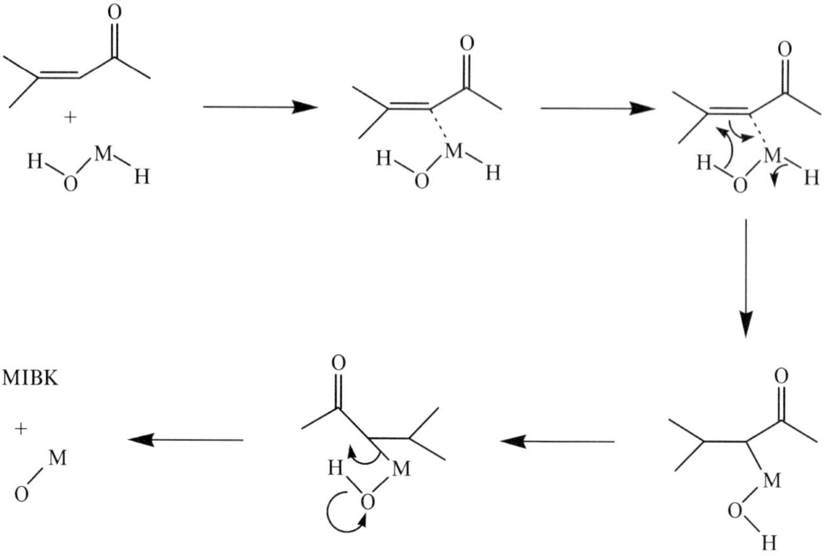

Scheme 21.4

benzoate reaction. In the transesterification of dimethyl terephthalate, a selectivity to bis(2-hydroxyethyl) terephthalate of 88% at a 100% conversion of dimethyl terephthalate was obtained. Once again the catalyst was not characterized to confirm the presence of Cs_2O, although the superbasic nature of the material suggests that the oxide was formed. Note that the support hydroxyl population will have been dramatically reduced and, as has been mentioned before, it is likely that the hydroxyl population will influence the strength and nature of the alkali metal basic site.

21.2.2
Alkaline Earth Metal Oxides

The alkaline earth metal oxides have been extensively studied in terms of their base behavior, for example Reference [39], with perhaps most attention centering upon MgO. In general, base strength increases down the group as anticipated, with the order being BaO > SrO > CaO > MgO, for example, in their surface-area-normalized reaction rates in the aldol condensation of acetone [40]. High-temperature pre-treatment is generally required in order to activate the oxides and they are particularly sensitive to rehydration and recarbonation. Figure 21.7, taken from Hattori's review [4], shows the equilibrium pressure for the decomposition of the carbonates where the increasing severity of necessary pre-treatment is evident, with decomposition equilibria being negligible at ambient temperature. Similar considerations apply to the metal hydroxides. Figure 21.8 shows the evolution of sites of varying strength as a function of the temperature applied in the thermal pre-treatment of $Mg(OH)_2$.

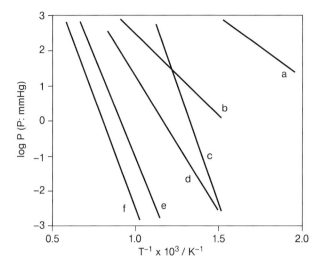

Equilibrium pressure for decomposition

Figure 21.7 The decomposition characteristics of various alkaline earth metal oxide precursors as a function of temperature (a) SrO_2, (b) BaO_2, (c) $MgCO_3$, (d) $CaCO_3$, (e) $SrCO_3$ and (f) $BaCO_3$. (Reproduced from [5] with permission).

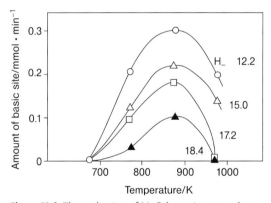

Figure 21.8 The evaluation of MgO base site strength as a function of $Mg(OH)_2$ decomposition temperature. (Reproduced from [5] with permission).

In alkaline earth metal oxides, the (100) surface termination plane, which exposes equal numbers of anions and cations, is prevalent and, as illustrated in Figure 21.2, it can be anticipated that an entire family of different co-ordination sites, of different basicity, can be exhibited. Furthermore, it would be expected that this would lead to a dependence upon crystallite morphology and/or particle size.

Figure 21.9 TEM of MgO prepared by burning magnesium in air and collecting the ribbon residue. (Reproduced from [41] with permission).

Figure 21.10 TEM of MgO prepared by thermal decomposition of $Mg(OH)_2$ at 800 °C. (Reproduced from [41] with permission).

The morphology of, for example, MgO is known to be a critical function of the preparative procedure adopted [41, 42]. Figures 21.9 to 21.11 are transmission electron microscopy (TEM) images of MgO samples prepared from a number of different routes [41].

Whilst all the samples display (100)-terminated crystallites, decomposition of the basic carbonate yields higher index mean planes generated by the aggregation of (100)-terminated microstructures as shown in Figure 21.12 [43]. The high-resolution TEM (HRTEM) image shown in Figure 21.13 displays terraced structures on a smoke sample recalcined after prolonged storage in air [44]. Further considerations to be made are that decomposition of $Mg(OH)_2$ at intermediate temperatures can lead to the generation of metastable (111)-terminated crystallites as shown in Figure 21.14 [45], which are believed to be stabilized by surface

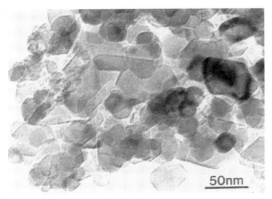

Figure 21.11 TEM of MgO prepared by thermal decomposition of magnesium hydroxycarbonate. (Reproduced from [41] with permission).

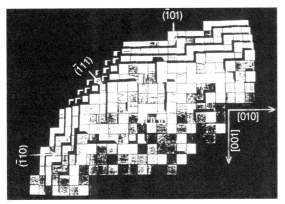

Figure 21.12 High index mean planes exhibited by MgO prepared by the decomposition of magnesium basic carbonate precursor. (Reproduced from [43] with permission).

hydroxyl groups. To our knowledge, the role of morphology and crystallite size on basicity has been little studied in relation to base catalysis and it would be anticipated that low co-ordination sites lead to higher base site strengths. However, caution in making such comparisons is necessary, since some precursors, in the case of MgO for example, can lead to high impurity levels in the resultant samples [41].

With the advent of synthetic procedures leading to the production of highly dispersed metal oxides [46, 47], access to a wide range of surface area is now possible (that is $<3\,m^2g^{-1}$ to $>300\,m^2g^{-1}$ for MgO) An alternative approach to the generation, and stabilization, of high surface area MgO samples is the use of a support, for example MgO/γ-Al$_2$O$_3$ [48].

Addition of alkali metal ions can also be used to modify base strength/defect structure as required, for example Reference [49], although this is not always

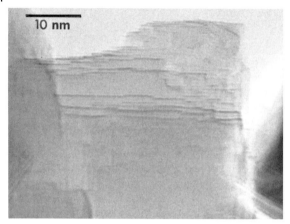

Figure 21.13 TEM of commercial MgO smoke following calcination after prolonged storage in air. (Reproduced from [44] with permission).

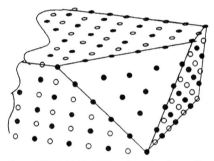

Figure 21.14 MgO (111) polar microfacet. (Reproduced from [45] with permission).

beneficial in terms of catalytic performance [50] and it can lead to substantial loss in surface area. The deposition of alkali metals themselves leads to the generation of "superbasicity", for example References [20, 51]. It has recently been reported that CaO can be made superbasic by appropriate treatment with ammonium carbonate [52].

The reaction of methanol and acetonitrile to form acrylonitrile has been studied over solid base catalysts derived from MgO [53]:

$$CH_3OH + CH_3CN \rightarrow CH_2=CHCN + H_2 + H_2O$$

Along with acrylonitrile (ACN), propionitrile (PPN) was also produced. Heating to 400 °C in nitrogen activated the catalysts. High surface area MgO produced the best yields of acrylonitrile and propionitrile but favored the hydrogenated nitrile. The low surface area MgO was considerably less effective at hydrogenation and gave a high ACN:PPN ratio. The main role of the modifiers (Cr, Mn) was to inhibit hydrogenation of acrylonitrile to propionitrile; however, the Cr-promoted MgO showed slightly higher specific activity than MgO, suggesting a promotional effect.

Studies on the use of SrO as a base catalyst are much less common than those using MgO. However, Mohri and coworkers [54] investigated the use of SrO for the base-catalyzed isomerization of 1-butene. The isomerization of alkenes by base catalysts has been known for many years [55] and usually requires a strong base. Mohri and coworkers found that activity was only detected after heating SrO *in vacuo* to 650 °C and that a temperature treatment of 1000 °C was required to achieve maximum activity. Selective poisoning revealed that both the Sr^{2+} and the O^{2-} ions had discrete roles with the *trans*-2-butene yield associated with the Sr^{2+}, while the *cis*-2-butene yield was associated with the O^{2-}.

21.2.3
Hydrotalcites

Hydrotalcites, or layered double hydroxides (LDH), are a class of material related to the mineral hydrotalcite which has the formula $[Mg_6Al_2(OH)_{16}]CO_3 \cdot 4H_2O$. They are layered double hydroxides with the general formula $M^{2+}_x M^{3+}_{1-x}(OH)_{2x} A_m \cdot zH_2O$ which have the trivalent and divalent cations contained within the octahedral sites of positively charged hydroxide sheets, in a structure akin to that of brucite, $Mg(OH)_2$. The anions, A_m, and water are contained between the layers of sheets and their intercalation and removal can be monitored by the position of the (003) reflection in their powder diffraction patterns [56]. A schematic of the structure is shown in Figure 21.15 [56].

Hydrotalcites are of interest because of their ability to function as base catalysts, for example References [57–61]. In a very topical application, IFP have developed a biofuels process using Al, Zn and Ti mixed oxides [62]. The process is being commercialized by Axens. Figueras and coworkers [63]) have studied Mg-Al

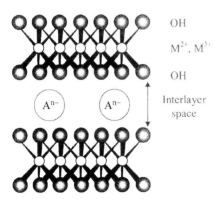

Figure 21.15 Schematic of the hydrotalcite structure. (Reproduced from [56] with permission).

hydrotalcites for the transesterification of methyl acetoacetate. The uncalcined material was as expected inactive; however, calcined and OH-exchanged materials were active with yields of 55 and 77% respectively. The best catalyst reported was a *tert*-BuO-exchanged hydrotalcite which gave 98% yield in a shorter reaction time and could be recycled. This Mg-Al-O^{t}Bu hydrotalcite was also used for a range of other reactions, which suggested a strong base although no measurement was made.

In addition, substitution of variable oxidation state metal ions such as Co^{2+}, Fe^{3+} and Cu^{2+} generates materials with redox catalytic properties, for example References [57, 64]. Hydrotalcites themselves have been used as basic supports for metals, leading to bifunctional catalysts. An example of this is provided in the study of de Jong and coworkers in which palladium–hydrotalcite combinations were reported to be active for single-stage synthesis of MIBK from acetone and hydrogen [65].

Although hydrotalcites are relatively stable (up to circa 500 °C), they are also of potential application as precursors of mixed metal oxide catalysts, for example Reference [66]. Dehydration–rehydration equilibria account for the switching between hydrotalcites and mixed/supported metal oxides, which is sometimes termed the "*memory effect*" [67–69]. Recent advances have seen attempts to prepare highly dispersed LDH systems, such as those dispersed within mesoporous carbon [70]. Owing to widespread interest in their application, hydrotalcite catalysts have been the subject of a number of reviews, for example References [71–75]. Other layered-based systems have also attracted attention for application in catalysis, for example Reference [76].

21.2.4
Rare Earth Oxides

Although comparisons are often drawn between the chemistry of alkaline earth and rare earth elements, rare earth oxides appear to have been less well studied than alkaline earth metal oxides as base catalysts. The activity of the rare earth oxides and preparation variables, with particular attention being paid to activity in catalytic pyrolysis, has been reviewed elsewhere [77]. Figure 21.5 shows the differential heats of CO_2 adsorption for some rare earth oxides. Rare earth oxides are generally tested as sesquioxides (M_2O_3), notable exceptions being CeO_2, PrO_2 and Tb_4O_7, and arguably most attention has centred upon La_2O_3. The bifunctional nature of these materials has been described in relation to Ho_2O_3 [78] where acid–base properties have been elucidated with a combination of temperature programmed isopropanol decomposition and FTIR spectroscopic studies of pyridine adsorption. Schuth and coworkers [79] have demonstrated that it is possible to disperse high loadings of yttria on SBA-15, which may open up new possibilities for the application of rare earth oxides. Furthermore, the inclusion of rare earth oxide components in other host matrices, such as yttria-stabilized zirconia, results in materials with catalytic activity, for example References [80, 81], although this has been ascribed to Lewis acidity. A comprehensive study [82] detailing the influence of preparation route, calcination condition and atmosphere (often neglected) on the basicity and catalytic activity of single rare earth and alkaline earth oxides as well as mixed rare earth–alkaline earth oxides has been published recently.

21.2.5
Basic Zeolites

The occlusion of highly dispersed oxides of alkali metals leads to zeolites possessing pronounced base strengths. Whilst basicity can be generated by exchange of alkali metal ions, for example Reference [83], zeolites containing occluded highly dispersed alkali metal oxide clusters, which possess higher base strength, have attracted most attention, for example References [84–87]. The most widely studied systems have been based upon cesium, where loading is typically achieved by impregnation with cesium acetate followed by thermal decomposition to yield cesium oxide clusters. However, variation of the general procedure can modify the nature of the occluded species. For example, Doskocil and Mankidy [88] have reported that it is possible to obtain occluded sodium metal within NaX by thermal decomposition of loaded sodium azide. Sodium oxide clusters were produced via the oxidation of the sodium clusters or, alternatively, loaded sodium acetate. Zhu and coworkers [89] have reported that dispersion and decomposition of KNO_3 over KL zeolite leads to superbasicity. In addition to alkali metal modification, shape-selective base catalysts generated by the nitridation of zeolites and aluminophosphates have been reported by Ernst and coworkers [90].

21.2.6
Zirconia Superbases

Solid superbases, which have been defined as materials with base strengths, $H_- \geq 26$, represent a highly desirable class of solid materials. This is because they offer the possibility of activation of relatively unreactive species under mild reaction conditions.

Whilst the application of superbases is reasonably well established in organic synthesis, for example References [91–93], the field of heterogeneous catalytic superbases is less well developed. However, it is important to acknowledge that, as stated earlier, $Na/NaOH/Al_2O_3$, which is applied by Sumitomo on an industrial scale for the isomerization of 5-vinyl-2-norbornene to 5-ethylidene-2-norbornene, is reported to have a base strength of $H_- > 37$. Whilst a number of nitrides, have been claimed to be superbasic, for example Reference [94] we restrict our attention to oxide-based examples in this section. In the sections above, mention was made of the possibility of preparing superbases by alkali metal modification of alkaline earth oxides. The possibility of the generation of superbase sites via the controlled thermal decomposition of KNO_3/ZrO_2 has attracted some attention. Wang and coworkers [95, 96] reported that sites with a strength $H_- = 27$ could be generated in such systems providing attention was paid to the loadings and thermal decomposition temperatures. Samples were prepared by grinding KNO_3 and ZrO_2 together followed by calcination. Furthermore, catalytic activity for the isomerization of *cis*-but-2-ene at 0 °C was reported and the samples were reported to be coke resistant in the high-temperature decomposition of isopropanol. It was argued that vacant sites in the structure of ZrO_2 were crucial to the development of such high base site strengths and this was used to explain the maximum activity observed at 14 wt% KNO_3/ZrO_2. In a study investigating the influence of zirconia and potassium precursor compounds upon the activity of potassium-doped zirconia for gas-phase and liquid-phase reactions by Kemnitz and coworkers [97], demonstrated that potassium leaching occurred for liquid-phase reactions. In addition, their studies indicated that employing hydrous zirconia precursors led to materials with a broader range of base sites with higher strength. Superbase sites have also been reported to be exhibited by KOH-modified ZrO_2 used to support Ru in ammonia decomposition catalysts for hydrogen generation [98]. Related systems, other than those employing ZrO_2, have also been identified as superbases, for example those derived from KNO_3 supported on MgO-modified SBA-15 [99]. Yamaguchi and coworkers [100] have reported that decomposition of carbonate and nitrate precursors, such as $CsCO_3$ and KNO_3, over alumina and zirconia supports can lead to the generation of catalysts with strong basicity. It was reported that both silica and titania were ineffective as supports and the generation of basicity was argued to be a consequence of the liberation of free oxygen species on thermal decomposition of the carbonate and nitrate salts (yielding CO_2 and O^{2-} and NO_2 and O^- respectively).

21.3
Outlook

The study of metal oxide-catalyzed base catalysis is becoming more widespread, although it still lags far behind acid catalysis. In this chapter, we have outlined some of the general considerations to be made in the characterization of base catalysts, have briefly described some of the common classes of base oxide materials applied as catalysts and have described a selection of the more commercially relevant reactions. Table 21.5, which is not intended to be exhaustive, summarizes some of the other base-catalyzed reactions that have been reported in the more recent literature. Attention has been restricted to oxides, although other basic materials such as oxynitrides, for example References [124, 125], may offer advantages, for example in terms of the ability to tune base site strength through control of the level of nitrogen incorporation. A survey of the literature demonstrates the vast scope of base-catalyzed reactions that have been studied with metal oxide catalysts. However, it is clear that in many cases fundamental studies, such as the elucidation of structure–activity relationships, are lacking. Ideally, the development of such understanding will lead to knowledge facilitating the tailored design of effective catalysts.

Table 21.5 Some examples of base-catalyzed reactions.

Reaction	Catalyst(s)	Ref.
Transesterification	CaO superbase	[52]
Transesterification of soybean oil	CaO	[101, 102]
Transesterification of rapeseed oil	Li/CaO, Na/CaO, K/CaO, Li/MgO	[103]
Transesterification of methyl benzoate and dimethyl terephthalate with ethylene glycol	Na/NaOH/γ-Al$_2$O$_3$ and Cs$_x$O/γ-Al$_2$O$_3$	[38]
Transesterification of glyceryl tributyrate with methanol	MgAl hydrotalcites	[56]
Transesterification of ethyl acetate with methanol	NaOH, KOH, LiOH supported on mesoporous smectites	[104]
Transesterification of soybean oil	ZrO$_2$, ZnO or γ-Al$_2$O$_3$ supported alkaline earth oxides	[105]
Synthesis of phytosterol esters	La$_2$O$_3$	[106]
Self-condensation of propanol	MgO	[50]
Dimerization of ethanol	MgO, CaO, BaO, Na/Al$_2$O$_3$, K/Al$_2$O$_3$, Cs/Al$_2$O$_3$, Mg/SiO$_2$	[107]
Isopropanol conversion to MIBK	CuMI(MII)O$_x$	[108]
Cyclohexanol conversion to cyclohexanone/cyclohexene	CaO/TiO$_2$, SrO/TiO$_2$, BaO/TiO$_2$	[109]
Cyanoethylation of methanol	BaO, MgO, KOH/Al$_2$O$_3$, KF/Al$_2$O$_3$	[110]
Cyanoethylation of alcohols	MgO, CaO, SrO, BaO, Mg(OH)$_2$, Ca(OH)$_2$, Sr(OH)$_2\cdot$8H$_2$O, Ba(OH)$_2\cdot$8H$_2$O, La$_2$O$_3$, KOH/Al$_2$O$_3$, KF/Al$_2$O$_3$	[111]
Synthesis of dimethyl carbonate from propylene carbonate and methanol	MgO, CaO	[112]
Synthesis of ethylene carbonate from urea and ethylene glycol	CaO, La$_2$O$_3$, MgO, ZnO, ZrO$_2$ and Al$_2$O$_3$	[113]
Synthesis of propylene carbonate from urea and propane-1,2-diol	CaO, MgO, ZnO, La$_2$O$_3$	[114] review

Table 21.5 Continued

Reaction	Catalyst(s)	Ref.
Synthesis of 1,3-dialkylureas from ethylene carbonate and amine	CaO, MgO, ZnO, ZrO_2, CH_3ONa	[115]
Synthesis of dimethyl carbonate from propylene oxide, carbon dioxide and methanol	NaOH, KOH, LiOH supported on mesoporous smectites	[104]
Propylene–ethylene conversion to pentenes and heptenes	MgO, K/MgO	[116]
Aldol condensation of citral with acetone	MgO, Li/MgO, Na/MgO, K/MgO, Cs/MgO	[117]
Self-condensation of acetone	MgO, Li/MgO, Na/MgO, K/MgO, Cs/MgO	[118]
Knoevenagel condensation of benzaldehyde with ethyl cyanoacetate	NaOH, KOH, LiOH supported on mesoporous smectites	[104]
Methylation of catchecol	Cs_2O/TiO_2, Cs_2O/Al_2O_3, Cs_2O/SiO_2	[119]
Benzaldehyde reduction	MgO, CaO, BaO, SrO	[120]
Ligand-free Heck reaction	Pd/MgO, Pd/CaO, Pd/SrO, Pd/BaO	[121]
Naphthenic acid decarboxylation	MgO	[122]
N-alkylation of aniline	Cs_2O	[123]

References

1 Pines, H. and Stalick, W.M. (1977) *Base-Catalyzed Reaction of Hydrocarbons and Related Compounds*, Academic Press, New York, pp. 240–308.
2 Tanabe, K. and Holderich, W.F. (1999) *Applied Catalysis A: General*, **181**, 399.
3 Tanabe, K. (1970) *Solid Acids and Bases*, Kodansha, Tokyo.
4 Hattori, H. (1995) *Chemical Reviews*, **95**, 537.
5 Hattori, H. (1988) *Materials Chemistry and Physics*, **18**, 533.
6 Ono, Y. and Baba, T. (1997) *Catalysis Today*, **38**, 321.
7 Pines, H. and Schaap, L.A. (1960) *Advanced Synthesis Catalysis*, **12**, 117.
8 Iglesia, E., Batron, D.G., Biscardi, J.A., Gines, M.J.L. and Soled, S.L. (1997) *Catalysis Today*, **38**, 339.
9 Rosynek, M.P., Fox, J.S. and Jensen, J.L. (1981) *Journal of Catalysis*, **71**, 64.
10 Angeles-Aramendia, M., Borav, V., Jiminez, C., Marinas, J.M., Marinas, A., Porras, A. and Urbano, F.J. (1999) *Journal of Catalysis*, **183**, 240.
11 Martra, G., Gianotti, E. and Coluccia, S. (2008) The application of UV-visible-NIR spectroscopy to oxides, in *Metal Oxide*

Catalysis, Vol. **1** (eds S.D. Jackson and J.S.J. Hargreaves), Wiley-VCH Verlag, Weinheim, pp. 51–94.

12. Coluccia, S. and Tench, A.J. (1981) *Proceedings of the 7th International Congress on Catalysis, Tokyo, 1980* (eds K. Tanabe and T. Sieyarna), Kodansha/Elsevier, Tokyo/Amsterdam, p. 1160.

13. Chizallet, C., Bailly, M.L., Costentin, G., Lauron-Pernot, H., Krafft, J.M., Bazin, P., Saussey, J. and Che, M. (2006) *Catalysis Today*, **116**, 196.

14. Kantam, M.L., Ranganathan, K.V.S., Mahendar, K., Chakrapani, L. and Choudary, B.M. (2007) *Tetrahedron Letters*, **48**, 7646.

15. Xu, C.L., Bartley, J.K., Enache, D.I., Knight, D.W. and Hutchings, G.J. (2005) *Synthesis – Stuttgart*, **19**, 3468.

16. King, F. and Kelly, G.J. (2002). *Catalysis Today*, **73**, 75.

17. Hathaway, P.E. and Davis, M.E. (1989) *Journal of Catalysis*, **116**, 263.

18. Tatibouet, J.M. (1997) *Applied Catalysis A: General*, **148**, 213.

19. Brown, A.S.C. and Hargreaves, J.S.J. (1999) *Green Chemistry*, **1**, 17.

20. Gorzawski, H. and Holderich, W.F. (1999) *Journal of Molecular Catalysis A – Chemical*, **144**, 181.

21. Canning, A.S., Jackson, S.D., McLeod, E. and Parker, G.M. (2007) *Science and Technology in Catalysis 2006*, Vol. **172** (eds K. Eguchi, M. Machida and I. Yamanaka), Studies in Surface Science and Catalysis, Elsevier, Tokyo, p. 401.

22. Auroux, A. and Gervasini, A. (1990) *Journal of Physical Chemistry*, **94**, 6371.

23. Martin, D. and Duprez, D. (1997) *Journal of Molecular Catalysis A – Chemical*, **118**, 113.

24. Vinek, H., Noller, H., Ebel, M. and Schwarz, K. (1977) *Journal of the Chemical Society – Faraday Transactions*, **1**, 73–743.

25. Tanaka, K., Yanashima, H., Minobe, M. and Suzukamo, G. (1997) *Applied Surface Science*, **121/122**, 461.

26. Huang, M., Adnot, A. and Kaliaguine, S. (1992) *Journal of Catalysis*, **137**, 322.

27. Shen, J., Tu, M., Hu, C. and Chen, Y. (1998) *Langmuir*, **14**, 2756.

28. Dingerdissen, U., Guenter, L. and Steuerle, U. (1996) DE Patent 19514146, assigned to BASF AG.

29. Tuneki, H., Yano, H., Shimasaki, Y. and Ariyoshi, K. (1993) US Patent 5231189, assigned to Nippon Catalytic Chem Ind.

30. Gortsema, F.P., Sharkey, J.J., Wildman, G.T. and Beshty, B.S. (1992) EP481674, assigned to Merck & Co. Inc.

31. Jackson, S.D., Johnson, D.W., Scott, J.D., Kelly, G.J. and Williams, B.P. (2001) European Patent EP1073517, assigned to INEOS Acrylics UK Ltd.

32. Hagen, G.P. and Montag, R.A. (1991) US Patent 4990662, assigned to Amoco Corp.

33. Tai, J. and Davis, R.J. (2007) *Catalysis Today*, **123**, 42.

34. Hamilton, C.A., Jackson, S.D. and Kelly, G.J. (2004) *Applied Catalysis A: General*, **263**, 63.

35. Canning, A.S., Gamman, J.J., Jackson, S.D. and Urquart, S. (2007) *Catalysis of Organic Reactions* (ed. S.R. Schmidt), Taylor & Francis, Boca Raton, FL, p. 67.

36. Canning, A.S., Jackson, S.D., McLeod, E. and Vass, E.M. (2005) *Applied Catalysis A: General*, **289**, 59.

37. Canning, A.S., Jackson, S.D., McLeod, E. and Parker, G.M. (2007) Studies in surface science and catalysis, in *Science and Technology in Catalysis 2006*, Vol. **172** (eds K. Eguchi, M. Machida and I. Yamanaka), Elsevier, Tokyo, p. 401.

38. Gorzawski, H. and Holderich, W.F. (1999) *Applied Catalysis A: General*, **179**, 131.

39. Wei, T., Wang, M.H., Wei, W., Sun, Y.H. and Zhang, B. (2003) *Fuel Processing Technology*, **83**, 175.

40. Zhang, G., Hattori, H. and Tanabe, K. (1988) *Applied Catalysis A: General*, **36**, 198.

41. Hargreaves, J.S.J., Hutchings, G.J., Joyner, R.W. and Kiely, C.J. (1992) *Journal of Catalysis*, **135**, 576.

42. Burrows, A., Coluccia, S., Hargreaves, J.S.J., Joyner, R.W., Kiely, C.J., Martra, G., Mellor, I.M. and Stockenhuber, M. (2005) *Journal of Catalysis*, **234**, 14.

43. Moodie, A.F. and Warble, C.E. (1971) *Journal of Crystal Growth*, **10**, 26.

44. Martra, G., Cacciatori, T., Marchese, L., Hargreaves, J.S.J., Mellor, I.M., Joyner, R.W. and Coluccia, S. (2001) *Catalysis Today*, **70**, 121.

45. Pantazidis, A., Burrows, A., Kiely, C.J. and Mirodatos, C. (1998) *Journal of Catalysis*, **177**, 325.

46 Knozinger, E., Diwald, O. and Sterrer, M. (2000) *Journal of Molecular Catalysis A – Chemical*, **162**, 83.
47 Utamapanya, S., Klabunde, K.J. and Schlup, J.R. (1991) *Chemistry of Materials*, **3**, 175.
48 Jiang, D., Zhao, B., Xie, Y., Pan, G., Ran, G. and Min, E. (2001) *Applied Catalysis A: General*, **219**, 69.
49 Wang, J.X. and Lunsford, J.H. (1986) *Journal of Physical Chemistry*, **90**, 5883.
50 Ndou, A.S. and Colville, N.J. (2004) *Applied Catalysis A: General*, **275**, 103.
51 Matsuhashi, H., Oikawa, M. and Arata, K. (2000) *Langmuir*, **16**, 8201.
52 Zhu, H., Wu, Z., Chen, Y., Zhang, P., Duan, S. and Liu, X. (2006) *Chinese Journal of Catalysis*, **27**, 391.
53 Jackson, S.D., Kelly, G.J., Hamilton, C.A. and Davies, L. (2003) *Reaction Kinetics and Catalysis Letters*, **79**, 213.
54 Mohri, M., Tanabe, K. and Hattori, H. (1974) *Journal of Catalysis*, **32**, 144.
55 Pines, H. and Haag, W. (1958) *Journal of Organic Chemistry*, **23**, 328.
56 Cantrell, D.G., Gillie, L.J., Lee, A.F. and Wilson, K. (2003) *Applied Catalysis A: General*, **287**, 183.
57 Palmores, A.E., Lopez-Nieto, J.M., Lazaro, F.J., Lopez, A. and Corma, A. (1999) *Applied Catalysis B – Environmental*, **20**, 257.
58 Corma, A., Fornes, V., Martin-Aranda, R.M. and Rey, F. (1998) *Journal of Catalysis*, **134**, 58.
59 Climent, M.J., Corma, A., Iborra, S. and Primo, J. (1995) *Journal of Catalysis*, **151**, 60.
60 Guida, A., Lhouty, M.H., Tichit, D., Figueras, F. and Gineste, P. (1997) *Applied Catalysis A: General*, **164**, 251.
61 Velu, S. and Swamy, C.S. (1996) *Applied Catalysis A: General*, **145**, 225.
62 Hillion, G., Delfort, B., Lendresse, C. and Le Pennec, D. (2005) French Patent 2,869,612, IFP.
63 Choudary, B.M., Lakshmi Kantam, M., Venkat Reddy, C., Aranganathan, S., Lakshmi Santhi, P. and Figueras, F. (2000) *Journal of Molecular Catalysis A – Chemical*, **159**, 411.
64 Lamonier, J.F., Boutoundou, A-B., Gernequin, C., Perez-Zurita, M.J., Siffert, S. and Aboukais, A. (2007) *Catalysis Letters*, **118**, 165.
65 Winter, F., Wothers, M., van Dillen, A.J. and de Jong, K.P. (2006) *Applied Catalysis A: General*, **307**, 231.
66 Liu, Y., Lotero, E., Goodwin, J.G., Jr and Mo, X. (2007) *Applied Catalysis A: General*, **331**, 138.
67 Cavani, F., Trifiro, F. and Vaccari, A. (1991) *Catalysis Today*, **11**, 173.
68 Rey, F., Fornes, V. and Rojo, J.M. (1992) *Journal of the Chemical Society – Faraday Transactions*, **88**, 2233.
69 Palomares, A.E., Prato, J.G. and Corma, A. (2004) *Journal of Catalysis*, **221**, 62.
70 Dubey, A. (2007) *Green Chemistry*, **9**, 424.
71 Kannan, S. (2006) *Catalysis Surveys from Asia*, **10**, 117.
72 Sels, B.F., de Vos, D.E. and Jacobs, P.A. (2001) *Catalysis Reviews – Science and Engineering*, **43**, 443.
73 Evans, D.G. and Duan, X. (2006) *Chemical Communications*, 485.
74 Tichit, D. and Coq, B. (2003) *CATTECH*, **7**, 206.
75 Figueras, F. (2004) *Topics in Catalysis*, **29**, 189.
76 Centi, G. and Perathoner, S. (2008) *Microporous and Mesoporous Materials*, **107**, 3.
77 Hussein, G.A.M. (1996) *Journal of Analytical and Applied Pyrolysis*, **37**, 111.
78 Mekhemer, G.A.H. (2004) *Applied Catalysis A: General*, **275**, 1
79 Schuth, F., Winger, A. and Sauer, J. (2001) *Microporous and Mesoporous Materials*, **44**, 465.
80 Kumar, P., Pandey, R.K., Bodas, M.S. and Dongare, M.K. (2001) *Synlett*, **2**, 206.
81 Pandey, R.K., Deshmukh, A.N. and Kumar, P. (2004) *Synthetic Communications*, **34**, 1117.
82 Ivanova, A.S. (2005) *Kinetics and Catalysis*, **46**, 620.
83 Walton, K.S., Abney, M.B. and LeVan, M.D. (2006) *Microporous and Mesoporous Materials*, **91**, 78.
84 Hathaway, P.E. and Davis, M.E. (1989) *Journal of Catalysis*, **116**, 279.
85 Barthomeuf, D. (1996) *Catalysis Reviews – Science and Engineering*, **38**, 521.
87 Rodriguez, I., Cambon, H., Brunel, D. and Lasperas, M. (1998) *Journal of Molecular Catalysis A – Chemical*, **130**, 195.
88 Doskocil, E.J. and Mankidy, P.J. (2003) *Applied Catalysis A: General*, **252**, 119.

89 Zhu, J.H., Chun, Y., Wang, Y. and Xu, Q.H. (1997) *Material Letters*, **33**, 207
90 Ernst, S., Hartmann, M., Sauerbeck, S. and Bongers, T. (2000) *Applied Catalysis A: General*, **200**, 117.
91 Pibre, G., Chaumant, P., Fleurzard, E. and Cassagnam, P. (2008) *Polymer*, **49**, 234.
92 Ozeryanskii, V.A., Milov, A.A., Minku, V.L. and Pozharskii, A.E. (2006) *Angewandte Chemie – International Edition in English*, **45**, 1453.
93 Mordini, A., Peruzzi, D., Russo, F., Valacchi, M., Reginato, G. and Brandi, A. (2005) *Tetrahedron*, **61**, 3349.
94 Kaskel, S. and Slichke, K. (2001) *Journal of Catalysis*, **201**, 270.
95 Wang, Y., Huang, W.Y., Wu, Z., Chun, Y. and Zhu, J.H. (2000) *Material Letters*, **46**, 198.
96 Wang, Y., Huang, W.Y., Chun, Y., Xia, J.R. and Zhu, J.H. (2001) *Chemistry of Materials*, **13**, 670.
97 Li, Z.-J., Prescott, H.A., Deutsch, J., Trunschke, A., Lieske, H. and Kemnitz, E. (2006) *Catalysis Letters*, **92**, 175.
98 Yin, S.F., Xu, B.Q., Wang, S.J. and Au, C.T. (2006) *Applied Catalysis A: General*, **301**, 202.
99 Wu, Z.Y., Jiang, Q., Wang, Y.M., Wang, H.J., Sun, L.B., Shi, L.Y., Xu, J.H., Wang, Y., Chun, Y. and Zhu, J.H. (2006) *Chemistry of Materials*, **18**, 4600.
100 Yamaguchi, T., Komatsu, M. and Okawa, M. (2002) *Catalysis Surveys from Japan*, **5**, 81.
101 Liu, X.J., He, H.Y., Wang, Y.J., Zhu, S.L. and Piao, X.L. (2008) *Fuel*, **87**, 216.
102 Kouzu, M., Kasuno, T., Tajika, M., Yamanaka, S. and Hidaka, J. (2008) *Applied Catalysis A: General*, **334**, 357.
103 MacLeod, C.S., Harvey, A.P., Lee, A.F. and Wilson, K. (2008) *Chemical Engineering Journal*, **135**, 63.
104 Fujita, S.-I., Bhanage, B.M., Aoki, D., Ochiai, Y., Iwasa, N. and Arai, M. (2006) *Applied Catalysis A: General*, **313**, 151.
105 Zhenqiang, Y. and Wenlei, X. (2007) *Fuel Processing Technology*, **88**, 631.
106 Valange, S., Beauchard, A., Barrault, J., Gabelica, Z., Daturi, M. and Can, F. (2007) *Journal of Catalysis*, **251**, 113.
107 Ndou, A.S., Plint, N. and Coville, N.J. (2003) *Applied Catalysis A: General*, **251**, 337.
108 Di Cosimo, J.I., Torres, G. and Apesteguia, C.R. (2002) *Journal of Catalysis*, **208**, 114.
109 Reddy, B.M., Ratnam, K.J., Saika, P. and Thrimurthulu, G. (2007) *Journal of Molecular Catalysis A – Chemical*, **276**, 197.
110 Kabashima, H. and Hattori, H. (1997) *Applied Catalysis A: General*, **161**, L33.
111 Kabashima, H. and Hattori, H. (1998) *Catalysis Today*, **44**, 277.
112 Wei, T., Wang, M., Wei, W., Sun, Y. and Zhang, B. (2003) *Fuel Processing Technology*, **83**, 175.
113 Li, Q., Zhang, W., Zhao, N., Wei, W. and Sun, Y. (2006) *Catalysis Today*, **115**, 111.
114 Qibiao, L., Ning, Z., Wei, W. and Yuhan, S. (2007) *Journal of Molecular Catalysis A – Chemical*, **270**, 44.
115 Fujita, S., Bhanage, B.M., Kanamaru, H. and Arai, M. (2005) *Journal of Molecular Catalysis A – Chemical*, **230**, 43.
116 Sun, N. and Klabunde, K. (1999) *Journal of Catalysis*, **185**, 506.
117 Diez, V.K., Apesteguia, C.R. and Di Cosimo, J.I. (2006) *Journal of Catalysis*, **210**, 235.
118 Di, J.I. and Apesteguia, C.R. (1998) *Journal of Molecular Catalysis A – Chemical*, **130**, 177.
119 Vishwanathan, V., Ndou, A.S., Sikhwivhilu, L., Plint, N., Raghavan, K.V. and Coville, N.J. (2001) *Chemical Communications*, 893.
120 Saadi, A., Rassoul, Z. and Bettahar, M.M. (2006) *Journal of Molecular Catalysis A – Chemical*, **258**, 59.
121 Chen, F., Lin, I., Li, H., Gan, G.J., Toh, K. and Tham, L. (2007) *Catalysis Communications*, **8**, 2053.
122 Zhang, A., Ma, Q., Wang, K., Liu, X., Shuler, P. and Tang, Y. (2006) *Applied Catalysis A: General*, 303.
123 Sivasanker, S. (2003) *Science and technology in catalysis 2002*, Studies in Surface Science and Catalysis, Vol. 85, Elsevier, p. 145.
124 Delsarte, S., Florea, M., Mauge, F. and Grange, P. (2006) *Catalysis Today*, **116**, 216.
125 Hasni, M., Prado, G., Rouchard, J., Grange, P., Devillers, M. and Delsarte, S. (2006) *Journal of Molecular Catalysis A – Chemical*, **247**, 116.

Index

a

A tensor 16
aberration 448
absorbance 98 ff, 186
absorption 51, 107
– AEO 82
– anhydrous oxides 111
– EPR 19
– TEM 453
– UV-Vis-NIR 59 f
absorption coefficient
– EXAFS 304
– IR spectroscopy 57, 67, 99
– XAS 301
acceptors 55
accuracy 309, 454
acetaldehyde production 548, 554
acetone 600, 624, 828
acetonitrile 155
– base catalysis 834
– surface basicity 166
– titanium silicalite-1 742
– vanadium phosphate catalysts 507
acid–base properties
– CuMgAl 321
– IR spectra 167
– metal oxides 617
– mixed metal oxides 415 f
– NMR 195
– POMs 567, 578
– surface 227
– thermal analysis 392
– vanadium pentoxides 431
– vanadium phosphates 507
– zeolites 311
acid-catalyzed reactions 677
acid probes 586
acid sites
– IR spectra 139, 156

– superacidic metal oxides 675, 682
– thermal analysis 404
acid zeolites 310 ff
acidic oxides 406
acidic systems 496
acidity
– IR spectra 147
– POMs 572
– surface properties 227
– thermal analysis 402 f
acoustic modes 124
acrylic acid oxidation 281 f
acrylic fibre ammoxidation 771
acrylonitrile–butadiene–styrene (ABS) 771
acrylonitrile synthesis 771–818
actinide elements 540
activation
– base catalysis 829
– IR spectra 134, 156
– thermal analysis 399, 405
– vanadium phosphate catalysts 500
active centers 69
active sites
– ammoxidation 776
– EXAFS 314
– thermal analysis 392
– titanium silicalite-1 720, 740 f
– uranium oxides 551
– vanadium phosphate catalysts 529
activity
– base catalysis 820
– chromium oxides 597
– IR spectroscopy 107
– photocatalytic 762
– POMs 564
– superacidic metal oxides 673
– uranium oxides 545
– vanadia catalysts 605

Metal Oxide Catalysis. Edited by S. David Jackson and Justin S. J. Hargreaves
Copyright © 2009 WILEY-VCH Verlag GmbH & Co. KGaA, Weinheim
ISBN: 978-3-527-31815-5

– vanadium oxides   210
– vanadium phosphate catalysts   504
addenda atoms   561, 569, 581
adiponitrile (ADN)   803
adsorbate–adsorbate interactions   398
adsorbed probe molecules   133 f
adsorption   394
– base catalysis   823
– Lewis/Brønsted sites   146
– metal oxide catalysts   645
– microcalorimetry   393
– photocatalysis   760
– superacidic metal oxides   681
– titanium silicalite-1   705, 740 f
adsorption–desorption heats   401 f
aerobic partial oxidation   642
aggregation   477
alcohols
– dehydration   618
– methoxide solutions   623
– oxidation   570, 727
– surface basicity   166
– titanium silicalite-1   708
– TS-1 catalyzed   316, 724
– vanadium phosphate catalysts   510
aldehydes   316, 724
aldol condensation   554, 830
aliphatic nitriles   154
aliphatic oximes   733
alkali-exchanged zeolites   428
alkali halides   101
alkaline earth oxides (AEO)
– base catalysis   826 f, 830
– IR spectra   61, 153
– PL spectra   81 f
– powder EPR   33
– promoted vanadium phosphates   523
alkaline earth salts   423
– destruction   547
– hydroxylation   707 ff
– NMR   195
– oxidation   499
– oxygenation   641
alkane–alkene conversion   605
alkanes
– olefins   178
– oxidative   416
– vanadium oxides   210
– vanadium/chromium oxides   595–612
alkenes
– epoxidation   570, 576
– metathesis   650
– TS-1 catalyzed   316
alkenyl mechanism   524

alkoxide mechanism   527
alkyl acid anhydride   693
alkyl amines   147
alkyl chains   480
alkyl radicals   47
alkylaromatic ammoxidation   791
alkylaromatic side-chain alkylation   819
alkylation   618
alkylbenzene ammoxidation   791
allyl alcohol oxidation   725
allyl chloride epoxidation   726
alumina supports
– surface structure   228, 231
– vanadium catalysts   214, 177–196
– vanadium oxide   211, 433, 602
aluminas   121
– bulk structure   347 f
– CP-MAS NMR   230
– differential heats   406
– electronic structures   347, 351
– hydroxyl groups   141
– IR spectroscopy   115, 159, 311
– Lewis acid sites   159 f
– metal oxide catalysts   614
– MQMAS   202
– NMR   195
– SATRAS   203
– surface oxo-anions   136
– surfaces   356, 359 f
– thermal analysis   408
– XPS   260
aluminosilicates   425
aluminum containing systems   204, 208 f, 780
aluminum hydroxides   209, 341 f, 347
aluminum layers   355
aluminum oxides   221 f, 348
aluminum supports   177–196, 221 f
aluminum zeolites   124, 300
amide ammoxidation   793
amines   316, 733, 793
ammonia
– adsorption   408, 418 f
– POMs   567, 586
– surface basicity   166, 402
ammonium salts   565
ammoxidation
– hydrocarbons/mixed oxides   771–818
– propane   281
– titanium silicalite-1   730 f
amorphous material   110 f, 129 f, 505
amorphous silica/aluminas   228, 425
amorphous structures   200
amphoteric oxides   406

amplitude reduction factor (AFAC) 302
amplitudes 446
analcime 478
anatase 118
Anderson structures 562
angular momentum 2, 7, 53, 197
anharmonic oscillator 77
anhydrides 693
anhydrous oxides 111 f
aniline oxidation 733
anion–cation distance 65
anion doping 760
anisotropic fine term 14
anthracene 13
anthrahydroquinone autoxidation (AO) 736
antiferromagnetic AF_2 state 372
antimonate catalysts 786
antimony 549
antiphases 446 f
antisymmetrized wavefunctions 326
aperture 446
applications
– base catalysis 840
– EXAFS 309 f
– IR spectroscopy 96 f
– metal oxide catalysts 641
– oxide systems 32 f
– partial oxidation reactions 568
– POMs 566 ff
– superacidic metal oxides 698
– surface chemistry 96 f
– UV-Vis-NIR spectroscopy 51–94
– vanadium oxide 256 f
aqueous routes 509
aqueous solutions 628
aquocomplexes 132
aragonite 123
argon probe molecule 147, 678 ff
aromatic acids ammoxidation 793
aromatic compounds hydroxylation 712
aromatic nitriles synthesis 792
aromatic oximes 733
aromatics Friedel–Crafts acylation 692 f
atomic displacements 99
atomic layer deposition (ALD) 409, 431
atomic layers 455
atomic number effect 453
atomistic models 34
attenuated multiple total internal reflection technique (ATR) 103
Auger electron process 249, 265, 305
augmented-wave pseudopotentials 341
Aurivillius phases 456
aziridine synthesis 827

b

background function 306
backscattering 301
Baeyer–Villiger oxidation 570
band gaps 68, 756
band structures 51 f, 344, 377, 762
band theory 332 f, 613
base-catalyzed Claisen–Schmidt condensation 652
base indicators 666
base metal oxide catalysis 819–844
base properties 617, 822
baseline correction 307
basic oxide crystal structures 454 f
basic oxides 406
basic principles
– electronic structure methods 336 f
– EPR 2 f
– EXAFS 299 f
– heterogeneous photocatalysis 756 f
– XAS 300 f
– XPS 244 f
basic probe molecules 160, 586
basic strength method 152
basic zeolites 837
basicity 402 f
bayerite 116, 341, 348
beam damage 256
Becke–Lee–Yang–Parr (BLYP) approach 331, 346
Beer–Lambert law 99
Bellamy–Hallam–Williams relation 150
bending magnets 304
bending mode 120
benzaldehyde production 544, 552
benzene
– ammoxidation 805
– hydroxylation 716
– oxidation 544 ff, 570
– XPS 263
benzoate-like intermediates 551
benzoic anhydride 694
benzoylation 692
benzylic derivative oxidation 571
Bi–Mo mixtures 553
bifunctional catalysis 567
bimolecular intermediates 682
binary mixed metal oxides 415 f
binary oxides 135 f
binary systems 418
binary vanadium oxides 269
binding energy 59
– alumina 336
– base catalysis 824 f

– vanadium oxides 257
– VPO XPS 270
– XAS 300
– XPS 244, 248 ff, 285
binominal expansion 12
biodiesel production 643
bismuth molybdate 417, 775
bismuth oxide-based solid solutions 455, 460 f
bismuth promoted vanadium phosphate catalysts 523
bismuth uranates 551
Bloch function 336
boehmite, Al-NMR 222, 229, 347 ff
Bohr frequency condition 4
Boltzmann factor 234
Boltzmann law 98
Boltzmann–Maxwell law 5
bond lengths/angles 97
bonding 340
– metal oxide catalysts 614, 631
– oxidation reactions 487 ff
borated superacidic metal oxides 674
borated surface oxo-anions 136
boria 409, 420
boron trifluoride 166
bottom-up preparation 619
boundary conditions 354, 616
Bragg angles 305
Bragg reflections 446, 450
branched alkane hydroxylation 709
Bravais cell 108, 112, 122
bridging 139
bright-field image 511
broadening 262
bromothymol blue 821
Brønsted acid sites
– IR spectra 139 f
– oxidation reactions 495
– silica-aluminas 145
– solid-state NMR 229 f
– superacidic metal oxides 666, 683, 689
– thermal analysis 392, 398, 404
– vanadium phosphate catalysts 507
– zeolites 427
Brønsted acidity
– metal oxides 146 f, 153 f
– POMs 567, 578
– protonic zeolites 153 f
Brønsted basic–OH surface groups 652
brucite-type layers 130, 133
buckling 372
building blocks 57, 619
bulk catalyst structure 229

bulk energy 335
bulk excitons 82
bulk heteropolyacids 422 f
bulk oxide catalysts 404 f
bulk phases 519
bulk properties 196, 614
bulk structures 200, 323, 347 f
bulk termination 357
bulk-type catalysis 564
bulk V-P-O ammoxidation 791
Burstein–Moss effect 69
butadiene ammoxidation 798
butane–butene fractions 568
butane conversion 643, 690
butane dehydrogenation 210, 260 f, 266 f, 598
butane desorption 609
butane isomerization 682, 685 f
butane oxidation 269 f, 499, 507, 511, 524
butane reaction profile 690
butanol oxidation 729, 742
butanone 42
butene 263
butylamine 233

c
C_2 hydrocarbons ammoxidation 807
C_4/C_5 olefins epoxidation 719
C_4 hydrocarbons ammoxidation 797
C–H bond
– activation 221
– breaking 492
– hydroxylation 708 ff
– POMs 575
– vanadium phosphate catalysts 499, 507
– XPS 255
calcination
– aluminas 115
– base catalysis 834
– ceramic acids 695
– metal oxide catalysts 627, 638
– spikelet echo spectra 219
– superacidic metal oxides 668 ff
– uranium oxides 556
– vanadium phosphate catalysts 515
calcite 123
calibration
– electronic structures 340
– TEM 453
– volumetric-calorimetric line 395
– XPS 252, 256 f
calorimetric methods 391–442
capillary pressure 623
carbon–carbon bonds 265, 403

carbon deposition 598
carbon dioxide 166, 402, 624
carbon disulfide 166
carbon formation 264
carbon monoxide
– chemisorption 384
– IR spectra 161 f
– oxidation 547, 552
– redox properties 403
– surface acidity 147
– surface basicity 166
carbon monoxide adsorption 419
carbon-supported heteropolyacid catalysts 423
carbonates 133 ff, 629
carbonyl compounds ammoximation 730
carboxylic acids 511
carcinogenic properties 649
case studies
– surface adsorbed NO_2 30 f
– vanadia catalysts 429 f
– XPS 256 f
catalyst precursors see precursors
catalysts
– anatase 121
– base catalysis 825 f
– IR spectra 52, 61, 132
– metal oxide 613–664
– NMR 195
– oxidation reactions 487–498
– partial oxidation 561–594
– powder EPR 33
– solid-state NMR 195–242
– spin number 24
– thermal behavior 393 f
– uranium oxides 543
– vanadium phosphate 499–538
– XAS 299–322
catalytic activity
– MoO_3/ZrO_2 695
– superacidic metal oxides 665–704
– vanadium oxides 210
– $WO_3/SnO_2/Al_2O_3$ 696 f
catalytic ammoxidation 771–818
catalytic cracking 195
catechol hydroxylation 734
cation exchanged zeolites 165 f
cation ordering 455 f
cationic centers 161 f
cationic radicals 759
CATOFIN process 596
cavities 24 ff, 158 ff
CdSe/ZnS quantum dots 477
ceramic acid 695

ceria 384, 632 f
ceria-based catalysts 420
ceria–lanthana coprecipitated mixed oxides 415
ceria–zirconia solid solutions 415
cerium-promoted vanadium phosphate catalysts 519 ff
character tables 107
characterization see properties
charge carrier lifetimes 761
charge-coupled devices (CCD) 470
charge density difference 373
charge localization 378
charge migration 756
charge separation 458
charge transfer transitions 54 f, 87, 573
charge trapping 761
charging contributions 252
chemical feedstocks 260
chemical properties
– metal oxide catalysts 616, 620
– POMs 566
– UV-Vis-NIR spectroscopy 51
chemical shift
– CRAMPS 209
– NMR 197, 203
– vanadium oxides 212
– XPS 250
chemical vapor deposition (CVD) 409
chemisorption 604
chlorinated compound removal 755
chloroaniline 821
chlorocarbon adsorption 649
chloroform 166
chloro-substituted xylene 794
chromia–alumina catalyst 596
chromium oxides
– alkane dehydrogenation 595–612
– combustion methods 639
– TEM 472
chromium promoted vanadium phosphate catalysts 519, 523
Claisen–Schmidt condensation 652
clay catalysts 424 f
cleaning method 256
close-packed structures 465
clusters
– DFT calculations 34, 278
– heteropolyoxometallate catalysts 561
– photocatalysis 763
– POMs 587
– TEM 477
– uranium oxides 542
– vanadium phosphate catalysts 503

coal convertion 546
coating 463
cobalt metal ions 413
cobalt promoted vanadium phosphate catalysts 519, 522
coherent electron waves 448
coke 265, 598–608
color centers 36
combination modes 99
combined rotation/multiple pulse spectroscopy (CRAMPS) 209 f
combustion methods 544, 638
complex interactions 165 f
complex precipitates 132
Compton scattering 306
concerted mechanism 529
condensation 128, 622 f
conduction band 344
– heterogeneous photocatalysis 756
– UV-Vis-NIR spectroscopy 57
– XPS 255
confinement effects 153
conjugated polyenes 182
consecutive alkenyl/alkoxide mechanism 524 ff
contamination 247
continuous-flow microwave reactor 632
convergent beam electron diffraction (CBED) 452
conversion data 52, 552
coordinates 325
coordination number 70, 302
coordination sites 65, 820
copper addenda elements 572
copper-exchanged ETS-10 catalysts 428
copper metal ions 28, 413
copper oxides 74
coprecipitation 415, 627 ff
core levels
– electronic structures 341
– XAS 300
– XPS 248, 259, 277
core polarization 245, 249
core–shell nanoparticles 637
core–shell quantum dots 464
core–shell structures 476
corners 562, 615
correlation methods 327 f
corundum sesquioxides 142
corundum structure 347
Coulomb interactions 244, 370, 480
Coulomb operator 324
counter-cations 422, 572
counter ions 561

covalent bonding 615
coverage profiles 396, 399
cracking 195, 435
critical points 624, 633
cross-polarization (CP) NMR 206
crotonaldehyde production 554
crystal defects see defects
crystal field 18, 53
crystal growth 475
crystal structures
– aluminas 362
– divalent elements 113
– metal oxide catalysts 615
– NMR 200
– TEM 443
– ternary oxides 123
– uranium oxides 541
– UV-Vis-NIR spectroscopy 62
– vanadium oxides 211
– XPS 249
crystalline aluminosilicates 310, 425
crystalline anhydrous oxides 111 f
crystalline complex 121 f
crystalline nanoparticles 488
crystalline phases 318
crystalline solids 108 f
crystallites 614
crystallization 511, 706
crystallographic cell vector 354
Cs-corrected HRTEM 449
Cs_2Cu_x reduction 572 f
Cu–Cu scattering 313
cubic close packing (ccp) structures 443
cubic fluorite structure 541
cubic phases 467
cubic unit cell 356
CuO (tenorite) 115
cyano-pyridine synthesis 795
cycle time 596
cyclic amines 147
cyclization 727
cycloalkane isomerization 686
cyclohexanol oxidation 729
cyclohexanol/hexane ammoxidation 800 ff
cyclohexanone ammoxidation 730 ff, 800
cyclohexene epoxidation 576
cylindrical internal reflection technique (CIR) 104

d

d bands 53, 72 f, 370 f
D tensor 17
Dawson structure 561
dead time 453

dealumination 425
Debye–Waller factor 302
decay curves 79
decomposition
– base catalysis 824, 831
– IR spectra 133
– metal oxides 645
defects 598
– metal oxide catalysts 614 ff
– oxides 465 f
– transition metal oxides 375 f
– UV-Vis-NIR spectroscopy 51
defocus 448
deformation modes 151
degeneracy 8, 197
deglitching 307
dehydrated state 178 f
dehydration 578, 672
dehydrogenation
– base catalysis 825 f
– NMR 195, 209
– vanadia catalysts 604
– vanadium/chromium oxides 595–612
– XPS 260 ff
delocalization 344, 584
density functional theory (DFT)
– electronic structures 323, 328 f, 371
– oxidation reactions 494
– vanadia catalysts 180
– XPS 255, 278
density of states (DOS)
– aluminas 364
– electronic structures 342 f
– NiO 374
– TiO_2 surface 378
– vanadyl pyrophosphate 278
desorption 394
– thermal analysis 393
– uranium oxides 548
– vanadium phosphate catalysts 526
destructive adsorption 649
deuteroxyls 169
diaspore structures 130
dielectric properties 25, 59, 102
difference electronic charge density map 381
differential charging 253
differential heats 394 ff, 406
differential pumping 246
differential scanning calorimetry (DSC) 392
differential thermal analysis (DTA) 392, 586
differential thermogravimetry (DTG) 392, 586
diffraction techniques 443–486

diffuse reflectance Fourier transform (DRIFT) 105, 133, 647
diffuse reflectance spectroscopy (DRS) 59 f, 104 f, 134
– oxidation reactions 490
– polycrystalline AEO 63
diffusivity 156 f
dihydrofuran 526
diluted self-supporting disks 101
dimeric compounds 262
dimeric condensed tetrahedral anions 127
dimethyl amine probe molecules 155
dimethylcyclopentane hydroxylation 710
dinitrosyl species 46
diphenols 316
diphenylpropionitrile 155
dipolar coupling 197, 206
dipole interaction 32, 209
dipole operators 16
direct dehydrogenation 260 f
direct hydrogen peroxide synthesis 737
direct methanol fuel cell (DMFC) 646
direct propane oxidation 281 f
dislocations 617
disordered planes 507
dispersed metal oxide catalysts 487, 613–664
dispersion 69, 334
displacement 367
disproportionation 682
distortions 25, 462
divalent element oxides 112 f
domains 466
donors 55
doping 408 f, 525, 760
double bonds 135 f, 195
double resonance (DR) NMR 206 f
double rotation (DOR) 202
doublets 12
drying 620, 623, 673
dyes adsorption 760
dynamic angle spinning (DAS) 202

e
edge X-ray absorption fine structure (EXAFS) 299 f
edges 562, 615
effective potentials 329
Eigen functions 8, 308
elastic scattering 306
elastic strain 479
electrochemistry 585
electron backdonation 162
electron delocalization 584

electron diffraction 450 f
electron–electron interactions 18, 78, 323, 370
electron energy loss spectra (EELS) 371
electron–hole pairs 82, 756, 761
electron paramagnetic resonance (EPR) 1–50, 596
electron populations 370
electron–proton interactions 34
electron radiation 446
electron–solid interactions 445 f
electron spectroscopy–chemical analysis (ESCA) 249
electron spin resonance (ESR) 584
electron trapping 35
electron wavelength 445 f
electron Zeeman interaction 2 f
electronegativity
– metal oxide catalysts 616
– oxidation reactions 487
– promoted vanadium phosphate catalysts 521
– UV-Vis-NIR spectroscopy 56
electronic perturbation 157
electronic structures 324 f
– heterogeneous photocatalysis 756
– oxidation reactions 490
electronic transitions 52 f
electrostatic stability 355 f
emission IR spectroscopy (IRES) 105
emission technique 51, 82, 87, 105 f
emittance 105
energy dispersive X-ray spectroscopy (EDX) 445, 452 f
energy gap 56
energy levels 4, 197
energy state transitions 51
enthalpy diagram 398
environmental influences 367 f
epoxidation
– olefins 717
– POMs 576
– titanium silicalite-1 316, 705
equilibrium constants 400, 665
Erythrosin B 760
escape depth 253
esterification 618
ethanol
– critical point parameters 624
– dimerization 839
– oxidation 546, 549, 729
ethene oxidation 413
ethers 147
ethylbenzene dehydrogenation 555

ethylcyclopropane hydroxylation 711
ethylene polymerization 488
Ewald sphere 450
excess energy 306
exchange-correlation potential 329
exchange energy 324 f
exchange splitting 373
excitation
– AEO 82
– IR spectroscopy 51, 98
– photocatalysis 760
– PL spectroscopy 80
– XPS 249, 274
excitonic surface states 59 ff
exfoliation 480, 511
exhaust gas removal 548
exothermic phenomenon 393
experimental techniques
– EPR powder spectra 24 f
– EXAFS 303 f
– IR methods 97 f
– thermal analysis 393 f
extended X-ray absorption fine structure (EXAFS) 490
external cationic sites 165 f
extra-framework aluminum (EFAL) 425

f
F centers 615
factor group analysis 125
FAU 158, 307
faujasite (USY) 310, 425
FDU-12 470
feed composition, ammoxidation 777
feedstocks 260, 502–514
FER 154, 158
Fermi contact term 31, 40
Fermi energy 345
Fermi level 252, 290
Fermi–Dirac statistics 344
ferric oxides *see* iron oxides
field gradients 199
flame processes 638
flow adsorption microcalorimetry 399, 426
fluid catalytic cracking (FCC) 310
fluid solutions 10 ff
fluidized-bed reactor 774 ff
fluorescence
– EXAFS 306
– PL spectroscopy 77
– TEM 453
– UV-Vis-NIR spectroscopy 60
– vanadia catalysts 179, 601
fluorite structure 461, 540, 554

Fock operator   324
formaldehyde oxidation   553
Fourier filtering   307
Fourier transform   208, 337
framework silicates   128
Franck–Condon principle   76 f
free energies   350
free fatty acids (FFA)   643
freedom degrees   107, 327
Freon   116 624
frequencies   4, 27, 100
Friedel–Crafts acylation   692 f
F-test values   308
FTIR spectra see IR spectra
fuel oil oxidation, uranium oxides   543
full width at half maximum (FWHM)   248, 256 f, 273
fumaric acids   499
functionality   554, 776
functionalization agents   566
fundamental transitions   99
furan production   266 f, 526, 555

g

g factor   3
g tensor   14
gallium-containing catalysts   220
gallium oxides
– electronic structures   351
– IR spectra   118, 135, 157 ff
– NMR   225
– polymorphs   142
– TEM   635
– thermal analysis   407 ff
gaps   354
gas flow thermal analysis   399 f
gas-phase synthesis   513
gas–solid interfaces   45
gas–solid transformations   619
gas/vapor manipulation   134
Gaussian image plane   448
Gaussian lineshape   24
Gaussian smearing   381
Gaussian target function   337
gels   671
geminal silanols   142
generalized gradient approximation (GGA)   330, 341, 359
germania   160
gibbsite structures   222, 341, 347
  see also aluminas
goethite structures   130
gold-based catalysts   554
gradient-corrected DFT technique   335

grain boundaries   446, 616
graphitic paracrystals   599
green catalytic processes   588
group approximation   110
group II oxides   33
gyromagnetic ratio   197

h

H-FER zeolite   156
H-method calculations   681
Hamiltonians   1 ff, 197
Hammett indicators   667–704, 820
harmonic rejection   304
harmonics   99
Hartmann–Hahn condition   206
Hartree–Fock (HF) approximation   323 ff
heat diagram   398
heat flow calorimetry   392 f
heat treatment   620
Heck reaction   840
helium system   304
hemihydrate precursors   508
hemispherical analyzer (HSA)   245
Henry constants   100, 714
heterocyclic amines   147
heterogeneous photocatalysis   42
heterogroups, alkylaromatic   795
heterolytic hydroxylation   715
heteropolyacids (HPA)   422 f, 583, 791
heteropolycompounds   416
heteropolyoxometallate catalysts   561–594
hexagonal close packing (hcp)   443
hexagonal layered structure   541
hexane ammoxidation   802
Hf sulfated metal oxides   669
hierarchical structures   562
high T_c superconducting oxides   446, 463 ff
high-order Laue zone (HOLZ) diffraction   450
high-pressure setup   245 f
high-resolution transmission electron microscopy (HRTEM)   443 f
high-surface-area oxides   487
highly dispersed metal oxide catalysts   613–664
highly dispersed oxide materials   51
highly dispersed supported oxo-species   69 f
highly dispersed transition metal ions   85 f
highly siliceous zeolites   128
hindered basic probes   154
hindered nitriles   165
hindered rotations/translations   110
Hohenberg–Kohn theorem   328
hole trapping   43

HOMO–LUMO structures   280
homolytic hydroxylation   715
host oxides   456
hybrid functionals   376
hybridization ratio   32
hybridized orbitals   366
hybrids   331
hydrated compounds   132 f
hydrazine   477
hydrocarbons
– catalytic ammoxidation   771–818
– cracking   618
– IR spectra   133
– oxidative dehydrogenation   267 f
– POMs   576
– surface acidity   147
hydrocarbon oxidation
– EXAFS   315
– selective   429
– uranium oxides   544
– vanadium phosphate catalysts   499
hydrochloric acid   509
hydrodesulfurization/denitrogenation   218
hydrogen   82, 147
– abstraction   526
– bonding   147, 398, 405, 615
– formation   260
– reduction   182 f, 597
hydrogen peroxide   569, 575 f, 588
– hydroxylation   713
– oxidation   575 ff
– processing   705, 737
– use problems   736
hydrogen silicates   136
hydrogenation   195, 829
hydrolysis   620
hydroquinone hydroxylation   734
hydrotalcites   132, 420 f, 835
hydrothermal processes   524, 633, 706
hydroxide coprecipitation products   628
hydroxides   132, 620 f
hydroxy chloride paratacamite   130
hydroxy silicoaluminum compounds   424
hydroxyl coverage   406
hydroxyl groups   61, 139 ff, 392, 377
hydroxyl radicals   759
hydroxylamines   316, 731
hydroxylation   707 ff
hyperfine parameters   7 f, 34, 39
HYSCORE techniques   14

i

ilmenites   122
image contrast   445 f, 472

image wave function   448
imine ammoxidation   793
impregnation   409, 826
impurities   81, 133 f, 467
incident electron wave   448
incommensurate superstructures   460 f
indicators   820
indium oxide   411, 635
indium promoted vanadium phosphate
   catalysts   523
inelastic scattering   95, 579
infrared reflection absorption spectroscopy
   (IRRAS)   103
infrared spectroscopic methods   95–176
inorganic dye adsorption   760
inorganic radicals   38 f
inorganic solids   107 f
insulating oxides
– AEO   81 f
– electronic structures   345
– UV-Vis-NIR spectroscopy   61
– XPS   258
integral heat   394 ff, 406
intensity
– Raman spectroscopy   186
– TEM   448
– XPS   247
interband transitions   58
intercalating compounds   511
intercept   58
interfaces   617
interlayer coupling   480
intermediates
– epoxide solvolysis/rearrangement   726
– propane ammoxidation   789 f, 793 ff
– superacidic metal oxides   682, 687
– titanium silicalite-1   743 f
– uranium oxides   551
– vanadium phosphate catalysts   507
internal cationic sites   165 f
internal vibrations   110
interstitials   467, 616
intersystem crossing (ISC)   79
intravalence   56
ion coordinations   113
ion implantation   761
ion–scattering spectroscopy   276
ionization potential   56, 829
IR spectra
– gaseous CO_2, CO, NO, CH_4 108
– MgO monocrystal   103
– POMs   581 f
– superacidic metal oxides   672
iron addenda elements   572

iron containing zeolite catalysts   313
iron doped vanadium phosphate catalysts
    519, 522
iron hydroxides   670
iron oxides   142, 308, 635
– heterogeneous photocatalysis   757
– nanoclusters   314
– TEM   477
iron sulfated metal oxides   669
isobutyronitrile (IBN)   155 f
isoelectric point   87
isolated molecular species   107 f
isolated surfaces   357
isomerization   83, 682, 685 ff
isopolyanions   561 ff
isosteric heat   394
isothermal measurements   401
isotherms   396
isotopic substitution   40
isotopically labeled molecules   168 f
isotropic hyperfine coupling   9 f
isotropic spectra analysis   10 f

j
Jahn–Teller effect   53, 73

k
K-edge absorption   300, 318
k-point   335, 342, 355
K–V–O catalysts   216
Keggin-type
– heteropolyoxometallates   561
– heteropoly compounds   422
– heteropolyacids   562
– heteropolysalts   129
– POMs   569, 585
ketones   147, 316, 570, 708
kinetic energy   244, 250, 300
Kirchoff law   105
KIT-6   471, 474
Knoevenagel condensation   821, 840
Kohn–Sham orbitals   329 ff, 338, 345
Kubelka–Munk (KM) equation   60, 104

l
L-edge   262, 300
L-method calculations   681
labeled molecules   168 f
Lambert–Beer law   248
Langmuir adsorption   680
lanthanides   225, 462
Laporte (orbital) selection rule   54, 75
lattice distortion   462
lattice oxygen
– ammoxidation   776
– DR UV-Vis spectra   71
– POMs   568
– thermal analysis   433
– vanadium phosphate catalysts   526
– VPO XPS   279
lattice parameters   57, 362, 460
lattice vacancies   542
lattice vectors   332, 354
lattice vibrations   110, 123
layer silicates   128, 424 f
layered defects   465 f
layered double hydroxides (LDH)   132 f,
    420, 834
layered structures   541
layers   60, 355 f, 455, 508
Lewis acid sites
– aluminas/SAs   159 f
– base catalysis   820
– CP-MAS NMR   233
– superacidic metal oxides   682
– thermal analysis   392, 405
– vanadium phosphate catalysts   507
– zeolites   427
Lewis acid–base, metal oxide catalysts   613
Lewis acidity
– aluminas   364
– highly covalent oxides   161 f
– ionic oxides   160 f
– metal oxide catalysts   617
– protonic zeolites   161 f
– thermal analysis   416
Lewis-to-Brønsted acidity ratio (L/B ratio)
    508
ligand field theory   613
ligand to metal charge-transfer transition
    (LMCT)   55, 69 f
ligands   70
Lindqvist structure   561
line broadening   200, 208, 310
linear alkane hydroxylation   708
linear olefin epoxidation   718
linewidths   24, 200
linkages   70
liquid–solid interface   25, 45
liquid–solid transformations   619
liquid superacids   667
liquid–vapor interface   623
lithium niobate structures   122
local density approximation (LDA)   330,
    341 ff, 346
localized electron populations   370
longitudinal optical mode   99, 102
Lorentzian lineshape   24

low-dimensional oxide crystals 476 f
low energy electron diffraction (LEED) 353
low energy ion scattering spectroscopy
  (LEISS) 489
low temperature coprecipitation 628
lower valence band (LVB) 344
lowest unoccupied molecular orbital
  (LUMO) 279
Lowry–Brønsted acidity definition 146
LPG dehydrogenation process 596 f
luminescence center 81
Lummus/Houdry CATOFIN process 596 f

m

Madelung potential 65, 82, 617
magic acids 666
magic angle spinning (MAS) 200 f
magnesia 408
– active sites 645
– differential heats 406
– DOS plots 342
– DR UV-Vis-NIR 61 ff, 81 ff
– IR spectra 102, 112, 131
– metal oxide catalysts 614
– Miller planes 354
– powder EPR 34
– surface calculations 359 f
– TEM 625, 832
magnesium supports 231
magnesium vanadates 434
magnetic field 200
magnetic moment 3, 197
magnetization transfer 206
magnetogyric ratio 7
maleic anhydride conversion
– metal oxide catalysts 643
– selective oxidation/VPO 268 f
– vanadium phosphate catalysts 499, 513, 524
manganese polymorphs 119
Mars–van Krevelen mechanism 196, 317, 375
mass –thickness contrast 445, 470
matrix isolation 42
Maxwell–Boltzmann law 5
MCM-41 materials 314, 472
mean free path 302
mechanical mixing 690
mechanochemical activation 516, 526
mesoporous materials 428
mesoporous silica 470 f, 764
mesoporous silica-aluminas 145 f
metal–alkyl bonds 600
metal antimonates 786
metal catalysts 309

metal-centered transitions 53 f
metal–cupferron decomposition 635
metal hydroxides 129 f
metal ion doped zeolites 29
metal ion implantation 761
metal oxide catalysts 613–664
– oxidation reactions 487–498
– solid-state NMR 195–242
metal oxide
– base catalysis 819–844
– supports 221 f
– XPS 255
metal–oxygen bonds 111, 132 f, 561
metal–oxygen–metal bonding 443
metal phosphates 791
metal-to-insulator transition 460
metal-to-ligand charge-transfer transition
  (MLCT) 55
methacrolein oxidation 568
methacrylic acid oxidation 568
methane 33, 553
methanol
– adsorption/decomposition 645
– base catalysis 834
– critical point parameters 624
– oxidation 493, 729
– titanium silicalite-1 729, 742
– transesterification 644
– uranium oxides 553
methyl-1-propanol oxidation 729
methylamine 155
MFI 154, 158, 310
micelles 636
Michael addition 821
microcalorimetry 586
microemulsion-mediated hydrothermal
  (MMH) method 634 f
microheterogeneity 19
microporous silica-aluminas 145 f
microporous zeolites 468 f
microtwin defects 466
microwave-assisted coprecipitation 631
microwave power 24, 29
Mie theory 59
Miller index 369
Miller plane 354
mirage effect 106, 121, 304
mixed metal oxide catalysts 690
mixed metal oxides 266 f
mixed metallic insulating behavior 252
mixed oxides 121 f, 771–818
modified oxides 408 f
modulation amplitudes 24
molar integral heat 394

molecular configurations   211
molecular oxygen oxidation   570
molecular probes   147, 166
molecular structure   487–498
molecular vibrational modes   107 f
Møller–Plesset perturbation methods   326
molybdate   137
molybdate catalysts   775
molybdated superacidic metal oxides   674
molybdenum addenda atoms   561, 572
molybdenum oxide catalysts   218, 379
molybdenum promoted vanadium phosphate catalysts   519
molybdenum trioxide   335, 379, 674
molybdenum–uranium oxide catalyst   553
momentum vector   57
monatomic non-absorbing gases   105
monoclinic fluorite structure   541
monolayer oxides   137, 488
monolayer surface coverage   490 f
monomeric compounds   262
monomolecular intermediate   682
mono-oxygen donors   707
monosulfate structure   688
Monte Carlo (QMC) simulations   330, 359
mordenite   426, 588, 680
Morse potential   76
Mott–Hubbard compounds   255
Mott–Hubbard model   371
MoV-based mixed metal oxides   281 f
Mo-V-Nb-Te-O system   780 ff
Mo–V–Te mixed-metal oxide catalysts   416
MO_x redox active systems   496
Mulliken analysis   339
Mulliken charges   255
multicomponent analysis   318
multicomponent catalysts   775 ff
multicomponent molybdate   791
multi-electron calculations   262
multi-electron wavefunction   325
multifrequency measurements   27 f
multimetal molybdates (MMM)   775
multiple adsorption sites   159
multiple product option   780
multiple quantum magic angle spinning (MQMAS)   202 f
multiple scattering   301 f, 312, 451
multiplet splitting   249, 256
multiwavelength excitation   177, 185
mutual exclusion rule   110

n

N–H stretching   825
nanobelts/rods/wires   477 f

nanocrystallites   563, 626
nanomaterials   69, 614, 619
nanoparticles
– metal oxide catalysts   614 f, 619, 625 ff
– powder EPR   33
– X-ray spectroscopy   476 f
naphtha steam reforming   557
near-edge X-ray absorption fine structure (NEXAFS)   301
negative differential resistance (NDR)   574
nephelauxetic support series   75
neutron diffraction   353, 582
Newton's laws   350
NH_3 adsorption   405
NH_3 probes   507
Ni/Co/Fe/Mg molybdates catalysts   775
nickel ions   413
nickel oxides   372
nickel–uranium oxide catalysts   555
NiMo/ASA catalyst   206
niobium oxides   413
nitrile–butadiene–rubber (NBR)   771
nitriles   147, 771
nitroaniline   821
nitrochlorobenzene indicators   666
nitrocompounds oxidation   732
nitrogen dioxide   30 f, 38
nitrogen monoxide   165 f, 403
nitrogen probe molecule   684
nitromethane   166
nitrotoluene indicators   666
nitrous oxide   624
nitroxide radical   27
NMR techniques
– oxidation catalysts   195–242
– POMs   579, 580 f
– vanadia catalysts   601
normal modes   123 ff
NO_x reduction   555
nuclear magneton   8
nuclear spins   198
nuclear Zeeman interaction   6
nucleophilic reactions   280, 622

o

O–H bonds   130, 139, 348, 526
occupancy   344
octanol oxidation   729
olefins   178 f
– diffusion   742
– epoxidation   717
– isomerization   618
– metathesis   488

– oxidation 549, 554, 717
– polymerization 153
oligocations 424
operando spectroscopy 96, 169
optic axis 448
optical properties 51, 56 f, 109
orbitals 55, 244, 301, 324, 526
ordered mesoporous silicas 470
organic dyes adsorption 760
organic phosphates 233
organic solvents 510
organometallic derivatives 566
organoperoxy radicals 42
orthotoluonitrile 165
oscillations 302
outer shell electronic transitions 52
overlayer-type contamination 247
overoxidation 521
overtone modes 99, 151
oxalates 629
oxidation
– hydrocarbons 773
– IR spectra 164
– NMR 195–245, 218
– superacidic metal oxides 672
– supported metal oxide catalysts 487–498
– titanium silicalite-1 740 f
oxidation catalysts 299–322
oxidation states
– cationic centers 161 f
– chromium 597
– heteropolyoxometallate catalysts 561
– uranium oxides 540
– vanadium phosphate catalysts 500 f
oxidative dehydrogenation 260–269, 416
oxide catalysts 613–664
– cracking 196
– IR spectra 52, 132
– solid-state NMR 195 f
– thermal analysis 404 f
– XAS 299–322
– X-ray spectroscopy 472 f
oxide materials 476 f, 755–770
oxide nanotubes 479 f
oxide supports 225 f
oxides/zeotype-systems 85 f
oximes oxidation 316, 733
oxoaluminum structure 62
oxoanions 561–594
oxo-salts 122 f
oxyfunctionalization 571
oxygen
– binding energies 253, 275
– defect structures 377
– lattices 351
– liberation 376
– mobility 416
– PL spectroscopy 82
– removal 604
– superstructures 459 f
– vacancies 376 f, 383, 456, 466
– vibrational modes 112
– X-ray absorption 309, 312
oxygen-rebound mechanism 711
oxygenation 641, 727

p

palladium catalysts 414
parabolic energy valley dependence 57
paraffin isomerization 668
paramagnetic centers 1
paratacamite 116
partial oxidation 403, 561–594
partial pressure 368
partially filled d bands 370 f
participating catalytic active site number 494
partition coefficients 740
patterns 11 f, 20, 452
Pauli's exclusion principle 57
pentane isomerization 682
pentanol oxidation 729
Perdew–Burke–Ernzerhof (PBE) approach 57, 330, 341, 346
Perdew–Wang (PW91) approach 331, 346
periodic electronic structures 323–390
periodic quantum chemistry 332 f
perovskites 122, 167, 455–467
peroxo intermediates 571
peroxyacyl radicals 42
perturbation methods 328
petroleum fractions 543
phase contrast 445
phase diagrams 370, 540
phase shifts 302, 307, 446 f
phenol
– hydroxylation 712, 734
– oxidation 544, 570
– TS-1 catalyzed 316
phenylethylamine (PEA) 417
phonon absorption/emission 66
phosphine/-oxides 147, 233
phosphorescence 77 f
phosphorus enrichment 277
phosphorus-modified aluminas 224
phosphorus-to-vanadium ratio (P/V) 276, 504

photoacoustic spectrometry (PAS) 106 f, 582 f
photocatalysts 68, 225, 755–770
photocatalytic processes
– AEO 83
– oxides 755–770
– powder EPR 42
– superacidic metal oxides 698
photocurrent 244
photoelectrochemical cell 768
photoelectron process 302, 305
photoelectron spectroscopy (PES) 243–298
photoemission cross-sections 276
photoionization cross-sections 287
photoirradiation 42
photoluminescence spectroscopy 60, 76 f, 80 f
photonic sponges 763
photons 57, 244, 300
photothermal beam deflection spectroscopy (PBDS) 106 f
physical mixtures 690
physical photocatalytic activity enhancement 763
physical properties 26 f, 392
– metal oxide catalysts 613–620
– POMs 566
physical transformations 619
physisorption 398, 406
phytosterol esters synthesis 839
picoline ammoxidation 795
pillared clays (PILC) 424 f
pitchblende 539
Planck constant 98, 339, 445, 451
plane wave basis set 338
point defects 33, 467 f, 616
point symmetry 36
poisoning 260, 600
polarization 100, 245
polyatomic chemical species 97 f
polymeric tetrahedral oxo-ions 128
polyoxometallates (POMs) 561–594
pore structure 156, 310, 564
porous materials 468 f, 472 f
potassium hexaniobate 480
potassium modified uranium oxides 556
potential-assisted photocatalysis 766
powder spectra (EPR) 19 f, 24 ff
powder X-ray diffraction 551
precipitation 132, 627 f
precursors
– Al-NMR 222
– base catalysis 831
– IR spectra 51, 129 f, 132

– metal oxide catalysts 617, 634
– spin-echo mapping 205
– uranium oxides 550, 556
– vanadium phosphate catalysts 500–515
– XPS 265
pre-edges 309, 317
preparation
– EXAFS 314
– photocatalysts 763
– superacidic metal oxides 665–704
– thermal analysis 415
– uranium oxides 543
– vanadia coverage 186
– vanadium phosphate catalysts 500, 513
pressure 396
primary structures 563
probing/probe molecules 160, 401 ff
process integration 739
product distribution 686 f
promoted catalysts 519 f, 525, 689 f
propane
– ammoxidation 777 f
– chemisorption 495
– conversion 547, 691
– dehydrogenation 213, 598
– oxidation 281 f
propanol oxidation 729, 742
propene oxidation 718, 774 ff, 777
propene oxide synthesis (HPPO) 735, 739
properties
– highly dispersed metal oxide catalysts 613–664
– MoVTeNb XPS 282
– oxide catalysts 81 f
– sulfated superacidic metal oxides 671
– superacidic metal oxides 675
  see also chemical, physical properties
propylene ammoxidation 549
propylene oxidation 318
proton affinity 402
protonic acidity 435
protonic zeolites
– Brønsted acidity 153 f
– IR spectra 158
– Lewis acidity 161
– surface hydroxyl groups 144 f
pseudobohemite 116
pseudoliquid bulk type catalysis 564
pseudopotentials 340 f
pulsed flow calorimetry 399
purification steps 777
PW91 approach 360
pyridine
– POMs 586

– adsorption 84, 427
– CP-MAS NMR 233
– IR spectra 135, 153
– surface basicity 166, 402
– vanadium phosphate catalysts 507
pyrolysis 640
pyrophosphate groups 277
pyrrole 166

q

Q–size effect 68
quadrupolar broadening 225
quadrupolar interaction 14, 199
quadrupolar phase adjusted spinning sideband (QPASS) 226
quadrupolar vanadium isotopes 212
quadrupole moment 198
quantifications 24 f, 186 f, 247 f
quantum chemistry calculations 323
quantum dots 476
quantum efficiency 79 f
quantum mechanics 3, 197
quantum mechanics/molecular mechanics (QM/MM) 323
quaternary metal oxides 415 f
quencher molecules 80

r

Racah parameter 53
radicals 12, 29, 38 f, 376
radio frequencies 202
Raman spectroscopy 95, 177–196
– butane dehydrogenation 606
– coke 599
– CrO_x/alumina catalyst 597
– polyatomic chemical species 97 f
– skeletal 111
– vanadia catalysts 489, 601
rapid reduction 574
rare earth oxides 836
Rayleigh theory 100
reactive intermediates 743
real space 451
reciprocal space 332, 451
recombination 47, 756, 761
recrystallization 463
recycle processes 779
redox properties 45
– heterogeneous photocatalysis 756
– metal oxide 613, 619 f, 227
– oxidation reactions 495, 491
– POMs 567, 578, 585
– thermal analysis 392, 403 f
– vanadium phosphate catalysts 516

– heterogeneous photocatalysis 756
– vanadia catalysts 178 f
– vanadium phosphate catalysts 530
reduced states 183 f
reducing agents 508, 707
reduction 554 f, 585, 604
refinery-based C_4 stream 650
reflectance
– divalent elements 112
– IR spectroscopy 59, 98, 101
– Raman spectroscopy 186
reflection techniques 101 f
refractive index 60, 102
regeneration 596, 609, 745
relativistic correction 445
relaxation 5 f, 245, 251, 360, 366
reoxidation 185, 260
resolution 354, 448 f
resonance absorption 21
resonance enhancement 184, 189
resonance line width 6
resonance Raman spectroscopy 177–196
resonant frequency 198
restricted Hartree–Fock (RHF) 326, 346
reststrahlen effect 102
rhombic symmetry 21
rock salt structure 63, 102, 143 f, 356
rocking mode 120
rosary pattern 472
rotation effects 25 f, 107, 201
rutile 118, 342 757, 786

s

S-compound oxidation 734
salt-free production 734
sample preparation 100, 134, 309
sample tumbling effects 25 f
satellite transition magic angle spinning (SATRAS) 200 f, 256 f
saturation effects 29
SBA-15 472, 650
scaling factor 203
scanning electron microscopy (SEM) 445, 512
scanning tunneling microscopy (STM) 462
scattering 59
– POMs 579
– TEM 448, 451
– XAS 302
Scheelite-type structure 775
Scherzer equation 449
Schiff base amines 571
Schottky pairs 616
Schrödinger equation 244

Schuster–Kubelka–Munk (SKM) model 59
second generation TiO_2 photocatalysts 68
secondary structures 563
segregation 276, 617
selected area electron diffraction (SAED) 450
selection rules 98, 103
selective n-butane dehydrogenation 260 f
selective oxidation
– EXAFS 315 f
– maleic anhydride/VPO 268 f
– uranium oxides 548
– XPS 267 f
selectivity
– toluene oxidation 552
– uranium oxides 556
– vanadium phosphate catalysts 504, 519
selectivity–activity relationships 487–498
self-absorption 187
self-interstitials 467
semiconductors
– electronic structures 345
– photocatalysis 756, 760
– transition metal oxide surfaces 381
– UV absorption bands 66 f
sensitivity
– IR spectra 134
– oxidation catalysts 198
– photocatalysis 760
– VPO 276
sensors 698
sesquioxides 121, 142 f, 836
Shannon's tabulation 355
shear planes 318
shells 452
shifting 182
shrinkage 474, 623
Si–(OH)–Al bridges 144
Si–O–Si bridges 119, 160, 228
side-chain alkylation 819
signal-to-noise ratio 60, 134, 245
silanol groups 141, 144
silica
– hydroxyl groups 141
– optical absorption 62
– polymorphs 119
– supports 227, 433, 553
– thermal analysis 408
– X-ray spectroscopy 470 f
silica-aluminas (SAs)
– CP-MAS NMR 206, 229 f
– Lewis acid sites 159 f
– supports 206, 231
– surface Brønsted acid sites 150

– surface hydroxyl groups 145 f
– thermal analysis 418
silica-titania/zirconia 417
silicalites 316, 428
silicates 136, 424 f
silicon 206, 757
silicon sulfated metal oxides 669
silver catalysts 413, 519
single crystals 19, 26
single-site active centers 69
singlet state 77
singularities 21
site chains 383
site symmetry 24
skeletal IR spectra 111 f, 129 f
skeletal isomerization 685
slab models 353, 357 f
Slater determinants 326
smectites 424
smoothing function 344
Sn sulfated metal oxides 669
SnO_2
– heterogeneous photocatalysis 757
– support 231
– thermal analysis 407
SO_2 adsorption 405
SO_4/ZrO_2 preparation 672
soft pseudopotentials 340
SOHIO process 774 ff
sol–gel precursors 634
sol–gel processing 415, 620, 672
solid acids 679
solid solutions 121 f
solid–gas interface 393, 404
solid–liquid interface 400
solid-state Hamiltonian 14
solid-state NMR 195–242
solid-state TEM 443
solid superacids 667 f
solids energy band transitions 56
solvents 622 f, 741
solvolysis/rearrangement 726
solvothermal techniques 633
sonochemical coprecipitation 630
space group 454, 471
space vector 332
spacings 450
spectral properties 13, 19, 51
spectroscopic detection 97 f, 139–159
spherical aberration 448
spikelet-echo technique 208
spin density 41, 384
spin-echo mapping 204 f, 216
spin Hamiltonian 1 ff, 44 ff

spin–lattice relaxation 6
spin–orbit coupling 79
spin quantum number 197
spin states 78, 325, 371
spin–spin coupling Hamiltonian 18, 199
spin–spin relaxation 6, 208
spinel model 352
spinel-type structure oxides 115 f, 121, 142 f
spinning 209, 223
splitting 372
– EPR 10, 23, 27
– IR spectroscopy 53, 110, 144
– POMs 581
– XPS 249
spontaneous ionization 33
spray pyrolysis 640
sructural defects 616
stabilization 504, 633
stacking faults 465, 617
stacking sequence 349, 355, 469
stannia gel 669
state multiplicity 99
static calorimetry–volumetry coupling 394 f
steady-state kinetics 493
steam cracker-based C_4 stream 650
steam reforming 556 f
Stefan's law 105
Stern–Volmer equation 80
stoichiometric range 276, 540
Stokes shift 78
strengths 147 f
stretching
– IR spectra 111, 123, 139, 161
– POMs 581
– vanadia catalysts 181
structure–activity relationships 190, 195, 487–498
structure calculations 323–390
structural characterization 69 ff, 221
structural transformations 517
structures
– Al_2O_3-supported vanadia catalysts 178 f
– boehmite 349
– bulk oxides 209 f
– heteropolyoxometallate catalysts 561 f
– hydrotalcites 835
– hydroxyl groups 140
– M1 XPS 281
– metal oxide catalysts 614
– oxide crystals 454 f
– sulfated zirconia 687
– surface defects 35
– surface properties 227

– Ti–OOH species 742
– Ti–zeolites 707
– uranium oxides 540 f
– USb_3O_{10} 550
styrene–acrylonitrile (SAN) 771
styrene dehydrogenation 607
subshell photoionization 287
substituted alkanes 709
substituted alkylbenzenes 791
substituted aromatic nitriles 793
substituted benzenes 717
substituted olefins 726
sulfated metal oxide catalysts 669 ff
sulfated oxides 143 f
sulfated silicas 671
sulfated titanias 408
sulfated zirconias 234, 408, 673, 685 ff
sulfates 133, 137
sulfation 670
sulfides/sulfoxides 316
sulfur dioxide 166, 408, 432, 516
sulfurous compound removal 755
superacidic metal oxides 665–704
superbase 824 ff, 837 ff
supercells 336, 372
superconducting oxides 455
superhyperfine interactions 35
superoxide radicals 759
superoxo species 528
superstructures 121, 455 f
supported boria catalysts 409
supported catalysts 252
supported heteropolyacids 422 f, 791
supported metal oxide catalysts 408 f, 487–498
supported transition metal oxides 54 f, 496
supported vanadium oxide catalysts 210 f
– oxidation reactions 487–498
– Raman spectroscopy 177–196
supports 195 ff, 210, 221 f
Suprasil 81
surface acid–base properties 404
surface acid sites 495
surface acidity
– IR spectra 147
– NMR 196, 229 f
– thermal analysis 402 f
surface active sites 392
surface adsorption 30 f, 323
surface anion vacancies 35
surface area 564, 617
surface basicity 166
surface calculations 359 f
surface catalysis 564

surface chemistry 96 f, 133 f
surface coverage 490 f
surface defects 33 f
surface density 180
surface electromagnetic wave spectroscopy (SEWS) 104
surface element–oxygen bonds 134 f
surface energy 355, 368
surface equilibration 401
surface hydroxyl groups 142 ff
surface M–O–M bridges 134 ff
surface OH groups 142
surface oxo-anions 136 f
surface paramagnetic centers 1
surface profile imaging 462 f
surface properties
– metal oxides 227 f, 614
– NMR 196
– oxides 402 ff
surface reactivity 81, 492 f
surface relaxation 112
surface-sensitive techniques 244 f
surface states 61
surface structure 206 f, 227 f, 353 f
surface tension 479
surface-to-volume ratio 478
surfactants 641
symmetrical hydrogen bonding 150
symmetry 157, 301
symmetry point group 124, 136
synchrotron 245
synergistic interactions 433, 575
synthesis, metal oxide catalysts 613–664, 706 f

t
Tanabe–Sugano diagrams 53
tapered element oscillating microbalance (TEOM) 604
tartaric acids 499
tellurium cations capping 587
tellurium dioxide, XPS 287
temperature dependence 392 ff
– adsorption–desorption heats 401 f
– powder EPR 26 ff
– reduced states 183
temperature programmed desorption (TPD) 399
– base catalysis 823
– superacidic metal oxides 675 ff
– uranium oxides 548
– vanadium phosphate catalysts 526
temperature programmed oxidation (TPO) 399, 619

temperature programmed reduction (TPR)
– chromia 596
– metal oxide catalysts 619
– superacidic metal oxides 678
– vanadium phosphate catalysts 526
temperature programmed surface reaction (TPSR) 489 ff
templates 473, 640
temporal analysis of products (TAP) 503, 548
tensors 14 ff
terminal planes 464
terminating layer 253
ternary phases 121 f
ternary systems 418
tertiary structures 563
tetraethylorthosilicate (TEOS) 523
tetravalent oxides 118 f, 143 f
thermal analysis 391–442, 579
thermal decomposition 46, 638
thermal desorption 399
– POMs 581
– XPS 255
thermal treatment 550, 567 f
thermodynamic equilibrium constant 665
thermogravimetric analysis (TGA) 391, 417, 579, 586
thermokinetic parameters 398
thermolysis 586
thioethers oxidation 734
thiols 166
Ti-MCM 73
Ti–OOH species 741 f
time constants 398
tin oxides 410
titania
– (110) surface 366
– DOS plots 342
– heterogeneous photocatalysis 757
– metal oxide catalysts 614, 625, 634
– nanotubular material 479
– photocatalysis 757, 763
– polymorphs 118, 143
– supported vanadium oxide 433
– surface calculations 359 f
– surface oxo-anions 136
– thermal analysis 408
see also rutile
titanium dioxide 42
titanium oxide incorporation 764
titanium promoted vanadium phosphate catalysts 521
titanium silicalite-1 315, 705–754
titanium sulfated metal oxides 669

titanium(IV) oxide 755 f, 763 ff
titanosilicate ETS-10 428
TNU complex zeolites 468
tolerance factor 460
toluene 310, 544, 552, 692
top-down preparation 619
total electron yield (TEY) 247
total oxidation 547
toxic properties 649
transesterification 644, 839
transfer of population in double resonance (TRAPDOR) 207
transient radical intermediates 42
transition aluminas 221 f, 351, 404
transition metal ions (TMI) 45 f, 53, 69 f, 85 f
transition metal oxides 613
– oxidation reactions 496
– partially filled d bands 370 f
– periodic electronic structures 323, 351
transition metal substituted polyoxometallates (TMSP) 568
transition metal zeolites 312 ff
transitions
– energy bands 56 f
– EPR 9
– TEM 452
– XAS 301
– XPS 248
translations 107
transmission electron aberration-corrected microscope (TEAM) 450
transmission electron microscopy (TEM) 443–486
transmission/absorption IR technique 100 f, 133
transmittance 60, 98, 304
transverse optical mode 99, 102
trapping 761
triacetin transesterification 644
trialkyl-substituted phenols 571
triethylamine 402
triglycerides 643
trigonal-planar oxo-anions 123
trilayer units 363
trimer structural model 689
trimethylamine 155, 402
trimethylborate 166
triplet configurations 12, 78, 371
trivalent element oxides 115 f
tungstate 137
tungstated aluminas 697
tungstated oxides 143 f
tungstated stannia 696
tungstated superacidic metal oxides 674

tungsten addenda atoms 561
tungstic heteropolyacids 422
turnover frequency 493, 605
twin defects 466 f
twisting modes 132
two-spin Hamiltonian 9

u

ultra-high vacuum (UHV) system 36, 245, 260
ultra-soft pseudopotentials 340
ultra-stable (USY) zeolites 310, 425
undulators 304
uniaxial symmetry 20
unit cell 332, 451, 563
unrestricted Hartree–Fock (UHF) 326
unsaturated alcohols epoxidation 724
upper valence band (UVB) 342
uranium oxides heterogeneous catalysis 539–560
UV absorption bands 61 f, 66 f
UV irradiation 43
UV photoemission spectroscopy (UPS) 376
UV-visible radiation 756
UV-visible spectroscopy 583
UV-vis-NIR spectroscopy 51–94

v

V–Mg–O catalysts 216
V–Mo–Nb–Te oxide catalysts 779
V–Mo–O catalyst ammoxidation 804
V–O–M bonds 487 ff
V–O–S bonds 178
V–O-support bond 494
V–O–V bonds 267
V–P–Ti compound 214
V–Sb–O-based catalysts 789 f
vacancies
– electronic structures 352, 376
– metal oxide catalysts 614 ff
– TEM 452, 466
– transition metal oxide surfaces 376
– uranium oxides 542
vacuum system 304
vacuum treatment 72
valence band 372
– heterogeneous photocatalysis 756
– UV-Vis-NIR spectroscopy 57
– VPO 273
valence charges 255, 259, 286
valence electrons 244, 613
valence states 340
– EPR 45
– heteropolyoxometallates 561

- vanadia catalysts   178
- XPS   248 ff
van der Waals interaction   405
vanadates   137, 184, 216 f
vanadia catalysts   429 f, 488–498, 601
vanadia coverage   186
vanadium   86
- cation capping   587
- heteropoly catalysts   578
- impregnation   227
- TiO$_2$ supported   791
vanadium oxides
- alkane dehydrogenation   595–612
- oxidation reactions   487 ff
- oxidative dehydrogenation   266 f
- PL spectroscopy   86
- Raman spectroscopy   177–196
- solid-state NMR   210 f
- XPS   255
vanadium pentoxides   430
vanadium phosphate catalysts   499–538
vanadium phosphorus oxides (VPO)   499–538
- selective dehydrogenation   268
- solid-state NMR   216 f
- spin-echo mapping   204
- thermal analysis   416
vanadyl groups   87, 259
vanadyl phosphates   500, 508
Vanderbilt potentials   340
vapor condensation methods   639
vegetable oils   643
vibrational spectroscopies   76, 95–126
visible light sensitization   760
VO$_2$ P$_2$O$_7$, HREM   500
VO$_x$/Al$_2$O$_3$ catalysts   230
V$_2$O$_5$–Al$_2$O$_3$ catalysts   202, 489
V$_x$O$_y$/alumina   260 f
volatile organic compounds (VOCs)   547
volumetric-calorimetric adsorption   394
VPA route   509

w

wagging modes   132
water
- addition   516, 547
- adsorption   132 f, 139 f, 376
- critical point   624
- oil micro-emulsions   636
- solvent   742
- splitting   755
- vapor adsorption   64
wavefunctions   16, 332
wavelengths   52, 445

wavenumber   56, 98, 134, 521
Weiss temperature   204
wet chemical transformations   619
wigglers   304
WO$_3$ photocatalysis   757
WO$_3$/ZrO$_2$ preparation   674
work function   244
wurtzite hexagonal-phases   477

x

X-band frequency   4
X-ray absorption fine structure spectroscopy (XAFS)   601
X-ray absorption near-edge structure (XANES)   301, 308 f, 490
X-ray absorption spectroscopy (XAS)   247, 299–322
X-ray diffraction (XRD)   332, 353, 503, 512
X-ray fluorescence process   305
X-ray photoelectron spectroscopy (XPS)   248 f
- chromia   596
- monolayer surface coverage   489
- superacidic metal oxides   672
- surface sensitive technique   243 ff
- vanadia catalysts   601
X-ray transitions   248 f
xylene ammoxidation   794

y

YBa$_2$Cu$_3$O$_{7-\delta}$ 463, 467
YBaCo$_2$O$_{5+x}$ 459
YSrCo$_2$O$_5$ 448

z

ZAF calibration   453
Zeeman interaction   2 ff
Zeeman shift   197
Zeeman states   207
zeolite acid catalysis   154
zeolites   29, 47, 62, 85 f
- adsorption microcalorimetry   402
- ammoxidation   810
- analcime   478
- base catalysis   837
- cation exchanged   165 f
- crystallites   469
- EXAFS   310 ff
- H-FER   300
- IR spectra   121, 124, 143, 165
- photocatalysis   763
- POMs   588
- surface Brønsted acid sites   150
- thermal analysis   425 f

– titanium silicalite-1   707 ff
– X-ray spectroscopy   468 f
– ZSM-5   310 ff, 425
zero field interactions   18, 27
zero point of charge (PZC)   87
zero valent state   47
Ziegler–Natta catalysts   47
zinc blende structure   333
zinc oxides   635
zinc promoted vanadium phosphate catalysts   519
zirconia
– IR spectra   143, 160
– metal oxide catalysts   614
– NMR   208, 234
– polymorphs   118
– precipitation methods   629
– sulfated   669 ff
– superacidic metal oxides   669
– superbases   837
– surface oxo-anions   139
– thermal analysis   408
zirconium promoted vanadium phosphate catalysts   519, 522
ZnO   115, 334, 477
ZnS photocatalysis   757